ENGINEERING MECHANICS

ENGINEERING MECHANICS

THIRD EDITION

PART II: DYNAMICS

FERDINAND L. SINGER

New York University

HARPER & ROW, PUBLISHERS

New York Evanston San Francisco London

Sponsoring Editor: John A. Woods
Project Editor: Lois Wernick
Designer: T. R. Funderburk
Production Supervisor: Will C. Jomarrón

ENGINEERING MECHANICS, Third Edition, Part II: Dynamics

Library of Congress Cataloging in Publication Data
Singer, Ferdinand Leon, Date-
Engineering mechanics.
CONTENTS: pt. 1. Statics—pt. 2. Dynamics
1. Mechanics, Applied. I. Title.
TA350.S532 1975b 620.1 74-16275
ISBN 0-06-046233-7 (v. 2)

To my cherished wife Evelyn
and our three daughters,
Joan, Karen, and Lucy

CONTENTS

PART II DYNAMICS

PREFACE TO THE THIRD EDITION

The previous editions endeavored to show how a few basic concepts—the relations between a force and its components, the principle of moments, and Newton's laws of motion—may be combined and applied to a wide variety of practical situations that are encountered by engineers. Another purpose was to help the student develop the logical, orderly processes of thinking that characterize an engineer. Both of these objectives have been emphasized to an even greater extent in this third edition.

This book has been completely rewritten and expanded to include the suggestions received from many users of the previous editions. There are two very significant changes: a reorganization that coordinates usually disparate topics into unified entities in both statics and dynamics and an integrated treatment of scalar-geometric analysis and vector analysis.

No previous experience in vector analysis is necessary because each application of vector notation is preceded or accompanied by detailed explanations that emphasize its geometric significance. The use of force multipliers permits vectors to be expressed in a simple manner rather than the decimalized notation so common in other books. In addition, force multipliers also simplify the transition from scalar-geometric notation to vector notation and vice versa.

Neither vector notation nor the scalar-geometric method is used to the exclusion of the other; instead each is employed where it is most appropriate. For two-dimensional analysis, the scalar-geometric method is retained as the simplest and most direct approach. For three-dimensional analysis and the development of general concepts, vector notation provides the most direct approach, particularly for explaining the effects of a change in direction of a body's motion. But primarily the book emphasizes how to analyze and comprehend the various areas of mechanics irrespective of the mathematical method.

Other noteworthy changes and additions in Part I, Statics, include a unified and coordinated treatment of plane and space statics; an expansion of the analysis of structures to incorporate the criteria for stability; a more organized approach to frames containing multiforce members; an optional section on space trusses; and additional material on friction, shear and bending moment in beams, and virtual work.

Part II, Dynamics, has been reorganized and rewritten to take full advantage of vector notation. Especially noteworthy is the chapter on the general principles of dynamics that is preceded and followed, respectively, by the kinematics and kinetics of a particle. The immediate application of kinematics to the kinetics of a particle contributes to understanding their interaction and reinforces each. Similarly, the kinematics of plane motion immediately precedes its application to kinetics. These fundamental approaches to the force-acceleration method are followed by chapters on the work-energy and impulse-momentum methods. Much thought was given to adding the particle motions of these chapters to particle kinetics but was discarded in favor of concentrating on basic dynamics instead of a multiplicity of approaches. However, the book is written so that those portions of energy and momentum methods that are pertinent to particle motion can be combined with particle dynamics, if so desired. After a chapter on spatial dynamics, the concluding chapter is an introduction to mechanical vibrations, which seems to be the natural place to discuss simple harmonic motion and various types of pendulums without interrupting the continuity of other basic developments.

Each portion of the book contains more material than can reasonably be covered in a semester. Sections on advanced or specialized topics are identified by an asterisk. These sections are independent of the rest of the book so that flexibility of course outlines or objectives may be attained.

It is hoped that these topics, as well as the others in the text, are presented in a manner that will relieve instructors of the burden of detailed explanations, thereby leaving them free to expand and develop those details that they wish to emphasize. Not only are the theoretical aspects rigorously presented, but their practical applications are also emphasized. Numerous illustrative problems (most of them new) show in detail how principles are applied. The explanations are complete—nothing is taken for granted. As each equation or principle is applied, it is stated in brackets at the left-hand side of the page. Adjacent to it, values are substituted in the respective order in which the symbols appear in the equation. This procedure enables students to readily follow the various steps of the solution without continually referring to the body of the text.

But primarily, the students' points of view and their special problems have been kept in mind. The prose style is almost conversational, and, without being too brief or lengthy, should be conducive to self-study. The physical significance of fundamental concepts and the assumptions and limitations made in developing them are carefully discussed so that understanding is achieved and memorization reduced to a minimum. The summaries appended to each chapter are intended to give the student a concise statement of key elements, which should be useful in review and post-college work.

There is a completely revised set of more than 1200 problems that have been carefully prepared to give wide variety in type and scope. It is anticipated that an instructor can make an adequate selection without repetition for four or five years. The problems have been arranged approximately in their order of difficulty and answers to two-thirds of them have been given; the others may on occasion be used for quizzes or to demonstrate mastery of concepts. No problem is presented that is not preceded by an adequate text discussion of the concepts involved, and usually numerical values that simplify the arithmetical computations are selected.

The numbering plan used in this revision enables one to locate quickly any cross reference. With this plan, all figures, equations, tables, and problems are preceded by the chapter and section numeral in which they appear and are numbered consecutively throughout each section. Figures for assigned problems are given the same number of the problem to which they refer in order to simplify correlation of a problem figure with the corresponding problem data.

The author wishes to acknowledge his indebtedness to his colleagues all over the nation for the many valuable suggestions that they so generously offered him for this revision. To identify them individually would make too lengthy a list (with possibly an inadvertent omission), but each has received my thanks. Particular mention, however, must go to the publisher's reviewers who painstakingly noted inconsistencies and offered suggestions for changes that were very helpful, especially those of Professor Andrew Pytel of The Pennsylvania State University and Professor William G. Plumtree of California State University, Los Angeles. Although great care was taken to eliminate errors, it is inevitable that some will still be found. The author will appreciate being informed about them and will welcome any comments that readers may care to offer.

January, 1975 Ferdinand L. Singer

LIST OF SYMBOLS AND ABBREVIATIONS*

A	area, amplitude of vibration
$\mathbf{a}$, a	acceleration
$\bar{\mathbf{a}}$, $\bar{a}$	acceleration of mass center
$\mathbf{a}_n$, a_n	normal acceleration
$\mathbf{a}_t$, a_t	tangential acceleration
b	breadth, width
$\mathbf{C}$	couple
C	centroid, instant center
d	depth, diameter, distance, moment arm
E	modulus of elasticity in tension or compression
$\hat{\mathbf{e}}$	unit vector along axes whose orientation changes
e	coefficient of restitution, natural base of logarithms
e_{st}	static elongation
$\mathbf{F}$	frictional resistance
f	coefficient of friction, frequency
G	modulus of elasticity in shear, gravity or mass center
g	gravitational acceleration (32.2 fps^2)
$\mathbf{H}$	angular momentum
h	height
I	moment of inertia
$\bar{I}$	centroidal moment of inertia
$\hat{\mathbf{i}}$, $\hat{\mathbf{j}}$, $\hat{\mathbf{k}}$	unit vectors along coordinate axes

* With very few exceptions, these symbols and abbreviations agree with those approved by the American Standards Association.

J	polar moment of inertia
$\overline{J}$	centroidal polar moment of inertia
KE	kinetic energy
k	radius of gyration, spring constant
L	length
$\mathbf{M}$	moment of force
M	bending moment
m	mass (W/g)
$\hat{\mathbf{n}}$	general unit vector
n	revolutions per minute, subscript denoting normal direction
$\mathbf{P}, \mathbf{Q}, \mathbf{F}$	forces or loads, concentrated
P_{uv}, P_{xy}	products of inertia
$\mathbf{p}$	linear momentum
$\mathbf{R}$	reaction, resultant force
$\mathbf{r}$	absolute position vector
r, R	radii
RW	resultant work
s	magnitude of displacement, arc length
t	thickness, time, subscript denoting tangential direction
U	work
u, v, w	rectangular coordinates
V	volume, transverse shearing force
$\mathbf{v}, v$	velocity
$\overline{\mathbf{v}}, \overline{v}$	velocity of mass center
$\mathbf{W}, W$	weight, load
w	load per unit length
XYZ	inertial reference frame
xyz	moving reference frame
x, y, z	rectangular coordinates, moving or fixed
$\overline{x}, \overline{y}, \overline{z}$	coordinates of centroid, center of gravity, or mass center
$\boldsymbol{\alpha}, \alpha$	angular acceleration
$\alpha, \beta, \gamma \ldots$	angles
β	phase angle, angle of contact for belt friction
γ	weight per unit volume (specific weight)
$\delta r, \delta\theta$	virtual displacement (linear or angular)
$\delta s, \delta\theta$	magnitude of virtual displacement (linear or angular)
δU	virtual work
θ	angular coordinate, second Eulerian angle
$\boldsymbol{\rho}$	relative position vector
ρ	mass density, radius of curvature, variable radius
ρ, θ	polar coordinates
τ	period, periodic time

ϕ	angle of friction
φ	first Eulerian angle
ψ	third Eulerian angle
$\boldsymbol{\omega}$, ω	angular velocity
ω or ω_n	natural circular frequency $(2\pi f)$
$\boldsymbol{\Omega}$	angular velocity of moving reference frame
Ω	angular velocity of precession
$\hat{}$	circumflex denoting unit vector
deg	degrees
FBD	free-body diagram
ft	feet
fps	feet per second
fps^2	feet per second per second
hp	horsepower
hr	hour
in.	inches
ips	inches per second
ips^2	inches per second per second
K	kip (1000 lb)
kw	kilowatt
lb	pounds
mph	miles per hour
psf	pounds per square foot
psi	pounds per square inch
rad	radians
rev	revolutions
rpm	revolutions per minute
rps	revolutions per second
sec	seconds
shm	simple harmonic motion

II

DYNAMICS

9

KINEMATICS OF A PARTICLE

9-1 INTRODUCTION

In this and in succeeding chapters, we shall develop rigorously the principles of dynamics. At this place, however, we present an overall preview of what these principles are and how they are interrelated. First recall that in Chapter 2 we showed that any force system could be reduced to a resultant force-couple system. In statics, both the resultant force and the resultant couple were zero, thus establishing the equations of static equilibrium. In dynamics, however, the resultant force-couple system is not zero and causes a change in the state of motion of the body on which they act.

Usually the body is rigid and the resultant force-couple system is applied at the center of gravity of the body. If the resultant of the applied force system consists only of a single force passing through the gravity center of a body starting from rest as in Fig. 9-1.1*a*, the body will move in the direction of the resultant **R,** but it will not rotate. If the direction of **R** is constant, the motion of the body is along a straight path and is called rectilinear translation. If the direction of **R** varies, although continuing to pass through the gravity center, so will the motion of the body, resulting in a curved path motion known as curvilinear translation.

If the resultant of the applied force system is a couple **M** as

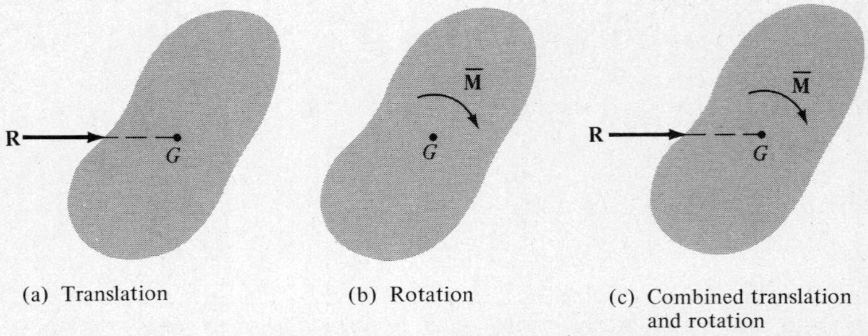

Figure 9-1.1 Types of rigid-body motion as determined by different resultants of applied forces.

in Fig. 9-1.1*b*, the body will spin about an axis passing through its center of gravity and directed perpendicular to the plane of the couple, but the center of gravity will remain stationary. All particles will describe circular paths about the axis of rotation. This type of motion is called centroidal rotation.

When the resultant of the applied force system consists of both a centroidal force and a couple as in Fig. 9-1.1*c*, the body has a motion consisting of a combined translation and a centroidal rotation. Further, if the applied forces always lie in the same plane, the resulting motion will be coplanar; otherwise, the motion will be spatial or three-dimensional.

The converse of the preceding discussion is also true; namely, the type of motion specifies the resultant of the applied force system. For example, a body constrained to have a motion of translation requires that the applied forces be so distributed that their resultant is a single force passing through the center of gravity. Or if a body is constrained so that it can only rotate about its centroidal axis, the resultant of the applied force system must be a couple.

The correlation between applied forces and the motion of a body is governed by only two basic equations which we shall develop later but merely state now. They are

$$\mathbf{R} = m\bar{\mathbf{a}} = \frac{W}{g}\bar{\mathbf{a}} \tag{9-1.1}$$

which relates the resultant force $\mathbf{R}$ to the mass m of the body and the acceleration $\bar{\mathbf{a}}$ of its mass center. (Frequently, the alternate use of W/g for mass m is a useful strategem.)

The other basic equation is

$$\Sigma\bar{\mathbf{M}} = \dot{\bar{\mathbf{H}}} \tag{9-1.2}$$

which relates the moment sum $\Sigma\bar{\mathbf{M}}$ of the applied forces about the mass center to the rate of change $\dot{\bar{\mathbf{H}}}$ of the angular momentum $\bar{\mathbf{H}}$ of the body about its mass center. This equation is valid for all angular motion, but it reduces to the simpler form

$$\Sigma\bar{\mathbf{M}} = \bar{I}\boldsymbol{\alpha} \qquad (9\text{-}1.2a)$$

for coplanar motion of a rigid body which is symmetrical about a plane of motion containing its mass center. The term $\bar{I}$ is the mass moment of inertia of the body about a centroidal axis perpendicular to the plane of motion while $\boldsymbol{\alpha}$ is the angular acceleration of the body.

These equations are valid only at a given instant of the body's motion. They are used to determine either the instantaneous forces acting when the motion of a body is known or conversely, when the forces are specified, they determine the accelerations from which the motion of the body may be calculated. Note that either $\mathbf{R}$ or $\Sigma\bar{\mathbf{M}}$ or both may be constant or variable. If variable, they may depend on position, velocity, or time or even a combination of these items. Such combinations may affect the complexity of the mathematical solution in a given situation, but should not obscure the fact that all problems in dynamics depend on only these two basic equations plus the correlation among acceleration and the motion. Such correlation is known as the time-geometry of motion and constitutes the area known as kinematics. It defines the motion of a particle or a body without consideration of the forces causing the motion. Since kinematics is the starting point for the study of dynamics, it is necessary that we begin with a clear understanding of what is meant by displacement, velocity, and acceleration and how they are related to each other.

9-2 MOTION OF A PARTICLE

The natural way of describing the motion of a particle is by lines of sight from an arbitrary origin to the successive positions of the particle. These lines of sight are called position vectors. Usually the position vectors emanate from an origin that is *fixed* in space, resulting in what is called absolute motion. With respect to a *moving* origin, the position vectors define what is known as relative motion. In most engineering situations, any origin on the earth's surface constitutes a fixed origin. For orbiting satellites, however, because of the rotation of the earth's surface, the fixed origin should be at the earth's center. For interplanetary motion, the origin may be taken at the mass center of the solar system, while for astronomical observations, reference axes are referred to the so-called fixed stars.

Although the following discussion is valid with respect to either a fixed or moving origin, for the present we take the origin as fixed. Later in Section 12-9, we shall discuss the correlation between motions with respect to fixed and moving origins. In Fig. 9-2.1, the position vectors $\mathbf{r}_A$ and $\mathbf{r}_B$ completely define positions A and B of a moving particle. Their difference $\Delta\mathbf{r}$ denotes the change in position (also known as displacement) occurring in the elapsed time Δt. Observe that the change in position $\Delta\mathbf{r}$ is independent of the choice of an origin as shown by the dashed position vectors to A and B from any other fixed origin O_1. Another observer at O_1 simultaneously sees the same displacement $\Delta\mathbf{r}$ as an observer at O. Since the elapsed time interval is the same for both observers, both record the same displacement, velocity, and acceleration of the moving particle albeit in terms of different position vectors.

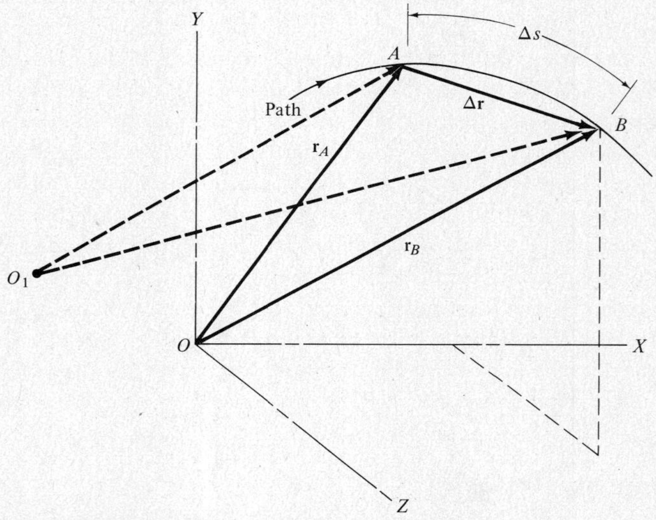

Figure 9-2.1 Relation between path and position vectors.

Observe carefully the distinction between a change in position and the distance traveled along the path. The change in position $\Delta\mathbf{r}$ is a vector involving both magnitude and direction whereas the arc distance Δs is a scalar measuring only the length of the path. The difference between the vector displacement (i.e., change in position) and the scalar distance is very evident if we assume the particle to move from A to B and back to A. The change in position, $\Delta\mathbf{r}$, would then be zero, but the distance traversed would be the accumulated length from A to B and back to A. This distinction is important because all the kinematic relations we shall develop involve vector change of position and *not* scalar distance.

VELOCITY

We define velocity as the time rate of change of position. If Δt is the elapsed time between A and B, the average velocity between these points will be

$$\mathbf{v}_{\text{ave}} = \frac{\Delta \mathbf{r}}{\Delta t}$$

and the instantaneous velocity is the limit of this ratio as Δt approaches zero, or

$$\mathbf{v} = \underset{\Delta t \to 0}{\text{limit}} \frac{\Delta \mathbf{r}}{\Delta t} = \frac{d\mathbf{r}}{dt} \tag{9-2.1}$$

The geometric significance of this result is that in Fig. 9-2.1 as Δt approaches zero, B gets closer to A and chord $\Delta \mathbf{r}$ coincides more completely with the arc Δs so that, in the limit, $d\mathbf{r}$ coincides with ds and consequently the velocity $\mathbf{v}$ is tangent to the path.

We can carry this interpretation further by rewriting Eq. (9-2.1) as

$$\mathbf{v} = \underset{\Delta t \to 0}{\text{limit}} \frac{\Delta \mathbf{r}}{\Delta s} \frac{\Delta s}{\Delta t} = \frac{d\mathbf{r}}{ds} \frac{ds}{dt}$$

The ratio $\dfrac{\Delta \mathbf{r}}{\Delta s}$ is the chord of an arc divided by the length of the arc. In the limit, the chord and the arc coincide and $\dfrac{d\mathbf{r}}{ds}$ becomes a vector of unit length directed tangent to the path. Denoting this unit tangent vector by $\hat{\mathbf{e}}_t$ and recognizing that $\dfrac{ds}{dt}$ denotes the *speed* (i.e., magnitude of velocity or rate of change of distance) of the particle along the path, we obtain

$$\mathbf{v} = \frac{ds}{dt} \hat{\mathbf{e}}_t = v \hat{\mathbf{e}}_t \tag{9-2.2}$$

ACCELERATION

We define acceleration as the time rate of change of velocity. If $\Delta \mathbf{v}$ is the change in velocity during the time Δt, the average acceleration will be

$$\mathbf{a}_{\text{ave}} = \frac{\Delta \mathbf{v}}{\Delta t}$$

and the instantaneous acceleration is the limit of this ratio as Δt approaches zero, or

$$\mathbf{a} = \underset{\Delta t \to 0}{\text{limit}} \frac{\Delta \mathbf{v}}{\Delta t} = \frac{d\mathbf{v}}{dt} \tag{9-2.3}$$

Since $\mathbf{v} = \dfrac{d\mathbf{r}}{dt}$, the instantaneous acceleration may also be written as

$$\mathbf{a} = \frac{d\mathbf{v}}{dt} = \frac{d\left(\dfrac{d\mathbf{r}}{dt}\right)}{dt} = \frac{d^2\mathbf{r}}{dt^2} \tag{9-2.3a}$$

Since time is so frequently the independent variable in dynamics, it will often simplify notation if we use the dot convention devised by Newton to signify differentiation with respect to time. This consists of a single dot above a variable to denote its first derivative with respect to time, while two dots above a variable denote its second time derivative. When we use this convention, velocity and acceleration may be expressed as

$$\mathbf{v} = \frac{d\mathbf{r}}{dt} = \dot{\mathbf{r}} \tag{9-2.1}$$

$$\mathbf{a} = \frac{d\mathbf{v}}{dt} = \dot{\mathbf{v}} = \ddot{\mathbf{r}} \tag{9-2.3}$$

Let us now discuss more explicitly how the change in the velocity vector is correlated to acceleration. The change in a vector is easy to see if the vectors radiate from the same reference point. While this is natural for the position vectors as in Fig. 9-2.2a, it must be contrived for the velocity vectors. We do this by representing the velocities as free vectors, radiating from a common origin as in Fig. 9-2.2b where the path, for ease of representation, is assumed to be in the plane of the paper. A visual comparison between the simultaneous changes in the position and velocity vectors is achieved by drawing the position vector **r** to the particle at *equal*

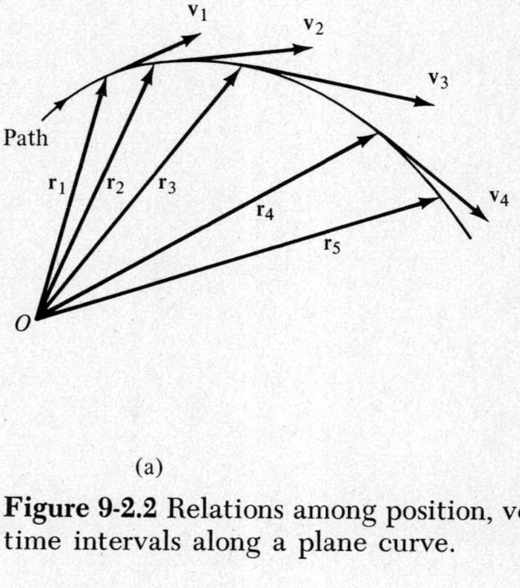

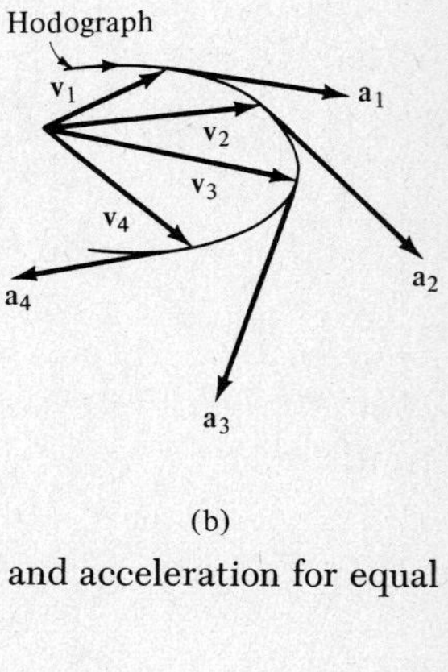

Figure 9-2.2 Relations among position, velocity, and acceleration for equal time intervals along a plane curve.

successive intervals of time. Then as shown in Fig. 9-2.2a, the velocity is tangent to the path and its magnitude is proportional to the distance from one position vector head to the next.

We now draw the successive velocity vectors from a common origin as in Fig. 9-2.2b. Then it is obvious that the change in velocity (drawn from one velocity vector head to the next) is caused by both the swinging and stretching of the velocity vector; that is, by changes in both the direction and the magnitude of the velocity vector. In addition, just as the path of the particle is generated by the head of the position vector, we can imagine a curve being generated simultaneously by the head of the velocity vector. This curve is called the *hodograph.* A chord of the hodograph, drawn from one velocity vector head to the next, is the change in velocity during the corresponding time interval. In the limit as the time interval approaches zero, this chord coincides with the hodograph so that the acceleration vector is tangent to the hodograph in the analogous manner that the velocity is tangent to the corresponding point of the actual path of the particle. As Fig. 9-2.2b demonstrates, the velocity and acceleration have different directions so that the acceleration is *not* tangent to the path of the particle except for the special case of straight-line motion.

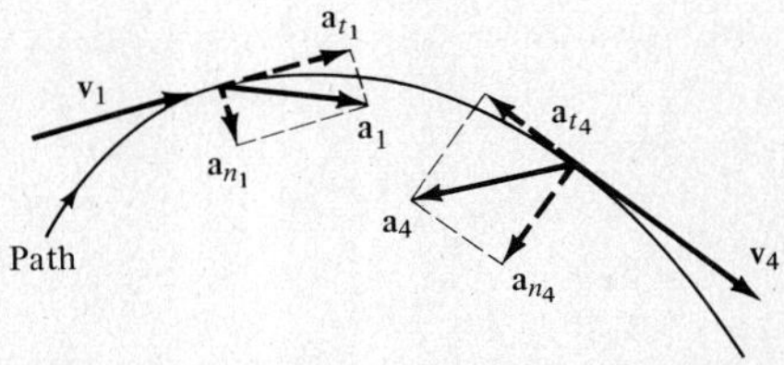

Figure 9-2.3 Velocity and acceleration related to path. Particle is speeding up at position 1, but slowing down at 4.

The correlation between velocity and acceleration is explained further in Fig. 9-2.3 where we have drawn the instantaneous velocity and acceleration vectors on the path itself for positions 1 and 4. At each position, the acceleration vector may be resolved into components that are tangent and normal to the path. In every instance, the normal component of acceleration will be directed inward toward the instantaneous center of curvature of the path. As we shall show later in Section 9-7, this normal component of acceleration occurs because of the changing *direction* of the velocity. The tangential component of acceleration coincides with the velocity and denotes the changing *magnitude* of the velocity. Indeed, if we consider the particle to be an automobile moving along a curving road, the car's speedometer measures the *speed* or magnitude of velocity while the tangential acceleration represents the variation in speed, being positive as we increase speed and vice versa.

UNITS

The units used for displacement, velocity, and acceleration depend on the units chosen to represent length and time, such as foot, centimeter, and mile for length; and second, minute, and hour for time. Accordingly, since displacement is synonymous with length, velocity with change of length per unit time, and acceleration with change of velocity per unit time, the common units for these terms are:

Displacement: foot, inch, centimeter, mile
Velocity: feet per second (fps), inches per second (ips), centimeters per second (cm/sec), miles per hour (mph), etc.
Acceleration: feet per second per second (fps^2), inches per second per second (ips^2), miles per hour per hour (mph^2), etc.

9-3 RECTILINEAR MOTION

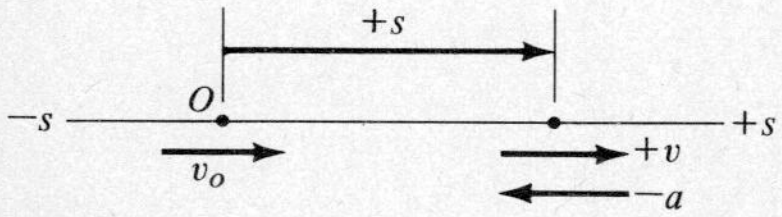

Figure 9-3.1

In rectilinear motion, the particle travels along a straight line. Since the motion is unidirectional, the sense of the position, velocity, and acceleration vectors may be denoted by a plus or minus sign as in Fig. 9-3.1. We may therefore rewrite the vector differential equations of motion in terms of the position coordinate s, measured along the rectilinear path, rather than the position vector $\mathbf{r}$ to obtain

$$v = \frac{ds}{dt} \tag{9-3.1}$$

$$a = \frac{dv}{dt} \tag{9-3.2}$$

$$a\,ds = v\,dv \tag{9-3.3}$$

where (9-3.3) has been obtained by eliminating dt from the others.

SIGNS

A convenient sign convention is to take the initial direction of motion as the positive sense of s, v, and a. Hence, a negative value of velocity obtained in applying the equations would mean that the velocity is directed oppositely to the initial direction of motion. A negative value for displacement would indicate that the position of the moving particle is to be measured back from the origin of displacement. Finally, if a moving particle returns to its starting position, the displacement s will be zero, *not* the distance actually traversed by the particle. Read the discussion of Fig. 9-2.1 on p. 330 again.

We now turn to the solution of the differential equations of motion. Our problem is to determine the relations among s, v, a, and t when a relation between any two of them is specified. Basically, we have three principal variables related by a common parameter t. Each of these principal variables may be known in terms of the time, or they may be specified in terms of each other or even a combination of the others. Let us consider here the simpler combinations indicated in the following box.

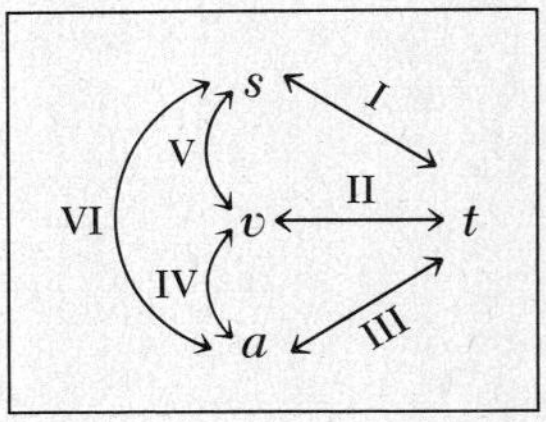

Case I. Given the displacement in terms of the time, i.e., $s = f(t)$, to find the v-t and a-t relations.

This is the simplest case and is easily solved by two successive differentiations of the s-t relation; thus, $v = \dfrac{ds}{dt}$ and $a = \dfrac{dv}{dt} = \dfrac{d^2s}{dt^2}$.

Case II. Given the velocity in terms of the time, i.e., $v = f(t)$, to find the a-t and s-t relations.

Applying the definition $a = \dfrac{dv}{dt}$ gives the a-t relation directly, while separating the variables in $v = \dfrac{ds}{dt}$ gives $\int ds = \int f(t)\, dt$ which may be integrated to find the s-t relation.

Case III. Given the acceleration in terms of the time, i.e., $a = f(t)$, to find the v-t and s-t relations.

From $a = f(t) = \dfrac{dv}{dt}$, we separate the variables v and t to obtain $dv = f(t)\, dt$ which may be integrated to determine the v-t relation, whence we proceed as outlined in Case II to find the s-t relation.

When the principal variables are not given as functions of the time, direct differentiation or integration cannot be performed without a preliminary treatment which we now discuss.

Case IV. Given $a = f(v)$, to find the a-t, v-t, and s-t relations.

Starting with $a = f(v) = \dfrac{dv}{dt}$, we separate the variables v and t to obtain $\dfrac{dv}{f(v)} = dt$ which is integrated to determine the v-t relation. Then we can proceed as outlined above in Case II.

Case V. Given $v = f(s)$, to find the s-t, v-t, and a-t relations.

Here, as in Case IV, one of the principal variables is given in terms of an adjacent variable. From the definition of velocity, we have $v = f(s) = \dfrac{ds}{dt}$ whence separating the variables s and t, we obtain $\dfrac{ds}{f(s)} = dt$ which upon integration determines the s-t relation. Thus, we have Case I and two successive differentiations yield the v-t and a-t relations.

Case VI. Given $a = f(s)$, to find the s-t, v-t, and a-t relations.

Here the principal variables are not adjacent as in the two previous cases. In this case, we substitute the given relation in $v\,dv = a\,ds$ to obtain $v\,dv = f(s)\,ds$ which is integrated and solved for $v = f(s)$. We now proceed as outlined in Case V.

The preceding outline is only a guide to mathematical analysis and should by no means be memorized or used as a set of rules. While the procedures are straightforward, some of the integrations sometimes may be complex enough to require reference to your calculus textbook.

UNIFORM ACCELERATION

One variation of Case III is so often used that we discuss it separately. This is the case of straight-line motion in which the acceleration is constant. As will be seen later (Section 11-2), this condition arises when a body is acted upon by forces which remain constant in magnitude and direction.

The usual assumptions are that at time $t = 0$, the initial velocity is v_o and $s = 0$. Applying $dv = a\,dt$ and integrating between the given limits, we have

$$\int_{v_o}^{v} dv = a\int_{0}^{t} dt \qquad \text{or} \qquad \boxed{v = v_o + at} \tag{9-3.4}$$

Note that a is placed outside the integral sign because it is assumed constant.

We now replace the variable v, just found in terms of the time, in the differential equation $ds = v\,dt$ and again proceed to integrate between the given limits. This gives

$$\int_{0}^{s} ds = \int_{0}^{t} (v_o + at)\,dt \qquad \text{or} \qquad \boxed{s = v_o t + \tfrac{1}{2}at^2} \tag{9-3.5}$$

Finally, we use $v\,dv = a\,ds$ and again proceed to integrate between definite limits to obtain

$$\int_{v_o}^{v} v\,dv = a\int_{0}^{s} ds \qquad \text{or} \qquad \boxed{v^2 = v_o^2 + 2as} \tag{9-3.6}$$

Remember that these equations can only be used when the acceleration is known to be constant. A common error is the attempt to apply them to all types of motion.

For bodies falling near the surface of the earth under the influence of gravity only, the acceleration may be assumed constant at the value $g = 32.2$ fps^2 directed downward. In any motion involving freely falling bodies, therefore, the equations of motion for constant acceleration may be applied directly by merely replacing a by g. The most convenient sign convention to use is that in which the initial direction of motion is the positive sense for displacement, velocity, and acceleration.

ILLUSTRATIVE PROBLEMS

9-3.1. As shown in Fig. 9-3.2, a stone is thrown vertically into the air from a tower 100 ft high at the same instant that a second stone is thrown upward from the ground. The initial velocity of the first stone is 50 fps and that of the second stone is 75 fps. When and where will the stones be at the same height from the ground?

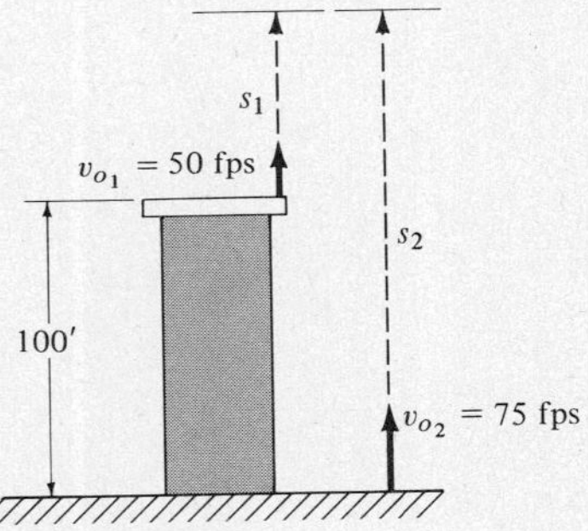

Figure 9-3.2

Solution

The initial direction of motion for each stone is upward which we therefore take as the positive sense for s, v, and a. Applying Eq. (9-3.5) and noting that the acceleration for freely falling bodies is $g = 32.2$ fps^2 directed downward and therefore negative, we obtain

$[s = v_o t + \frac{1}{2}at^2]$ For stone 1: $s_1 = 50t - 16.1t^2$ (*a*)

For stone 2: $s_2 = 75t - 16.1t^2$ (*b*)

From Fig. 9-3.2, $s_2 - s_1 = 100$ ft. Hence, subtracting Eq. (*a*) from Eq. (*b*) gives

$$s_2 - s_1 = 100 = 25t \quad \text{or} \quad t = 4 \text{ sec}$$

Substituting this value of t in Eqs. (*a*) and (*b*), we have

$$s_1 = 50(4) - 16.1(4)^2 = 200 - 257.6 \qquad s_1 = -57.6 \text{ ft} \qquad \textit{Ans.}$$

$$s_2 = 75(4) - 16.1(4)^2 = 300 - 257.6 \qquad s_2 = +42.4 \text{ ft} \qquad \textit{Ans.}$$

Hence, the stones pass each other 57.6 ft below the top of the tower, or 42.4 ft from the ground. Note that although we assumed that they would pass above the top of the tower, the negative sign of s_1 indicates otherwise. Since the terms involved in the equations are vector quantities, an incorrect assumption of direction merely results in a negative sign.

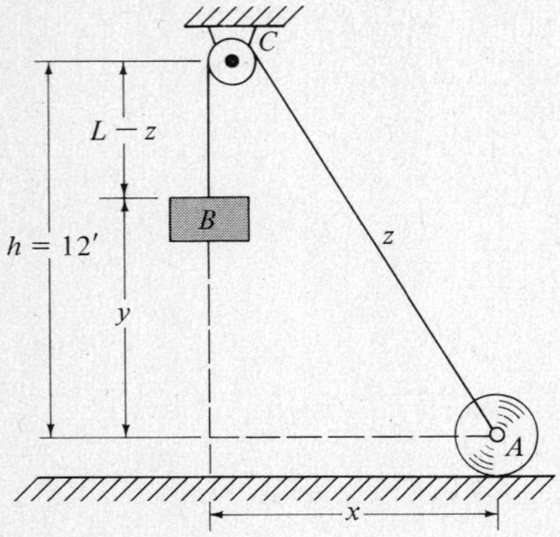

Figure 9-3.3

9-3.2. A rope of length L connects the wheel A and the weight B by passing over a pulley of negligible size at C as shown in Fig. 9-3.3. At the instant when $x = 9$ ft, the center of wheel A has a velocity $v_A = 10$ fps and an acceleration $a_A = 4$ fps^2, both rightward. What is then the velocity and acceleration of B?

Solution

If we denote the variable distance AC by z, then the vertical length $CB = L - z$ and hence

$$h = L - z + y \tag{a}$$

From the figure, we also have

$$z^2 = x^2 + h^2 \tag{b}$$

By eliminating z from these relations, y may be expressed directly in terms of x so that successive differentiation with respect to time will relate the velocity and acceleration of B to that of A. However, a preferable method is to retain z as a common parameter and proceed as follows:

Using the dot notation to indicate differentiation with respect to time, and noting that $\dot{y} = v_B$, we obtain from Eq. (a),

$$0 = -\dot{z} + \dot{y} \quad \text{or} \quad v_B = \dot{z} \tag{c}$$

This result is not surprising if we note that the change in length of z determines the rise (or fall) of B.

We next differentiate Eq. (b) with respect to the time which gives

$$2z\dot{z} = 2x\dot{x} \quad \text{or} \quad zv_B = xv_A \tag{d}$$

Another differentiation of Eq. (d) yields

$$\dot{z}v_b + z\dot{v}_B = \dot{x}v_A + x\dot{v}_A$$

or

$$v_B{}^2 + za_B = v_A{}^2 + xa_A \tag{e}$$

By substituting the given data in Eqs. (d) and (e) and noting that $z = 15$ ft when $x = 9$ ft, we obtain

[From Eq. (d)] $\quad 15v_B = 9(10) \qquad v_B = 6$ fps up ***Ans.***

[From Eq. (e)] $\quad (6)^2 + 15a_B = (10)^2 + 9(4) \qquad a_B = 6.67$ fps^2 up ***Ans.***

An interesting aspect of this example is that to determine the accelerations here, the velocities must first be found.

9-3.3. The rectilinear motion of a particle is given by $s = v^2 - 9$ where s is in ft and v in fps. When $t = 0$, $s = 0$ and $v = 3$ fps. Determine the s-t, v-t, and a-t relations.

Solution

The given data suggest the procedure outlined in Case V; i.e., $v = f(s)$. However, rather than follow this outline by expressing v in terms of s, we differentiate the given relation directly with respect to time, and obtain

$$\frac{ds}{dt} = 2v\frac{dv}{dt} \quad \text{or} \quad v = 2va \quad \text{or} \quad a = \frac{1}{2}\text{ fps}^2 \qquad \textit{Ans.}$$

which is the a-t relation and shows a to be constant.

Since the acceleration is constant, we proceed directly to apply Eqs. (9-3.4) and (9-3.5) which are valid for constant acceleration. We therefore obtain

$$[v = v_o + at] \qquad v = 3 + \tfrac{1}{2}t \qquad \textit{Ans.}$$

$$[s = v_o t + \tfrac{1}{2}at^2] \qquad s = 3t + \tfrac{1}{4}t^2 \qquad \textit{Ans.}$$

9-3.4. The motion of a particle is defined by the relation $a = 2t$ where a is in fps^2 and t is in seconds. It is known that $s = 4$ ft and $v = 2$ fps when $t = 1$ sec. Find s and v at $t = 4$ sec.

Solution

Since a is given in terms of t, we start with $a = dv/dt$ in which we substitute $a = 2t$ to obtain $2t = dv/dt$. Separating the variables and integrating between the given limits, we have

$$[dv = a\,dt] \qquad \int_2^v dv = \int_1^t 2t\,dt$$

whence

$$v - 2 = t^2 - 1 \quad \text{or} \quad v = t^2 + 1 \qquad (a)$$

We now replace the variable v, just found in terms of the time, in the definition of velocity written in the form $ds = v\,dt$ and again proceed to integrate between the given limits. This gives

$$[ds = v\,dt] \qquad \int_4^s ds = \int_1^t (t^2 + 1)\,dt$$

whence

$$s = \frac{t^3}{3} + t + \frac{8}{3} \qquad (b)$$

Finally, substituting $t = 4$ sec in these v-t and s-t relations, we obtain

$$[\text{From Eq. } (a)] \qquad v = (4)^2 + 1 = 17 \text{ fps} \qquad \textit{Ans.}$$

$$[\text{From Eq. } (b)] \qquad s = \frac{(4)^3}{3} + 4 + \frac{8}{3} = 28 \text{ ft} \qquad \textit{Ans.}$$

A common error to be avoided is to substitute in $ds = v\,dt$ the particular value of v at $t = 4$ sec instead of using the v-t relation expressing v as a variable in terms of the time.

PROBLEMS

9-3.5. An automobile is driven at 30 mph for 12 min, then at 40 mph for 20 min, and finally at 50 mph for 8 min. What is its average speed over this interval?

9-3.6. How fast must the automobile of the previous problem move in the last 8 min to obtain an average speed of 35 mph?

9-3.7. Car A at a gasoline station stays there for 10 min after a car B passes at an average speed of 40 mph. How long will it take car A moving at an average speed of 50 mph to overtake car B?

40 min **Ans.**

9-3.8. On a certain stretch of track, trains run at 60 mph. How far back of a stopped train should a warning torpedo be placed to signal an oncoming train? Assume that the brakes are applied at once and retard the train at the uniform rate of 4 fps^2.

9-3.9. A stone is thrown vertically upward and returns to earth in 5 sec. How high does it go?

$h = 100.6$ ft **Ans.**

9-3.10. A ship being launched slides down the ways with a constant acceleration. She takes 4 sec to slide the first foot. How long will she take to slide down the ways if their length is 900 ft?

$t = 2$ min **Ans.**

9-3.11. A ball is thrown vertically into the air at 120 fps. After 3 sec, another ball is thrown vertically. What initial velocity must the second ball have to pass the first ball at 100 ft from the ground?

$v_o = 84.9$ fps **Ans.**

9-3.12. A stone is dropped down a well and 5 sec later the sound of the splash is heard. If the velocity of sound is 1120 fps, what is the depth of the well?

353 ft **Ans.**

9-3.13. A stone is dropped into a well with no initial velocity and 4.5 sec later the splash is heard. Then a second stone is thrown downward into the well with an initial velocity v_o and the splash is heard in 4.0 sec. If the velocity of sound is constant at 1120 fps, determine the initial velocity of the second stone.

9-3.14. A train moving with constant acceleration travels 24 ft during the 10th sec of its motion and 18 ft during the 12th sec of its motion. Find its initial velocity.

$v_o = 52.5$ fps **Ans.**

9-3.15. An automobile starting from rest speeds up to 40 fps with a constant acceleration of 4 fps^2, runs at this speed for a time, and finally comes to rest with a deceleration of 5 fps^2. If the total distance traveled is 1200 ft, find the total time required.

9-3.16. An auto A is moving at 20 fps and accelerating at 5 fps^2 to overtake an auto B which is 384 ft ahead. If auto B is moving at 60 fps and decelerating at 3 fps^2, how soon will A pass B?

$$t = 16 \text{ sec} \qquad \textit{Ans.}$$

9-3.17. A balloon rises from the ground with a constant acceleration of 3 fps^2. Five seconds later, a stone is thrown vertically up from the launching site. What must be the minimum initial velocity of the stone for it to just touch the balloon? Note that the balloon and the stone have the same velocity at contact.

$$v_o = 66.4 \text{ fps} \qquad \textit{Ans.}$$

9-3.18. The rectilinear motion of a particle is governed by the equation $s = r \sin \omega t$ where r and ω are constants. Show that the acceleration is $a = -\omega^2 s$.

9-3.19. The motion of a particle along a straight line is defined by $s = \frac{1}{3}t^3 - 36t$. (a) Find the average acceleration during the fourth second. (b) When the particle reverses its direction, what is its acceleration?

$$\text{(a) } 7 \text{ fps}^2; \text{ (b) } 12 \text{ fps}^2 \qquad \textit{Ans.}$$

9-3.20. A ladder of length L moves with its ends in contact with a vertical wall and a horizontal floor. If the ladder starts from a vertical position and its lower end A moves along the floor with a constant velocity v_A, show that the velocity of the upper end B is $v_B = -v_A \tan \theta$ where θ is the angle between the ladder and the wall. What does the minus sign mean? Is it physically possible for the upper end B to remain in contact with the wall throughout the entire motion? Explain.

9-3.21. In the previous problem, find the acceleration of the upper end B of the ladder as a function of θ.

9-3.22. The velocity of a particle moving along the X axis is defined by $v = kx^3 - 4x^2 + 6x$ where v is in fps, x is in ft, and k is a constant. If $k = 1$, compute the value of the acceleration when $x = 2$ ft.

$$a = 8 \text{ fps}^2 \qquad \textit{Ans.}$$

9-3.23. In the previous problem, find the smallest value of k that will make the acceleration equal to 16 fps^2 at $x = 3$ ft.

9-3.24. Because of the resistance exerted by a fluid, the rectilinear motion of a particle is given by $a = -kv$ where k is a constant. When $t = 0$, $s = 0$ and $v = v_o$. Determine the particle's velocity as a function of (a) the time t and (b) its position s. (c) What is the maximum distance the particle will move?

$$\text{(c) max. } s = v_o/k \qquad \textit{Ans.}$$

9-3.25. The motion of a particle in rectilinear motion is defined by

$a = 6\sqrt{v}$ where a is in fps^2 and v in fps. When $t = 2$ sec, $v = 36$ fps and $s = 30$ ft. Determine the value of s at $t = 3$ sec.

$s = 87$ ft **Ans.**

9-3.26. The rectilinear motion of a particle is governed by $a = -8s^{-2}$ where a is in fps^2 and s is in feet. When $t = 1$ sec, $s = 4$ ft, and $v = 2$ fps. Determine the acceleration of the particle at $t = 2$ sec.

$a = -0.237$ fps^2 **Ans.**

9-3.27. The straight line motion of a particle is defined by $a = 3 + \frac{1}{2}t$. When $t = 0$, $s = 2$ ft and $v = -4$ fps. Find s at $t = 6$ sec.

$s = 50$ ft **Ans.**

9-3.28. The rectilinear motion of a particle is governed by $a = 12t - 6t^2$. It starts from rest when $t = 0$. Determine its velocity when it returns to its starting position.

*9-4 MOTION CURVES

The use of motion curves which show the variation of s, v, and a with time frequently provides a preferred method of solving the problems considered before. The method is especially useful when the motion has distinct phases, each requiring its associated set of equations. The method also provides a means of using experimental data to determine the s-t, v-t, or a-t curves when any one of them is known. Especially useful is the a-t curve showing the variation of acceleration with time since it alone is sufficient to determine values of velocity and displacement at any instant of the entire motion.

Consider the a-t curve in Fig. 9-4.1a and assume that the velocity v_1 and displacement s_1 are known at the time t_1. The velocity v_2 at any other time t_2 is found by writing the definition of acceleration in the form $dv = a\,dt$ and integrating between the corresponding limits. This gives

$$\Delta v = \int_{v_1}^{v_2} dv = \int_{t_1}^{t_2} a\,dt$$

The geometric significance of the right-hand term is apparent from Fig. 9-4.1a. During the infinitesimal time interval dt, the acceleration a may be considered constant. Obviously $a\,dt$ is the area of the unshaded vertical strip. Since $\int_{t_1}^{t_2} a\,dt$ means the summation of such strips, we conclude that the shaded area under the a-t curve between the times t_1 and t_2 represents the velocity change Δv during this time interval, or

$$v_2 - v_1 = \Delta v = (\text{Area})_{a\text{-}t} \qquad (9\text{-}4.1)$$

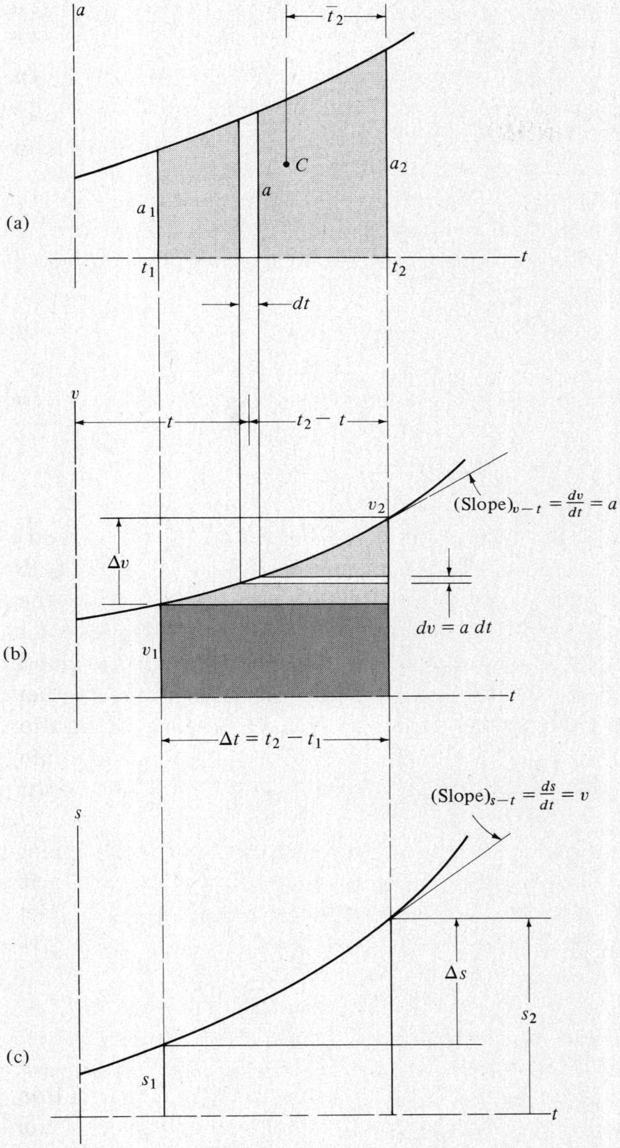

Figure 9-4.1 Relations among *a-t*, *v-t*, and *s-t* curves.

Similarly, writing the definition of velocity in the form $ds = v\,dt$ and integrating between corresponding limits, we obtain

$$\int_{s_1}^{s_2} ds = \int_{t_1}^{t_2} v\,dt \qquad \text{or} \qquad s_2 - s_1 = \Delta s = (\text{Area})_{v\text{-}t} \qquad (9\text{-}4.2)$$

where in similar fashion in part (b) the shaded area under the *v-t* curve represents the corresponding change in displacement during the time interval from t_1 to t_2.

Instead of using a summation of vertical elements $v\,dt$ to determine the area under the *v-t* curve, the alternate subdivision shown in part (b) for this area leads to a very useful result. Thus, let the shaded area of part (b) be the sum of the rectangle of area $v_1(t_2 - t_1)$ and the sum of the unshaded horizontal strips of area $dv(t_2 - t)$. Note that v_1 is the velocity at the start of the time interval $(t_2 - t_1) = \Delta t$, and that dv is the increment of velocity at any time t during this interval. Then we have

$$s_2 - s_1 = \Delta s = (\text{Area})_{v\text{-}t} = v_1(\Delta t) + \int_{v_1}^{v_2} dv(t_2 - t)$$

Substituting $dv = a\,dt$ in the integral, we obtain

$$\Delta s = v_1(\Delta t) + \int_{t_1}^{t_2} (a\,dt)(t_2 - t)$$

To interpret the meaning of the second right-hand term, we note that $a\,dt$ is the area of an elemental strip under the *a-t* curve and that $(t_2 - t)$ is the moment arm of this elemental strip about an ordinate through t_2. The integral therefore is the summation of moments of area of all such strips and is equivalent to the moment of area about the ordinate at t_2 of the area under the *a-t* curve included in the time interval $\Delta t = t_2 - t_1$. Hence, we obtain

$$\Delta s = v_1(\Delta t) + (\text{Area})_{a\text{-}t}(\bar{t}_2) \qquad (9\text{-}4.3)$$

in which $\bar{t}_2$ is the moment arm to the centroid C of the shaded area in part (a). In applying this equation, note that v_1 is the velocity at the start of the time interval and that the moment of the area under the *a-t* curve during the time interval is taken about an ordinate at the end of the time interval. When the area under the *a-t* curve is resolved into parts like those listed in Table 9-4.1, the total moment of area is the sum of the moments of area of its composite parts.

The use of Eq. (9-4.3) in conjunction with Eq. (9-4.1) is especially convenient as only the *a-t* curve is necessary to find the changes in displacement and in velocity.

Next consider the shape of the *v-t* curve. Since $a = dv/dt$, the slope dv/dt of the *v-t* curve at any instant such as t_2 is determined by the corresponding ordinate of the *a-t* curve. Since the shown values of a are positive and increasing, the corresponding slopes of the *v-t* curve are positive (i.e., up to the right) and increasingly steeper.

In similar fashion, the shape of the *s-t* curve is determined by $v = ds/dt$. Since ds/dt is the slope of the *s-t* curve, it has increasingly

Table 9-4.1. Properties of Areas

Equation	Graph	Area	Location of centroid
(zero degree) $y = h$	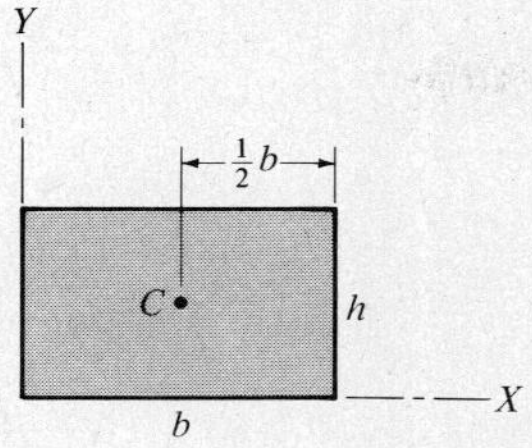	$\frac{1}{1}(bh)$	$\frac{1}{2}b$
(1st degree) $y = mx$	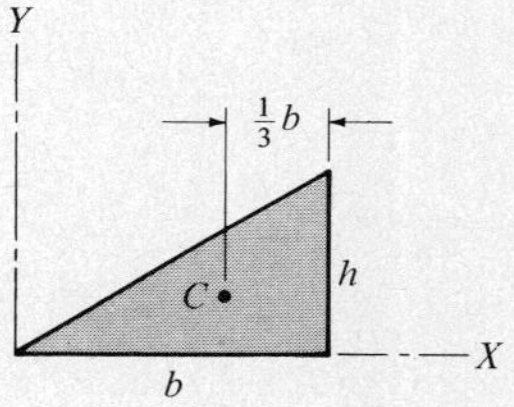	$\frac{1}{2}(bh)$	$\frac{1}{3}b$
(2nd degree) $y = kx^2$	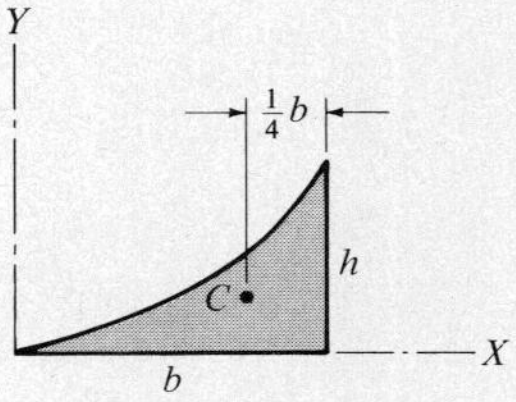	$\frac{1}{3}(bh)$	$\frac{1}{4}b$
(3rd degree) $y = kx^3$	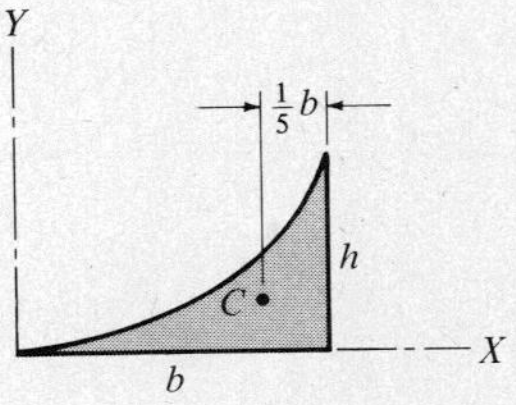	$\frac{1}{4}(bh)$	$\frac{1}{5}b$
(nth degree) $y = kx^n$	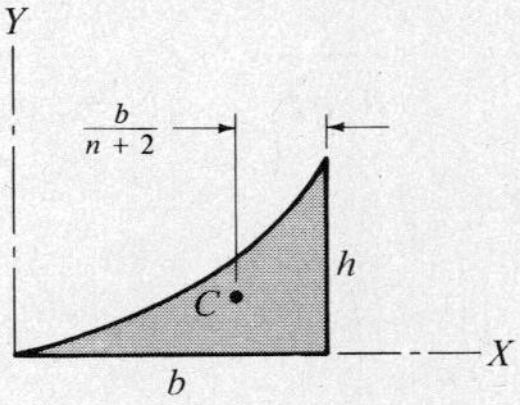	$\frac{1}{n+1}(bh)$	$\frac{1}{n+2}(b)$

positive steeper slopes to agree with the corresponding larger and positive ordinates of the v-t curve.

If you have studied or will study the relations between load, shear, and moment diagrams (see Chapter 8), you should observe that they are related in exactly the same manner as are the a-t, v-t, and s-t diagrams for motion curves. Except for a change in notation, one is exactly analogous to the other as shown in the following comparison:

Shear and Moment Diagrams	*Motion Curves*
$w = \frac{dV}{dx} = (\text{Slope})_{\text{shear}}$	$a = \frac{dv}{dt} = (\text{Slope})_{v\text{-}t}$
$V = \frac{dM}{dx} = (\text{Slope})_{\text{moment}}$	$v = \frac{ds}{dt} = (\text{Slope})_{s\text{-}t}$
$\Delta V = (\text{Area})_{\text{load}}$	$\Delta v = (\text{Area})_{a\text{-}t}$
$\Delta M = (\text{Area})_{\text{shear}}$	$\Delta s = (\text{Area})_{v\text{-}t}$

Two other motion curves are of minor interest. For the a-s curve in which acceleration is plotted against displacement, application of $\int_{s_1}^{s_2} a\, ds = \int_{v_1}^{v_2} v\, dv$ results in

$$\tfrac{1}{2}(v_2{}^2 - v_1{}^2) = (\text{Area})_{a\text{-}s} \tag{9-4.4}$$

For the v-s curve showing the variation of velocity with displacement, writing $a = v(dv/ds)$ shows that the acceleration can be found by multiplying v at any instant by the corresponding slope of the v-s curve.

ILLUSTRATIVE PROBLEMS

9-4.1. A particle, starting with an initial velocity of 60 fps, has a rectilinear motion with a constant deceleration of 10 fps^2. Determine the velocity and displacement at the end of 9 sec by sketching the a-t, v-t, and s-t curves and using the relations between them.

Solution

This elementary problem is easily solved by using the equations of constant acceleration, but its very simplicity will develop confidence in applying the relations between motion curves to more complex cases. Since the acceleration is constant but negative, the slope of the v-t curve is likewise constant and negative or directed down to

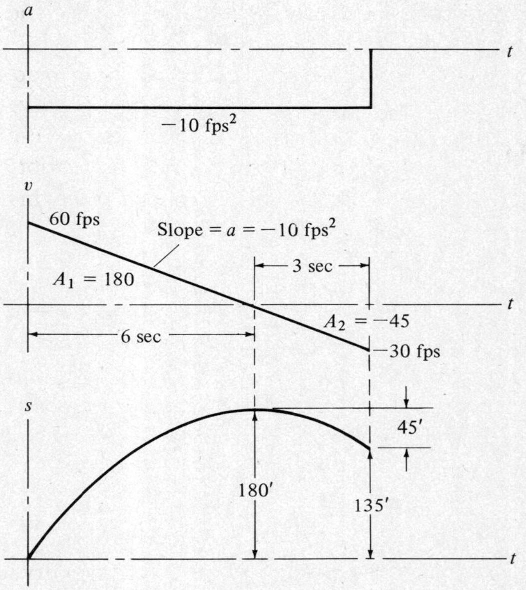

Figure 9-4.2

the right as shown in Fig. 9-4.2. The velocity change in the 9-sec interval is

$$[\Delta v = (\text{Area})_{a\text{-}t}] \qquad \Delta v = -10(9) = -90 \text{ fps}$$

which decreases the initial velocity of 60 fps to −30 fps.

Evidently the velocity is zero at $t_1 = 6$ sec so that the change in displacement can be computed as the algebraic sum of the positive and negative areas A_1 and A_2 under the v-t curve. Hence,

$$[\Delta s = (\text{Area})_{v\text{-}t}] \; \Delta s = \tfrac{1}{2}(60)(6) - \tfrac{1}{2}(30)(3) = 180 - 45 = 135 \text{ ft} \quad \textbf{\textit{Ans}}$$

Alternatively, Eq. (9-4.3) can be used to obtain Δs in the 9-sec interval. This gives

$$[\Delta s = v_o(\Delta t) + (\text{Area})_{a\text{-}t}(\bar{t}_2)]$$
$$\Delta s = 60(9) + (-10)(9)(\tfrac{9}{2}) = 135 \text{ ft} \qquad \textbf{\textit{Check}}$$

which, in this case, is identical to applying $s = v_o t + \frac{1}{2}at^2$.

The shape of the s-t curve is determined by noting that the velocity ordinate equals the slope at the corresponding ordinate of the s-t curve. Thus, at $t = 0$, the tangent to the s-t curve is up to the right, becoming less steep and eventually horizontal at $t = 6$ sec as the corresponding velocity ordinates gradually reduce to zero. Thereafter the tangents to the s-t curve are increasingly steeper down to the right as the velocity ordinates become increasingly negative.

Observe that here the s-t curve is a symmetric parabola with its vertex at $t = 6$ sec because at equal intervals to either side of this instant, the velocities are numerically equal but of opposite sign, thereby producing equal slopes but of opposite inclinations.

Finally, we observe that the shape of the s-t curve indicates that the moving particle reaches a maximum rightward displacement of 180 ft at $t = 6$ sec after which it returns leftward. The total *distance* traveled is the sum of these travels and is given by the sum of A_1 and A_2 or $180 + 45 = 225$ ft.

9-4.2. Draw the v-t and s-t curves for the two stones having the motions described in Illus. Prob. 9-3.1. The first stone is thrown vertically upward from a tower 100 ft high with a velocity of 50 fps at the same instant the second stone is projected upward from the ground with a velocity of 75 fps. When and where are the stones at the same level?

Solution

Both stones move with the same downward or negative acceleration of $g = -32.2$ fps^2. Consequently, from $a = (\text{Slope})_{v\text{-}t}$, both v-t curves have the same slope. But since the stones start with different velocities, their v-t curves are parallel as shown in Fig. 9-4.3. It is apparent that the stones have a constant *relative* velocity of $75 - 50 = 25$ fps.

Choosing the initial position of stone 2 as a common origin for displacement, we draw the s-t curves as shown. Their shapes are

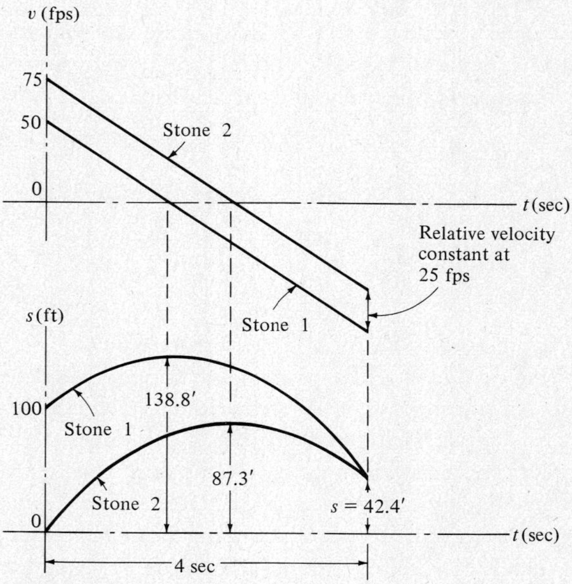

Figure 9-4.3

determined from making their slopes proportional to their corresponding velocity ordinates. Since the initial *relative* displacement of 100 ft between the stones is being reduced at the constant relative velocity of 25 fps, the stones meet in 4 sec. This checks the value previously computed in Prob. 9-3.1 and explains the physical significance of that computation. The common height s at this instant is most easily computed from

$$s = v_o t - \tfrac{1}{2}gt^2 = 75(4) - \tfrac{1}{2}(32.2)(4)^2 = 42.4 \text{ ft} \qquad \textbf{\textit{Ans.}}$$

9-4.3. The a-t curve for a particle having rectilinear motion is shown in Fig. 9-4.4. At $t = 0$, the velocity is 8 fps and the particle is 60 ft to the left of the origin of displacement. Draw the v-t and s-t curves, specifying values of v and s at $t = 4$ sec, 7 sec, and 13 sec.

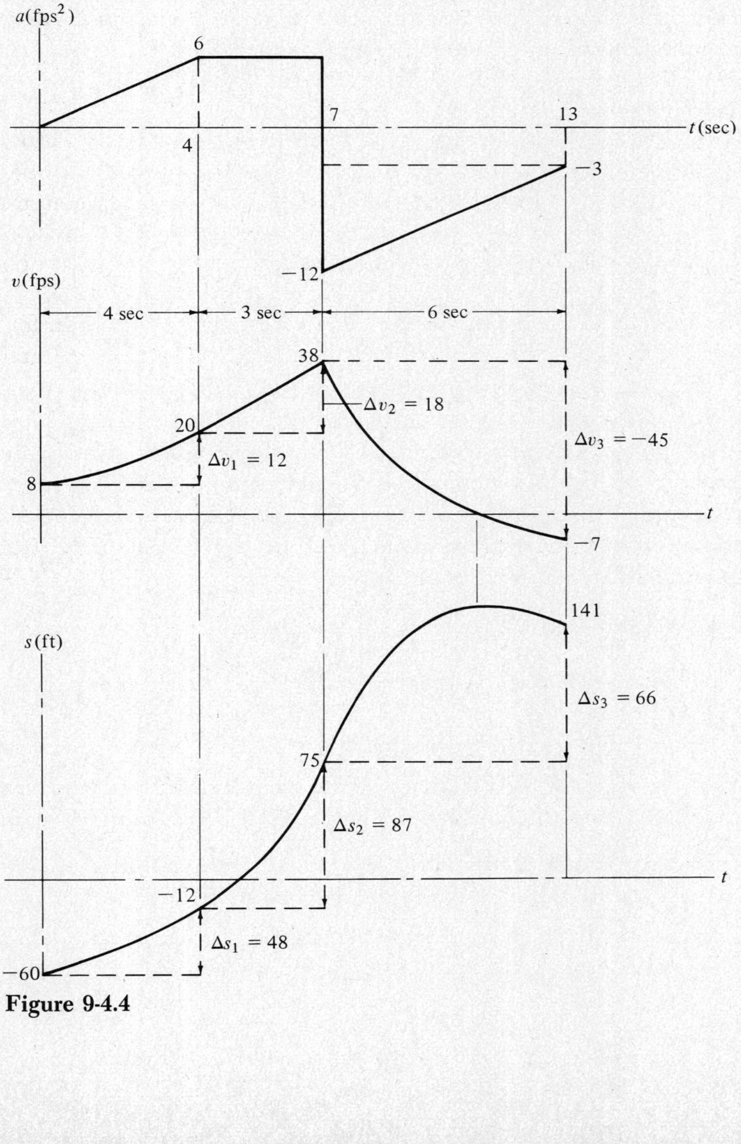

Figure 9-4.4

Solution

The changes in velocity during the specified time intervals of 4 sec, 3 sec, and 6 sec are found by applying Eq. (9-4.1) to the area under the *a-t* curve as follows:

$$[\Delta v = (\text{Area})_{a\text{-}t}] \qquad \begin{aligned} \Delta v_1 &= \tfrac{1}{2}(4)(6) = 12 \text{ fps} \\ \Delta v_2 &= (3)(6) = 18 \text{ fps} \\ \Delta v_3 &= -3(6) - \tfrac{1}{2}(9)(6) = -18 - 27 = -45 \text{ fps} \end{aligned}$$

The velocity ordinates are found by adding $\Delta v_1 = 12$ fps to the initial velocity of 8 fps to obtain 20 fps at $t = 4$ sec; to this result is added $\Delta v_2 = 18$ fps to give 38 fps at $t = 7$ sec; from this value we subtract $\Delta v_3 = 45$ fps to obtain -7 fps at $t = 13$ sec. The *v-t* curve is drawn through these points, its shape being determined by making the slope of the *v-t* curve at any instant proportional to the corresponding value of acceleration. Observe that each segment of the *v-t* curve is one degree higher than its corresponding part of the *a-t* diagram.

Now we may apply Eq. (9-4.2) to the area under this *v-t* curve to obtain the following changes in displacement. Observe that in the first 4-sec interval, the area is composed of a rectangle surmounted by a second-degree curve while in the second interval of 3 sec, the area consists of a rectangle plus a triangle.

$$[\Delta s = (\text{Area})_{v\text{-}t}] \qquad \begin{aligned} \Delta s_1 &= 8(4) + \tfrac{1}{3}(12)(4) = 48 \text{ ft} \\ \Delta s_2 &= 20(3) + \tfrac{1}{2}(18)(3) = 87 \text{ ft} \end{aligned}$$

The last increment of displacement Δs_3 cannot easily be found from the area under the *v-t* curve. Instead we find it by applying Eq. (9-4.3) to the last 6-sec interval. The velocity at the start of this interval is 38 fps. The moment of area under the *a-t* curve is found as the sum of the moments of areas of its rectangular and triangular parts, moments being taken about the time ordinate at the end of the interval. Hence, we obtain

$$[\Delta s = v_1(\Delta t) + (\text{Area})_{a\text{-}t}(\bar{t}_2)]$$

$$\Delta s_3 = (38)(6) + (-3)(6)\left(\frac{1}{2} \times 6\right) + \frac{(-9)(6)}{2}\left(\frac{2}{3} \times 6\right)$$

$$= 228 - 54 - 108 = 66 \text{ ft}$$

If desired, we could also have found the increments of displacement Δs_1 and Δs_2 directly from the *a-t* curve. Using this approach gives

$$\Delta s_1 = (8)(4) + \tfrac{1}{2}(6)(4)(\tfrac{1}{3} \times 4) = 32 + 16 = 48 \text{ ft} \qquad \textbf{\textit{Check}}$$

$$\Delta s_2 = (20)(3) + (6)(3)(\tfrac{1}{2} \times 3) = 60 + 27 = 87 \text{ ft} \qquad \textbf{\textit{Check}}$$

The required values of displacement are obtained by adding $\Delta s_1 = 48$ ft to the initial displacement of -60 ft to give $s = -12$ ft at $t = 4$ sec; then adding $\Delta s_2 = 87$ ft to this result to obtain $s = 75$ ft at $t = 7$ sec; finally adding $\Delta s_3 = 66$ ft to this value to obtain $s = 141$ ft at $t = 13$ sec. The shape of the *s-t* curve which is drawn through these points is determined by making the slope of the *s-t* curve at any instant proportional to the corresponding value of velocity.

PROBLEMS

9-4.4. The elevators in a modern office building are designed to accelerate or decelerate at a constant rate to or from 1800 fpm in 6 sec. With the aid of a *v-t* diagram, determine the minimum time for the elevator to rise 1200 ft.

46 sec **Ans.**

9-4.5. An auto is accelerated from rest to a top speed of 60 mph and then immediately decelerated to a stop. If the total elapsed time is 20 sec, determine the distance covered. The acceleration and deceleration are both constant but not necessarily of the same magnitude.

9-4.6. An auto starts from rest and comes to a stop $\frac{1}{4}$ mile away. If its acceleration and deceleration are limited to 10 fps^2 and 20 fps^2, respectively, what maximum speed does it reach if the elapsed time is to be as small as possible?

$v = 90.6$ mph **Ans.**

9-4.7. A car and a truck are both moving at 45 mph along a highway. To pass the truck safely, the car must move 200 ft relative to the truck. Determine the minimum time to do this if the car can accelerate at 6 fps^2 but must not exceed a speed of 60 mph.

9-4.8. An auto is to travel a distance from *A* to *B* of 1800 ft in exactly 40 sec. The auto accelerates and decelerates at 6 fps^2, starting from rest at *A* and coming to a stop at *B*. Find its maximum speed in fps.

$v = 60$ fps **Ans.**

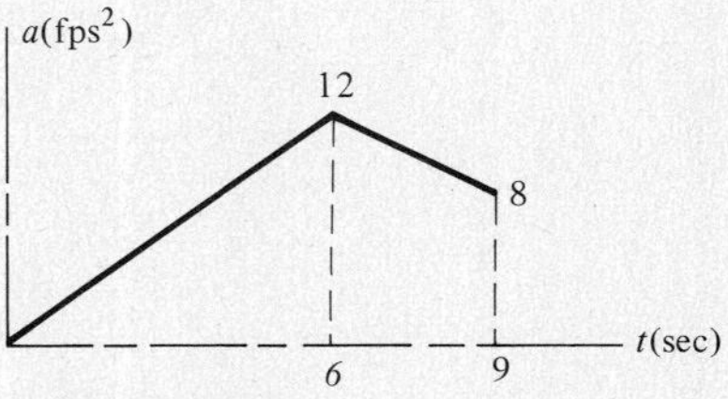

Figure P-9-4.9

9-4.9. The motion of a particle starting from rest is governed by the *a-t* curve shown in Fig. P-9-4.9. Sketch the *v-t* and *s-t* curves. Determine the displacement at $t = 9$ sec.

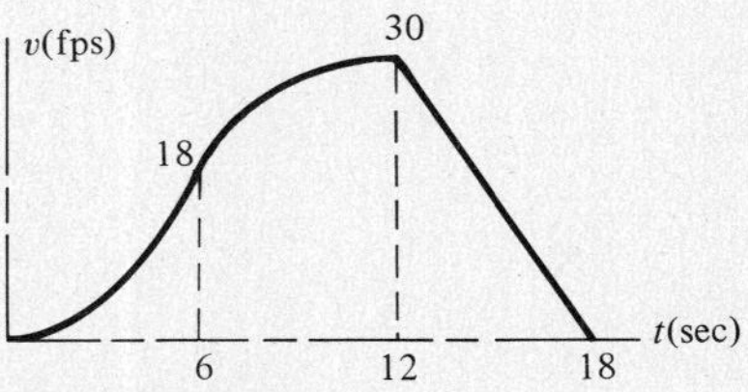

Figure P-9-4.10

9-4.10. The curved portions of the *v-t* curve shown in Fig. P-9-4.10 are second-degree parabolas with horizontal slope at $t = 0$ and $t = 12$ sec. Sketch the *a-t* and *s-t* curves if s_o is zero. Check values of *s* using both Eqs. (9-4.2) and (9-4.3).

$s = 282$ ft at $t = 18$ sec **Ans.**

9-4.11. One cycle of the *s-t* curve for a certain machine is shown in Fig. P-9-4.11 (p. 352). The curves are second-degree parabolas that are tangent at the indicated points of inflection and have zero slope at $t = 0$, 8, and 12 sec. Sketch the *v-t* and *a-t* curves and compute the maximum velocity and maximum acceleration.

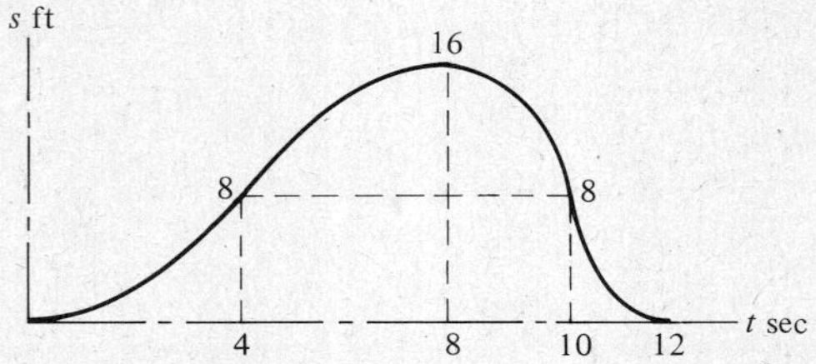

Figure P-9-4.11

9-4.12. The rate of change of acceleration, called "jerk," of an elevator is constant at ± 2 fps^3. Find the shortest time for the elevator, starting from rest, to rise 32 ft and stop.

$t = 8$ sec *Ans.*

9-4.13. The acceleration of a particle is given by $a = 18 - 3t$ where a is in fps^2 and t is in seconds. If the particle starts from rest, determine its velocity when it has returned to its initial position.

9-4.14. The motion of a particle is defined by $a = 10t - t^2$ where a is in fps^2 and t is in seconds. How far will it move from rest before starting to reverse its direction of motion?

$s = 1406$ ft *Ans.*

9-4.15. An auto starts from rest and reaches a speed of 60 fps in 15 sec. The acceleration increases uniformly from zero for the first 9 sec after which the acceleration reduces uniformly to zero in the next 6 sec. Compute the displacement in this 15-sec interval.

9-4.16. The acceleration of an object moving along a straight path decreases uniformly from 10 fps^2 to zero in 12 sec, at which time its velocity is 6 fps. Find its initial velocity and the change in position during the 12-sec interval.

$\Delta s = -168$ ft *Ans.*

9-4.17. An object attains a velocity of 48 fps with an acceleration which varies uniformly from 4 fps^2 to 12 fps^2 in 9 sec. Compute its initial velocity and the change in position during the 9-sec interval.

9-4.18. The velocity of a particle changes from -8 fps to 58 fps during a 12-sec interval during which its acceleration increases uniformly with time from an initial value of 3 fps^2. Determine the displacement covered in the 12-sec interval.

240 ft *Ans.*

9-5 INTRODUCTION TO VECTOR CALCULUS

In all motions except rectilinear motion, the vector quantities involved change in direction as well as in magnitude. Any change in the direction of a vector results in its derivative having a direction different from that of the original vector. This directional consideration is automatically accounted for in vector calculus; indeed, it is the basic difference between vector and scalar calculus. We shall emphasize and explain this distinction in subsequent applications.

For the present, we need only the following rules for the differentiation of the sums and products of two vectors. They are almost exactly the same as those for scalar calculus:

$$\frac{d}{dt}(\mathbf{A} + \mathbf{B}) = \frac{d\mathbf{A}}{dt} + \frac{d\mathbf{B}}{dt} \qquad (9\text{-}5.1)$$

$$\frac{d}{dt}(\mathbf{A} \cdot \mathbf{B}) = \frac{d\mathbf{A}}{dt} \cdot \mathbf{B} + \mathbf{A} \cdot \frac{d\mathbf{B}}{dt} \qquad (9\text{-}5.2)$$

$$\frac{d}{dt}(\mathbf{A} \times \mathbf{B}) = \frac{d\mathbf{A}}{dt} \times \mathbf{B} + \mathbf{A} \times \frac{d\mathbf{B}}{dt} \qquad (9\text{-}5.3)$$

These rules will be very useful in theoretical derivations. Notice that is is essential to maintain the order of the terms in differentiating the cross product.

One of the most useful methods of distinguishing between changes in magnitude and direction of a vector is given by the following rule for differentiating the product of a scalar and a vector, both of which are functions of t:

$$\frac{d}{dt}(n\mathbf{A}) = \frac{dn}{dt}\mathbf{A} + n\frac{d\mathbf{A}}{dt} \qquad (9\text{-}5.4)$$

To integrate a vector function, we reverse the process of differentiation. Thus, if

$$\mathbf{A} = \frac{d\mathbf{B}}{dt}$$

the integral of **B** is

$$\mathbf{B} = \int \mathbf{A}\, dt + \mathbf{C} \qquad (9\text{-}5.5)$$

where the constant of integration **C** is a vector, constant in both magnitude and direction. One of the most useful methods of performing the integration is to express the integrand in terms of constant unit vectors directed along fixed coordinate directions. For example, if

$$\mathbf{A} = A_x\hat{\mathbf{i}} + A_y\hat{\mathbf{j}} + A_z\hat{\mathbf{k}}$$

where A_x, A_y, A_z are scalar functions of t, the integral becomes

$$\begin{aligned} \mathbf{B} &= \int \mathbf{A}\, dt = \int (A_x\hat{\mathbf{i}} + A_y\hat{\mathbf{j}} + A_z\hat{\mathbf{k}})\, dt + \mathbf{C} \\ &= \hat{\mathbf{i}}\int A_x\, dt + \hat{\mathbf{j}}\int A_y\, dt + \hat{\mathbf{k}}\int A_z\, dt + \mathbf{C} \end{aligned} \qquad (9\text{-}5.6)$$

in which the constant unit vectors have been extracted from the integral sign and **C** is a vector constant of integration. Additional concepts will be developed in subsequent sections as they are needed.

9.6 RECTANGULAR COMPONENTS OF CURVILINEAR MOTION

We now consider the motion of a particle along a space curve generated by the tip of a variable position vector. With respect to a fixed set of reference axes as shown in Fig. 9-6.1, the position vector **r** of the moving particle P may be expressed as

$$\mathbf{r} = x\hat{\mathbf{i}} + y\hat{\mathbf{j}} + z\hat{\mathbf{k}} \tag{9-6.1}$$

where the rectangular coordinates x, y, z are functions of time. When we successively differentiate this expression to obtain velocity and acceleration, note that each of the right-hand terms are a product

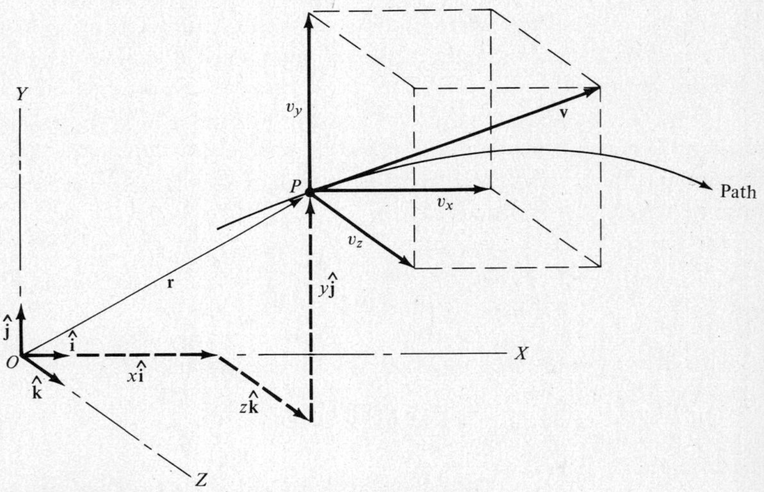

Figure 9-6.1 Rectangular components of velocity.

of a scalar and a vector. Hence, applying the rule expressed by Eq. (9-5.4), the time derivative of **r** is

$$\frac{d\mathbf{r}}{dt} = \frac{dx}{dt}\hat{\mathbf{i}} + x\frac{d\hat{\mathbf{i}}}{dt} + \frac{dy}{dt}\hat{\mathbf{j}} + y\frac{d\hat{\mathbf{j}}}{dt} + \frac{dz}{dt}\hat{\mathbf{k}} + z\frac{d\hat{\mathbf{k}}}{dt}$$

However, here the unit vectors $\hat{\mathbf{i}}$, $\hat{\mathbf{j}}$, $\hat{\mathbf{k}}$ are constant in magnitude as well as direction so that their derivatives are zero.

Hence, we find the velocity to be

$$\left.\begin{aligned}\mathbf{v} = \frac{d\mathbf{r}}{dt} &= \frac{dx}{dt}\hat{\mathbf{i}} + \frac{dy}{dt}\hat{\mathbf{j}} + \frac{dz}{dt}\hat{\mathbf{k}} \\ &= v_x\hat{\mathbf{i}} + v_y\hat{\mathbf{j}} + v_z\hat{\mathbf{k}}\end{aligned}\right\} \tag{9-6.2}$$

and the acceleration to be

$$\left.\begin{aligned} \mathbf{a} &= \frac{d\mathbf{v}}{dt} = \frac{dv_x}{dt}\hat{\mathbf{i}} + \frac{dv_y}{dt}\hat{\mathbf{j}} + \frac{dv_z}{dt}\hat{\mathbf{k}} \\ &= \frac{d^2\mathbf{r}}{dt^2} = \frac{d^2\mathbf{x}}{dt^2}\hat{\mathbf{i}} + \frac{d^2y}{dt^2}\hat{\mathbf{j}} + \frac{d^2z}{dt^2}\hat{\mathbf{k}} \\ &= a_x\hat{\mathbf{i}} + a_y\hat{\mathbf{j}} + a_z\hat{\mathbf{k}} \end{aligned}\right\} \qquad (9\text{-}6.3)$$

The significance of these results is that the motion of a particle is equivalent to the simultaneous projection of the motion along fixed rectangular coordinate axes. Each such projection is then a rectilinear motion, the details of which have been presented in Section 9-3.

When the sense and direction of a vector component is completely defined by its subscript (as is the situation here), it is often useful to replace formal vector notation with the following alternative form:

$$\begin{aligned} r &= x \mathrel{+\!\!\!\rightarrow} y \mathrel{+\!\!\!\rightarrow} z \\ v &= v_x \mathrel{+\!\!\!\rightarrow} v_y \mathrel{+\!\!\!\rightarrow} v_z \\ a &= a_x \mathrel{+\!\!\!\rightarrow} a_y \mathrel{+\!\!\!\rightarrow} a_z \end{aligned} \qquad (9\text{-}6.4)$$

where the symbol $\mathrel{+\!\!\!\rightarrow}$ is used to denote that we are dealing with the geometric sum of quantities whose magnitudes have known directions. The positive directions of these quantities are in the positive directions of the coordinate axes.

At other times, it may be convenient to ignore vector notation completely and consider only the scalar relations among components. Thus, for a moving particle whose position is defined by rectangular components x, y, and z which vary with time, the components of velocity and acceleration may be written as

$$v_x = \frac{dx}{dt}, \quad v_y = \frac{dy}{dt}, \quad v_z = \frac{dz}{dt}$$

and (9-6.5)

$$a_x = \frac{dv_x}{dt} = \frac{d^2x}{dt^2}, \quad a_y = \frac{dv_y}{dt} = \frac{d^2y}{dt^2}, \quad a_z = \frac{dv_z}{dt} = \frac{d^2z}{dt^2}$$

We shall use each of these notations from time to time because each has its own particular advantages. The formal vector notation provides a method of deriving complex results in a compact manner. When it comes to solving numerical problems, however, the vector representations must be expressed in scalar form, and Eqs. (9-6.4) and (9-6.5) are convenient ways of keeping track of the details. Later we shall discuss other scalar representations of a vector quantity, but these will be just different ways of examining the same physical entity.

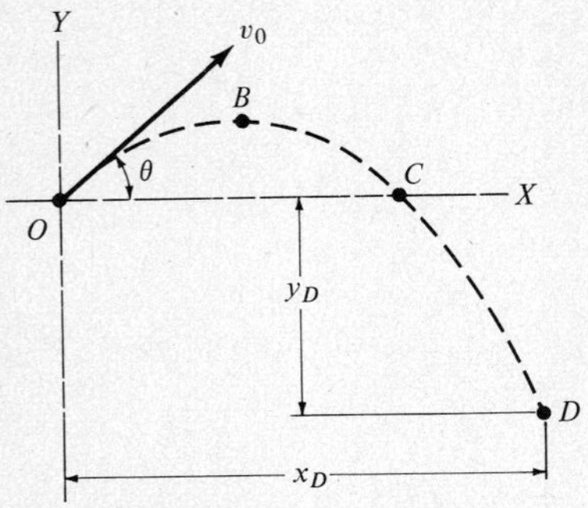

Figure 9-6.2 Flight of projectile.

FLIGHT OF PROJECTILES, AIR RESISTANCE NEGLECTED

Consider the motion of a projectile in which such factors as wind velocity, air resistance, and rotation of the earth, which affect the actual flight of the projectile, will be neglected. Then the flight path of the projectile will be a plane curve such as *OBCD* shown in Fig. 9-6.2. It is convenient to resolve the curvilinear motion along the path into rectilinear motions along *X* and *Y* axes having an origin at the initial point of flight. The initial velocity is denoted by v_o directed at an angle θ with the *X* axis.

From the initial direction of motion, rectilinear displacements will be positive both rightward and upward. Since the only force assumed to be acting on the projectile is its own weight, its total acceleration at all positions is due to gravity and is directed vertically downward with the value g. Hence, the rectangular components of this acceleration are constant at $a_x = 0$ and $a_y = -g$.

Since these components of acceleration are constant, we may use the equations of rectilinear motion with constant acceleration to determine the corresponding rectangular components of the curvilinear motion. This procedure is explained in the accompanying table which should be clear enough to eliminate memorizing the results.

Flight of Projectiles[a]

Rectilinear motion with constant acceleration	X component of flight ($a_x = 0$, $v_{o_x} = v_o \cos\theta$)	Y component of flight ($a_y = -g$, $v_{o_y} = v_o \sin\theta$)
$v = v_o + at$	$v_x = v_{o_x} + a_x t$ or $v_x = v_o \cos\theta$	$v_y = v_{o_y} + a_y t$ or $v_y = v_o \sin\theta - gt$
$s = v_o t + \frac{1}{2}at^2$	$x = v_{o_x} t + \frac{1}{2}a_x t^2$ or $x = (v_o \cos\theta)t$	$y = v_{o_y} t + \frac{1}{2}a_y t^2$ or $y = (v_o \sin\theta)t - \frac{1}{2}gt^2$

[a] If air resistance, etc., were not neglected, a_x and a_y would become variable quantities, and the differential equations of motion—Eqs. (9-3.1), (9-3.2), and (9-3.3)—would have to be solved to obtain the proper equations for the flight of the projectile.

If, as in Fig. 9-6.2, the time of flight is less than that required to reach *C*, the projectile will be above its initial position and values of the *Y* displacement will be positive. If the time of flight is more than that required to reach *C*, the projectile will be on the path *CD* and values of the *Y* displacement will then be negative. At the topmost point of flight *B*, the value of v_y will be zero.

ILLUSTRATIVE PROBLEMS

9-6.1. In Fig. 9-6.3, the pivoted rigid link AB carries a pin P whose position is controlled by the horizontal slotted bar. At the instant that $y = 2$ in., the slotted bar is moving upward at a constant velocity of 4 ips. What is then the x components of velocity and acceleration of the pin P?

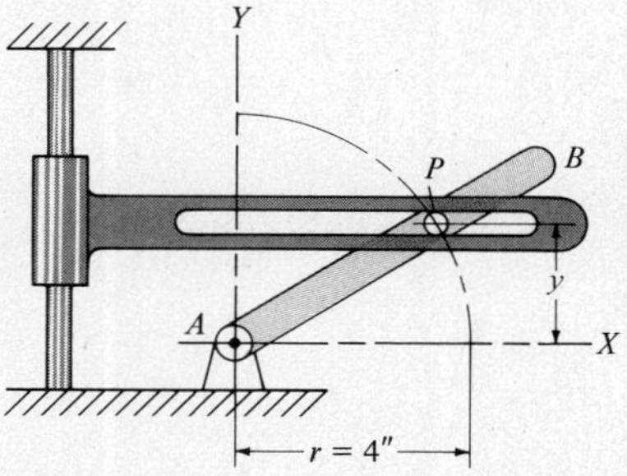

Figure 9-6.3

Solution

The motion of P is common to the upward motion of the bar and the circular path $x^2 + y^2 = 16$ where at $y = 2$ in., $x = 3.46$ in. As determined by the bar, $\dot{y} = 4$ ips and $\ddot{y} = 0$. On the circular path, the velocity and acceleration of P are determined by two successive time differentiations of the path to be

$$x\dot{x} + y\dot{y} = 0 \qquad (a)$$
$$x\ddot{x} + \dot{x}^2 + y\ddot{y} + \dot{y}^2 = 0 \qquad (b)$$

Into Eq. (a), we substitute the known values of x, y, and $\dot{y}$ to obtain

$$3.46\dot{x} + 2(4) = 0 \quad \text{whence} \quad \dot{x} = -2.31 \text{ ips} \qquad \textit{Ans.}$$

With $\dot{x}$ known, Eq. (b) now gives

$$3.46\ddot{x} + (2.31)^2 + 2(0) + (4)^2 = 0$$

whence

$$\ddot{x} = -6.16 \text{ ips}^2 \qquad \textit{Ans.}$$

9-6.2. A projectile is fired from the top of a cliff 300 ft high with a velocity of 1414 fps directed at 45° to the horizontal. Find the range on a horizontal plane through the base of the cliff.

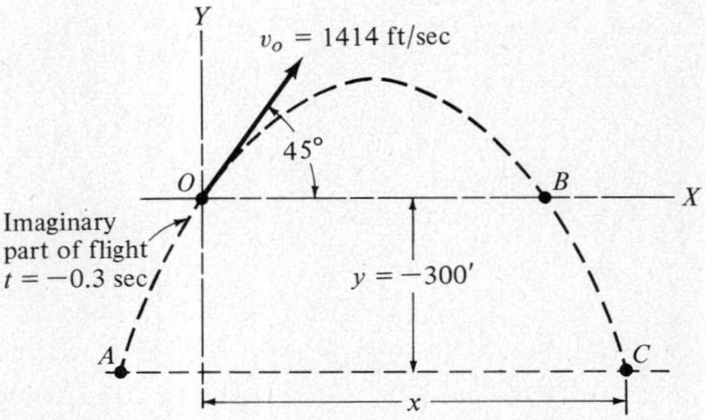

Figure 9-6.4

Solution

Figure 9-6.4 represents the conditions of the problem. The initial components of velocity are v_{o_x} = components. Then the components of dis-

$v_{o_y} = 1000$ fps directed rightward and upward. When we choose these directions as the positive senses of all vector components, Y displacement is positive upward and the downward acceleration g is negative; furthermore, the final position of the projectile is below the origin and hence negative. Therefore, we have

$$[y = v_{o_y}t - \tfrac{1}{2}gt^2] \qquad -300 = 1000t - 16.1t^2 \qquad (a)$$

$$t = 62.4 \text{ sec} \quad \text{or} \quad t = -0.3 \text{ sec}$$

Using the positive value of t, we obtain

$$[x = v_{o_x}t] \qquad x = 1000(62.4) = 62{,}400 \text{ ft} \qquad \textbf{\textit{Ans.}}$$

The negative value of t obtained from Eq. (a), i.e., $t = -0.3$ sec, may be interpreted as the time required for the projectile to leave the base of the cliff at A and rise to the origin O. This is also the time required for the projectile to travel from B to C. This observation may be checked by finding the time required to travel distance OB from

$$[y = v_{o_y}t - \tfrac{1}{2}gt^2] \qquad 0 = 1000t - 16.1t^2 \quad \text{or} \quad t = 62.1 \text{ sec}$$

Adding 0.3 sec to this value gives the total time of flight to be 62.4 sec as before.

9-6.3. Determine the position at which a ball thrown up to the right will strike the inclined surface shown in Fig. 9-6.5. The initial velocity of the ball is 100 fps directed at $\theta = \tan^{-1}\frac{4}{3}$ with the horizontal.

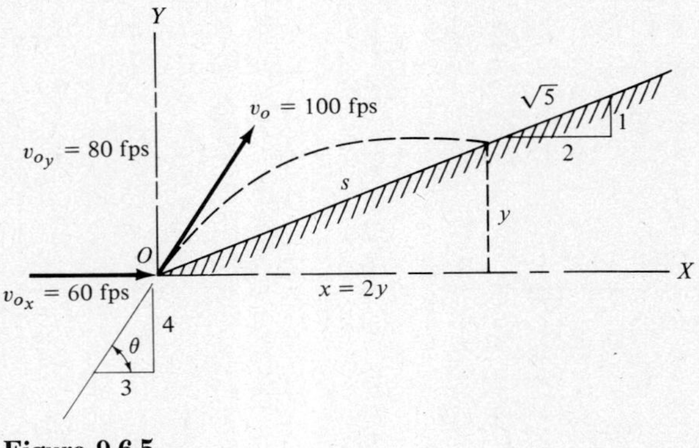

Figure 9-6.5

Solution

Because of the slope of the incline, the components of displacement are $x = 2y$. The initial components of velocity are $v_{o_x} = 60$ fps and $v_{o_y} = 80$ fps. Their directions establish the positive sense of all vector placement are

$$[x = v_{o_x} t] \qquad x = 2y = 60t \qquad (a)$$

$$[y = v_{o_y} t - \tfrac{1}{2}gt^2] \qquad y = 80t - 16.1t^2 \qquad (b)$$

Multiplying Eq. (*b*) by -2 and adding it to Eq. (*a*) eliminates y to give

$$0 = -100t + 32.2t^2 \quad \text{or} \quad t = 3.11 \text{ sec}$$

Substituting this value of t in Eq. (*a*) gives

$$x = 60(3.11) = 186.6 \text{ ft}$$

from which the distance s along the incline is found to be

$$\left[\frac{s}{\sqrt{5}} = \frac{x}{2}\right] \qquad \frac{s}{\sqrt{5}} = \frac{186.6}{2} \qquad s = 208 \text{ ft} \qquad \textit{Ans.}$$

If the incline had sloped down to the right instead of up, the final coordinate of y displacement would be below the origin and hence negative. Therefore, by inserting a minus sign before y in Eq. (*b*), the time of flight down the incline will become $t = 6.84$ sec. The corresponding values of x and s become $x = 410$ ft and $s = 458$ ft.

PROBLEMS

9-6.4. A golf ball is hit from an elevated tee to a green; the distance, horizontally, is 120 yards. If the initial velocity of the ball is 100 fps at 53.1° to the horizontal, how high is the tee above the green?

9-6.5. A golf ball is driven with the initial velocity of 161 fps upward at 30° to the horizontal from an elevated tee 64.4 ft above a level fairway. What horizontal distance will the ball travel before hitting the fairway?

226 yards ***Ans.***

9-6.6. A projectile is fired with an initial velocity of v_o fps upward at an angle of $\theta°$ with the horizontal. Find the horizontal distance covered before the projectile returns to its original level. Also determine the maximum height attained by the projectile.

$$x = \frac{v_o^2 \sin 2\theta}{g}; \; h = \frac{v_o^2 \sin^2 \theta}{2g} \qquad \textit{Ans.}$$

9-6.7. As an example of poetic license, consider the following passage from Longfellow's *The Song of Hiawatha:*

Swift of foot was Hiawatha:
He could shoot an arrow from him
And run forward with such swiftness

That the arrow fell behind him!
Strong of arm was Hiawatha:
He could shoot ten arrows upward,
Shoot them with such strength and swiftness
That the tenth had left the bowstring
Ere the first to earth had fallen.

Assuming that 1 sec elapses between the discharge of each arrow and that Hiawatha shot at his greatest range each time, how fast must he have been able to run?

9-6.8. A stunt man is to drive an auto across the water-filled gap shown in Fig. P-9-6.8. Determine the auto's minimum take-off velocity and the angle θ of the landing ramp.

$$v_o = 22.4 \text{ mph}; \ \theta = 45° \qquad \textbf{\textit{Ans.}}$$

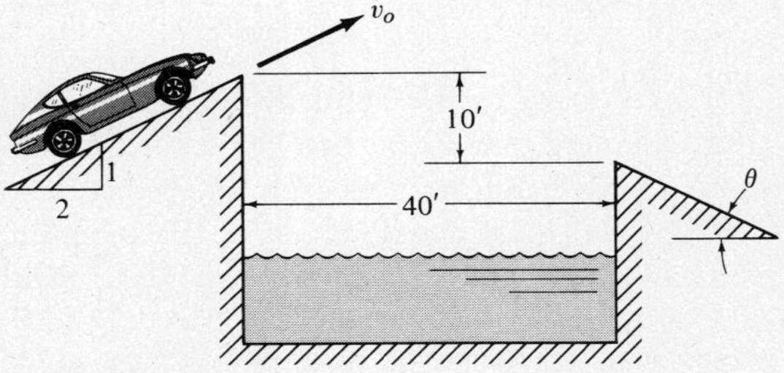

Figure P-9-6.8

9-6.9. A ball is thrown so that it just clears a 25-ft wall 100 ft away. If it left the hand 5 ft above the ground and at an angle of 60° to the horizontal, what was the initial velocity of the ball?

$$v_o = 75.4 \text{ fps} \qquad \textbf{\textit{Ans.}}$$

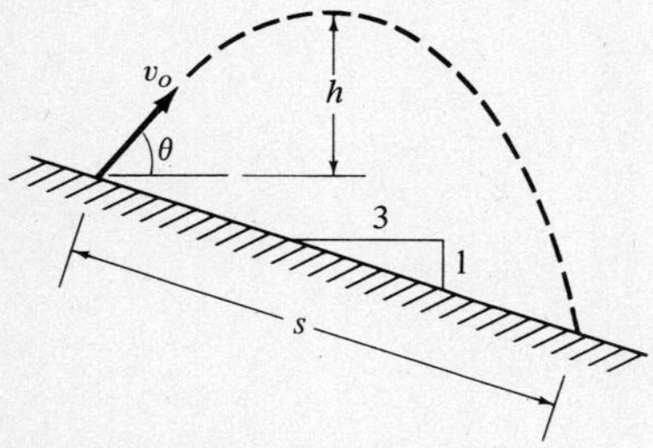

Figure P-9-6.10

9-6.10. In Fig. P-9-6.10, a ball thrown down the incline strikes it at a distance $s = 254.5$ ft. If the ball rises to a maximum height $h = 64.4$ ft above the point of release, compute its initial velocity and inclination θ.

9-6.11. A particle has an initial velocity of 100 fps up to the right at 30° with the horizontal. The components of acceleration are constant at $a_x = -4 \text{ fps}^2$ and $a_y = -20 \text{ fps}^2$. Compute the horizontal distance covered until the particle reaches a point 60 ft below its original elevation.

$$x = 448 \text{ ft} \qquad \textbf{\textit{Ans.}}$$

9-6.12. A particle moves along the path $y = x^2 - 4x + 100$, starting with an initial velocity of $\mathbf{v}_o = (4\hat{\mathbf{i}} - 16\hat{\mathbf{j}})$ fps. If v_x is constant, determine v_y and a_y at $x = 16$ ft.

9-6.13. If the velocity of a particle is defined by $\mathbf{v} = (2t + 1)\hat{\mathbf{i}} + 3\hat{\mathbf{j}}$ fps, and its position vector at $t = 1$ sec is $\mathbf{r} = 4\hat{\mathbf{i}} + 3\hat{\mathbf{j}}$ ft, determine the path of the particle in terms of its x and y coordinates.

$$9x = y^2 + 3y + 18 \qquad \textbf{\textit{Ans.}}$$

9-6.14. The trajectory of a space probe makes an angle of 60° with the horizontal when it is moving at 10,000 mph. A course correction is to be made by thruster engines when the trajectory is at 30° to the horizontal. Assuming gravitational acceleration is constant at 30 fps^2 during this unpowered flight, what time interval should elapse before the thruster engines are ignited? What is the increase in altitude in this interval?

282 sec; 453 miles **Ans.**

9-6.15. A rocket is released from a jet fighter flying horizontally at 750 mph at an altitude of 8000 ft above its target. The rocket thrust gives it a constant horizontal acceleration of 0.6 g. Determine the angle between the horizontal and the line of sight to the target.

9-6.16. If the initial velocity of an object is 40 fps, determine the horizontal distance it can cover without rising more than 10 ft.

48.7 ft **Ans.**

9-6.17. The telescoping rod shown in Fig. P-9-6.17 forces the pin P to move along the fixed path $9y = x^2$ where x and y are in inches. At any time t, the x coordinate of P is given by $x = t^2 - 5t$. Determine the y components of velocity and acceleration of P at $x = 6$ in.

$v_y = 9.33$ ips; $a_y = 13.56$ ips^2 **Ans.**

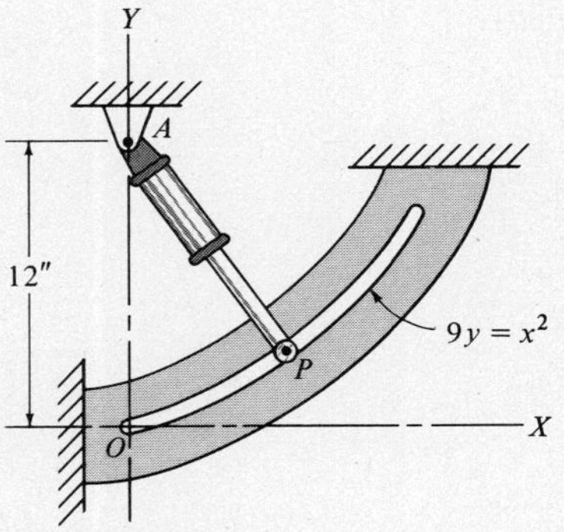

Figure P-9-6.17

9-6.18. A particle moves in the XY plane so that its x coordinate is defined by $x = 5t^3 - 105t$ where x is in inches and t is in seconds. When $t = 2$ sec, the total acceleration is 75 ips^2. If the y component of acceleration is constant and the particle starts from rest at the origin when $t = 0$, determine its total velocity when $t = 4$ sec.

9-6.19. A particle is constrained to move upward to the right along the path $2y^2 = x^3 + 26$ where x and y are in inches. The x coordinate of the particle at any time is $x = t^2 - t + 4$. Determine the y components of velocity and acceleration at $x = 6$ in.

$v_y = 7.36$ ips; $a_y = 7.35$ ips^2 **Ans.**

9-6.20. The position of pin P in the circular slot shown in Fig. P-9-6.20 is controlled by the inclined guide which is moving rightward at the constant rate of 4 ips for an interval of its motion. Compute the velocity and acceleration of P at the given position.

Hint: By sketching the position of the guide a short time t after the given position, obtain the absolute coordinates of motion along the guide in terms of time. The absolute motion of P in the circular slot is equal to the geometric sum of the motion of the guide plus that of P along the guide.

$\dot{x} = 2.56$ ips; $\dot{y} = -1.92$ ips; $\ddot{x} = -1.23$ ips^2; $\ddot{y} = -1.64$ ips^2 **Ans.**

9-6.21. Solve Prob. 9-6.20 if at the given position the guide is moving leftward with a velocity of 5 ips and an acceleration of 4 ips^2.

$\ddot{x} = -4.48$ ips^2; $\ddot{y} = -0.64$ ips^2 **Ans.**

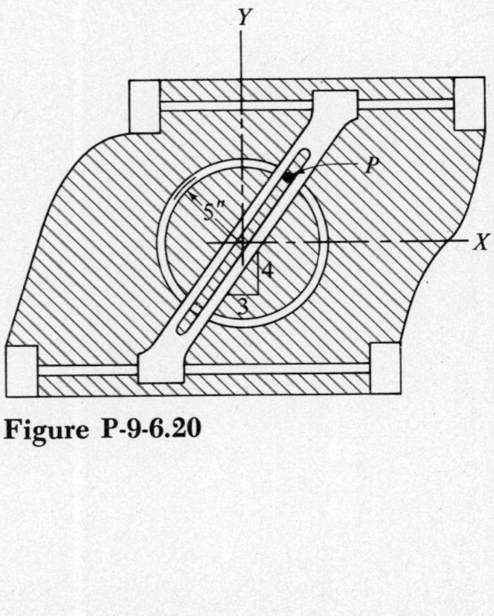

Figure P-9-6.20

9-7 NORMAL AND TANGENTIAL COMPONENTS OF ACCELERATION

In many situations, the most useful components of acceleration are those which are tangent and normal to the path because, as we shall show, these components separate and denote respectively the rate of change of *magnitude* and of *direction* of velocity. They are particularly useful when we need to relate velocity and acceleration directly with the path itself. The following discussion refers to motion in a plane, but it is also applicable to three-dimensional spatial motion.

In Section 9-2, we have already denoted velocity in terms of its magnitude and direction by the expression $\mathbf{v} = v\hat{\mathbf{e}}_t$. Since acceleration is the time derivative of velocity, we obtain

$$\mathbf{a} = \frac{d\mathbf{v}}{dt} = \frac{d}{dt}(v\hat{\mathbf{e}}_t) = \frac{dv}{dt}\hat{\mathbf{e}}_t + v\frac{d\hat{\mathbf{e}}_t}{dt} \qquad (a)$$

Here $\dfrac{dv}{dt} = \dfrac{d^2s}{dt^2} = a_t$ represents the change in the magnitude of the velocity, directed tangent to the path as indicated by $\hat{\mathbf{e}}_t$.

The directional change of $\mathbf{v}$ is expressed by $\dfrac{d\hat{\mathbf{e}}_t}{dt}$ which we now interpret. Unlike unit vectors associated with a fixed set of reference axes, $\hat{\mathbf{e}}_t$ is a unit vector whose direction changes with the path. Thus, in Fig. 9-7.1, unit tangent vectors at two positions A and B, separated by the time interval Δt, have the same magnitude of unity but are respectively perpendicular to the normals to the path at A and B. The directions of these normals differ by an angle of $\Delta\theta$ radians. The radius of curvature at A is designated as ρ and the radius of curvature at B will approach this value as $\Delta\theta \to 0$. When we draw the unit tangent vectors from a common origin as in part (b), they

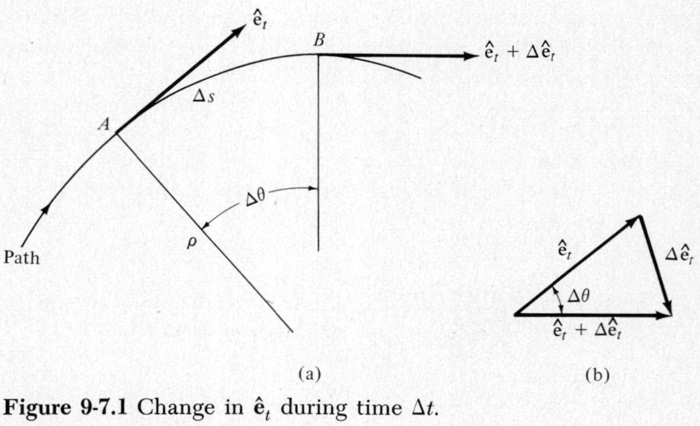

Figure 9-7.1 Change in $\hat{\mathbf{e}}_t$ during time Δt.

form an isosceles triangle whose vertex angle is $\Delta\theta$ and whose base $\Delta\hat{\mathbf{e}}_t$ denotes their directional change.

Consider now the geometric significance of permitting the time interval between positions A and B to approach zero as a limit. In Fig. 9-7.1a, B approaches to within a differential distance ds of A; angle $\Delta\theta$ becomes $d\theta$; the radius of curvature at B also becomes ρ. In part (b) of Fig. 9-7.1, as $\Delta\theta \to 0$, the base angles approach 90° and hence in the limit, the direction of $\Delta\hat{\mathbf{e}}_t$ is normal to the path at A. As a result of $\Delta t \to 0$, therefore, Fig. 9-7.1 is transformed into

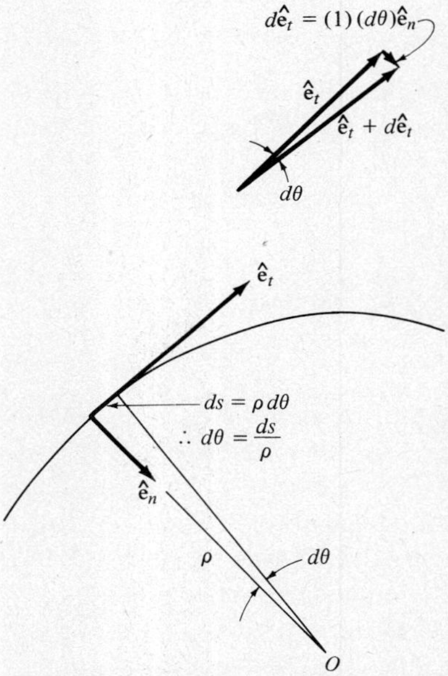

Figure 9-7.2 Change in $\hat{\mathbf{e}}_t$ as $\Delta t \to 0$.

Fig. 9-7.2. Note that in the time dt, the unit tangent vector $\hat{\mathbf{e}}_t$ swings through the angle $d\theta$ and its tip moves through the distance $|d\hat{\mathbf{e}}_t| = (1)\,d\theta = d\theta$ in a direction perpendicular to $\hat{\mathbf{e}}_t$; i.e., inward toward the center of curvature O in the direction defined by the unit normal vector $\hat{\mathbf{e}}_n$. Thus, we obtain

$$\frac{d\hat{\mathbf{e}}_t}{dt} = \frac{d\theta}{dt}\hat{\mathbf{e}}_n = \frac{1}{\rho}\frac{ds}{dt}\hat{\mathbf{e}}_n = \frac{v}{\rho}\hat{\mathbf{e}}_n \qquad (b)$$

Substituting this result in Eq. (a) then gives

$$\mathbf{a} = \frac{dv}{dt}\hat{\mathbf{e}}_t + \frac{v^2}{\rho}\hat{\mathbf{e}}_n = \mathbf{a}_t + \mathbf{a}_n \qquad (9\text{-}7.1)$$

where the magnitudes of the acceleration components are $a_t = \dfrac{dv}{dt}$

and $a_n = \dfrac{v^2}{\rho}$ and their directions are such that positive a_t is tangent to the path in the direction of increasing speed whereas a_n is always directed inward toward the center of curvature.

Note that for measurements along the path, the same relations exist among s, v, and a_t as discussed previously for rectilinear motion in Section 9-3. Observe that the tangential component of acceleration $a_t = \dfrac{dv}{dt}$ represents only the change in *magnitude* of velocity and that it will be zero if the speed is constant. On the other hand, the normal component of acceleration, $a_n = \dfrac{v^2}{\rho}$, is caused by the change in the *direction* of the velocity and will be zero only if $v = 0$ or if ρ is infinite as at a point of inflection of the path, or if the path is straight.

Alternate expressions for the magnitude of $\mathbf{a}_n$ are

$$a_n = v\omega = \rho\omega^2 = \frac{v^2}{\rho} \tag{9-7.2}$$

where $\omega = \dfrac{d\theta}{dt}$ represents the angular speed of the radius of curvature. One of the most useful applications of Eq. (9-7.2) occurs in motion along a circular path; for example, rotation of a rigid body about a fixed axis. Here ρ is the constant radius of a circular path described by a particle in the rotating body. For other curved paths, ρ and its angular speed ω are generally difficult to compute as a preliminary to finding the normal component of acceleration.

The correlation between the total acceleration $\mathbf{a}$ of a particle and its rectangular components or its normal and tangential components is shown in Fig. 9-7.3. Either set of components may be derived from the other by projecting one set upon the other. Doing this gives

$$\left.\begin{aligned} a_n &= a_x \sin\theta + a_y \cos\theta \\ a_t &= a_x \cos\theta - a_y \sin\theta \end{aligned}\right\} \tag{9-7.3}$$

or

$$\left.\begin{aligned} a_x &= a_n \sin\theta + a_t \cos\theta \\ a_y &= a_n \cos\theta - a_t \sin\theta \end{aligned}\right\} \tag{9-7.4}$$

Once you understand these correlations among the components of acceleration, you need not bother to remember them since they are so easily obtained from a diagram appropriate to the situation.

The preceding discussion of tangential and normal components of acceleration is also valid for three-dimensional spatial motion, but finds limited use because of the difficulty of specifying the directions of $\hat{\mathbf{e}}_t$ and $\hat{\mathbf{e}}_n$. The direction of the tangent to the spatial path is easily visualized, but there are an infinite number of perpendiculars to this

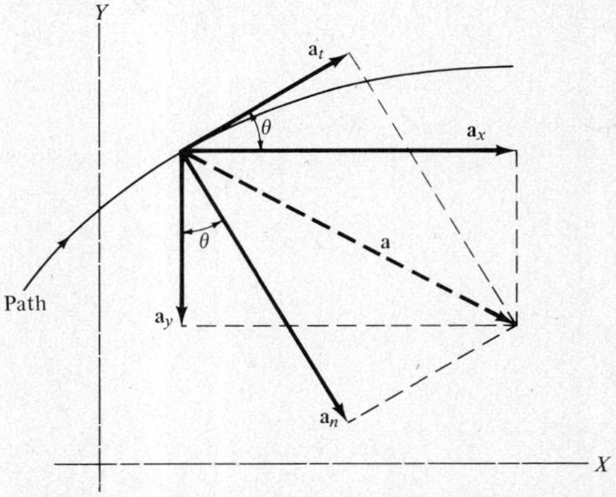

Figure 9-7.3 Relations between rectangular components and normal and tangential components of acceleration.

tangent. The direction of the *principal normal* along which $\hat{\mathbf{e}}_n$ is directed is that which passes through the center of curvature. Then $\hat{\mathbf{e}}_t$ and $\hat{\mathbf{e}}_n$ define a plane which is coincident with that described by the simultaneous stretching and swinging of the velocity vector. This plane is known as the *osculating plane*. The perpendicular to the osculating plane is the binormal direction. The unit binormal vector $\hat{\mathbf{e}}_b$, together with $\hat{\mathbf{e}}_t$ and $\hat{\mathbf{e}}_n$, forms a right-handed set of mutually perpendicular vectors, defined by $\hat{\mathbf{e}}_t \times \hat{\mathbf{e}}_n = \hat{\mathbf{e}}_b$, whose directions continually change in space with the motion of a particle along its path.

ILLUSTRATIVE PROBLEMS

9-7.1. A particle moves in the XY plane with $a_x = -6$ fps^2 and $a_y = -30$ fps^2. If its initial velocity is 100 fps directed at a slope of 4 to 3 as shown in Fig. 9-7.4, compute the radius of curvature of the path 2 sec later.

Solution

Instead of determining the radius of curvature by applying the familiar calculus formula to the equation of the path, a simpler procedure is to compute the radius of curvature from Eq. (9-7.2) after first finding the velocity v and normal acceleration a_n.

The components of the velocity after 2 sec are

$$[v = v_o + at] \qquad v_x = 100(\tfrac{3}{5}) - 6(2) = 48 \text{ fps}$$
$$v_y = 100(\tfrac{4}{5}) - 30(2) = 20 \text{ fps}$$

Combining these components yields the resultant velocity v and the inclination θ of the tangent to the path.

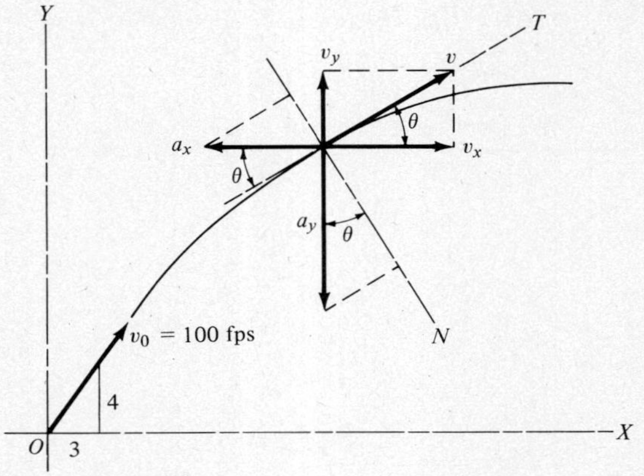

Figure 9-7.4

$$\left[v = \sqrt{v_x^{\,2} + v_y^{\,2}}\right] \qquad v = \sqrt{(48)^2 + (20)^2} = 52 \text{ fps}$$

$$\tan\theta = \frac{v_y}{v_x} = \frac{20}{48}; \qquad \theta = 22.6°$$

The normal component of acceleration is found by projecting the rectangular components of acceleration upon the normal to the path. As shown in the figure, these projections are oppositely directed so that a_n is equal to their difference.

$$[a_n = a_y \cos\theta - a_x \sin\theta] \qquad a_n = 30 \cos 22.6° - 6 \sin 22.6° = 25 \text{ fps}^2$$

Finally, by applying Eq. (9-7.2), we obtain

$$\left[a_n = \frac{v^2}{\rho}\right] \qquad 25 = \frac{(52)^2}{\rho} \qquad \rho = 108 \text{ ft} \quad \textbf{\textit{Ans.}}$$

9-7.2. The motion of a particle is defined by the position vector $\mathbf{r} = 5t\hat{\mathbf{i}} + 3t^2\hat{\mathbf{j}} + \frac{1}{3}t^3\hat{\mathbf{k}}$ where r is in feet and t is in seconds. At the instant when $t = 2$ sec, find the tangential and normal components of acceleration and the principal radius of curvature.

Solution

The velocity and acceleration at any instant are determined by successive differentiation of the position vector to be

$$\mathbf{v} = 5\hat{\mathbf{i}} + 6t\hat{\mathbf{j}} + t^2\hat{\mathbf{k}}$$
$$\mathbf{a} = 6\hat{\mathbf{j}} + 2t\hat{\mathbf{k}}$$

so that at $t = 2$ sec,

$$\mathbf{v} = 5\hat{\mathbf{i}} + 12\hat{\mathbf{j}} + 4\hat{\mathbf{k}}, \qquad v = \sqrt{185} = 13.6 \text{ fps}$$
$$\mathbf{a} = 6\hat{\mathbf{j}} + 4\hat{\mathbf{k}}, \qquad a = \sqrt{52} = 7.21 \text{ fps}^2$$

Since the velocity is tangent to the path, the unit tangent vector at $t = 2$ sec is

$$\hat{\mathbf{e}}_t = \frac{\mathbf{v}}{v} = \frac{1}{13.6}(5\hat{\mathbf{i}} + 12\hat{\mathbf{j}} + 4\hat{\mathbf{k}})$$

whence the tangential component of acceleration at this instant is

$$a_t = \mathbf{a} \cdot \hat{\mathbf{e}}_t = (6\hat{\mathbf{j}} + 4\hat{\mathbf{k}}) \cdot \frac{1}{13.6}(5\hat{\mathbf{i}} + 12\hat{\mathbf{j}} + 4\hat{\mathbf{k}})$$

$$= \frac{72 + 16}{13.6} = 6.47 \text{ fps}^2 \qquad \textit{Ans.}$$

Applying the concept that $\mathbf{a}_t$ and $\mathbf{a}_n$ are perpendicular components of $\mathbf{a}$, we find the magnitude of the normal component of acceleration from

$$a_n^2 = a^2 - a_t^2 = 52 - (6.47)^2 = 10.2; \quad a_n = 3.19 \text{ fps}^2 \qquad \textit{Ans.}$$

Finally, the principal radius of curvature is found from the relation

$$\rho = \frac{v^2}{a_n} = \frac{185}{3.19} = 58.0 \text{ ft} \qquad \textit{Ans.}$$

A general method of finding ρ in any coordinate system is to take the cross product of the velocity $\mathbf{v}$ and the acceleration $\mathbf{a}$. Thus,

$$\mathbf{v} \times \mathbf{a} = \mathbf{v} \times (\mathbf{a}_t + \mathbf{a}_n) = \mathbf{v} \times \mathbf{a}_n$$

where $\mathbf{v} \times \mathbf{a}_t = 0$ because $\mathbf{v}$ and $\mathbf{a}_t$ are collinear tangents to the path. Then the magnitude of $\mathbf{v} \times \mathbf{a}_n$ is equal to that of $\mathbf{v} \times \mathbf{a}$, or

$$|\mathbf{v} \times \mathbf{a}| = |\mathbf{v} \times \mathbf{a}_n| = v a_n \sin 90° = v a_n = v\frac{v^2}{\rho} = \frac{v^3}{\rho}$$

whence

$$\rho = \frac{v^3}{|\mathbf{v} \times \mathbf{a}|} \qquad (9\text{-}7.5)$$

To illustrate this general procedure, let us check the value of ρ previously obtained. Noting that at $t = 2$ sec, $\mathbf{v} = 5\hat{\mathbf{i}} + 12\hat{\mathbf{j}} + 4\hat{\mathbf{k}}$ and $\mathbf{a} = 6\hat{\mathbf{j}} + 4\hat{\mathbf{k}}$, we obtain

$$\mathbf{v} \times \mathbf{a} = \begin{vmatrix} 5 & 12 & 4 \\ 0 & 6 & 4 \\ \hat{\mathbf{i}} & \hat{\mathbf{j}} & \hat{\mathbf{k}} \end{vmatrix} = \hat{\mathbf{i}}(48 - 24) + \hat{\mathbf{j}}(-20) + \hat{\mathbf{k}}(30)$$

whence

$$|\mathbf{v} \times \mathbf{a}| = \sqrt{(24)^2 + (20)^2 + (30)^2} = 43.2 \text{ ft}^2/\text{sec}^3$$

Applying Eq. (9-7.5), we now obtain

$$\left[\rho = \frac{v^3}{|\mathbf{v} \times \mathbf{a}|}\right] \qquad \rho = \frac{(13.6)^3}{43.2} = 58.0 \text{ ft} \qquad \textit{Check}$$

Also observe that this general method determines the magnitude of $\mathbf{a}_n$ as follows:

$$[|\mathbf{v} \times \mathbf{a}| = va_n] \qquad 43.2 = 13.6a_n \qquad a_n = 3.19 \text{ fps}^2 \quad \textit{Check}$$

As a final comment, whenever the unit normal vector is needed to specify the direction of the normal acceleration, the following procedure may be used. Using the data previously found at $t = 2$ sec, we apply

$$\mathbf{a} = a_t\hat{\mathbf{e}}_t + a_n\hat{\mathbf{e}}_n$$

$$6\hat{\mathbf{j}} + 4\hat{\mathbf{k}} = 6.47\left(\frac{1}{13.6}\right)(5\hat{\mathbf{i}} + 12\hat{\mathbf{j}} + 4\hat{\mathbf{k}}) + 3.19\hat{\mathbf{e}}_n$$

which reduces to

$$3.19\hat{\mathbf{e}}_n = -2.48\hat{\mathbf{i}} + 0.28\hat{\mathbf{j}} + 2.097\hat{\mathbf{k}}$$

whence

$$\hat{\mathbf{e}}_n = -0.778\hat{\mathbf{i}} + 0.0888\hat{\mathbf{j}} + 0.658\hat{\mathbf{k}} \qquad \textit{Ans.}$$

The unit binormal vector is $\hat{\mathbf{e}}_b = \hat{\mathbf{e}}_t \times \hat{\mathbf{e}}_n$ as defined on p. 365. It may readily be determined now that $\hat{\mathbf{e}}_t$ and $\hat{\mathbf{e}}_n$ are known. In this example, we obtain

$$\hat{\mathbf{e}}_b = \frac{1}{13.6}\begin{vmatrix} 5 & 12 & 4 \\ -0.778 & 0.0888 & 0.658 \\ \hat{\mathbf{i}} & \hat{\mathbf{j}} & \hat{\mathbf{k}} \end{vmatrix} = 0.554\hat{\mathbf{i}} - 0.471\hat{\mathbf{j}} + 0.719\hat{\mathbf{k}}$$

PROBLEMS

9-7.3. A particle moves on a circular path of 40-ft radius so that its arc distance from a fixed point on the path is given by $s = 4t^3 - 10t$ where s is in feet and t is in seconds. Compute the total acceleration at the end of 2 sec.

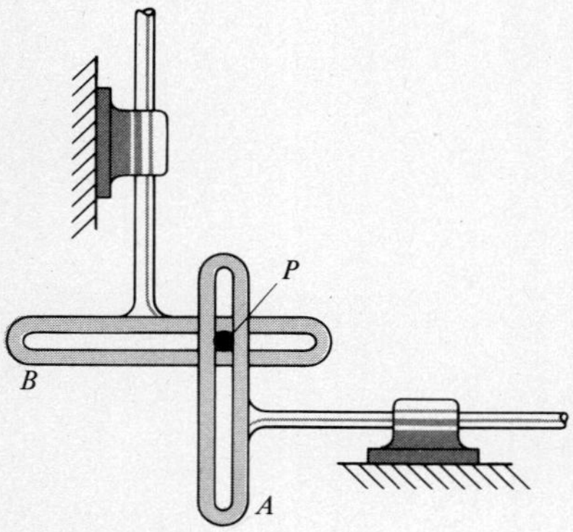

Figure P-9-7.4

9-7.4. The pin P moves along a curved path which is determined by the motions of two slotted links A and B. At the instant shown in Fig. P-9-7.4, A has a velocity of 12 ips and an acceleration of 10 ips^2 both to the right, while B has a velocity of 16 ips and an acceleration of 5 ips^2 both vertically downward. Find the radius of curvature of the path of P at this instant and sketch how the path curves.

80 in. ***Ans.***

9-7.5. A stone is thrown with an initial velocity of 100 fps upward at 60° to the horizontal. Compute the radius of curvature of its path at the position where it is 50 ft horizontally from its initial position.

251 ft ***Ans.***

9-7.6. A particle has an initial velocity of 200 fps up to the right at a slope of 0.75. The components of acceleration are constant at $a_x = -12$ fps^2 and $a_y = -20$ fps^2. Compute the radius of curvature at the start and at the top of the path.

9-7.7. A particle moves counterclockwise on a circular path of 400-ft radius. It starts from a position which is horizontally to the right of the center of the path and moves so that $s = 10t^2 + 20t$ where s is the arc distance in feet and t is in seconds. Compute the horizontal and vertical components of acceleration at the end of 3 sec.

$a_h = -22.2$ fps^2; $a_v = 12.7$ fps^2 ***Ans.***

9-7.8. The velocity of a particle is defined by $v_x = 100 - t^{3/2}$ and $v_y = 100 + 10t - 2t^2$ where v is in fps and t in seconds. Determine the radius of curvature at the top of its path.

9-7.9. Using the data of the preceding problem, determine the radius of curvature of the path of the particle at $t = 12$ sec.

280 ft ***Ans.***

9-7.10. The rectangular components of acceleration for a particle are $a_x = 3t$ and $a_y = 30 - 10t$ where a is in fps^2. If the particle starts from rest at the origin, find the radius of curvature of the path at $t = 2$ sec.

368 ft ***Ans.***

9-7.11. The telescoping rod shown in Fig. P-9-7.11 forces the pin P to move along the fixed path $9y = x^2$ where x and y are in inches. At any time t, the x coordinate of P is given by $x = t^2 - 5t$. At what rate is the speed of P changing at $t = 6$ sec? Solve without using the value of a_y.

12.05 fps^2 ***Ans.***

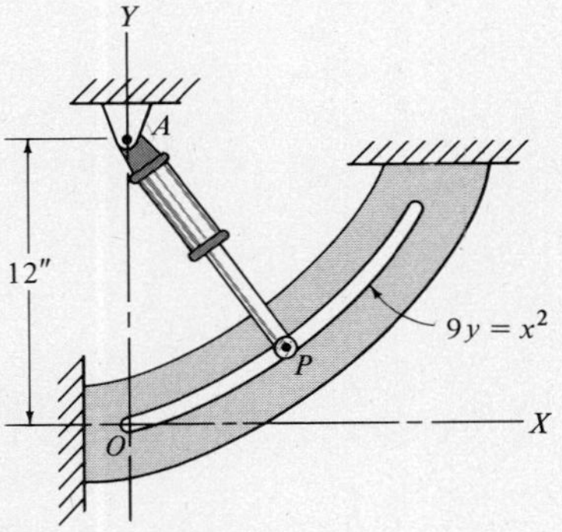

Figure P-9-7.11

9-7.12. The position vector of a particle is defined by $\mathbf{r} = (2t^3 - 12t)\hat{\mathbf{i}} + (6t^2)\hat{\mathbf{j}}$ where $\mathbf{r}$ is in feet and t is in seconds. Find the principal radius of curvature of the path of the particle at $t = 2$ sec.

9-7.13. The position vector of a particle is given by $\mathbf{r} = 2t^2\hat{\mathbf{i}} + 10t\hat{\mathbf{j}} + \frac{1}{3}t^3\hat{\mathbf{k}}$ where $\mathbf{r}$ is in feet and t is in seconds. Determine the normal and tangential components of acceleration and the principal radius of curvature of the path of the particle at $t = 3$ sec.

$a_n = 4.47$ fps^2; $a_t = 5.65$ fps^2; $\rho = 72.7$ ft ***Ans.***

9-7.14. Repeat Prob. 9-7.13 using $\mathbf{r} = 9t\hat{\mathbf{i}} - \frac{2}{3}t^3\hat{\mathbf{j}} + 3t^2\hat{\mathbf{k}}$ and $t = 2$ sec.

9-7.15. A particle starting from rest moves with the acceleration $\mathbf{a} = 4t\hat{\mathbf{i}} - 3t^2\hat{\mathbf{j}} + 6\hat{\mathbf{k}}$ fps^2. Determine the principal radius of curvature of its path at $t = 2$ sec.

$\rho = 40$ ft ***Ans.***

9-8 RADIAL AND TRANSVERSE COMPONENTS. CYLINDRICAL COORDINATES

Usually the problems of dynamics are solved by means of rectangular or normal and tangential coordinate systems. Less frequently, the motion of a particle is more conveniently described by polar coordinates and its three-dimensional extension to cylindrical coordinates. We now consider how velocity and acceleration are expressed in these coordinate systems.

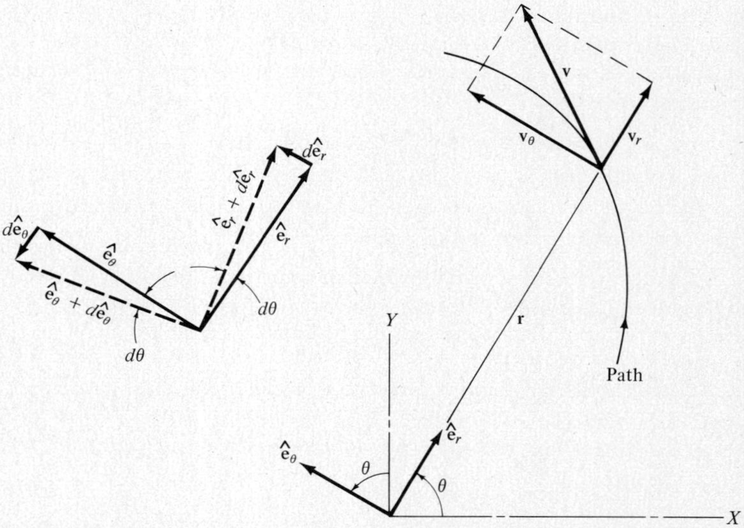

Figure 9-8.1 Radial and transverse components of velocity.

We start with plane curvilinear motion using polar coordinates to denote the position vector $\mathbf{r}$ in terms of its magnitude r and its angular coordinate θ measured from a fixed reference axis as shown in Fig. 9-8.1. Let the components of velocity be taken along and perpendicular to the position vector as shown, where $\mathbf{v}_r$ and $\mathbf{v}_\theta$ denote respectively the radial and transverse components of velocity. Note that the total velocity $\mathbf{v}$ is not perpendicular to $\mathbf{r}$ except for the special case when $\mathbf{r}$ is of constant length and the path therefore is circular.

Let us also establish unit vectors $\hat{\mathbf{e}}_r$ and $\hat{\mathbf{e}}_\theta$ acting in the positive directions of increasing r and θ; i.e., the unit vectors $\hat{\mathbf{e}}_r$ and $\hat{\mathbf{e}}_\theta$ are parallel to the positive senses of $\mathbf{v}_r$ and $\mathbf{v}_\theta$. Both these unit vectors will swing through the angle $d\theta$ as the position vector $\mathbf{r}$ swings through the angle $d\theta$ in the time dt. As shown in Fig. 9-8.1 (and explained in detail in Section 9-7), their tips move through the distances $|d\hat{\mathbf{e}}_r| = d\theta$ in the direction of $\hat{\mathbf{e}}_\theta$, and through $|d\hat{\mathbf{e}}_\theta| = d\theta$

in the direction of $-\hat{\mathbf{e}}_r$. Hence, their time rates of change (due to a swing but no stretch) are respectively

$$\left.\begin{aligned} \dot{\hat{\mathbf{e}}}_r &= \frac{d\hat{\mathbf{e}}_r}{dt} = \frac{d\theta}{dt}\hat{\mathbf{e}}_\theta = \dot{\theta}\,\hat{\mathbf{e}}_\theta \\ \text{and} \quad \dot{\hat{\mathbf{e}}}_\theta &= \frac{d\hat{\mathbf{e}}_\theta}{dt} = \frac{d\theta}{dt}(-\hat{\mathbf{e}}_r) = -\dot{\theta}\,\hat{\mathbf{e}}_r \end{aligned}\right\} \tag{9-8.1}$$

Another more direct method of obtaining these results is discussed later in Section 12-6 where we derive the omega theorem.

We may now proceed to obtain expressions for velocity and acceleration by successive differentiation of the position vector $\mathbf{r} = r\hat{\mathbf{e}}_r$. The velocity is found to be

$$\mathbf{v} = \frac{d\mathbf{r}}{dt} = \frac{d}{dt}(r\hat{\mathbf{e}}_r) = \dot{r}\hat{\mathbf{e}}_r + r\dot{\hat{\mathbf{e}}}_r$$

$$\mathbf{v} = \dot{r}\hat{\mathbf{e}}_r + r\dot{\theta}\hat{\mathbf{e}}_\theta \tag{9-8.2}$$

or

$$v = v_r \nrightarrow v_\theta$$

where $v_r = \dot{r}$ and $v_\theta = r\dot{\theta}$ are respectively the magnitudes of the radial and transverse components of velocity acting in the positive senses of $\hat{\mathbf{e}}_r$ and $\hat{\mathbf{e}}_\theta$.

The acceleration is the time rate of change of velocity which results in

$$\begin{aligned} \mathbf{a} &= \frac{d\mathbf{v}}{dt} = \frac{d}{dt}(\dot{r}\hat{\mathbf{e}}_r) + \frac{d}{dt}(r\dot{\theta}\hat{\mathbf{e}}_\theta) \\ &= \ddot{r}\hat{\mathbf{e}}_r + \dot{r}\dot{\hat{\mathbf{e}}}_r + \dot{r}\dot{\theta}\hat{\mathbf{e}}_\theta + r\ddot{\theta}\hat{\mathbf{e}}_\theta + r\dot{\theta}\dot{\hat{\mathbf{e}}}_\theta \end{aligned}$$

into which we substitute the values of $\dot{\hat{\mathbf{e}}}_r$ and $\dot{\hat{\mathbf{e}}}_\theta$ from Eq. (9-8.1) to obtain

$$\begin{aligned} \mathbf{a} &= \ddot{r}\hat{\mathbf{e}}_r + \dot{r}(\dot{\theta}\hat{\mathbf{e}}_\theta) + \dot{r}\dot{\theta}\hat{\mathbf{e}}_\theta + r\ddot{\theta}\hat{\mathbf{e}}_\theta + r\dot{\theta}(-\dot{\theta}\hat{\mathbf{e}}_r) \\ &= (\ddot{r} - r\dot{\theta}^2)\hat{\mathbf{e}}_r + (r\ddot{\theta} + 2\dot{r}\dot{\theta})\hat{\mathbf{e}}_\theta \end{aligned} \tag{9-8.3}$$

or

$$a = a_r \nrightarrow a_\theta$$

where $a_r = \ddot{r} - r\dot{\theta}^2$ and $a_\theta = r\ddot{\theta} + 2\dot{r}\dot{\theta}$ are respectively the magnitudes of the radial and transverse components of acceleration acting in the positive senses of $\hat{\mathbf{e}}_r$ and $\hat{\mathbf{e}}_\theta$.

An alternate form of the transverse component is

$$a_\theta = \frac{1}{r}\frac{d}{dt}(r^2\dot{\theta})$$

which is useful when it is necessary to integrate a_θ. This equivalence may be verified by direct differentiation of the alternate form.

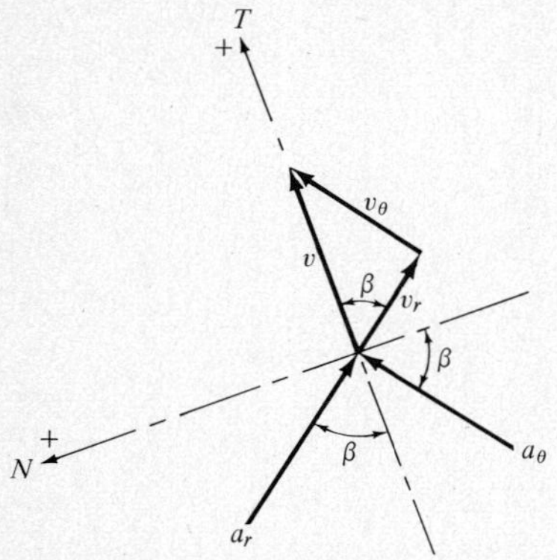

Figure 9-8.2 Algebraic summation of projections of a_r and a_θ along N and T determines a_n and a_t.

On occasion, it may be required to transform the radial and transverse components of acceleration into normal and tangential components. Since either set of components defines the same acceleration vector, this transformation is readily accomplished by projecting the r-θ components upon the normal and tangential directions. Thus in Fig. 9-8.2, the tangential direction (T) is along the velocity vector $\mathbf{v}$ while the normal direction (N) is perpendicular to $\mathbf{v}$. The algebraic sums of the projections of a_r and a_θ along the T and N directions then gives

$$a_t = a_r \cos\beta + a_\theta \sin\beta$$
$$a_n = a_\theta \cos\beta - a_r \sin\beta$$

which, since the relations among the velocity components is seen to give $\cos\beta = v_r/v$ and $\sin\beta = v_\theta/v$, readily reduces to

$$\left.\begin{aligned} a_t &= \frac{a_r v_r + a_\theta v_\theta}{v} \\ a_n &= \frac{a_\theta v_r - a_r v_\theta}{v} \end{aligned}\right\} \qquad (9\text{-}8.4)$$

These results are not important enough to memorize; they are discussed only to illustrate the geometric relations among the several sets of components and to emphasize a method of reasoning whereby they can be easily developed when needed.

A more elaborate approach which yields the same results, although not as direct, is to first express $\hat{\mathbf{e}}_t$ and $\hat{\mathbf{e}}_n$ in terms of $\hat{\mathbf{e}}_r$ and $\hat{\mathbf{e}}_\theta$ and then apply

$$a_t = \mathbf{a} \cdot \hat{\mathbf{e}}_t = (\mathbf{a}_r + \mathbf{a}_\theta) \cdot \hat{\mathbf{e}}_t$$
$$a_n = \mathbf{a} \cdot \hat{\mathbf{e}}_n = (\mathbf{a}_r + \mathbf{a}_\theta) \cdot \hat{\mathbf{e}}_n$$

This procedure is left to be done as an exercise in Prob. 9-8.4.

It is instructive to observe the geometric significance of the radial and transverse components of velocity and acceleration by discussing what happens if only $\mathbf{r}$ varies while θ is held constant, and then if only θ varies while $\mathbf{r}$ has a constant length. Permitting $\mathbf{r}$ to vary while θ is held constant is equivalent to rectilinear motion along the fixed direction of θ. This results in $\mathbf{v} = \dot{\mathbf{r}}$ and $\mathbf{a} = \ddot{\mathbf{r}}$. On the other hand, if $\mathbf{r}$ is of constant length while θ varies, we have a rotation along a circular path in which $v = r\dot{\theta} = r\omega$ and $a = a_n +\!\!\!\!\rightarrow a_t = (r\dot{\theta}^2 = r\omega^2) +\!\!\!\!\rightarrow (r\ddot{\theta} = r\alpha)$. In the equivalent expressions enclosed in the parentheses, $\omega = \dot{\theta}$ is the common symbol for angular velocity and $\alpha = \ddot{\theta}$ represents angular acceleration.

CYLINDRICAL COORDINATES

The extension of radial and transverse coordinates to three-dimensional space is accomplished by adding a fixed Z coordinate axis to the polar coordinate axes. The position vector is then given by $\mathbf{r} = r\hat{\mathbf{e}}_r + z\hat{\mathbf{k}}$ where $\hat{\mathbf{k}}$ is a unit vector in the Z direction. Note that the positive Z direction is that which makes the unit vectors $\hat{\mathbf{e}}_r$, $\hat{\mathbf{e}}_\theta$, and $\hat{\mathbf{k}}$ form a right-handed triad so that $\hat{\mathbf{e}}_r \times \hat{\mathbf{e}}_\theta = \hat{\mathbf{k}}$. Since $\hat{\mathbf{k}}$ is constant in direction as well as magnitude, $\dot{\hat{\mathbf{k}}} = 0$. Hence, the successive derivatives of $z\hat{\mathbf{k}}$ are merely $\dot{z}\hat{\mathbf{k}}$ and $\ddot{z}\hat{\mathbf{k}}$. By adding these terms to Eqs. (9-8.2) and (9-8.3), we obtain the following expressions for velocity and acceleration in cylindrical coordinates:

$$\left.\begin{aligned} \mathbf{v} &= \dot{\mathbf{r}} = \dot{r}\hat{\mathbf{e}}_r + r\dot{\theta}\hat{\mathbf{e}}_\theta + \dot{z}\hat{\mathbf{k}} \\ \mathbf{a} &= \ddot{\mathbf{r}} = (\ddot{r} - r\dot{\theta}^2)\hat{\mathbf{e}}_r + (r\ddot{\theta} + 2\dot{r}\dot{\theta})\hat{\mathbf{e}}_\theta + \ddot{z}\hat{\mathbf{k}} \end{aligned}\right\} \tag{9-8.5}$$

ILLUSTRATIVE PROBLEMS

9-8.1. The plane curvilinear motion of a particle is defined in polar coordinates by $r = t^3/3 + 2t$ and $\theta = 0.3t^2$ where r is in inches, θ is in radians, and t is in seconds. At the instant when $t = 2$ sec, determine the magnitudes of velocity, acceleration, and the radius of curvature of the path.

Solution

Into successive time derivatives of r and θ, we substitute $t = 2$ sec to obtain

$$\left.\begin{array}{ll} r = \frac{1}{3}t^3 + 2t & \theta = 0.3t^2 \\ \dot{r} = t^2 + 2 & \dot{\theta} = 0.6t \\ \ddot{r} = 2t & \ddot{\theta} = 0.6 \end{array}\right|_{t=2} \qquad \begin{array}{ll} r = \frac{20}{3} \text{ in.} & \theta = 1.2 \text{ rad} = 68.8° \\ \dot{r} = 6 \text{ ips} & \dot{\theta} = 1.2 \text{ rad/sec} \\ \ddot{r} = 4 \text{ ips}^2 & \ddot{\theta} = 0.6 \text{ rad/sec}^2 \end{array}$$

These values are now substituted into Eqs. (9-8.2) and (9-8.3) to obtain the following components of v and a:

$$v_r = \dot{r} = 6 \text{ ips}$$
$$v_\theta = r\dot{\theta} = \tfrac{20}{3}(1.2) = 8 \text{ ips}$$
$$a_r = \ddot{r} - r\dot{\theta}^2 = 4 - \tfrac{20}{3}(1.2)^2 = -5.6 \text{ ips}^2$$
$$a_\theta = r\ddot{\theta} + 2\dot{r}\dot{\theta} = \tfrac{20}{3}(0.6) + 2(6)(1.2) = 18.4 \text{ ips}^2$$

from which we find the velocity and acceleration to be

$[v = v_r +\!\!\!\!\rightarrow v_\theta]$ $\qquad v = 6 +\!\!\!\!\rightarrow 8 = 10$ ips **Ans.**

$[a = a_r +\!\!\!\!\rightarrow a_\theta]$ $\qquad a = 5.6 +\!\!\!\!\rightarrow 18.4 = 19.25$ ips^2 **Ans.**

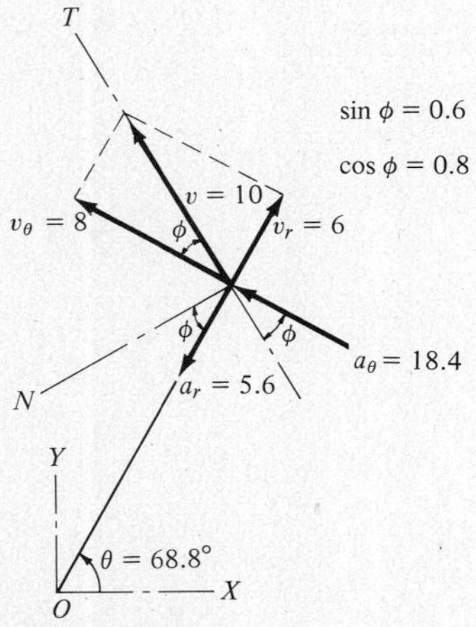

Figure 9-8.3

The directions of the velocity and acceleration components, with due regard for sign, are shown in Fig. 9-8.3. By projecting the acceleration components upon the normal N to the path (i.e., perpendicular to the tangential direction defined by v), we find the normal component of acceleration to be

$$a_n = \Sigma N = 5.6 \cos \phi + 18.4 \sin \phi$$
$$= 5.6(0.8) + 18.4(0.6) = 15.52 \text{ ips}^2$$

Using Eq. (9-7.1), we then find the instantaneous radius of curvature to be

$$\left[a_n = \frac{v^2}{\rho}\right] \qquad 15.52 = \frac{(10)^2}{\rho} \qquad \rho = 6.44 \text{ in.} \qquad \textit{Ans.}$$

9-8.2. The 6-in. crank OA of an oscillating arm quick-return mechanism is rotating counterclockwise at a constant rate of 10 rad/sec. When the mechanism is in the position shown in Fig. 9-8.4, determine the angular acceleration of arm BD.

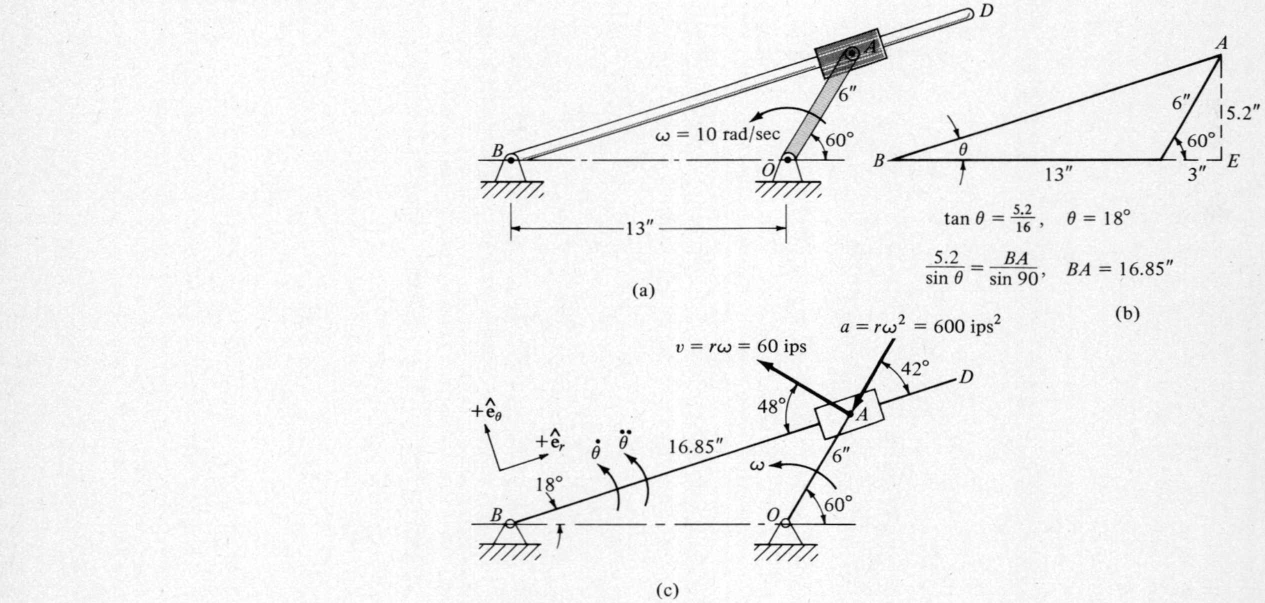

Figure 9-8.4

Solution

From the preliminary geometry indicated in part (b), we find $\theta = 18°$ and $BA = 16.85$ in. which we insert on the skeleton outline of the mechanism in part (c). The motion of A is common to both OA and BD. As part of OA, point A has pure rotation about O which causes the values of v and a shown. With respect to BD, however, point A is sliding toward B as BD rotates about B. It is natural, therefore, to describe the motion of A on BD in terms of r-θ components with an origin at B, the positive senses being as indicated.

By projecting v onto the radial and transverse directions, we obtain

$$v_r = \dot{r} = -60 \cos 48° = -40.1 \text{ ips}$$

and

$$v_\theta = r\dot{\theta} = 60 \sin 48°, \qquad 16.85\dot{\theta} = 60 \sin 48°, \qquad \dot{\theta} = 2.64 \text{ rad/sec}\circlearrowleft$$

Similarly, by projecting a upon the transverse direction, we obtain a_θ which enables us to find $\ddot{\theta}$ now that r, $\dot{r}$, and $\dot{\theta}$ are known. Thus, we get

$$[a_\theta = r\ddot{\theta} + 2\dot{r}\dot{\theta}] \quad -600 \sin 42° = 16.85\ddot{\theta} + 2(-40.1)(2.64)$$

$$\ddot{\theta} = \alpha = -11.2 = 11.2 \text{ rad/sec}^2\circlearrowright \qquad \textit{Ans.}$$

Later in Section 12-9, we shall discuss a more general approach to this type of problem by using a rotating coordinate system.

PROBLEMS

9-8.3. Derive Eq. (9-8.1) by the following method: Express $\hat{\mathbf{e}}_r$ and $\hat{\mathbf{e}}_\theta$ in terms of the unit vectors $\hat{\mathbf{i}}$ and $\hat{\mathbf{j}}$ along fixed X and Y axes, and then differentiate these expressions.

9-8.4. Show that the unit vectors $\hat{\mathbf{e}}_t$ and $\hat{\mathbf{e}}_n$ in terms of the unit vectors $\hat{\mathbf{e}}_r$ and $\hat{\mathbf{e}}_\theta$ are $\hat{\mathbf{e}}_t = \dfrac{v_r\hat{\mathbf{e}}_r + v_\theta\hat{\mathbf{e}}_\theta}{v}$ and $\hat{\mathbf{e}}_n = \dfrac{v_r\hat{\mathbf{e}}_\theta - v_\theta\hat{\mathbf{e}}_r}{v}$. Then verify Eq. (9-8.4) by using the procedure outlined on p. 372.

9-8.5. A pump impeller with straight radial vanes rotates at a constant rate of 180 rpm. Fluid particles move outward along the vanes with a constant speed of 20 ips. Find the acceleration of a particle when it is 4 in. from the shaft axis of the impeller.

$$a = 134 \text{ fps}^2 \qquad \textit{Ans.}$$

9-8.6. The motion of a particle is defined by the parametric equations $r = 4t^2$ and $\theta = 2t$ where r is in inches and t is in seconds. Find the total acceleration of the particle at $t = 2$ sec.

9-8.7. The polar coordinates of a particle moving on a plane curve are given by $r = t^3 - 3t + 10$ and $r = 2\theta$ where r is in inches, θ is in radians, and t is in seconds. Determine the acceleration of the particle at $t = 2$ sec.

$$a = 23.1 \text{ fps}^2 \qquad \textit{Ans.}$$

9-8.8. In a certain plane motion, it is known that when the radial distance from a fixed origin is $r = 20$ in., the r-θ components of velocity are $v_r = 60$ ips and $v_\theta = 80$ ips, while the components of acceleration are $a_r = -60 \text{ ips}^2$ and $a_\theta = +100 \text{ ips}^2$. At this instant, find the angular acceleration α (i.e., $\ddot{\theta}$) of the position vector and the component of acceleration normal to the path.

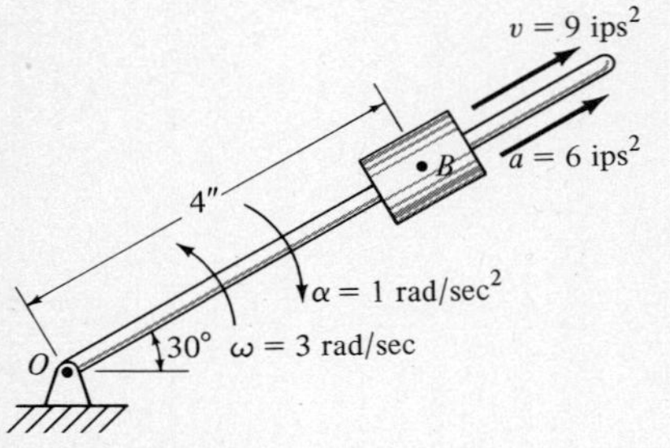

Figure P-9-8.9

9-8.9. At the position shown in Fig. P-9-8.9, the block B is sliding outward along the straight rod with the given values of velocity and acceleration relative to the rod. Simultaneously, the rod has the given values of angular velocity ω and angular acceleration α. Find the total acceleration of the block. What is the component of this acceleration normal to the path described by the block?

$a = 58.3 \text{ ips}^2$; $a_n = 54.0 \text{ ips}^2$ **Ans.**

9-8.10. A particle is constrained to move along the cardiod $r = 2(1 + \cos\theta)$ where r is in inches. If its position vector has a constant angular speed of 2 rad/sec counterclockwise, at what rate is the particle changing its speed at $\theta = 60°$?

Decreasing at 4 ips^2 **Ans.**

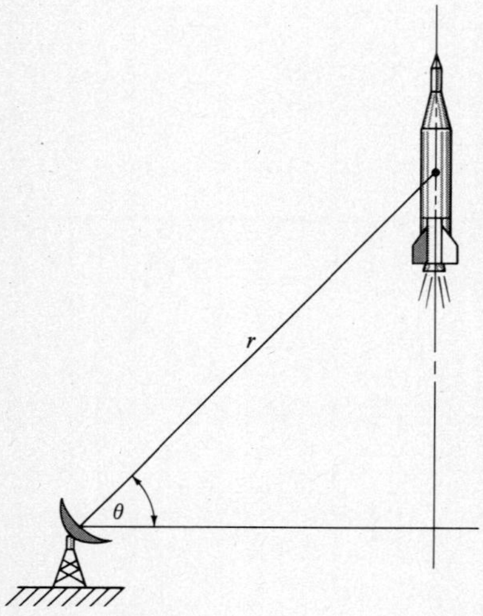

Figure P-9-8.11

9-8.11. A rocket is fired vertically and tracked by radar as shown in Fig. P-9-8.11. At the instant $\theta = 60°$, it is known that $r = 30{,}000$ ft, $\dot{\theta} = 0.02$ rad/sec, and $\ddot{\theta} = 0.002 \text{ rad/sec}^2$. Determine the velocity and acceleration of the rocket.

$v = 818$ mph; $a = 138.5$ mph/sec **Ans.**

9-8.12. A marine steering gear known as Rapson's slide is shown in Fig. P-9-8.12. The tiller AB is turned by the link DE acting upon the slider D. If link DE is moving rightward at a constant speed of 1 fpm, find the angular acceleration of the tiller at the given position.

$\alpha = 0.0241 \text{ rad/min}^2$ **Ans.**

9-8.13. In Fig. P-9-8.13, the motion of the pin P on a fixed circular path of radius r is controlled by the slotted rod OA. Between $\theta = \pm\pi/4$, the rod OA has a constant angular speed $\dot{\theta}$. Find the acceleration of P at any value of θ between the indicated limits.

9-8.14. The telescoping rod shown in Fig. P-9-8.14 forces the pin P to move along the fixed path $9y = x^2$. When $x = 6$ in., it is known that

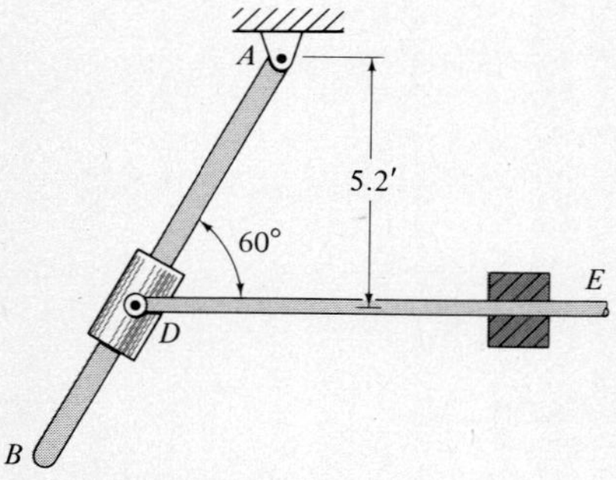

Figure P-9-8.12

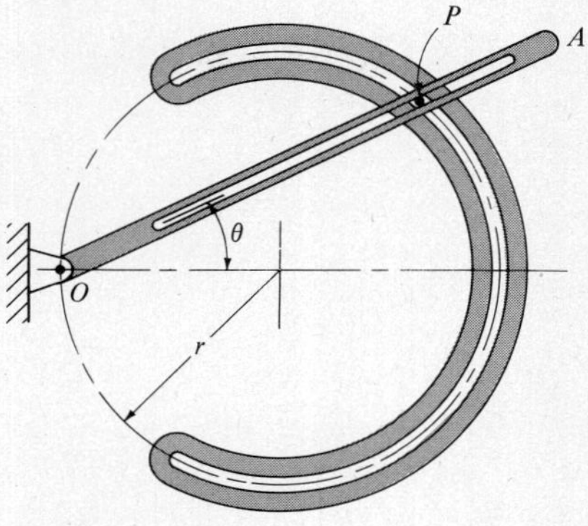

Figure P-9-8.13

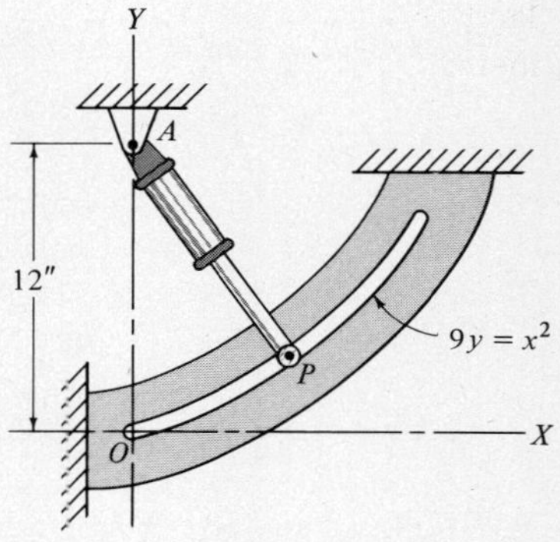

Figure P-9-8.14

the velocity and acceleration of P are respectively $\mathbf{v} = 12\hat{\mathbf{i}} + 16\hat{\mathbf{j}}$ ips and $\mathbf{a} = 10\hat{\mathbf{i}} + 20\hat{\mathbf{j}}$ ips^2. What is then the angular acceleration of the rod?

$$\alpha = 3.29 \text{ rad/sec}^2 \qquad \textbf{\textit{Ans.}}$$

9-8.15. The driving crank AB of the oscillating arm quick-return mechanism shown in Fig. P-9-8.15 rotates at a constant clockwise rate of 11.2 rad/sec. Compute the angular acceleration of the arm CD at the instant when crank AB is horizontal as shown.

$$\alpha_{CD} = 30.1 \text{ rad/sec}^2 \text{ counterclockwise} \qquad \textbf{\textit{Ans.}}$$

9-8.16. The elements of a quick-return mechanism have the given configuration at the instant shown in Fig. P-9-8.16. Find the angular acceleration of arm CD if the driving crank AB rotates at a constant clockwise rate of 5 rad/sec.

$$\alpha_{CD} = 2.51 \text{ rad/sec}^2 \text{ counterclockwise} \qquad \textbf{\textit{Ans.}}$$

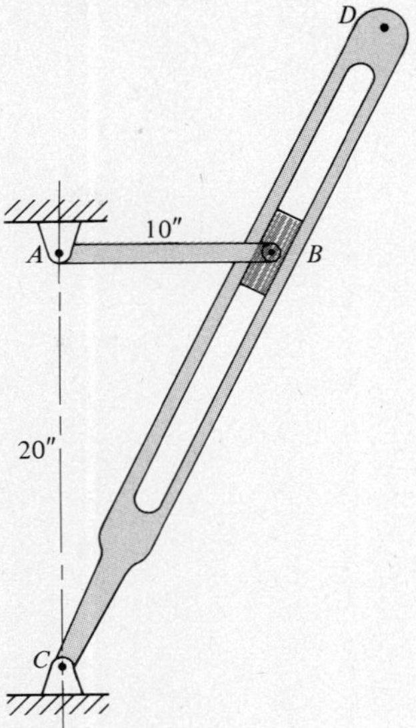

Figure P-9-8.15

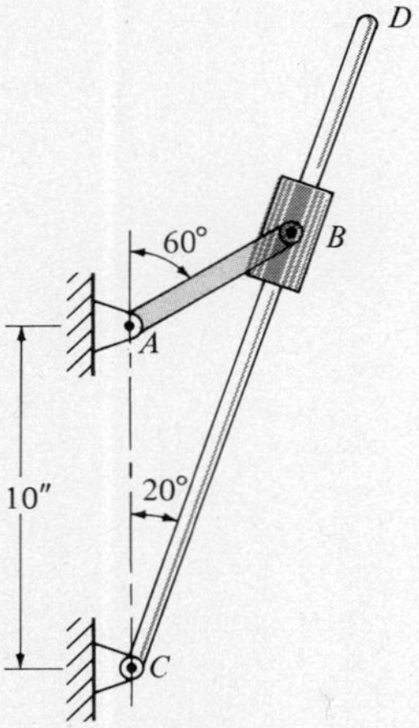

Figure P-9-8.16

SUMMARY

The displacement of a particle is the change in the position vectors from a fixed origin to the positions occupied by the particle on its path of travel. The absolute displacement (or change in position) is independent of the choice of a fixed origin. Velocity is defined as the time rate of change of displacement; acceleration is defined as the time rate of change of velocity. From these definitions, the vector differential equations of the kinematics of a particle are given by the equations

$$\mathbf{v} = \frac{d\mathbf{r}}{dt} = \dot{\mathbf{r}} \qquad (9\text{-}2.1)$$

and

$$\mathbf{a} = \frac{d\mathbf{v}}{dt} = \dot{\mathbf{v}} = \ddot{\mathbf{r}} \qquad (9\text{-}2.3)$$

For rectilinear motion in which displacement, velocity, and acceleration are unidirectional, the signed position coordinate s replaces the position vector $\mathbf{r}$ and the scalar differential equations of kinematics become

$$v = \frac{ds}{dt} \qquad (9\text{-}3.1)$$

$$a = \frac{dv}{dt} \qquad (9\text{-}3.2)$$

$$a\,ds = v\,dv \qquad (9\text{-}3.3)$$

In applying these equations, take the positive values of s, v, and a as corresponding to the positive sense of initial motion. When the motion can be expressed graphically (Section 9-4), the area under a v-t curve represents the change of displacement during the corresponding time interval, and the area under the a-t curve represents the corresponding change in velocity. The a-t curve can also be used to find the change in displacement during a time interval by means of the equation

$$\Delta s = v_1(\Delta t) + (\text{Area})_{a\text{-}t}(\bar{t}_2) \tag{9-4.3}$$

where v_1 is the velocity at the start of the time interval and the last term is the moment of the area under the a-t curve about an ordinate at the *end* of the time interval.

The curvilinear motion of a particle may be expressed in terms of rectangular components, normal and tangential components, or polar and cylindrical coordinates. In terms of rectangular components, we have

$$\mathbf{r} = x\hat{\mathbf{i}} + y\hat{\mathbf{j}} + z\hat{\mathbf{k}} \tag{9-6.1}$$

$$\mathbf{v} = v_x\hat{\mathbf{i}} + v_y\hat{\mathbf{j}} + v_z\hat{\mathbf{k}} \tag{9-6.2}$$

$$\mathbf{a} = a_x\hat{\mathbf{i}} + a_y\hat{\mathbf{j}} + a_z\hat{\mathbf{k}} \tag{9-6.3}$$

These formulations show that curvilinear motion may be considered equivalent to the simultaneous projection of the motion along fixed rectangular axes, and each projected motion can be treated as a rectilinear motion.

Tangential and normal components of acceleration separate and denote respectively the rate of change of *magnitude* and *direction* of velocity. Using unit vectors $\hat{\mathbf{e}}_t$ and $\hat{\mathbf{e}}_n$ to denote directions respectively tangent and normal to the path, we have

$$\mathbf{a} = \frac{dv}{dt}\hat{\mathbf{e}}_t + \frac{v^2}{\rho}\hat{\mathbf{e}}_n = \mathbf{a}_t + \mathbf{a}_n \tag{9-7.1}$$

where the magnitudes of the acceleration components are $a_t = \dfrac{dv}{dt}$ and $a_n = \dfrac{v^2}{\rho}$ and ρ is the instantaneous radius of curvature of the path. Their directions are such that positive a_t is tangent to the path in the direction of increasing speed whereas a_n is always directed inward toward the center of curvature.

In polar coordinates, the position vector $\mathbf{r}$ is expressed in terms of its magnitude r and its angular coordinate θ measured from a fixed reference axis. The radial and transverse components of velocity and acceleration are

$$\mathbf{v} = \dot{r}\hat{\mathbf{e}}_r + r\dot{\theta}\hat{\mathbf{e}}_\theta = \mathbf{v}_r + \mathbf{v}_\theta \tag{9-8.2}$$

and

$$\mathbf{a} = (\ddot{r} - \mathbf{r}\dot{\theta}^2)\hat{\mathbf{e}}_r + (r\ddot{\theta} + 2\dot{r}\dot{\theta})\hat{\mathbf{e}}_\theta = \mathbf{a}_r + \mathbf{a}_\theta \tag{9-8.3}$$

where $\hat{\mathbf{e}}_r$ and $\hat{\mathbf{e}}_\theta$ are unit vectors acting in the positive directions of increasing r and θ; i.e., the unit vectors $\hat{\mathbf{e}}_r$ and $\hat{\mathbf{e}}_\theta$ are directed parallel to the positive senses of $\mathbf{v}_r$ and $\mathbf{v}_\theta$.

The extension of polar coordinates to cylindrical coordinates is accomplished by adding a fixed Z axis perpendicular to the polar coordinate axes. The position vector is then $\mathbf{r} = r\hat{\mathbf{e}}_r + z\hat{\mathbf{k}}$ where $\hat{\mathbf{k}}$ is a unit vector in the positive Z direction. The positive Z direction is that which makes the unit vectors $\hat{\mathbf{e}}_r$, $\hat{\mathbf{e}}_\theta$, and $\hat{\mathbf{k}}$ form a right-handed triad so that $\hat{\mathbf{e}}_r \times \hat{\mathbf{e}}_\theta = \hat{\mathbf{k}}$.

10

GENERAL PRINCIPLES OF DYNAMICS

10-1 INTRODUCTION

Our objective in this chapter is to derive the two basic equations of dynamics that we stated in the preview of dynamics presented in Section 9-1. After discussing several associated concepts, we shall use kinematic principles to extend Newton's laws of motion for a single particle to bodies consisting of systems of particles. We concentrate here on the general principles of dynamics and in subsequent chapters consider their application to various types of rigid-body motion. But first it is appropriate to present the following brief background.

Dynamics is the branch of mechanics which deals with the study of bodies in motion. Compared with statics, dynamics is relatively new[1]; it is generally considered to have been begun by Galileo (1564–1642). Its development was greatly retarded by the lack of

[1] An excellent summary of the history of dynamics and those who contributed greatly to its development is given in Goodman and Warner's *Dynamics,* Wadsworth Publishing Co., pp. 1–8 and in Fanger's *Engineering Mechanics,* Merrill Publishing Co., pp. 501–512.

precise methods for measuring time. The experiments which form the foundation of dynamics require the use of three kinds of units: *force, length,* and *time.* Precise methods for measuring force and length are relatively simple and account in part for the early development of statics in which only these units of measurements are required. No accurate time-measuring devices, such as the pendulum clock developed by Huygens in 1657 and the balance wheel watch developed by Robert Hooke around 1666, were devised until after Galileo's death.

An understanding of dynamics was also retarded by the principles of natural philosophy which were promulgated by Aristotle and in Galileo's time were regarded as infallible. Galileo's experimental turn of mind led him to doubt these dogmas of abstract thought. For example, he did not accept the notion that heavy weights fall more rapidly than light ones. His experiments with dropping weights from the Leaning Tower of Pisa exploded this theory, but precipitated such bitter conflicts that he was forced to leave Pisa.[2]

Galileo's experiments with blocks sliding down inclined planes led to a relation between force and acceleration which Sir Isaac Newton (1642–1727) generalized and incorporated into the laws governing the motion of a particle that are named after him. Newton's laws of motion are the basis for extending the kinetics of a particle to a body composed of a system of particles. This extension is discussed in Sections 10-5 and 10-6.

At this point, definitions of the terms particle and body are pertinent. The term *particle* usually denotes an object of point size. The term *body* denotes a system of particles which form an object of appreciable size. The criterion of size is only relative however; the terms particle and body may apply equally to the same object. For example, in astronomical calculations the earth may be assumed to be a particle in comparison with the size of its path, whereas to earthbound observers, it is a body of great size. More realistically, a particle is a body so small that any differences in the motions of its parts can be ignored with negligible error. We shall also learn that any rigid body, regardless of its size, may be considered to be a particle if all of its parts move in identical parallel paths.

10-2 NEWTON'S LAWS OF MOTION FOR A PARTICLE

From his study of falling bodies, Galileo discovered the first two of what are commonly called Newton's laws of motion for a particle. Newton's name is associated with the laws of motion, however, because it was he who generalized them and, together with the law

[2] *The Star Gazer* by Z. Harsanyi (Putnam) is an excellent account of Galileo's life.

of universal gravitation, demonstrated their truth by astronomical predictions based on them.

Newton's laws of motion have been stated in a variety of ways. For our purposes, we shall phrase them as follows:

1. A particle acted upon by a balanced force system has no acceleration.

2. A particle acted upon by an unbalanced force system has an acceleration in line with and directly proportional to the resultant of the force system.[3]

3. Action and reaction forces between two particles are always equal and oppositely directed.

The reference frame or set of axes in which Newton's laws are valid is one having a fixed origin and fixed directions of the axes. It is called an *inertial, Newtonian,* or *Galilean frame of reference.* An inertial frame may also be a frame whose axes remain fixed in direction while its origin moves along a straight line with constant speed. Motion with respect to an inertial frame is known as absolute motion. The most accurate of such reference frames available to us is one having its origin at the mass center of the solar system and its axes directed toward the so-called fixed stars whose parallaxes are too small to be measured. Such a reference frame is suitable for celestial mechanics, but for orbiting satellites or long-range rockets, the inertial frame could be one fixed in the earth with its origin at the earth's center. With respect to most engineering applications, however, it is sufficiently accurate to use an inertial frame attached to the surface of the earth.

It should be observed that modern concepts of relativistic mechanics initiated by Albert Einstein (1879–1955) and of quantum mechanics developed by Max Planck (1858–1947), Niels Bohr (1885–1962), and Werner Heisenberg (1901–) establish the boundaries of Newtonian mechanics but do not invalidate it. Except for speeds approaching that of light or the motion of elements of the atom, Newton's laws will continue to be used by engineers and scientists. Their accuracy has been definitely proven by the Apollo space missions.

Another outstanding contribution to the study of mechanics is Newton's law of universal gravitation which states that any two particles of masses m_1 and m_2 separated by a distance r attract each other with equal, opposite forces that are directed from either particle toward the other. The magnitude of each of this pair of action

[3] An alternate version states that the resultant force on a particle is in line with and proportional to the time rate of change of momentum of the particle. Either statement is correct when applied to a particle of constant mass.

and reaction forces is given by

$$F = G\frac{m_1 m_2}{r^2} \tag{10-2.1}$$

where G is the universal constant of gravitation.[4] Its value has been shown to be 3.44×10^{-8} ft^4/lb-sec^4 in the ft-lb-sec system or 6.67×10^{-8} cm^3/gm-sec^2 in the cgs system.

Observe that Newton's law of gravitation does not define any of the physical quantities (force, mass, and length) contained in it. The defining relations among these quantities is expressed in Newton's second law. In Section 10-4 we shall discuss various systems of units that may be used. Of importance here is the concept of weight. The weight of a body is the gravitational force of attraction exerted on the body by the earth and depends on its position relative to the earth's center.

10-3 FUNDAMENTAL EQUATION OF KINETICS FOR A PARTICLE

Consider a particle of weight W acted upon by the forces shown in Fig. 10-3.1a. The resultant of these forces is obtained by a tip-to-tail addition as in Fig. 10-3.1b. Because all the forces on a particle are assumed to be concurrent (the size of a particle is a point), the direction and position of the resultant is indicated by the dashed

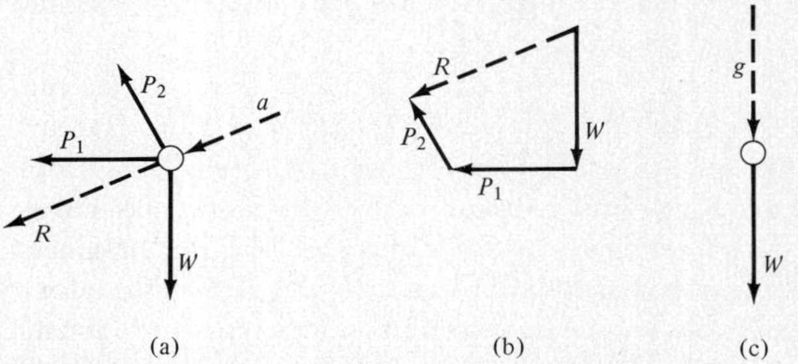

Figure 10-3.1 Resultant force on and acceleration of a particle.

vector **R** in Fig. 10-3.1a. From Newton's second law of motion, this resultant causes an acceleration **a** which is in line with **R** and directly proportional to it, or

$$\mathbf{R} = k\mathbf{a} \tag{a}$$

where k is some constant of proportionality.

If the same particle is now assumed to be in a vacuum, the

[4] Do not confuse G with the acceleration g due to gravity. Note that g denotes the acceleration of a freely falling body; its value varies with position but for most engineering applications has the value of 32.2 fps^2.

resultant force acting upon it is its weight W. By experiment, the acceleration produced by W is found to be the value of the gravitational constant g which acts in the direction of W, as shown in Fig. 10-3.1*c*. Again applying Newton's second law and assuming the same constant of proportionality, we have

$$W = kg \tag{b}$$

Dividing Eq. (*a*) by Eq. (*b*) gives

$$\mathbf{R} = \frac{W}{g}\mathbf{a} = m\mathbf{a} \tag{10-3.1}$$

Comparison with Eq. (*a*) discloses the value of the constant of proportionality to be W/g. This ratio of weight divided by the gravitational constant is often called the *gravitational mass m* of the particle since the weight W and the associated gravitational acceleration g vary with position on the earth, whereas inertial mass m is an invariant absolute property of the particle which is independent of position. However, the most refined experiments have failed to show any difference between gravitational mass and inertial mass and they are therefore used interchangeably in this book. One advantage of using gravitational mass is that, when applying Eq. (10-3.1), it results in a consistent set of dimensional units by automatically including the conversion factor g. Perhaps as good a concept of mass as any is to consider that it represents the inertia of a body, i.e., the resistance a body offers to a change in its motion.

EFFECTIVE FORCE ON A PARTICLE

The effective force on a particle is defined as the resultant force on a particle. Since $\mathbf{R}$ and $m\mathbf{a}$ are numerically equivalent, either may be said to be the effective force on a particle. The use of $m\mathbf{a}$ to represent this force is especially convenient in cases when the acceleration of the particle is known but the actual force system producing this acceleration is not known. This concept is used in succeeding sections to extend the laws of motion for a single particle to the motion of a body composed of a system of particles.

According to Newton's third law, for every force there is an equal but opposite reaction. In the case of a particle accelerated by a resultant force, this reaction is called the *inertia force* of the particle. It is convenient to think of the inertia force as a force numerically equal to $m\mathbf{a}$ but directed oppositely to the acceleration.

If the inertia force is considered to act on a particle together with the resultant force, the particle will be in a state of equilibrium. This will be called *dynamic equilibrium* to distinguish it from static equilibrium in which the particle is at rest or is moving with constant velocity.

10-4 GRAVITATIONAL AND ABSOLUTE SYSTEMS OF UNITS

In applying $\mathbf{R} = m\mathbf{a}$, any system of units can be used which will result in a force of one unit giving a unit acceleration to a unit mass. Four quantities are involved in this equation—force (F), mass (M), length (L), and time (T)—which are dimensionally related by

$$F = \frac{ML}{T^2}$$

Of these four quantities, L, T, and either F or M may be chosen as basic dimensions. Most engineers select F as a basic dimension and mass is the derived dimension expressed in terms of the others; namely, $M = FT^2/L$. This is known as a gravitational system since F is defined as the weight of a standard object. In the English system using foot-pound-second units, a unit mass (sometimes called a slug) has the dimensions of lb-sec^2/ft. In the metric system using meter-kilogram-second units, a unit mass has the dimensions of kg-sec^2/m.

In the absolute MKS system of units used by physicists and electrical engineers, inertial mass is a basic dimension and force is the derived dimension defined as $F = \dfrac{ML}{T^2}$. The unit of force, called a newton (nt), has the dimensions of kg-m/sec^2. The advantage to the electrical engineer of an absolute MKS system of units stems from the definition of a watt or volt-ampere as one newton-meter per second.

In general, either a gravitational or an absolute system of units may be used since each has its own advantage over the other. For example, in situations where the value of g may vary considerably as in rocket flight or space probes, it is preferable to consider mass

Table 10-4.1. Systems of Units

System	Length	Time	Force	Mass	Used by
Gravitational English (FPS)	foot (ft)	sec	pound (lb)	slug[a] (lb-sec^2/ft)	Engineers in English-speaking countries
Gravitational Metric (MKS)	meter (m)	sec	kilogram (kg)	metric slug[a] (kg-sec^2/m)	Engineers elsewhere
Absolute Metric (CGS)	centimeter (cm)	sec	dyne[a] (gm-cm/sec^2)	gram (gm)	Physicists rarely
Absolute Metric (MKS)	meter (m)	sec	newton[a] (kg-m/sec^2)	kilogram (kg)	Physicists generally

[a] Derived unit.

as a basic unit since it is a property independent of position. However, most engineers are concerned with problems in which the variation in g is negligible; this makes the common concept of the weight of a standard body as the unit of force convenient to use. Following this custom, in this book we shall use the English gravitational system of foot-pound-second units. A comparison of various systems of units is shown in Table 10-4.1.

10-5 D'ALEMBERT'S PRINCIPLE. MOTION OF THE MASS CENTER

D'Alembert's principle expresses the relation between the external forces applied to a system of particles and the effective force on each particle of the system. It is the basis for extending Newton's laws for a single particle to a system of particles. The principle may be stated as follows: *The resultant of the external forces applied to a body (rigid or nonrigid) composed of a system of particles is equal to the vector summation of the effective forces acting on all particles.* Because of the fundamental importance of this principle, we shall prove it both in terms of its physical meaning and again by a formal mathematical development.

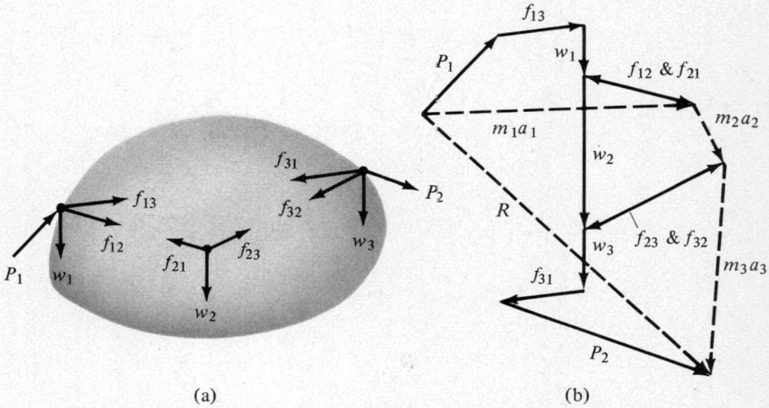

(a) (b)

$$R = W \nrightarrow P_1 \nrightarrow P_2 \nrightarrow \cdots = m_1a_1 \nrightarrow m_2a_2 \nrightarrow m_3a_3 \nrightarrow \cdots$$

Figure 10-5.1 Resultant of external forces is equivalent to that of the effective forces.

The physical significance of D'Alembert's principle is demonstrated in Fig. 10-5.1. The heavy outline denotes the boundary of a body which consists of a system of particles. The body may be rigid so that the particles remain at fixed distances from one another, or the body may be nonrigid like a gas or a liquid. The external forces acting on this body are its weight W and several forces of which two are shown, P_1 and P_2. For convenience and also for clarity, assume the body to consist of only the three particles shown. Each of these particles is acted upon by its own weight and by the internal forces exerted upon it by the other particles. In addition, the first particle at the left boundary is acted upon by the external force P_1 while the third particle at the right boundary is acted upon by the external force P_2.

In part (b) of the figure, the effective force $m_i a_i$ on each particle is shown as the tip-to-tail sum of the forces acting on each particle. Since the internal forces occur in equal opposite pairs, they cancel out of a vector summation of forces involving all particles.[5] Hence, as shown in the figure, we see that the resultant force R of the impressed forces P_1, P_2, and W (note that $W = w_1 + w_2 + w_3$) is equivalent to the vector sum of the effective forces. Hence, we obtain

$$\begin{aligned} R &= W \nrightarrow P_1 \nrightarrow P_2 \nrightarrow \cdots \\ &= m_1 a_1 \nrightarrow m_2 a_2 \nrightarrow m_3 a_3 \nrightarrow \cdots \end{aligned} \tag{10-5.1}$$

This expresses in equation form D'Alembert's principle that the resultant of the impressed forces acting on a body is equivalent to the vector sum of the effective forces acting on all the particles composing the body. In this development, there is no restriction that the body must be rigid.

Notice that if the effective forces $m_i a_i$ are reversed and considered to act on each respective particle of the system that the system will be in a state of balance known as dynamic equilibrium. As discussed in Section 10-3, these reversed effective forces are called the inertia forces of the particles. Thus, D'Alembert's principle may also be stated as follows: *The impressed forces acting on any body are in dynamic equilibrium with the inertia forces of the particles of the body.*

The preceding discussion should make it simple to follow a formal mathematical development of D'Alembert's principle and its extension to the motion of the mass center of any body. In a system of n particles, let $\mathbf{F}_i$ be the resultant of external forces acting on the ith particle and $\sum_{j=1}^{n} \mathbf{f}_{ij}$ be the resultant of internal forces exerted on particle i by all other particles. (Note that $\mathbf{f}_{ii} = 0$ since a particle does not exert an internal force upon itself.)

[5] In order to draw vector polygons of the forces acting on each particle so that the effective force $m_i a_i$ on each particle will be obvious, it is not possible to draw f_{13} and f_{31} as the equal, opposite, collinear vectors they really are. However, it should be obvious that f_{13} and f_{31} will cancel each other when we consider a tip-to-tail summation of all the forces shown.

Newton's second law for any particle may now be expressed by

$$\mathbf{F}_i + \sum_{j=1}^{n} \mathbf{f}_{ij} = m_i \mathbf{a}_i$$

Then the sum of the external and internal forces on all particles of a system is

$$\sum_{i=1}^{n} \mathbf{F}_i + \sum_{i=1}^{n} \sum_{j=1}^{n} \mathbf{f}_{ij} = \sum_{i=1}^{n} m_i \mathbf{a}_i$$

However, the internal forces occur in equal, opposite pairs ($\mathbf{f}_{ij} = -\mathbf{f}_{ji}$ by Newton's third law) so that the double sum is zero. Hence, we obtain

$$\sum_{i=1}^{n} \mathbf{F}_i = \sum_{i=1}^{n} m_i \mathbf{a}_i = \sum_{i=1}^{n} m_i \ddot{\mathbf{r}}_i \tag{a}$$

which is the mathematical statement of D'Alembert's principle.

To extend this discussion to determine the motion of the mass center of any system, we start by noting that the total moment of mass is the moment sum of the component masses, or

$$m\bar{\mathbf{r}} = \sum_{i=1}^{n} m_i \mathbf{r}_i$$

Note the use of the overbar in $\bar{\mathbf{r}}$ to denote the position vector to the mass center. When this relation is differentiated twice with respect to time, it becomes

$$m\ddot{\bar{\mathbf{r}}} = \sum_{i=1}^{n} m_i \ddot{\mathbf{r}}_i \tag{b}$$

Comparison of the right-hand terms of Eqs. (*a*) and (*b*) shows them to be identical; hence, we conclude that the left-hand terms must be equal, or

$$\sum_{i=1}^{n} F_i = m\ddot{\bar{\mathbf{r}}} \qquad \text{equivalent to} \qquad \boxed{\mathbf{R} = m\bar{\mathbf{a}} = \frac{W}{g}\bar{\mathbf{a}}} \tag{10-5.2}$$

This equation expresses the acceleration of the mass center of any body (rigid or nonrigid) in terms of the applied external forces. Since it is identical in form to Newton's second law for a single particle, we conclude that insofar as the relation between the resultant force and the acceleration of the mass center is concerned, *any system of particles may be considered equivalent to a single particle concentrated at the mass center and having the same mass as the mass of the system.* This concept as expressed by Eq. (10-5.2) is fundamental to the study of dynamics since it is valid for any body having any motion. We shall refer to it frequently in later chapters.

But there is an important difference between the system and the equivalent particle concept; the line of action of the resultant force on the system does not usually pass through the equivalent particle at the mass center of the system. The action line of **R** on a body or a system of particles must be found by applying the principle of moments; i.e., the moment of the resultant force is equal to the moment sum of its parts. We do this in the next section.

10-6 MOMENT EFFECT OF EXTERNAL FORCES

As we learned in statics, any force system can be reduced to a resultant force-couple system. Because the resultant force is independent of the choice of a reference center, we applied this in the previous section to show the equivalence of the force polygon of the applied forces on a body to the vector polygon of the effective forces. The magnitude of the resultant couple, however, depends on the position of a reference center. We now discuss the relation between the resultant couple and the motion of a system of particles. Of particular importance will be the rotational effect of the resultant couple on the angular motion of a rigid body.

Consider a swarm of particles (enclosed by the boundary shown in Fig. 10-6.1) which is acted upon by any set of forces. The moment effect $\Sigma \mathbf{M}_A$ of these forces (not shown) about any center A in or outside the system is equal to the moment sum of the equivalent force distribution expressed in terms of the effective forces. Since Newton's second law is only valid in an inertial frame, the effective force on a typical particle B is given by $m_i \mathbf{a}_i = m_i \ddot{\mathbf{r}}_i$ where $\mathbf{r}_i$ is the position vector of B from the origin O of the inertial reference frame. The moment of this typical effective force about the moment center A depends on the position vector $\boldsymbol{\rho}$ of B from A. Thus, we have

$$\Sigma \mathbf{M}_A = \Sigma \boldsymbol{\rho} \times m_i \ddot{\mathbf{r}}_i \tag{a}$$

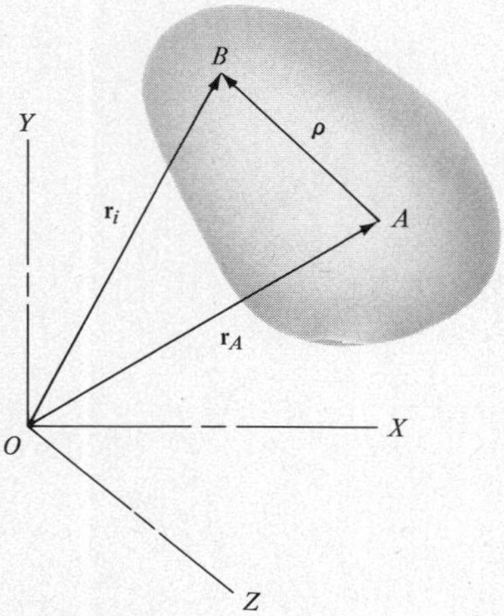

Figure 10-6.1

From the figure, we see that $\mathbf{r}_i = \mathbf{r}_A + \boldsymbol{\rho}$ so that two time differentiations result in $\ddot{\mathbf{r}}_i = \ddot{\mathbf{r}}_A + \ddot{\boldsymbol{\rho}}$ which we substitute into Eq. (a) to obtain

$$\Sigma \mathbf{M}_A = \Sigma \boldsymbol{\rho} \times m_i \ddot{\mathbf{r}}_A + \Sigma \boldsymbol{\rho} \times m_i \ddot{\boldsymbol{\rho}} \tag{b}$$

Now we simplify the first right-hand term by noting that $\ddot{\mathbf{r}}_A$ is independent of $\boldsymbol{\rho}$ so that we can take it out of the summation and rewrite it as

$$\Sigma \boldsymbol{\rho} \times m_i \ddot{\mathbf{r}}_A = (\Sigma \boldsymbol{\rho} m_i) \times \ddot{\mathbf{r}}_A = \bar{\boldsymbol{\rho}}_A m \times \ddot{\mathbf{r}}_A = \bar{\boldsymbol{\rho}}_A \times m \ddot{\mathbf{r}}_A \tag{c}$$

In the second of these equivalent terms, observe that $\Sigma \boldsymbol{\rho} m_i$ is the total moment of mass which is equivalent in the third term to the total mass m multiplied by the position vector $\bar{\boldsymbol{\rho}}_A$ from A to the center of mass. Finally we reassociated m with $\ddot{\mathbf{r}}_A$.

Consider next the second right-hand term of Eq. (*b*). Let $m_i\dot{\boldsymbol{\rho}}$ be defined as the linear momentum of a particle relative to A so that $\boldsymbol{\rho} \times m_i\dot{\boldsymbol{\rho}}$ is the moment of this relative momentum about A. Then using the symbol $\mathbf{H}_A$ to represent the total relative moment of momentum about A, we have

$$\mathbf{H}_A = \Sigma\boldsymbol{\rho} \times m_i\dot{\boldsymbol{\rho}} \tag{d}$$

whose time derivative is

$$\dot{\mathbf{H}}_A = \Sigma\dot{\boldsymbol{\rho}} \times m_i\dot{\boldsymbol{\rho}} + \Sigma\boldsymbol{\rho} \times m_i\ddot{\boldsymbol{\rho}} \tag{e}$$

in which the first right-hand term is zero, since it is the cross product of a vector with itself. Therefore, Eq. (*e*) reduces to

$$\dot{\mathbf{H}}_A = \Sigma\boldsymbol{\rho} \times m_i\ddot{\boldsymbol{\rho}} \tag{f}$$

We now return to Eq. (*b*) and replace its right-hand terms by their equivalents from Eqs. (*c*) and (*f*) to obtain the following fundamental moment equation for any system of particles:

$$\Sigma\mathbf{M}_A = \bar{\boldsymbol{\rho}}_A \times m\ddot{\mathbf{r}}_A + \dot{\mathbf{H}}_A \tag{10-6.1}$$

Usually, we select a moment center A that is either a fixed point O in space so that $\ddot{\mathbf{r}}_A = 0$, or else is the mass center G so that $\bar{\boldsymbol{\rho}}_A = 0$. Either of these choices of a moment center reduces Eq. (10-6.1) to the more convenient form

$$\Sigma\mathbf{M}_O = \dot{\mathbf{H}}_O \quad \text{or} \quad \Sigma\mathbf{M}_G = \dot{\mathbf{H}}_G \tag{10-6.2}$$

This simpler form also applies to the special case[6] when the acceleration of the reference point (i.e., $\ddot{\mathbf{r}}_A$) passes through the mass center, since then $\bar{\boldsymbol{\rho}}_A$ and $\ddot{\mathbf{r}}_A$ coincide and their cross product is zero.

In general, by adding an inertial force $-m\ddot{\mathbf{r}}_A$ acting at the mass center of the system and considering it as an additional external force acting on the system, all of the results derived by applying Newton's laws of motion relative to an inertial frame apply equally well with respect to a *translating* frame having the acceleration of the base point A.

The physical interpretation of Eq. (10-6.2) as shown in Fig. 10-6.2 will greatly enhance its universal application. This figure shows that at the mass center G, the resultant force-couple system of the forces acting on any system of particles is equivalent to a

[6] This occurs in a free-rolling body (see Section 12-5).

resultant effective force $m\bar{\mathbf{a}}$ at G plus the couple $\dot{\mathbf{H}}$ representing the time rate of change of the relative angular momentum $\mathbf{H}$ about G.

Some final observations are pertinent. We have now developed the two fundamental equations of dynamics; namely, Eqs. (10-5.2) and (10-6.2) which we merely stated in the preview of dynamics in Section 9-1. We shall discuss their application in subsequent chapters. At this time, though, consider the application of Eq. (10-6.2)

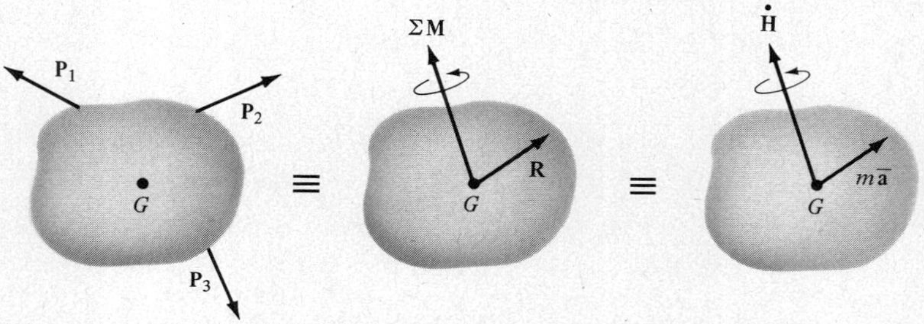

Figure 10-6.2 Equivalence of applied forces to their dynamic effects.

and its physical interpretation in Fig. 10-6.2 to a *translating* rigid body in which all particles travel along identical parallel paths. In this case, no angular motion of the body is involved and both the relative angular momentum $\mathbf{H}$ and its time derivative $\dot{\mathbf{H}}$ will be zero. We conclude, from the equivalences in Fig. 10-6.2, that the resultant of all forces on a translating rigid body reduces to a single force $\mathbf{R} = m\bar{\mathbf{a}}$ which passes through the mass center. This conclusion will be very useful when we apply it later in Section 11-4.

SUMMARY

The laws of motion for a particle (Section 10-2) are extended, by means of D'Alembert's principle (Section 10-5), to include a body (rigid or nonrigid) composed of a system of particles. D'Alembert's principle states that the resultant of the impressed forces acting on a body is equivalent to the vector sum of the effective forces acting on all the particles composing the body.

A useful variation of this principle states that the impressed forces acting on any body are in dynamic equilibrium with the inertia forces acting on the particles of the body.

By extending D'Alembert's principle to determine the motion of the mass center of any body, we find that the resultant of the applied external forces is equivalent to the product of the mass of the body and the acceleration of its mass center; it is expressed by the equation

$$\mathbf{R} = m\bar{\mathbf{a}} = \frac{W}{g}\bar{\mathbf{a}} \tag{10-5.2}$$

This equation is fundamental to dynamics; it is valid for both nonrigid and rigid bodies. However, the location of the resultant force is not specified by this equation.

The effect of the resultant couple acting on any body is given by

$$\Sigma\mathbf{M}_A = \bar{\boldsymbol{\rho}}_A \times m\ddot{\mathbf{r}}_A + \dot{\mathbf{H}}_A \tag{10-6.1}$$

where A is any center of moments and $\dot{\mathbf{H}}_A$ is the time derivative of the relative angular momentum about A. When the moment center is either a fixed point O in space or else is the mass center G, it reduces to

$$\Sigma\mathbf{M} = \dot{\mathbf{H}} \tag{10-6.2}$$

which is the second fundamental equation of dynamics. Its principle use is to determine the angular motion of rigid bodies which we do in subsequent chapters. In particular when applied to a *translating rigid body,* Eqs. (10-5.2) and (10-6.2) show that both the resultant of applied forces and the resultant of the effective forces passes through the mass center.

11

KINETICS OF PARTICLES

11-1 INTRODUCTION

In this chapter, we consider only particle motion. This restriction does not mean that we are limited to bodies of point size; it only means that we consider bodies so small that any differences in the motions of their parts can be ignored or that the bodies have no angular motion. The translational motion of rigid bodies is such a type.

Translation is defined as the motion of a *rigid* body in which a straight line passing through any two of its particles always remains parallel to its initial position. The condition that the body be rigid means that the distance separating any two particles remains constant. This, coupled with the requirement that a straight line in the body remain parallel to its initial direction, means that all particles travel the same or parallel paths. Thus, the outstanding *kinematic* characteristic of a translating rigid body is that all the particles have the same values of displacement, velocity, and acceleration. The motion is completely described by the motion of *any* particle of the body. The particle usually selected is the one at the mass center of the body. Furthermore, as shown on p. 391, the resultant of all forces acting upon the translating rigid body passes through its mass center.

Translation may be either rectilinear or curvilinear, depending on whether the path described by any particle of the body is straight or curved. This is illustrated in Fig. 11-1.1 where the motion in part (a) is called *rectilinear translation* and the motion in part (b) is called *curvilinear translation*. In either case, the pertinent kinetic equation is $\mathbf{R} = m\mathbf{a} = \dfrac{W}{g}\mathbf{a}$ obtained from Eq. (10-5.2). Note that the bar sign over $\mathbf{a}$ to identify the acceleration of the mass center can be omitted since all particles have the same motion.

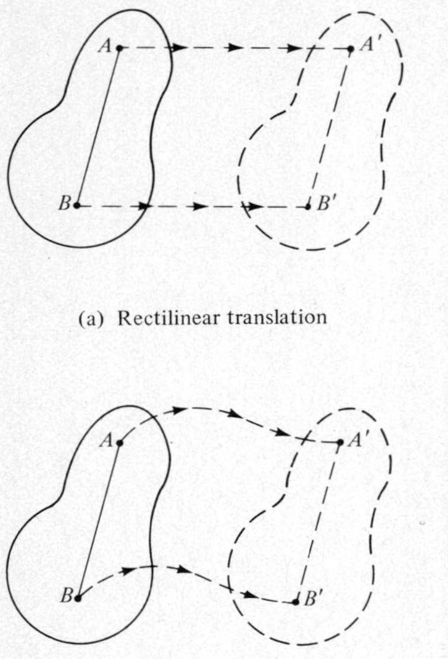

(a) Rectilinear translation

(b) Curvilinear translation

Figure 11-1.1 Translation of a rigid body.

The general problem in dynamics is to determine the relation between the motion of a body and the force system acting upon it. When the motion of a particle is specified, it is simple to determine the force causing the motion directly from $\mathbf{R} = m\mathbf{a}$. The converse problem of determining the motion of a particle when the force is specified is more involved. This requires solving $\mathbf{R} = m\mathbf{a}$ for the instantaneous acceleration $\mathbf{a}$ and then integrating one or more of the differential equations of kinematics (as in Section 9-3) to obtain the appropriate s-t, v-t, or v-s relation desired.

An alternate procedure is to integrate $\mathbf{R} = m\mathbf{a}$ directly, either in terms of the displacement or the time. The displacement integral leads to what is known as the work-energy relation, whereas the time integral yields the impulse-momentum relation. As we shall learn, the work-energy equation relates force, displacement, and velocity. Moreover, we shall see that it is a scalar equation which makes its use much simpler than the impulse-momentum relation which is a vector equation relating force, velocity, and time.

Each of these various methods—force-acceleration, work-energy, and impulse-momentum—and their respective advantages are discussed in subsequent chapters. Our consideration here of the kinetics of particle motion allows us to neglect temporarily the complications due to the angular motion of a general rigid-body motion.

11-2 TRANSLATION. ANALYSIS AS A PARTICLE

The kinetic equations for translation as a particle are obtained from the general equation governing the motion of the mass center (see Section 10-5). It was shown that any translating body could be treated as though it were a particle which had the same mass as the body and the same motion as the mass center of the body, i.e., $\mathbf{R} = \dfrac{W}{g}\bar{\mathbf{a}}$. When this equation is applied to the motion of translation, in which all particles have the same acceleration, the bar sign over $\bar{\mathbf{a}}$ can be omitted. Thus, we obtain $\mathbf{R} = \dfrac{W}{g}\mathbf{a}$ which is identical with Eq. (10-3.1) for the motion of a particle. The component forms to use for this equation depends on whether the path is straight or curved.

For rectilinear motion, it is convenient to select the line of motion as the X axis. *Let this axis be considered positive in the initial direction of motion.* Using this convention, we consider displacement, velocity, acceleration, and X components of forces as positive when directed in the initial direction of motion. Since the acceleration coincides with the X axis, we have $a_x = a$, and $a_y = a_z = 0$. Then the component forms of the kinetic equation for rectilinear motion become

$$\Sigma X = \frac{W}{g}a \qquad \Sigma Y = \Sigma Z = 0 \tag{11-2.1}$$

in which the force summations ΣX, ΣY, and ΣZ are the equivalent components of **R**.[1]

For curvilinear motion, the convenient components of acceleration to use are normal and tangential to the path, or on occasion the radial and transverse components. Calling the normal axis N and the tangential axis T, the corresponding scalar forms of $\mathbf{R} = \frac{W}{g}\mathbf{a}$ are

$$\Sigma N = \frac{W}{g}a_n = \frac{W}{g}\frac{v^2}{r} \qquad \Sigma T = \frac{W}{g}a_t \tag{11-2.2}$$

In applying these equations, the positive senses of ΣN and ΣT are in the directions of a_n and a_t.

In curvilinear motion, however, it is usually preferable to create dynamic equilibrium by applying inertia forces of magnitudes $\frac{W}{g}\frac{v^2}{r}$ and $\frac{W}{g}a_t$ acting through the mass center opposite in direction to the normal and tangential acceleration components respectively. These inertia forces are known as the centrifugal inertia force (or reversed normal effective force) and the tangential inertia force (or reversed tangential effective force). Recall that the normal component of acceleration is always directed inward toward the center of curvature of the path while the positive sense of the tangential component of acceleration will be in the direction of motion.

[1] Sometimes the force summations are represented by ΣF_x, ΣF_y, etc. We recommend using ΣX, ΣY, etc., since this is both easier to write and to read.

When motion along a curved path is defined by polar coordinates, it is convenient to use the radial and transverse components of acceleration as derived in Section 9-8. The corresponding components of $\mathbf{R} = \frac{W}{g}\mathbf{a}$ along radial and transverse direction then become

$$\Sigma F_r = \frac{W}{g}a_r = \frac{W}{g}(\ddot{r} - r\dot{\theta}^2)$$
$$\Sigma F_\theta = \frac{W}{g}a_\theta = \frac{W}{g}(r\ddot{\theta} + 2\dot{r}\dot{\theta}) \qquad (11\text{-}2.3)$$

In these equations, the positive senses are in the positive directions of increasing r and θ.

ILLUSTRATIVE PROBLEMS

11-2.1. Two blocks A and B are released from rest on a 30° incline when they are 60 ft apart. The coefficient of friction under the upper block A is 0.20 and that under the lower block B is 0.40. Compute the elapsed time until the blocks touch. After they touch and move as a unit, what will be the contact force between them?

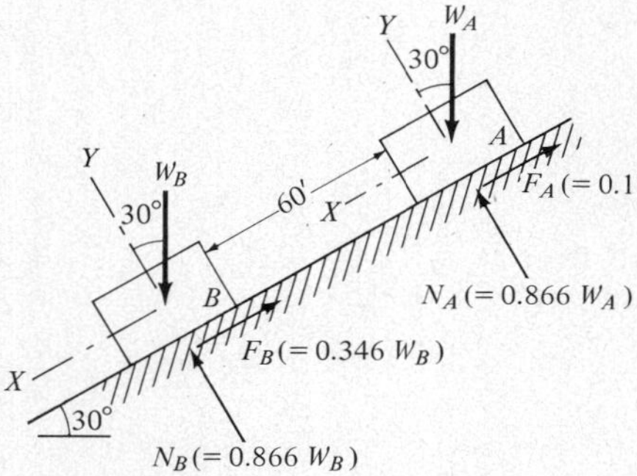

Figure 11-2.1

Solution

The FBD of each body is shown in Fig. 11-2.1. Select reference axes as shown with the X axis positive down the incline in the direction of motion. Applying $\Sigma Y = 0$ and $F = fN$, the normal and frictional components of the incline reaction are as shown. Applying $\Sigma X = (W/g)a$ to each block, we have

For A: $\quad 0.5W_A - 0.173W_A = \frac{W_A}{g}a_A, \qquad a_A = 0.327g$

For B: $\quad 0.5W_B - 0.346W_B = \dfrac{W_B}{g}a_B, \qquad a_B = 0.154g$

Thus, the relative acceleration with which A overtakes B is $a = a_A - a_B = 0.173g$ fps^2. Therefore, the time to traverse the relative distance between them of 60 ft at this constant acceleration is found from Eq. (9-3.5) to be

$[s = v_0t + \frac{1}{2}at^2] \qquad 60 = 0 + \frac{1}{2}(0.173)(32.2)t^2$

from which

$$t^2 = 21.5 \quad \text{and} \quad t = 4.64 \text{ sec} \qquad \textit{Ans.}$$

After the blocks touch, the common acceleration for the system is

$$\left[\Sigma X = \frac{W}{g}a\right]$$

$$0.5W_A - 0.173W_A + 0.5W_B - 0.346W_B = \frac{W_A + W_B}{g}a$$

which reduces to

$$\frac{a}{g} = \frac{0.327W_A + 0.154W_B}{W_A + W_B}$$

With this value of a/g, the contact force between the bodies may now be obtained from the FBD of either block. On block A, this contact force acts up the incline as shown in Fig. 11-2.2, whence we obtain

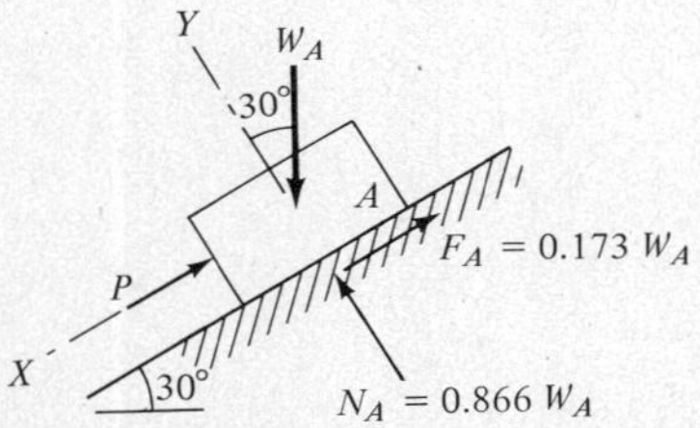

Figure 11-2.2 FBD of A after blocks touch.

$$\left[\Sigma X = \frac{W}{g}a\right]$$

$$0.5W_A - 0.173W_A - P = W_A\left(\frac{0.327W_A + 0.154W_B}{W_A + W_B}\right)$$

from which we find

$$P = \frac{0.173W_AW_B}{W_A + W_B} \qquad \textit{Ans.}$$

11-2.2. A rod ABC rotating at 20 rpm about a vertical axis through A supports a 200-lb ball at its lower end as shown in Fig. 11-2.3a. It is fixed in position by the rod BD. Neglecting the weights of rods AC and BD, compute the force F in rod BD. Is the force tensile or compressive? At what rpm will this force be zero?

Solution

The FBD of rod AC is shown in Fig. 11-2.3b. The ball may be considered as a particle moving in a horizontal circle of radius $r = 5 \sin 30° = 2.5$ ft. In order to eliminate the components of the

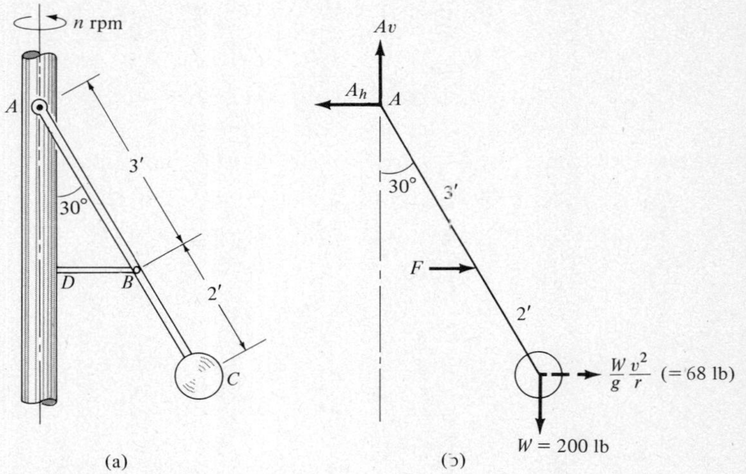

Figure 11-2.3

hinge reaction at A by using a moment summation about A, we create dynamic equilibrium by applying to the ball an inertia force $\frac{W}{g}\frac{v^2}{r}$ acting radially outward from the center of its path.

The constant magnitude of the velocity of the ball is found from

$$\left[v = \frac{s}{t} = \frac{2\pi rn}{t}\right] \qquad v = \frac{2\pi(2.5)(20)}{60} = 5.23 \text{ fps}$$

which then determines the centrifugal inertia force to be

$$\frac{W}{g}\frac{v^2}{r} = \frac{200(5.23)^2}{32.2(2.5)} = 68 \text{ lb}$$

Assuming the force F in BD to be compressive, it acts on AC as shown. Then a moment summation of dynamic equilibrium about A gives

$$[+\circlearrowleft \Sigma M_A = 0] \qquad F(3\cos 30°) + 68(5\cos 30°) - 200(2.5) = 0$$

$$F = 79 \text{ lb} \qquad \textbf{\textit{Ans.}}$$

Since F is positive, it acts as shown and confirms the assumption that BD is in compression.

To find the velocity v when $F = 0$ at the given position, we take another moment summation about A to obtain

$$[+\circlearrowleft \Sigma M_A = 0] \qquad W(5\sin 30°) - \frac{W}{g}\frac{v^2}{(2.5)}(5\cos 30°) = 0$$

whence

$$v^2 = 2.5(32.2)\tan 30° = 40 \quad \text{or} \quad v = 6.33 \text{ fps}$$

Finally, the rotational speed in rpm corresponding to this velocity is found to be

$$\left[v = \frac{2\pi rn}{t}\right] \qquad 6.33 = \frac{2\pi(2.5)n}{60} \qquad n = 24.2 \text{ rpm} \quad \textbf{\textit{Ans.}}$$

11-2.3. Determine the maximum speed v at which a car of weight W can make a horizontal turn around a highway curve of radius r that is banked (i.e., inclined) at $\theta°$ with the horizontal. The coefficient of static friction between the tires and the road is f and the corresponding angle of friction is ϕ.

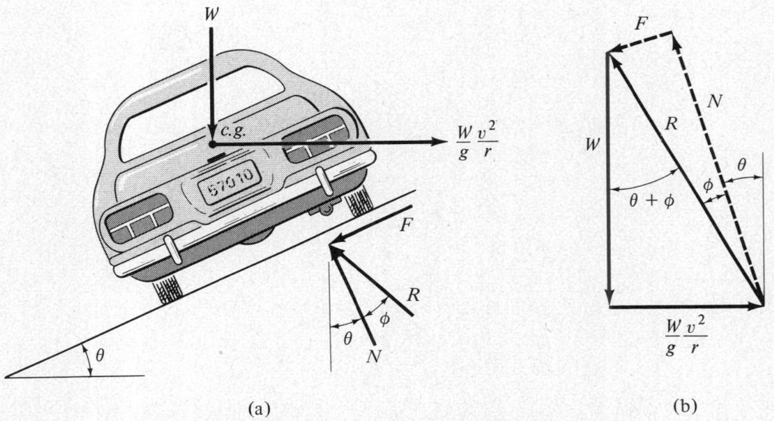

Figure 11-2.4 Maximum velocity, friction considered.

Solution

If the car were making a very sharp turn so that the dimensions of the car are not negligible compared with the radius of the turn, its motion would be an example of rigid-body rotation about the vertical axis of the turn. However, usually the radius of the turn is so large that no significant error is caused by analyzing the car as a particle.

A rear view of the FBD of the car is shown in Fig. 11-2.4a. The centrifugal inertia force $\dfrac{W}{g}\dfrac{v^2}{r}$ is applied through the mass center to create dynamic equilibrium; it acts horizontally outward from the vertical axis of the turn. If its upplane component exceeds the downplane component of the weight of the car, a downplane friction force is necessary to prevent skidding up the incline. When the car is traveling at maximum speed around the turn, all available static friction is required to prevent such skidding. Then the resultant R of the total road reaction is inclined at the angle of friction ϕ with the normal pressure N. Observe that since the car is considered as a particle, the total reaction R is directed through the *c.g.* of the car. Consequently, the outer wheels carry a greater load than the inner set of wheels.

From the force polygon shown in Fig. 11-2.4b, we obtain

$$\tan(\theta + \phi) = \frac{\frac{W}{g}\frac{v^2}{r}}{W} = \frac{v^2}{gr} \tag{11-2.4}$$

Note that this relation is independent of the weight. If the car is on the verge of slipping down the incline (because of insufficient speed), we would have

$$\tan(\theta - \phi) = \frac{v^2}{gr} \tag{11-2.5}$$

Two other situations are of interest. When a car is rounding a curve banked at such an angle θ with a speed v so that there is no tendency to slip up or down the road, θ is known as the *ideal angle of banking* and v as the *rated speed* of the curve. Under these conditions, we may apply the result above by taking $\phi = 0$. Note also that when the car is traveling at any intermediate speed between its rated speed and its maximum or minimum speed, the friction force does not act at its maximum value and cannot be determined from $F = fN$. The actual friction force, if desired in such cases, can only be found from a summation of the forces shown in Fig. 11-2.4a taken parallel to the incline.

As a specific application of this discussion, suppose the rated speed of a highway curve of 300-ft radius is 40 mph. If the coefficient of friction between the tires and the road is 0.6, what is the maximum speed at which a car can round the curve without skidding? The angle of banking for the rated speed (i.e., $\phi = 0$) is found from

$$\left[\tan\theta = \frac{v^2}{gr}\right] \qquad \tan\theta = \frac{\left(40 \times \frac{88}{60}\right)^2}{32.2(300)} = 0.356 \qquad \theta = 19.6^\circ$$

From the definition of the angle of friction, we have

$$[\tan\phi = f] \qquad \tan\phi = 0.6 \qquad \phi = 31^\circ$$

We may now apply Eq. (11-2.4) to get

$$\left[\tan(\theta + \phi) = \frac{v^2}{gr}\right] \qquad \tan(19.6^\circ + 31^\circ) = \frac{v^2}{32.2(300)}$$

whence

$$v_{max} = 108.5 \text{ fps} = 74 \text{ mph} \qquad \textbf{\textit{Ans.}}$$

PROBLEMS

11-2.4. A magnetic particle weighing 3.6 grams is pulled through a solenoid with an acceleration of 6 meters per sec^2. Compute the force in pounds acting on the particle.

Note: 1 lb = 454 grams and 1 in. = 2.54 cm.

$F = 0.00485$ lb **Ans.**

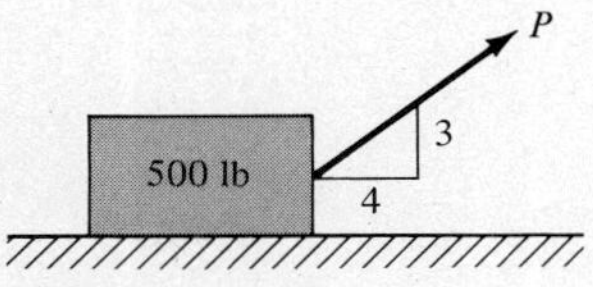

Figure P-11-2.5

11-2.5. Determine the force P that will give the body in Fig. P-11-2.5 an acceleration of $0.20g$ fps^2. The coefficient of kinetic friction is 0.20.

$P = 217.5$ lb **Ans.**

11-2.6. In Fig. P-11-2.5, assume $P = 200$ lb and the coefficient of kinetic friction is 0.5. Find the acceleration of the body under the following conditions: (a) the body is moving rightward; (b) the body is moving leftward; (c) the body starts from rest.

11-2.7. When a 644-lb boat is moving at 10 fps, the motor conks out. How much farther will the boat glide, assuming its resistance to motion is $2v$ lb where v is in fps?

$s = 100$ ft **Ans.**

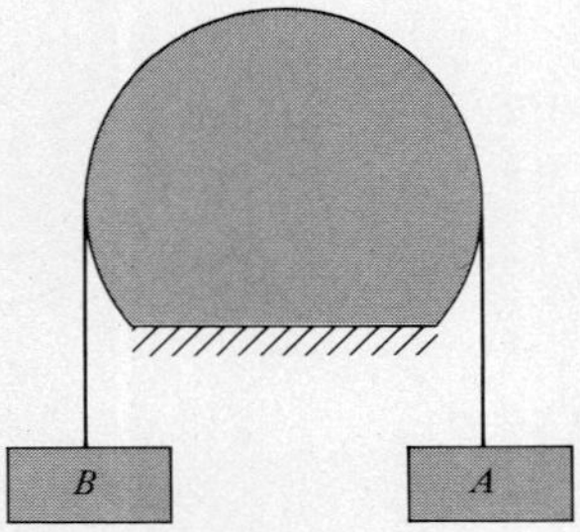

Figure P-11-2.8

11-2.8. Referring to Fig. P-11-2.8, assume A weighs 200 lb and B weighs 100 lb. Determine the acceleration of the bodies if the coefficient of kinetic friction is 0.10 between the cable and the fixed drum.

$a = 6.03$ fps^2 **Ans.**

11-2.9. Determine the acceleration of the bodies in Fig. P-11-2.9 if the coefficient of kinetic friction is 0.20 at all contact surfaces. Body A weighs 200 lb and B weighs 300 lb.

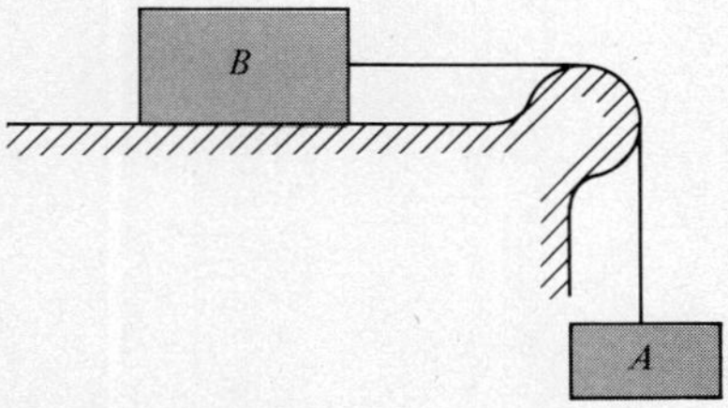

Figure P-11-2.9

11-2.10. The two bodies in Fig. P-11-2.10 are separated by a spring. Their motion down the incline is resisted by a force $P = 200$ lb. The coefficient of kinetic friction is 0.30 under A and 0.10 under B. Determine the force in the spring.

11-2.11. In Fig. P-11-2.11, the motion of block B on a smooth horizontal table is controlled by the rod which rotates about a vertical axis at O. At the position shown, B has the given values of velocity and acceleration

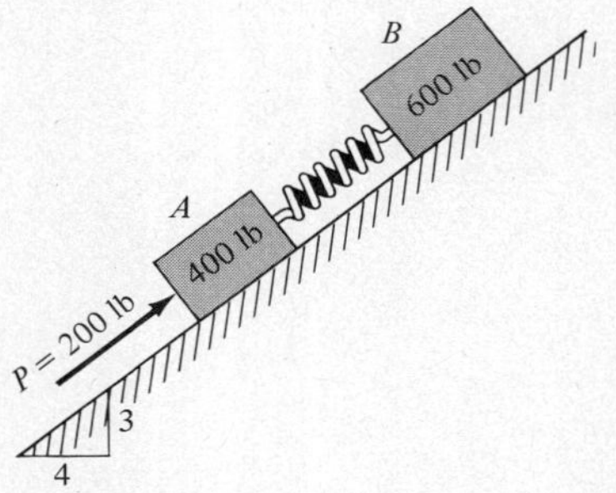

Figure P-11-2.10

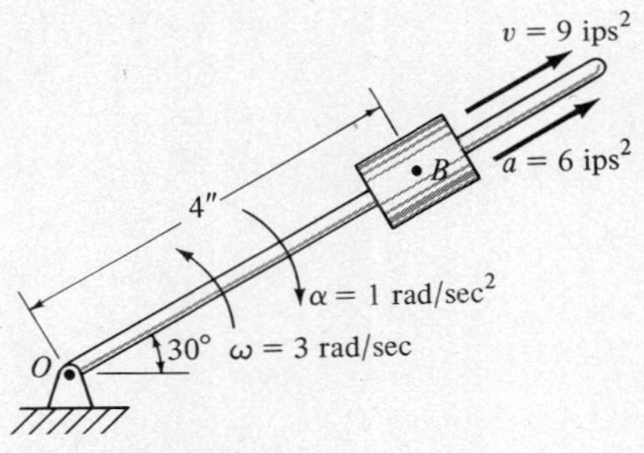

Figure P-11-2.11

relative to the rod which is rotating with the given values of angular velocity ω and angular acceleration α. If B weighs 12 lb, what moment does it exert about O?

$M = 6.21$ in.-lb **Ans.**

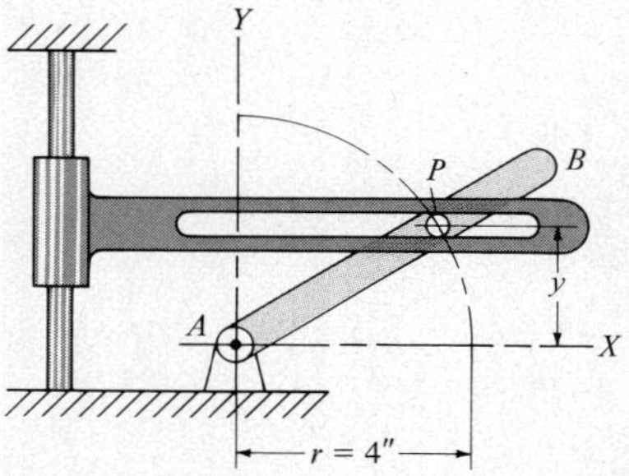

Figure P-11-2.12

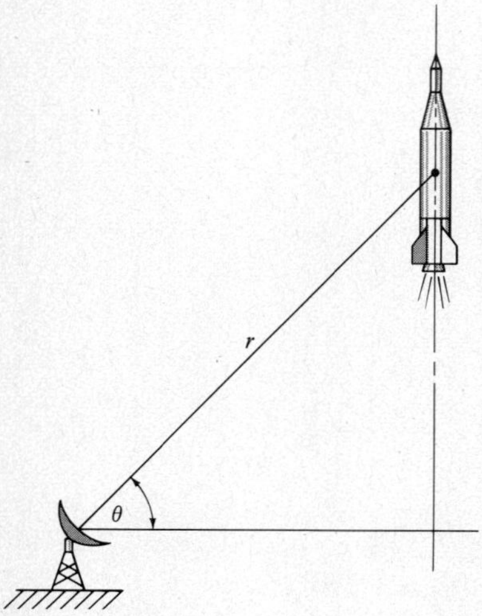

Figure P-11-2.13

11-2.12. The horizontal slotted bar in Fig. P-11-2.12 weighs 20 lb and slides with negligible friction along the vertical guide. Determine the force exerted on it by pin P when $y = 2$ in. if AB is then rotating at a constant angular velocity of 20 rad/sec.

21.4 lb **Ans.**

11-2.13. A rocket fired vertically is tracked by radar as shown in Fig. P-11-2.13. Determine the thrust P on the 10,000-lb rocket if when $\theta = 60°$, it is known that $r = 20{,}000$ ft, $\dot{\theta} = 0.02$ rad/sec, and $\ddot{\theta} = 0.003$ rad/sec^2. Assume g remains constant.

$P = 64{,}500$ lb **Ans.**

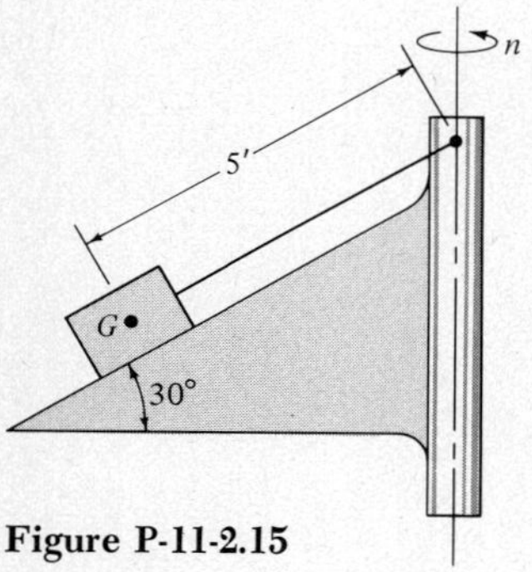

Figure P-11-2.15

11-2.14. In an amusement park ride, gondola cars of weight W are attached by rods 24 ft long to a bull wheel 16 ft in diameter. The bull wheel rotates in a horizontal plane about a vertical axis at its center. At how many rpm is it rotating when the supporting rods of the gondolas are inclined at 30° with the vertical? Consider each gondola to be a particle concentrated at the end of each 24-ft rod.

$n = 9.2$ rpm **Ans.**

11-2.15. A body of weight W rests on the smooth inclined surface of the frame shown in Fig. P-11-2.15. A peg attached to the frame forces the body to rotate with it about the vertical axis. Determine the speed in rpm at which the tension in the cord is equal to the weight of the body.

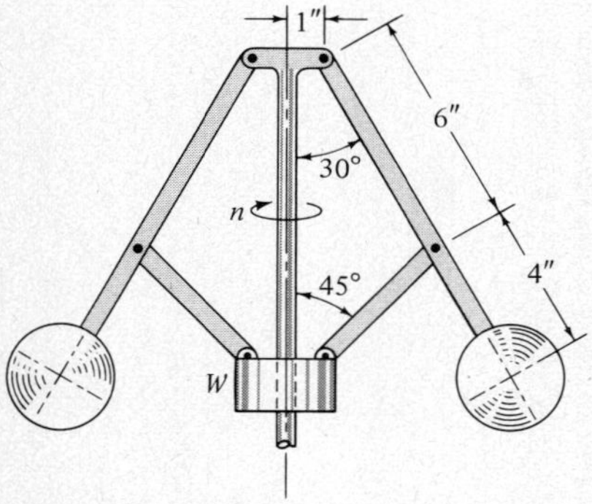

Figure P-11-2.16

11-2.16. What counterweight W will maintain the Corliss engine governor in the position shown in Fig. P-11-2.16 at a rotational speed of $n = 120$ rpm? Each flyball weighs 16.1 lb. Neglect the weight of the other links.

11-2.17. Why are railroad curves laid out in the form of a spiral with a gradually decreasing radius of curvature instead of a circle of constant radius?

11-2.18. An airplane makes a turn in a horizontal plane without side slip at 300 mph. At what angle must the plane be banked if the radius of the turn is 2 miles? If the pilot weighs 150 lb, what force does he exert on his seat?

11-2.19. A stuntman drives a motorcycle around a circular vertical wall 100 ft in diameter. The coefficient of friction between tires and wall is 0.60. What is the minimum speed that will prevent his sliding down the wall? At what angle will the motorcycle be inclined to the horizontal? What is the effect of traveling at a greater speed?

$v = 35.3$ mph; $\theta = 31°$ ***Ans.***

11-2.20. The superelevation of a railroad track is the number of inches that the outside rail is raised to prevent side thrust on the wheel flanges of cars rounding the curve at rated speed. Determine the superelevation e for a track having a gauge of 4 ft $8\frac{1}{2}$ in. of 2000-ft radius and a rated speed of 60 mph. What is the flange pressure P on the wheels of a 100,000-lb car that rounds the curve at 80 mph?

$e = 6.78$ in.; $P = 9230$ lb ***Ans.***

11-2.21. A car weighing 4000 lb rounds a curve of 200-ft radius banked at an angle of 30°. Find the friction force acting on the tires when the car is traveling at 60 mph. The coefficient of friction between the tires and the road is 0.60.

11-2.22. Find the angle of banking for a highway curve of 300-ft radius designed to accommodate cars traveling at 100 mph if the coefficient of friction between the tires and the road is 0.60. What is the rated speed of the curve?

$\theta = 34.8°$; $v = 56$ mph ***Ans.***

11-2.23. Transverse sections through the curves on an automobile speedway have the parabolic shape shown in Fig. P-11-2.23. Determine the y coordinate of an auto so that when it is moving at a speed of v fps there will be no tendency for side slip.

$y = v^2/2g$ ***Ans.***

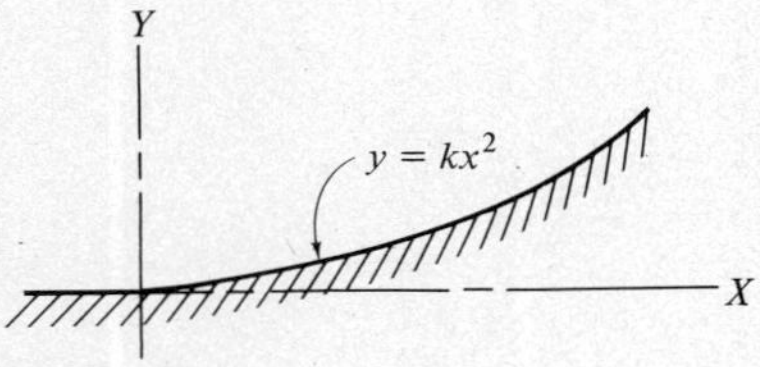

Figure P-11-2.23

11-2.24. The segment of road passing over the crest of a hill is defined by the parabolic curve $y = 0.4x - 0.01x^2$. A car weighing 3220 lb travels along the road at a constant speed of 30 fps. What is the normal pressure on the wheels of the car when it is at the crest of the hill where $y = 4$ ft? At what speed will the road pressure be zero?

Hint: The radius of curvature is defined by

$$\rho = \frac{\left[1 + \left(\frac{dy}{dx}\right)^2\right]^{3/2}}{\frac{d^2y}{dx^2}}$$

11-2.25. As a car passes over the crest of the hill described in the previous problem, the brakes are locked and the wheels skid. The tangential retardation produced is 0.4g. If the coefficient of friction between the tires and the road is 0.70, how fast is the car moving?

$v = 26.3$ fps **Ans.**

11-2.26. A satellite is put into a circular orbit at an altitude of 600 miles above the earth. The mean radius of the earth is 3960 miles. Determine the time to complete one orbit.

Hint: Refer to Eq. (10-2.1).

104.3 min. **Ans.**

11-3 FURTHER DISCUSSION OF PARTICLE KINETICS

The equations and methods developed in the preceding section are extended here to more complex situations involving the kinetics of particles; particularly, connected systems of particles. A general plan for the solution of such problems consists of the following guide:

1. Draw a free-body diagram for each body involved in the problem. Indicate thereon all forces, both known and unknown, representing the latter by an appropriate symbol. If the direction of any unknown (except friction) is incorrectly assumed, the solution will yield its correct magnitude but with a negative sign. Friction forces, however, must *always* be directed to oppose relative motion.

2. In order to ensure that friction forces oppose the motion, determine the direction of motion if not evident or specified, and indicate it by a dashed arrow near each free-body diagram. This dashed arrow will be a convenient reminder of the positive direction of the equation of motion.

3. Determine the kinematic relations between the bodies involved in a connected system.

4. Select the X axis as positive in the direction of initial motion and apply $\Sigma Y = 0$ and $\Sigma X = (W/g)a$ to each body. For curvilinear motion, create dynamic equilibrium.

5. Solve for the unknowns, using such additional equations of kinematics as may be required to determine relations among displacement, velocity, and time.

ILLUSTRATIVE PROBLEM

11-3.1. A rocket-propelled sled weighing 400 lb is at rest on a horizontal track for which the coefficient of static friction is 0.30 and the coefficient of kinetic friction is 0.20. A horizontal thrust of $P = 15t$ is applied for 20 sec and then shut off. P is in pounds and t is in seconds. What is the maximum velocity of the sled and how far will it be from its initial position when it comes to rest? Neglect the weight of expelled fuel.

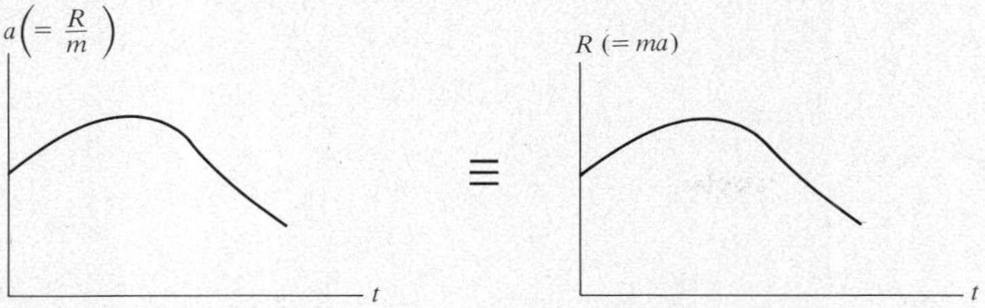

Figure 11-3.1 Correspondence between *a-t* and *F-t* curves.

Preliminary

In rectilinear motion, there is a direct relation between an acceleration-time (*a-t*) curve and the corresponding resultant force-time (*F-t*) curve as is shown in Fig. 11-3.1. Except for the factor m for mass, these two curves are equivalent. In other words, the area under the *a-t* curve is $1/m$ times the area under the *F-t* curve. Hence, in place of using $(\text{Area})_{a\text{-}t}$ as discussed in Section 9-4 to find velocity and displacement, we can use $(\text{Area})_{F\text{-}t}$. The relations corresponding to Eqs. (9-4.1) and (9-4.3) then become

$$v_2 - v_1 = \Delta v = \frac{1}{m}(\text{Area})_{F\text{-}t} \tag{11-3.1}$$

$$s_2 - s_1 = \Delta s = v_1(\Delta t) + \frac{1}{m}(\text{Area})_{F\text{-}t}(\bar{t}_2) \tag{11-3.2}$$

In this problem, we shall see the advantage of using these equations to determine the direct effect of forces on the motion. Actually, these equations are an elemental application of the impulse-momentum method mentioned in Section 11.1 which, by eliminating the acceleration, directly relates force, velocity, and time. A general discussion of the impulse-momentum method is presented later in Chapter 15.

Solution

A free-body diagram of the sled is shown in Fig. 11-3.2a. Applying $\Sigma Y = 0$ gives $N = 400$ lb. The maximum static friction force is obtained from

$$F = f_s N = 0.30(400) = 120 \text{ lb}$$

But this static value will act only when $P = 120$ lb or, since $P = 15t$, after 8 sec have elapsed. At any time between zero and 8 sec, the static friction will adjust itself to the net force tending to cause motion. As the time increases beyond 8 sec, the sled will slide and

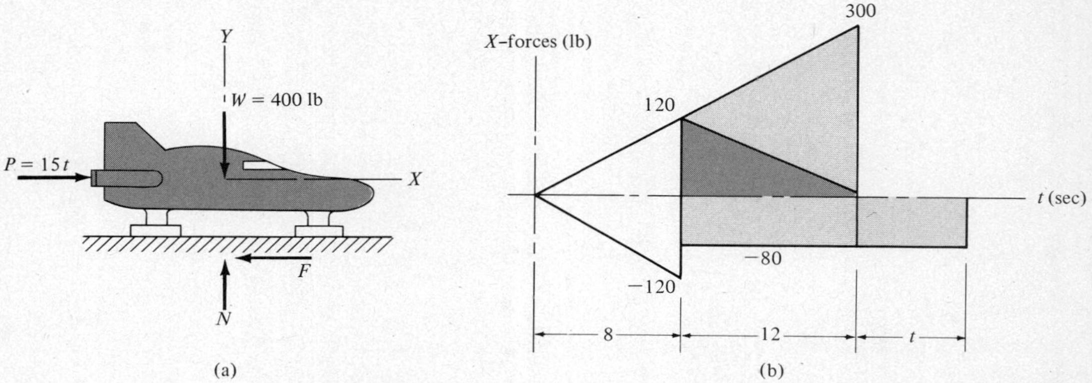

Figure 11-3.2

the kinetic friction force will act at its maximum value as found from

$$F = f_k N = 0.20(400) = 80 \text{ lb}$$

We shall assume that the transition from static friction to kinetic friction takes place instantaneously. Although this is not strictly true, it involves negligible error. With this assumption, a graphic representation of the X components of the forces acting, plotted against time, will be shown in Fig. 11-3.2b. Notice that it is convenient to show separately the variation of positive and negative X components. Note that during the initial 8 sec, the resultant force is zero but attains its maximum positive value during the next 12 sec, at the end of which the sled has its maximum velocity. This value may now be found by applying Eq. (11-3.1) to the shaded area during this 12-sec interval to obtain

$$\left[\Delta v = \frac{1}{m}(\text{Area})_{F\text{-}t}\right]$$

$$\max v = \frac{1}{400/32.2}\left[\left(\frac{300 + 120}{2}\right)(12) - 80(12)\right] = 125.5 \text{ fps} \quad \textbf{\textit{Ans.}}$$

After the rocket thrust is shut off, the sled will coast to a stop during a time t. This time may be found by applying Eq. (11-3.1) to the coasting interval since we now know the velocity at the beginning of this coasting interval, but as an independent approach, we will apply Eq. (11-3.1) to the entire motion over which the velocity change is zero. This gives

$$\left[\Delta v = \frac{1}{m}(\text{Area})_{F\text{-}t}\right] \qquad 0 = \frac{1}{m}\left[\left(\frac{300 - 200}{2}\right)(12) - 80(12 + t)\right]$$

$$t = 19.5 \text{ sec}$$

The required displacement can now be found by applying Eq. (11-3.2) to the entire shaded area of the F-t diagram. Remember to take the moment of each component area about the final time ordinate. Thus, we obtain

$$\left[\Delta s = v_1(\Delta t) + \frac{1}{m}(\text{Area})_{F\text{-}t}(\bar{t}_2)\right]$$

$$s = 0 + \frac{1}{400/32.2}\left[\left(\frac{300 \times 12}{2}\right)\left(\frac{12}{3} + 19.5\right) + \left(\frac{120 \times 12}{2}\right)\left(\frac{2}{3} \times 12 + 19.5\right) - 80(31.5)\left(\frac{31.5}{2}\right)\right]$$

which gives

$$s = 1802 \text{ ft} \qquad \textbf{\textit{Ans.}}$$

11-3.2. In the system of connected blocks shown in Fig. 11-3.3, the coefficient of kinetic friction under blocks A and C is 0.20. Compute the acceleration of each block and the tension in the connecting cable. The pulleys are assumed to be frictionless and of negligible weight.

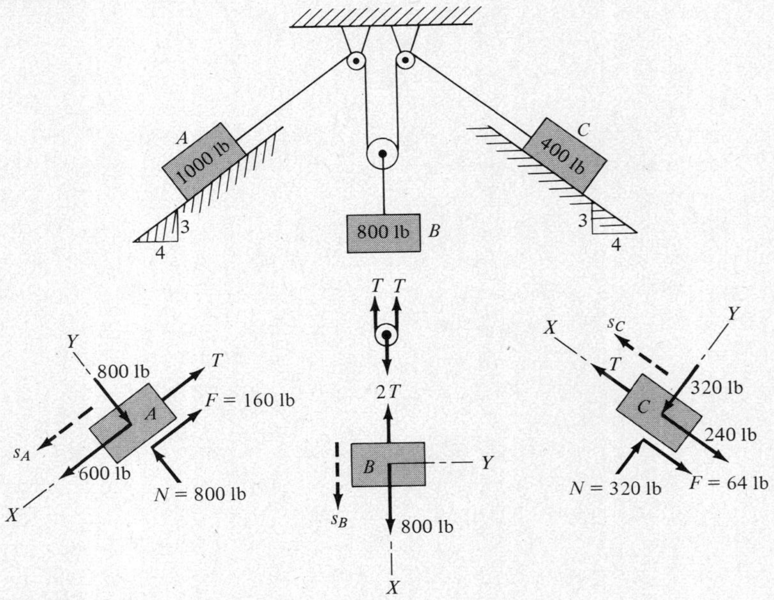

Figure 11-3.3

Preliminary

To simplify the computations, represent the weights of A and C by their components acting parallel and perpendicular to the inclines as shown on the free-body diagrams. To determine the direction of motion of the system, we begin by assuming that one part of the

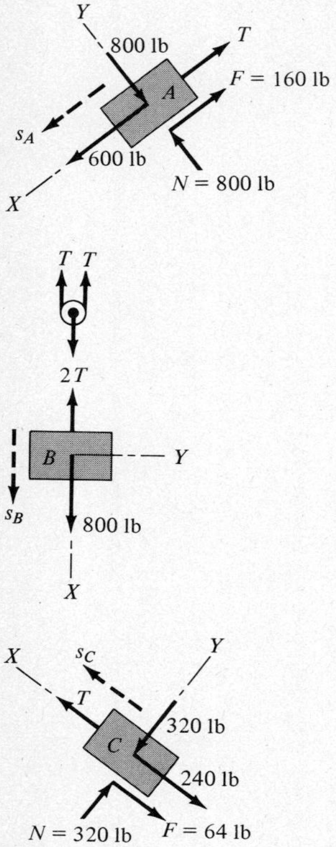

Figure 11-3.3

system does not move and then computing the tension necessary to keep it at rest. With this assumption, if any part of the system is found to be acted upon by an unbalanced force, that part will move in the direction of that force. When friction is involved, as in this problem, we must consider whether there is sufficient friction available to prevent motion.

For example, if we temporarily assume B is at rest, $2T = 800$ lb which gives $T = 400$ lb acting on A and C. Notice that since the floating pulley supporting B is of negligible weight, $\Sigma X = (W/g)a$, when applied to it reduces to $\Sigma X = 0$ as also is $\Sigma M = 0$. Hence, from the FBD of the floating pulley, the tension supporting B is $2T$ and the tension in the connecting cable is constant. From force summations parallel to the inclines and assuming $T = 400$ lb, the friction forces required to prevent motion of A and C are for A: $600 - 400 = 200$ lb and for C: $400 - 240 = 160$ lb. However, from $\Sigma Y = 0$ and $F = fN$, we find the available friction forces on A and C to be $F_A = 0.2(800) = 160$ lb and $F_C = 0.2(320) = 64$ lb. These are insufficient to keep A and C at rest and are the actual friction forces acting as shown to oppose the downward motion of A and the upward motion of C, the directions of motion being indicated by the dashed arrows adjacent to each FBD.

We do not yet know the motion of B because although the downplane motion of A tends to raise B, the upplane motion of C tends to lower B. We shall assume that B moves down. Since no friction acts on B, an incorrect assumption of its motion will be of no importance. The actual value of T will determine its motion. Notice that the FBD's of A and C require that T be less than $600 - 160 = 440$ lb to permit downward motion of A, but more than $240 + 64 = 304$ lb to cause upward motion of C. An estimated value of T will be the average of these limiting values or $\frac{1}{2}(440 + 304) = 372$ lb. Using this approximation on the FBD of B gives an unbalanced downward force on B and justifies the assumption that B moves down.

We must next determine the kinematic relations among the motion of the blocks. The simplest way to do this is to use a concept from the method of virtual work and that we redevelop later in the work-energy method: The total work done by internal connecting forces on a system is zero. Noting that work is the product of force and distance, and that positive work is done by a force acting in the direction of motion, we have as the sum of the work done by T on the system of connected blocks

$$Ts_C - Ts_A - 2Ts_B = 0$$

whence cancelling out T, the kinematic relation between the displacements is

$$s_C = s_A + 2s_B \tag{a}$$

on which successive differentiation with respect to time gives

$$v_C = v_A + 2v_B \tag{b}$$

and

$$a_C = a_A + 2a_B \tag{c}$$

An alternate approach to finding the kinematic relations among the blocks is the following method: The net motion of B is the difference between the partial motion s'_B of B caused by C while A is held at rest and the partial motion s''_B of B due to A while C is held at rest. Noting that these partial motions are only one-half the absolute motions of C and A, we obtain

with A at rest: $s'_B = \frac{1}{2} s_C$ (down)
with C at rest: $s''_B = \frac{1}{2} s_A$ (up)

Assuming that B moves down, its net motion is the difference of these partial motions or

$$s'_B - s''_B = s_B = \tfrac{1}{2} s_C - \tfrac{1}{2} s_A$$

which agrees with Eq. (*a*) above.

Solution

We are now ready to apply Eq. (11-2.1) to each body. Note carefully that in each case ΣX is applied as positive in the direction of motion indicated by the dashed displacement vectors adjacent to each FBD.

$\left[\Sigma X = \frac{W}{g} a\right]$ For A: $$600 - 160 - T = \frac{1000}{g} a_A \tag{d}$$

For B: $$800 - 2T = \frac{800}{g} a_B \tag{e}$$

For C: $$T - 64 - 240 = \frac{400}{g} a_C \tag{f}$$

When a_C is replaced by $a_A + 2a_B$ from Eq. (*c*), these equations are easily solved for their common term T by multiplying Eq. (*d*) by $-\frac{4}{10}$, Eq. (*e*) by -1, and adding them, whence we obtain

$$T = 377 \text{ lb}$$ ***Ans.***

and then by substitution,

$$a_A = 2.05 \text{ fps}^2$$
$$a_B = 1.89 \text{ fps}^2$$ ***Ans.***
$$a_C = 5.83 \text{ fps}^2$$

The positive values of these accelerations confirm our assumptions of the direction of motion of each block.

Some final comments are pertinent. If the coefficient of friction were increased to 0.30, the available friction force under A and C would become $F_A = 240$ lb and $F_C = 96$ lb. Under these conditions, enough friction is available to keep A at rest, but not sufficient to keep C at rest. It is essential to first determine which bodies *can* move and in which direction, or else much time will be wasted in getting impossible answers.

PROBLEMS

11-3.3. A 64.4-lb block is resting on a smooth horizontal surface. It is acted upon by a horizontal force P which varies according to the relation $\mathbf{P} = (12t - 3t^2)\hat{\mathbf{i}}$ where P is in pounds and t is in seconds. Determine the maximum positive velocity of the block, and the time when it reverses its motion.

11-3.4. A 60-lb body is free to slide on a horizontal smooth surface. It starts from rest and is struck three successive blows of 100 lb, 50 lb, and -100 lb at intervals of 2 sec. Assuming that each blow is along the line of motion and is constant for 0.1 sec, what is the velocity and displacement of the body after 8 sec?

Hint: Use an F-t diagram.

$v = 2.68$ fps; $s = 38.1$ ft **Ans.**

11-3.5. A 322-lb body moves under the action of a force given by the relation $\mathbf{P} = (18 - 3t)\hat{\mathbf{i}}$ where P is in pounds and t is in seconds. If the body starts from rest, in how many seconds will it return to its initial position? What then is its velocity?

11-3.6. If the force on the 322-lb body of the previous problem is changed to $\mathbf{P} = (10t - t^2)\hat{\mathbf{i}}$, how far will the body move from rest before it starts to reverse its direction?

$s = 140.8$ ft **Ans.**

11-3.7. Air resistance on a parachutist in free fall is given by kv^2. Find his terminal velocity if the parachute opens when his velocity is v_o fps.

$v = \sqrt{W/k}$ **Ans.**

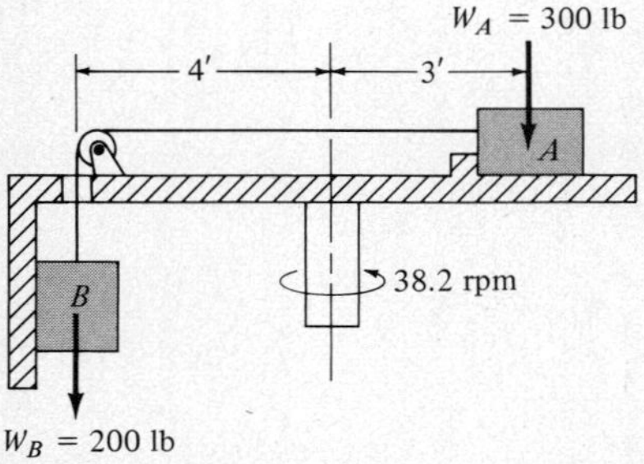

Figure P-11-3.9

11-3.8. A rocket weighing 10,000 lb is launched vertically with a constant thrust of 40,000 lb. Assuming air resistance to be given by $R = 20v$ where R is in pounds and v in fps, determine its height when its velocity is 1000 fps. Consider g and the weight of the rocket to be constant.

$h = 10{,}080$ ft **Ans.**

11-3.9. The frame shown in Fig. P-11-3.9 rotates about a vertical axis. The coefficient of friction under block A is 0.40. Compute the coefficient of friction at block B if B starts to rise when the frame rotates at 38.2 rpm.

11-3.10. In Prob. 11-3.9, block B starts to rise when the frame rotates at 38.2 rpm. Find the coefficient of friction under block A if the coefficient of friction at block B is 0.20.

$$f_A = 0.561 \qquad \textbf{Ans.}$$

11-3.11. Two blocks having the weights and positions shown in Fig. P-11-3.11 rest on a frame which rotates about its vertical axis at a constant speed. The coefficient of friction between the blocks and the frame is 0.20. Neglecting the weight and friction of the pulley, at what speed in rpm will the blocks start to slide? What is the tension in the cord at this instant?

$$n = 31.4 \text{ rpm}; \; T = 22.5 \text{ lb} \qquad \textbf{Ans.}$$

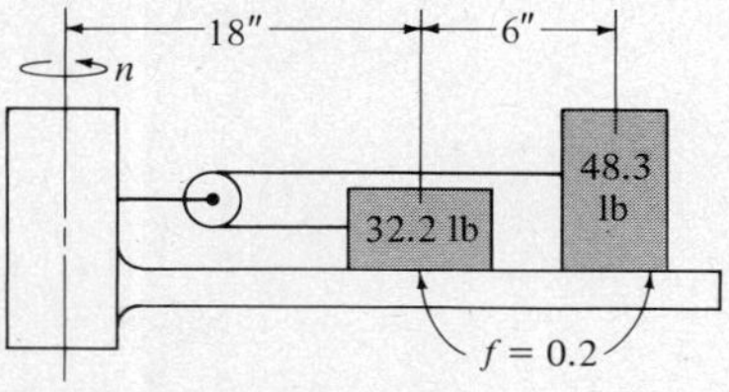

Figure P-11-3.11

11-3.12. Repeat Prob. 11-3.11 if the weights of the blocks are interchanged.

11-3.13. In the system of connected bodies shown in Fig. P-11-3.13, the pulleys are frictionless and of negligible weight. (a) Determine the weight of A to give B a downward acceleration of $0.6g$. (b) Can you find the weight of A to give B an upward acceleration of $0.6g$?

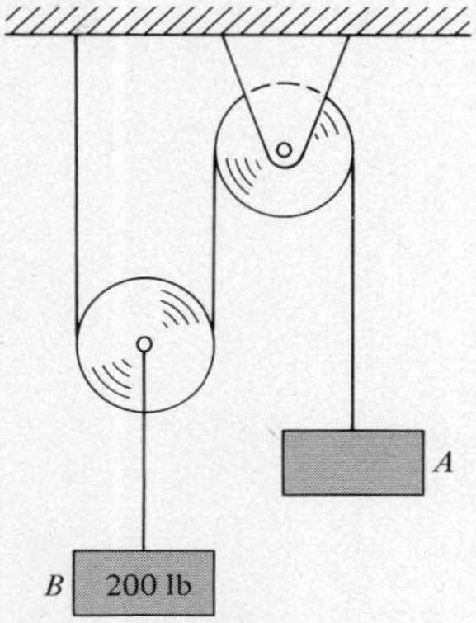

Figure P-11-3.13

11-3.14. Three bodies, each of weight W, are arranged as shown in Fig. P-11-3.14. The coefficient of kinetic friction between B and C is 0.30, and is 0.10 between C and the ground. How far does C move when B has moved s ft relative to C?

$$s_C = 0.4s \qquad \textbf{Ans.}$$

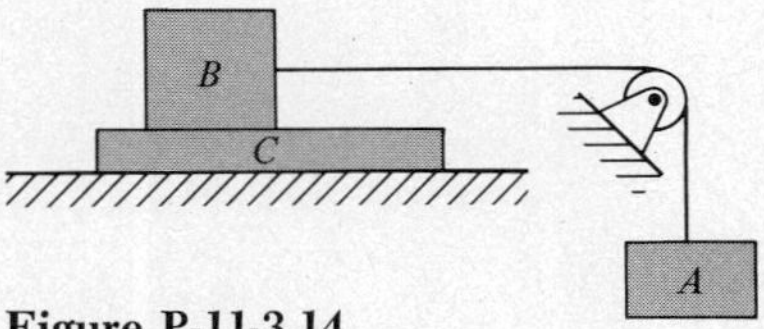

Figure P-11-3.14

11-3.15. Determine the acceleration of body A in Fig. P-11-3.15, assuming the pulleys to be frictionless and of negligible weight.

$$a_A = 2.61 \text{ fps}^2 \qquad \textbf{Ans.}$$

11-3.16. Find the acceleration of each body in the system shown in Fig. P-11-3.16.

Hint: First determine the direction of motion.

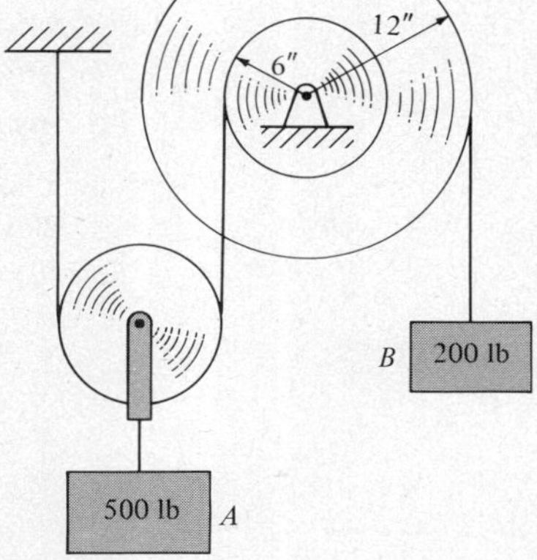

Figure P-11-3.15

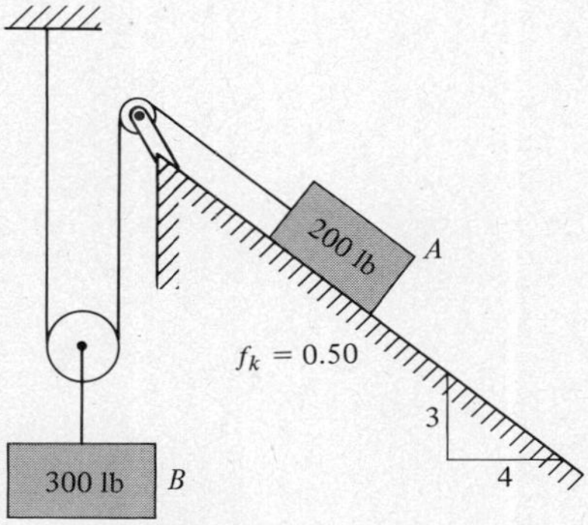

Figure P-11-3.16

11-3.17. Referring to Fig. P-11-3.17, compute the acceleration of body *B* and the tension in the cord supporting body *A*.

$$a_B = 5.03 \text{ fps}^2; \; T = 137.5 \text{ lb} \qquad \textit{Ans.}$$

11-3.18. In Fig. P-11-3.18, determine the acceleration of body *B* and the tension in the cord attached to it.

11-3.19. If the coefficient of kinetic friction is 0.25 under each body in the system shown in Fig. P-11-3.19, how far and in what direction will body *B* move in 5 sec starting from rest?

$$s_B = 26.9 \text{ ft down} \qquad \textit{Ans.}$$

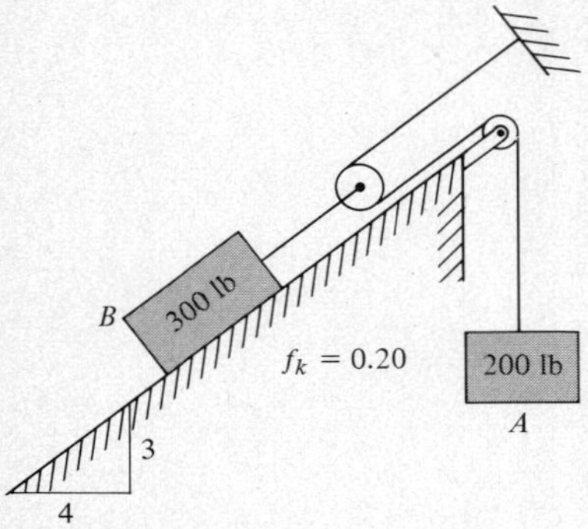

Figure P-11-3.17

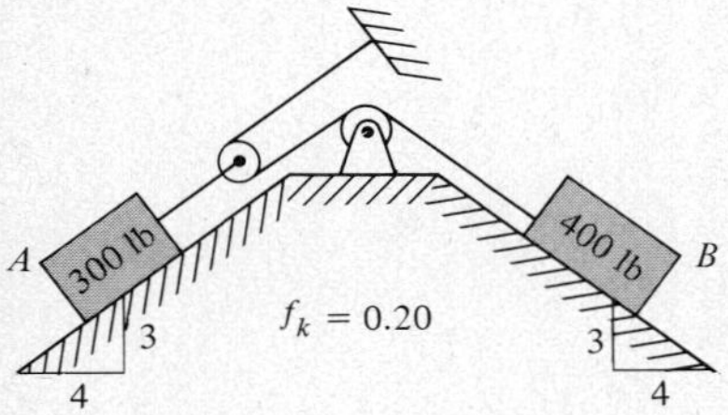

Figure P-11-3.18

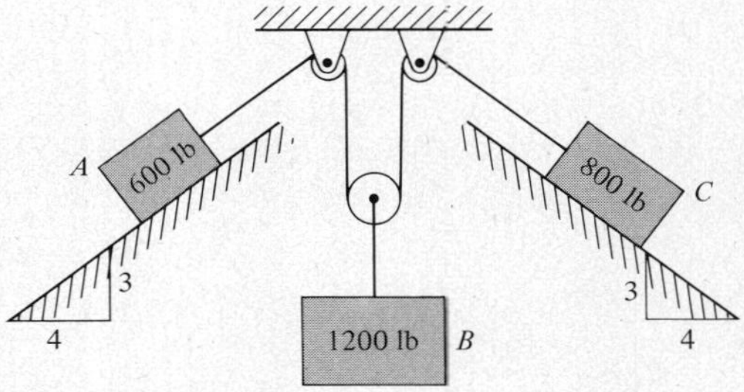

Figure P-11-3.19

11-3.20. Determine the acceleration of each body in Fig. P-11-3.20, assuming the pulleys to be frictionless and of negligible weight. The inclined plane is smooth.

11-3.21. Determine the tension in the cord supporting body *C* in Fig. P-11-3.21. The pulleys are frictionless and of negligible weight.

$$T = 212 \text{ lb} \qquad \textit{Ans.}$$

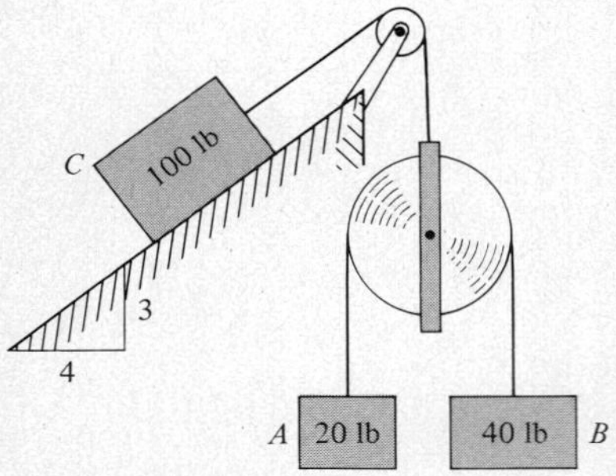

Figure P-11-3.20

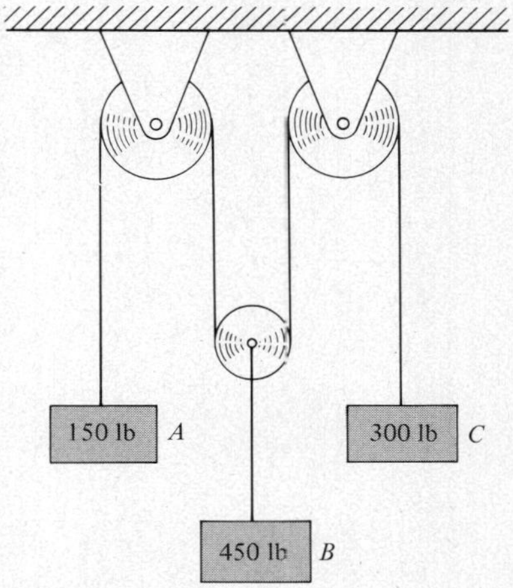

Figure P-11-3.21

11-3.22. The pulleys in Fig. P-11-3.22 are frictionless and of negligible weight. Determine the tension in the cable supporting body C.

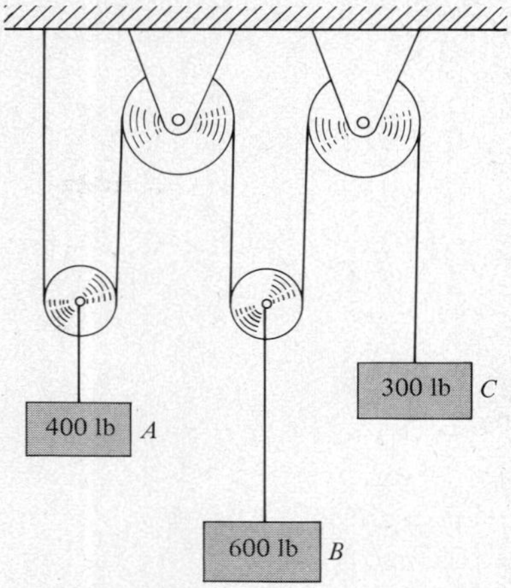

Figure P-11-3.22

11-3.23. In the system of connected bodies in Fig. P-11-3.23, the coefficient of kinetic friction is 0.20 under bodies B and C. Determine the acceleration of each body and the tension in the cord supporting A.

$$T = 348.2 \text{ lb};\ a_A = 4.18 \text{ fps}^2;\ a_B = 3.57 \text{ fps}^2 \qquad \textbf{Ans.}$$

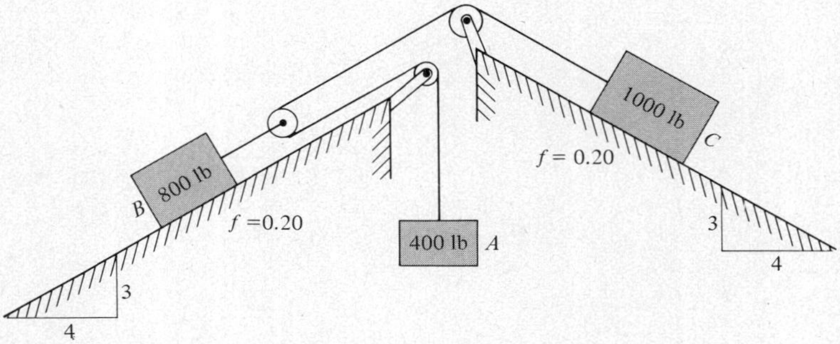

Figure P-11-3.23

11-3.24. Repeat the previous problem but change the weight of A to 600 lb, of B to 1000 lb, and of C to 500 lb.

11-3.25. Two bodies A and B, weighing 400 lb and 300 lb, respectively, are connected by a rigid bar of negligible weight and move along the smooth surfaces shown in Fig. P-11-3.25. If they start from rest at the given position, determine the acceleration of B at this instant.

$$a_B = 9.56 \text{ fps}^2 \qquad \textbf{Ans.}$$

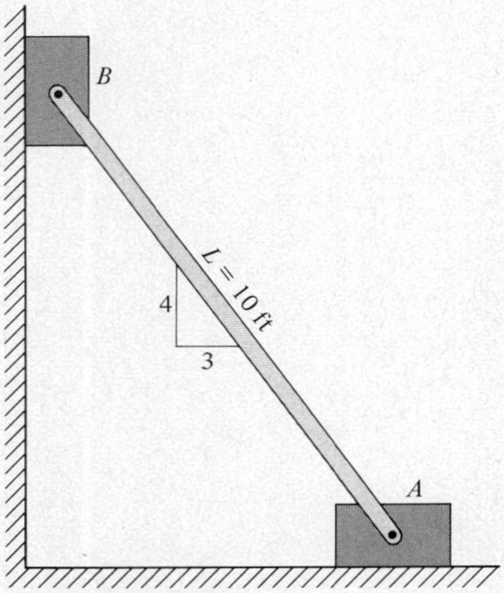

Figure P-11-3.25

11-3.26. Repeat the previous problem if a leftward horizontal force of 360 lb is applied to body A.

11-3.27. As shown in Fig. P-11-3.27, body A slides down the smooth surface of wedge B which is on a smooth horizontal floor. Their weights are W_A and W_B. Determine the acceleration of wedge B.

Hint: a_A equals the vector sum of a_B and the acceleration of A relative to B.

$$\frac{a_B}{g} = \frac{W_A \sin\theta \cos\theta}{W_B + W_A \sin^2\theta} \qquad \textbf{Ans.}$$

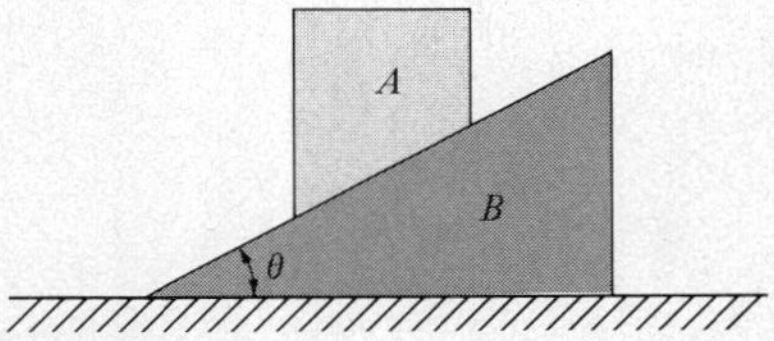

Figure P-11-3.27

11-4 TRANSLATION. ANALYSIS AS A RIGID BODY

In this section, we consider a group of problems which involve the magnitude and/or the position of forces that will constrain a body to be in translation. Our basic criterion is that, in a translating body, the resultant of all applied forces will pass through the mass center of the body (see Section 10-6). Then as shown in Fig. 11-4.1, the free-body diagram of the applied forces in part (a) is equivalent to part (b) where the resultant effective force $(W/g)a$ (equivalent to the resultant of the applied forces but expressed in terms of the mass and acceleration) acts through the mass center in the direction of the acceleration.

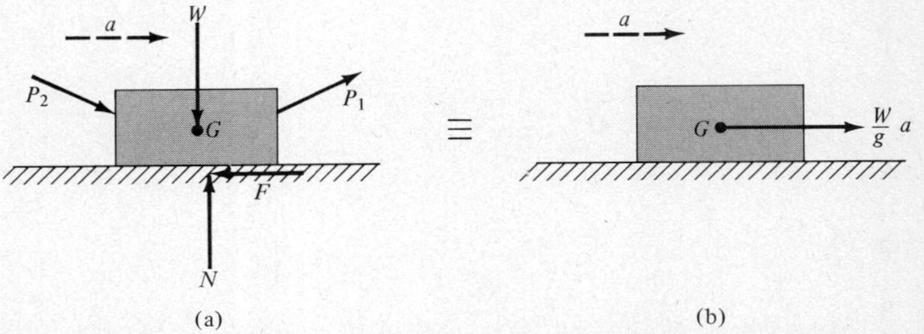

Figure 11-4.1 Equivalence of applied forces and resultant effective forces.

A force summation in any direction of part (a) is equal to the corresponding force summation of part (b); also, a moment summation about any center for part (a) is equal to the corresponding moment summation for part (b). Obviously moments about the mass center G are zero for either part, but both diagrams should be drawn when equating moment summations for both parts about centers other than G.

A useful stratagem that eliminates drawing two free-body diagrams is to add to each part a force equal to $(W/g)a$ reversed as shown in Fig. 11-4.2. There it is evident that part (b) then has a

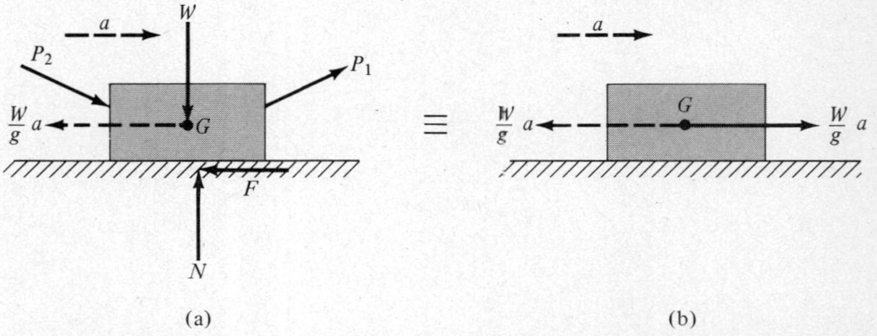

Figure 11-4.2 Creation of dynamic equilibrium by adding the inertia reaction $(w/g)a$.

zero resultant; hence, part (a) also has a zero resultant. Consequently, in Fig. 11-4.2a, force and moment summations are zero along any direction or about any axis. Thus, all the methods developed in statics may be applied to a single diagram upon which, in addition to the applied forces, there is included what may be called an inertia reaction[2] equal to $(W/g)a$ acting through the mass center in a direction opposite to the acceleration of the translating body. By adding the inertia reaction to the original force system, we create the condition known as *dynamic equilibrium*. The principal advantage of dynamic equilibrium is that a moment summation is zero about any convenient center such as at the intersection of two unknown forces.

ILLUSTRATIVE PROBLEMS

11-4.1. The frame of a certain machine accelerates rightward at $\frac{4}{5}g$ fps^2. As shown in Fig. 11-4.3, it carries a uniform bent bar ABC weighing 10 lb/ft pinned to it at C and braced by the uniform strut DE which weighs 50 lb. Determine the components of the force at pin D.

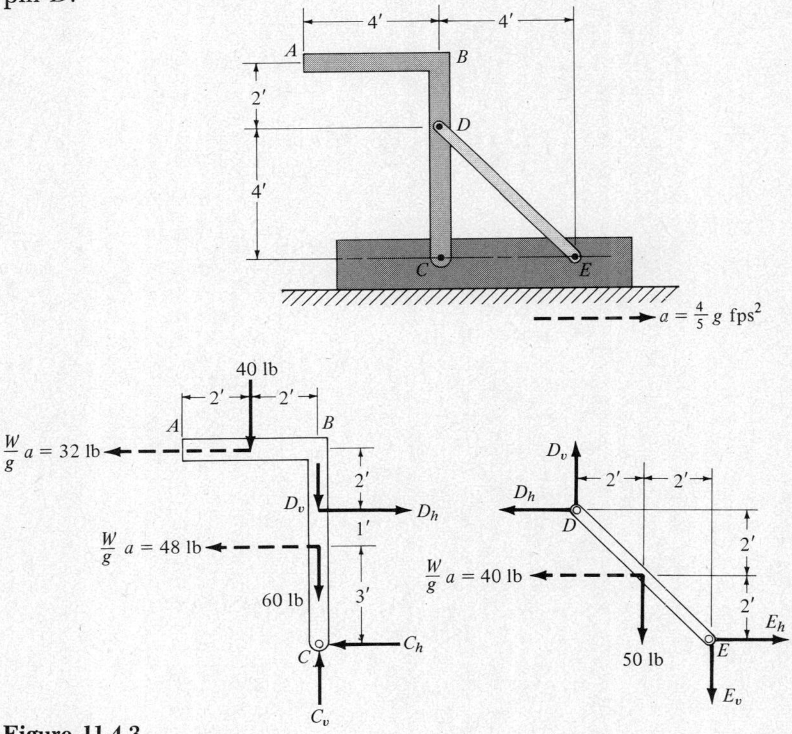

Figure 11-4.3

[2] The inertia reaction should be represented by a dashed vector to avoid any possible confusion with an applied force.

Solution

Except for the application of the inertia reactions, the analysis of this problem is exactly the same as for the method of members in Section 4-6. The free-body diagram of each member is placed in dynamic equilibrium by including the inertia reaction, acting through the gravity center of each part, directed opposite to the acceleration. Note that instead of locating the center of gravity of *ABC*, it is more convenient to apply the inertia reaction acting on each segment of its length. The inertia reactions are

$$\text{For } AB: \qquad \frac{W}{g}a = \frac{40}{g}\left(\frac{4}{5}g\right) = 32 \text{ lb}$$

$$\text{For } BC: \qquad \frac{W}{g}a = \frac{60}{g}\left(\frac{4}{5}g\right) = 48 \text{ lb}$$

$$\text{For } DE: \qquad \frac{W}{g}a = \frac{50}{g}\left(\frac{4}{5}g\right) = 40 \text{ lb}$$

Consider now the FBD of *ABC*. A moment summation about *C* eliminates three of the unknown forces and solves directly for D_h.

$$[\circlearrowleft + \Sigma M_C = 0] \qquad 4D_h - 48(3) - 32(6) - 40(2) = 0$$

$$D_h = 104 \text{ lb} \qquad \textit{Ans.}$$

Turning to the FBD of *DE* in which D_h is now known, we solve for D_v. A moment summation about *E* eliminates the unknown components acting at *E*.

$$[\circlearrowleft + \Sigma M_E = 0] \qquad 4D_v - (D_h = 104)(4) - 40(2) - 50(2) = 0$$

$$D_v = 149 \text{ lb} \qquad \textit{Ans.}$$

If desired, the components of the pin pressures at *C* and *E* can now be computed by applying a horizontal and vertical summation of forces first to the FBD of *ABC* and then to that of *DE*.

11-4.2. A homogeneous triangular plate weighing 64.4 lb is attached to a rigid vertical support by two links of negligible weight as shown in Fig. 11-4.4. At the given position, the plate has an upward velocity of 12 fps and the hydraulic cylinder exerts a downward force at *E* of $P = 135$ lb. Compute the total force acting on hinge *C*.

Solution

The parallel linkage constrains the plate to move without rotation; i.e., to be in curvilinear translation. The motion of its mass center is therefore the same as that of *A* or *B* which move in circular paths

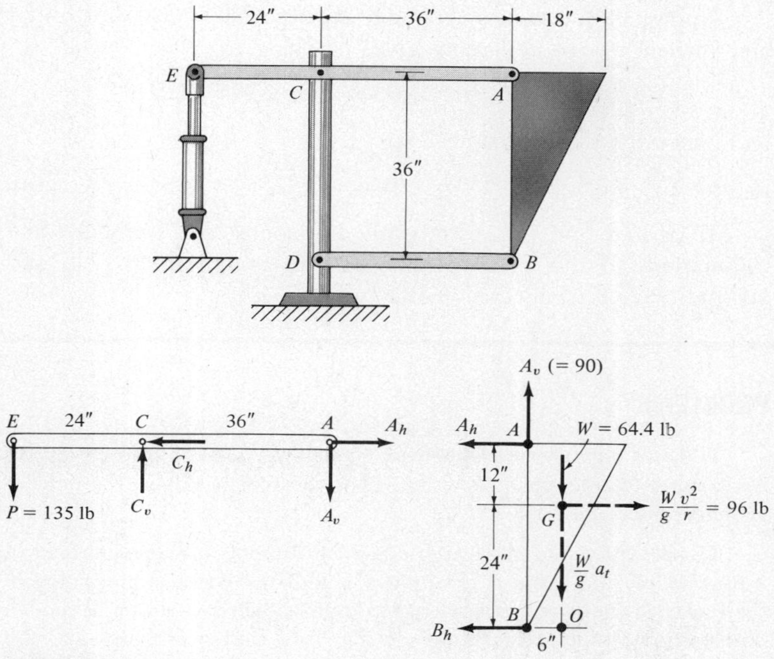

Figure 11-4.4

about C and D respectively. The acceleration of the mass center G thereby has the same normal and tangential components as A or B. These components determine the centrifugal inertia reaction and tangential inertia reaction which are applied as shown on the FBD of the plate to create dynamic equilibrium. The magnitude of the centrifugal inertia reaction is

$$\frac{W}{g}a_n = \frac{W}{g}\frac{v^2}{r} = \frac{64.4}{32.2}\frac{(12)^2}{36/12} = 96 \text{ lb}$$

but the tangential inertia reaction $(W/g)a_t$ is as yet unknown.

Since link BD has negligible weight, it acts as an axial force member and exerts only B_h on the FBD of the plate. However, link AE is a multi-force member having the FBD shown. The common unknowns acting on the FBD's of the link AE and the plate are A_v and A_h. Considering first AE, we have

$[\circlearrowleft + \Sigma M_C = 0] \qquad 36A_v - 135(24) = 0 \qquad A_v = 90 \text{ lb}$

Consider next the FBD of the plate where, with A_v known, a convenient moment center is at the intersection O of B_h and $(W/g)a_t$ which thereby determines A_h as follows:

$[+\circlearrowright \Sigma M_O = 0] \qquad 36A_h - 90(6) - 96(24) = 0 \qquad A_h = 79 \text{ lb}$

Applying these values of A_h and A_v to the FBD of AE, horizontal and vertical force summations give

$$C_h = A_h = 79 \text{ lb}; \qquad C_v = 135 + 90 = 225 \text{ lb}$$

and hence the total force at C is

$$[C = \sqrt{C_h^2 + C_v^2}] \qquad C = \sqrt{(79)^2 + (225)^2} = 238 \text{ lb} \qquad \textit{Ans.}$$

If desired, we may also apply horizontal and vertical force summations to the FBD of the plate to obtain $B_h = 17$ lb and $(W/g)a_t = 25.6$ whence $a_t = 12.8$ fps^2.

PROBLEMS

11-4.3. A juggler places the lower end of a vertical rod upon his finger. As it starts to tip, explain how he keeps the rod in balance by moving his finger back and forth.

11-4.4. The cable of a cargo crane can support a maximum load of 2 tons. While the crane is lowering a 1610-lb weight at uniform speed, the brake on the winch is applied too rapidly, thereby causing a sudden deceleration of the weight equal to 100 fps^2. The cable snaps and the weight falls, badly injuring a workman. For the purpose of establishing liability in this accident, is it likely that failure of the cable was due to its being weaker than its test strength of 2 tons? Why?

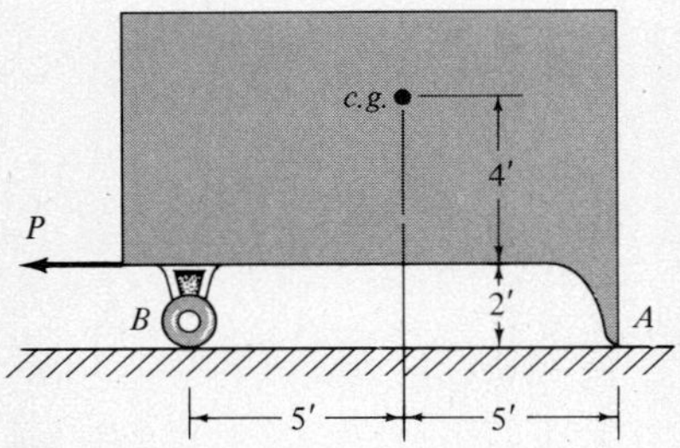

Figure P-11-4.6

11-4.5. A uniform box of weight W is 2 ft square and 6 ft high. It stands on end upon a truck with its sides parallel to the truck's motion. If the coefficient of friction between the box and the truck is 0.30, show whether the box will slide or tip first as the acceleration of the truck is increased. What happens if the truck is being braked?

11-4.6. The 440-lb body in Fig. P-11-4.6 is supported by wheels at B which roll freely without friction and by a skid at A under which the coefficient of friction is 0.50. Compute the value of P to cause an acceleration of $g/4$ fps^2.

$$P = 230 \text{ lb} \qquad \textit{Ans.}$$

11-4.7. If the value of P in Prob. 11-4.6 is 200 lb, compute the acceleration. If this value of P were applied at a higher position on the body, would the acceleration be changed in any way? Explain your conclusion.

11-4.8. An auto has a wheelbase of 10 ft and its $c.g.$ is 3 ft above the ground and 4 ft behind the front wheels. Which would give the greater acceleration on a level road—a front wheel or a rear wheel drive?

11-4.9. An auto has a wheelbase of 120 in. with its $c.g.$ located 60 in. ahead of the rear wheels and 30 in. above the pavement. If $f = 0.80$ at the tires and only the front wheel brakes are operating, compute the minimum distance in which the auto can be brought to rest from a speed of 60 mph, assuming the driver's reaction time before applying the brakes is $\frac{3}{4}$ sec.

$s = 306$ ft **Ans.**

11-4.10. The coefficient of friction under the sliding supports of the door in Fig. P-11-4.10 is 0.30 at A and 0.20 at B. What force P will give the 600-lb door a leftward acceleration of 8.05 fps²?

$P = 281$ lb **Ans.**

11-4.11. Determine the acceleration of the door in the preceding problem if $P = 250$ lb.

11-4.12. In Fig. P-11-4.12, find the angle θ at which a uniform bar of weight W will be maintained inside the smooth surface of a cylindrical drum translating and accelerating leftward at $0.8g$ fps².

$\theta = 38.7°$ **Ans.**

11-4.13. A bar weighing 2 lb/ft is bent at right angles into segments 26 in. and 13 in. long. It takes the position shown in Fig. P-11-4.13 when the frame F to which it is pinned at A is accelerated horizontally. Determine this acceleration and the total reaction at A.

$a = 8.93$ fps²; $A = 6.75$ lb **Ans.**

11-4.14. A 14-ft bar is bent in a right angle, as shown in Fig. P-11-4.14, so that the length $BC = 6$ ft. Compute the acceleration of the block to which the bar is freely pinned at A in order to maintain the bent bar in the given position. The bar weighs w lb/ft.

11-4.15. The uniform crate shown in Fig. P-11-4.15 weighs 200 lb. It is pulled up the incline by a counterweight W of 400 lb. Find the maximum and minimum values of d so that the crate does not tip over as it slides up the incline.

max. $d = 3.35$ ft; min. $d = 1.61$ ft **Ans.**

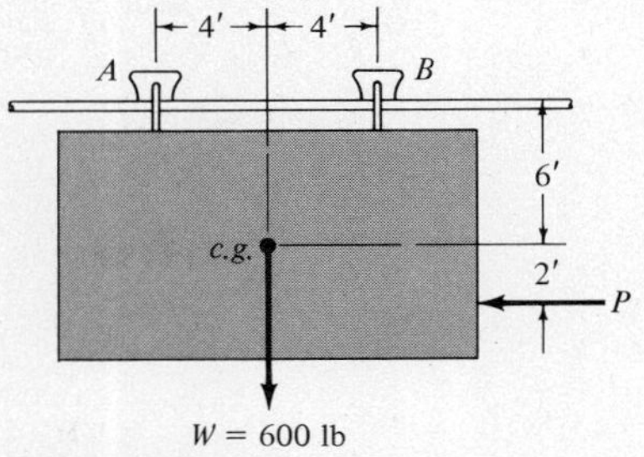

Figure P-11-4.10

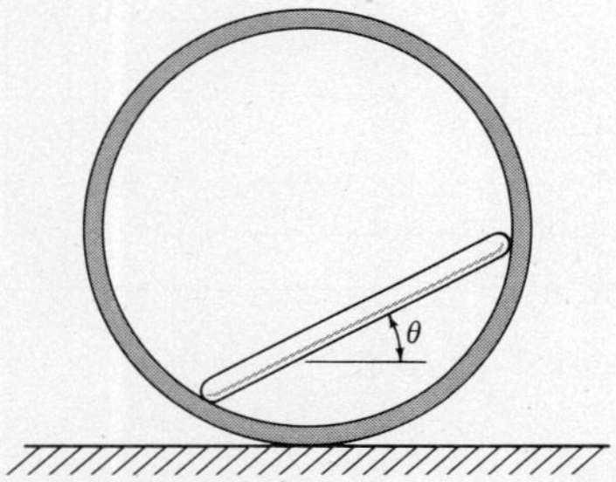

Figure P-11-4.12

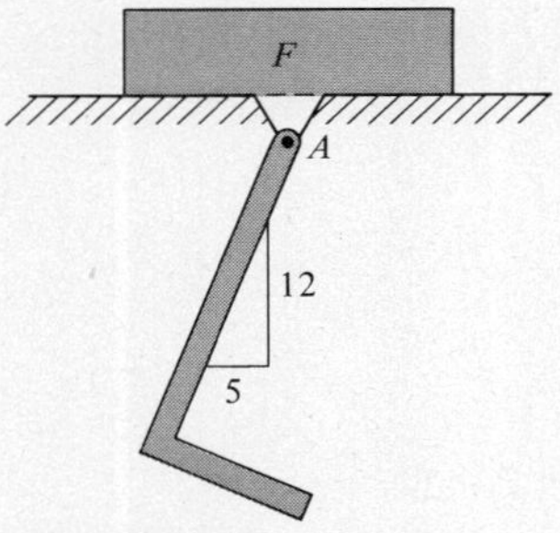

Figure P-11-4.13

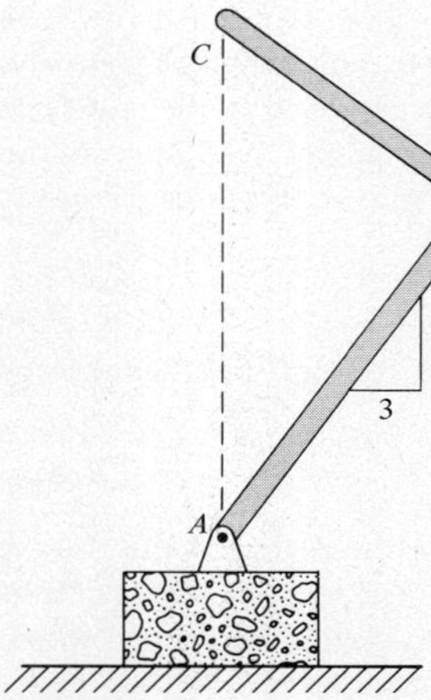

Figure P-11-4.14

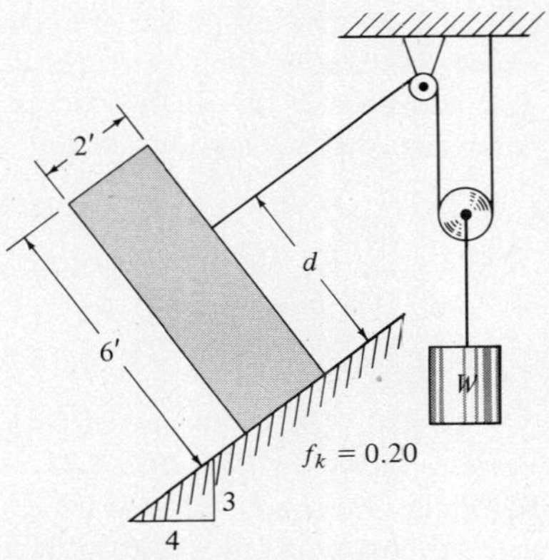

Figure P-11-4.15

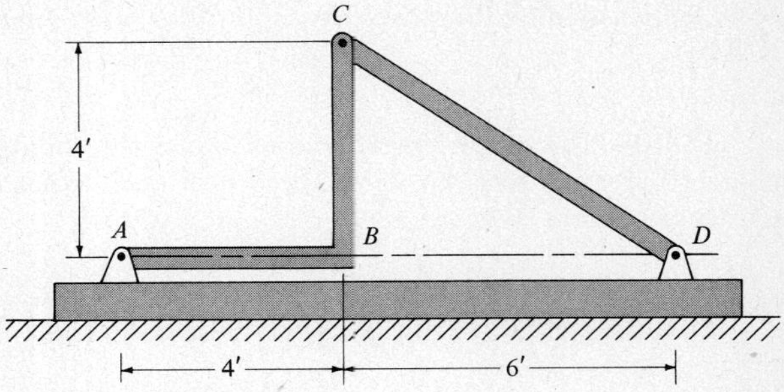

Figure P-11-4.16

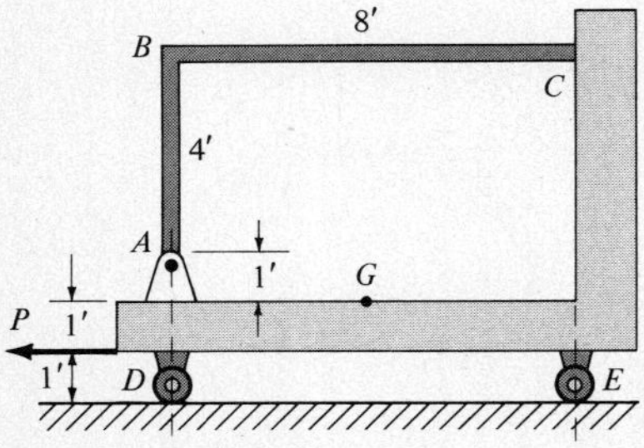

Figure P-11-4.18

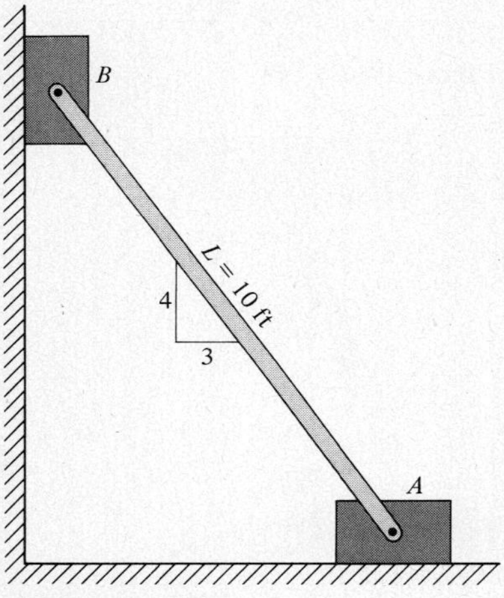

Figure P-11-4.20

11-4.16. The frame of a machine is accelerated leftward at $0.6g$ fps^2. As shown in Fig. P-11-4.16, it carries a uniform angle *ABC* weighing 80 lb which is braced by the uniform strut *CD* weighing 60 lb. Determine the components of the hinge force at *C* upon *CD*.

$C_h = 54$ lb right; $C_v = 18$ lb down **Ans.**

11-4.17. Solve the preceding problem if the frame of the machine is accelerated rightward at $0.8g$ fps^2.

11-4.18. The bent bar *ABC* weighing 10 lb/ft is mounted as shown in Fig. P-11-4.18 upon a carriage weighing 240 lb. The center of gravity of the carriage is at *G* midway between the wheels. If $P = 108$ lb and there is no frictional resistance at the wheels, find the wheel reactions and also the horizontal and vertical components of the hinge force at *A*.

$R_E = 193$ lb; $A_h = 74$ lb **Ans.**

11-4.19. Repeat the preceding problem if the magnitude and sense of *P* is such that the reaction of the carriage upon the bent bar at *C* is 40 lb leftward.

11-4.20. Two bodies *A* and *B*, weighing 300 lb and 400 lb respectively, are connected by a rigid bar of negligible weight attached to them at their gravity centers as shown in Fig. P-11-4.20. The angle of friction at the wall and floor is 15°. If the bodies start from rest at the given position, determine the acceleration of *B* at this instant.

$a_B = 4.61$ fps^2 **Ans.**

11-4.21. Repeat the preceding problem if a leftward horizontal force of 700 lb is applied to body *A*.

$a_B = 3.92$ fps^2 **Ans.**

11-4.22. The coefficient of friction between the road and the tires of the car shown in Fig. P-11-4.22 is 0.60. The car weighs 4000 lb. It is rounding a horizontal curve of 600-ft radius at the maximum speed at which side slipping impends. What is the value of the friction force acting under each wheel? How high above the road must the center of gravity be to

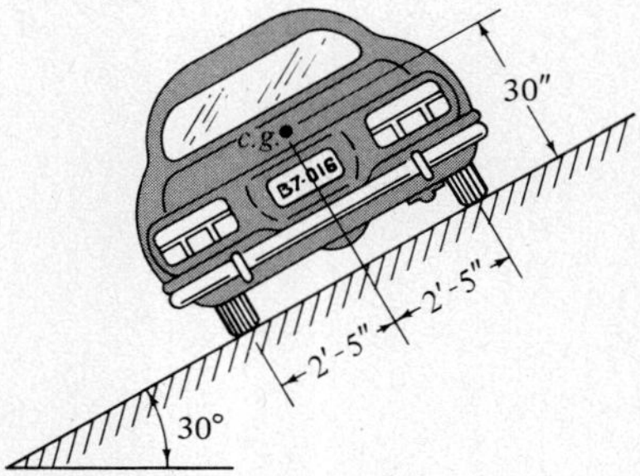

Figure P-11-4.22

limit this maximum speed by the tendency to overturn?

At inner wheels, $F = 804$ lb;
at outer wheels, $F = 3440$ lb; $h = 4.02$ ft **Ans.**

11-4.23. Repeat the preceding problem if the road is banked at 20° instead of 30° as shown in Fig. P-11-4.22.

11-4.24. The loading platform shown in Fig. P-11-4.24 is raised from rest at the given position by a leftward pull of 2500 lb exerted at *B* by the hydraulic cylinder. The weight of the platform and links are negligible compared with the 1000-lb weight of the crate whose center of gravity is at *G*. Determine the force at the hinge *E* at this instant. The coefficient of friction between the crate and the platform is 0.40.

$E = 825$ lb **Ans.**

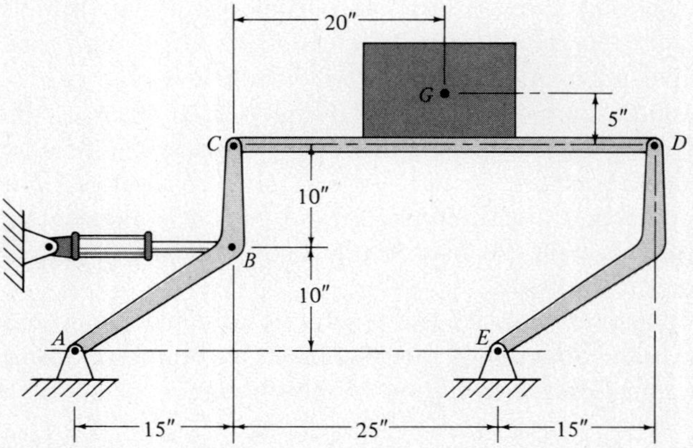

Figure P-11-4.24

11-4.25. Solve the preceding problem if, in place of the hydraulic cylinder, a counterclockwise torque of 500 ft-lb is applied to link *AC* at hinge *A*.

11-4.26. A homogeneous triangular plate weighing 100 lb is supported by two parallel links of negligible weight as shown in Fig. P-11-4.26 (p. 422). It moves under the action of the cable attached to a 200-lb weight. If the system starts from rest when $\theta = 90°$, find the force in link *CD*.

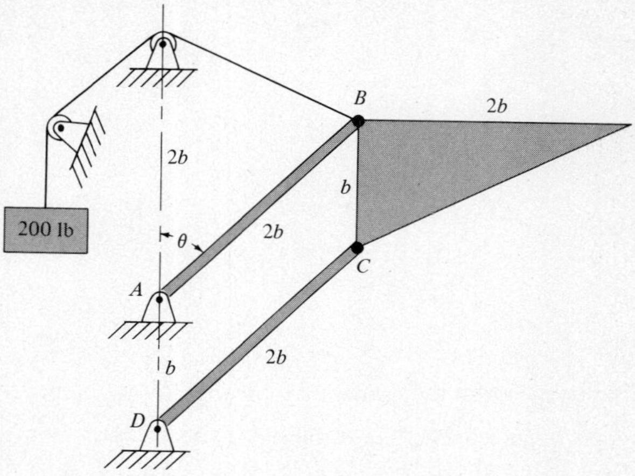

Figure P-11-4.26

11-4.27. Solve the preceding problem if the system starts from rest when $\theta = 60°$.

$$CD = 93.5 \text{ lb} \qquad \textbf{\textit{Ans.}}$$

SUMMARY

The rigid-body motion of translation is defined as motion in which a straight line passing through any two particles of a body always remains parallel to its initial position. All particles travel along the same or parallel paths and therefore have identical values of displacement, velocity, and acceleration. If the path is straight, the motion is called rectilinear translation; if curved, the motion is called curvilinear translation. Insofar as Newton's laws of motion for a particle are concerned, translation of a rigid body is equivalent to that of a particle having the mass of the body and the motion of the mass center of the body.

For rectilinear motion, let the X axis be directed along and positive in the initial direction of motion. Then the component forms of the kinetic equation for rectilinear motion become

$$\Sigma X = \frac{W}{g}a \qquad \Sigma Y = \Sigma Z = 0 \tag{11-2.1}$$

For curvilinear motion, the convenient components of acceleration to use are those normal and tangential to the path or, occasionally, the radial and transverse components. When we call the normal axis N and the tangential axis T, the corresponding scalar forms of $\mathbf{R} = \frac{W}{g}\mathbf{a}$ are

$$\Sigma N = \frac{W}{g}a_n = \frac{W}{g}\frac{v^2}{r}$$
$$\Sigma T = \frac{W}{g}a_t \qquad (11\text{-}2.2)$$

In applying these equations, be careful to take the positive senses of ΣN and ΣT in the positive directions of a_n and a_t. Usually, however, it is simpler to put curvilinear motion problems in dynamic equilibrium by applying inertia reactions of magnitudes $\frac{W}{g}\frac{v^2}{r}$ and $\frac{W}{g}a_t$ acting through the mass center which are directed opposite to the normal and tangential components of acceleration respectively.

When motion along a curved path is best defined by polar coordinates, the component forms of the kinetic equation along radial and transverse directions are

$$\Sigma F_r = \frac{W}{g}a_r = \frac{W}{g}(\ddot{r} - r\dot{\theta}^2)$$
$$\Sigma F_\theta = \frac{W}{g}a_\theta = \frac{W}{g}(r\ddot{\theta} + 2\dot{r}\dot{\theta}) \qquad (11\text{-}2.3)$$

In these equations, the positive senses are in the positive directions of increasing r and θ.

When considering the distribution of forces on a translating rigid body, it is often convenient to create dynamic equilibrium by applying the components of $\frac{W}{g}\mathbf{a}$ acting at the mass center of the body and directed opposite to the corresponding components of $\mathbf{a}$. The advantage of dynamic equilibrium is that all the methods of statics may be applied to a single free-body diagram which includes the components of the inertial reaction $\frac{W}{g}\mathbf{a}$. Especially useful is that with dynamic equilibrium a moment summation may be set equal to zero at a center which is at the intersection of two unknown forces.

12

KINEMATICS OF RIGID BODIES

12-1 INTRODUCTION. TYPES OF RIGID-BODY MOTION

In the preview of dynamics presented in Section 9-1, we discussed the correlation between the unbalanced resultant force-couple system acting on a body and its consequent motion. Two basic equations were stated; namely, $\mathbf{R} = (W/g)\overline{\mathbf{a}}$ and $\Sigma\mathbf{M} = \dot{\mathbf{H}}$. These were later derived in Sections 10-5 and 10-6. In addition to these basic equations, we noted the necessity for the correlation between acceleration and the motion which is the area known as kinematics.

For the same reasons that we began with a study of the kinematics of a particle as a prerequisite to the study of its kinetics, we similarly begin our study of rigid-body motion with the kinematics of rigid-body motion. The various types of motion to be discussed are:

1. Translation
2. Fixed-axis rotation
3. General plane motion
4. Rotation about a fixed point
5. General space motion

We have already shown in Section 11-2 that translation (either rectilinear or curvilinear) has the characteristic that all particles in the body have the same identical motion. This was the basis for treating a translating body as a particle having the mass of the body and the motion of any point of the body. More particularly, the motion of translation is characterized by the absence of any spin or angular motion of the body. Because of this, the kinematics of particle motion was sufficient to handle the translation of a rigid body, and therefore we included this type of motion in Chapter 11 as a simple extension of particle kinetics.

In the next section on fixed-axis rotation, we consider the concept whereby the position of a body is uniquely defined by its angular coordinate rather than by a linear coordinate which suffices for translation. In particular, we shall demonstrate an analogy between rectilinear motion and angular motion whereby much of our knowledge of rectilinear motion may be applied directly to rotational kinematics. Each of the remaining types of motion will be shown to be combinations of linear and angular motion for which our key objective will be to correlate the angular motion with the linear motion of the mass center. When this kinematic correlation is established, we will be ready to discuss the kinetics of the various types of rigid-body motions.

12-2 ANGULAR MOTION. FIXED-AXIS ROTATION

Fixed-axis rotation is defined as that motion of a rigid body in which the particles move in circular paths with their centers on a *fixed* straight line that is called the axis of *rotation*. The planes of the circles in which the particles move are perpendicular to the axis of rotation.

These concepts are illustrated in Fig. 12-2.1 where the motion of a typical particle P is determined by the change in its position vector $\mathbf{r}$ drawn from an origin O anywhere on the fixed axis of rotation. The condition that the body be rigid requires that the length of $\mathbf{r}$ and its inclination ϕ be constant; hence, P describes a circular path, perpendicular to the axis of rotation, centered about its projection B on this axis and of radius $BP = r \sin \phi$.

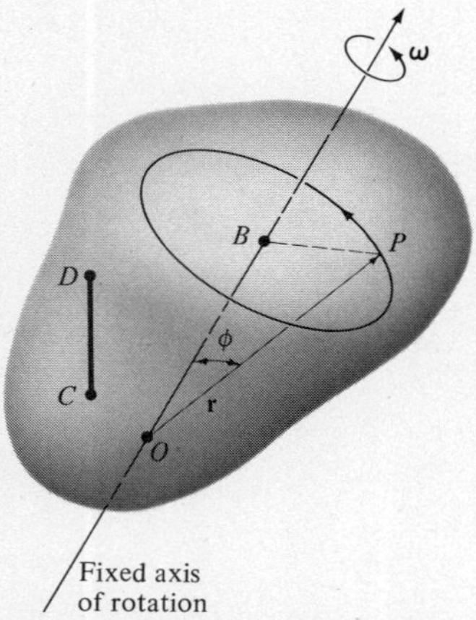

Figure 12-2.1 Motion of P in a fixed-axis rotation of a rigid body.

This motion, and that of any line CD in the body, is most conveniently described by projecting the rotating body upon a plane perpendicular to the axis of rotation as shown in Fig. 12-2.2. There O represents the fixed axis and CD is the projection of any line such as CD in Fig. 12-2.1. We see that if the radius to any point C is permitted to rotate through θ radians, point C moves through the arc distance $s_1 = r_1\theta$. Since the body is rigid, angle COD cannot change; hence, the radius to any other point D will also rotate through θ radians and point D will move through the arc distance $s_2 = r_2\theta$. From this we conclude that the radii to all particles of a rotating body have the same angular motion (i.e., the same θ)

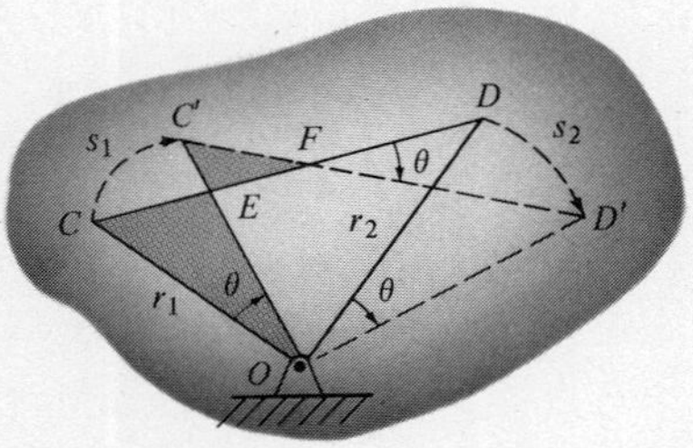

Figure 12-2.2 In fixed-axis rotation of a rigid body, all lines rotate through the same angular displacement.

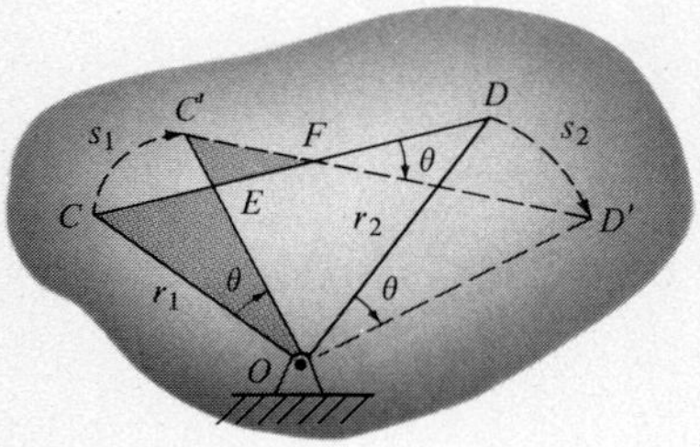

Figure 12-2.2

although their linear movements (i.e., s_1 and s_2) vary directly with their radial distances from the axis of rotation.

Of particular importance is the fact that the angle between the line CD and its subsequent position $C'D'$ is also equal to θ. We can easily prove this statement by noting that triangles CEO and $C'EF$ are similar because two of their angles are equal. Thus, angles CEO and $C'EF$ are equal vertical angles and angles ECO and $FC'E$ are the same apex angle of the congruent triangles COD and $C'OD'$. Consequently, the angles COE and EFC' are equal to each other and have the same value θ as must angle DFD' which is the vertical angle of EFC'. Our conclusion is that the projections of all line segments in a rigid body, such as CD in Fig. 12-2.1, upon a plane perpendicular to the axis of rotation all rotate through the same angle. We define this angle as the *angular displacement* of the rigid-body rotation. The unit of angular displacement may be radians, degrees, or revolutions, but radian measurement is preferred in order to correlate angular motion with linear motion.

For the case of fixed-axis rotation, it is permissible to define angular displacement $\boldsymbol{\theta}$ as a vector[1] directed along the axis of rotation, its sense being determined by the right-hand rule; e.g., the vector is directed along the axis of rotation as indicated by the extended

[1] For other than plane motion, $\boldsymbol{\theta}$ is not a vector in that the sequence of combining its components causes different results; i.e., $\boldsymbol{\theta} = \boldsymbol{\theta}_x + \boldsymbol{\theta}_y$ is not equal to $\boldsymbol{\theta}_y + \boldsymbol{\theta}_x$. For example, consider a top initially in the position

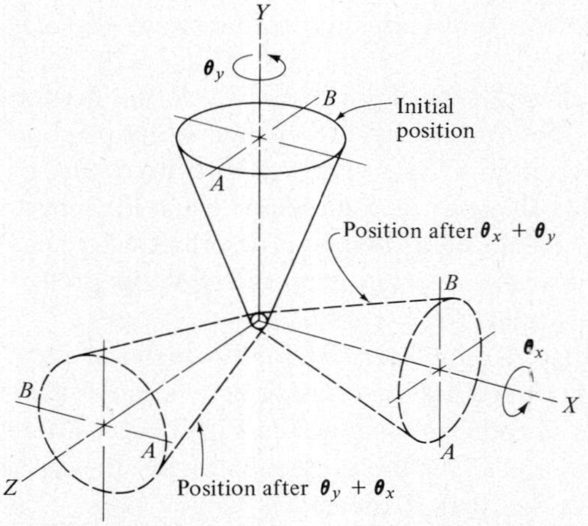

Figure 12-2.3 Effect of two different sequences of combining 90° rotations about X and Y axes.

shown in Fig. 12-2.3. The sequence in which two separate 90° rotations of the top about the X and Y axes are done will produce the two different dashed positions shown. However, although $\boldsymbol{\theta}$ is not a vector, its time derivatives $\boldsymbol{\omega}$ and $\boldsymbol{\alpha}$ are *always* vectors because any sequence of combining two *infinitesimal* rotations will produce the same result.

thumb of the right hand when the fingers are curled about the axis of rotation to correspond with the direction of rotation. The successive time derivatives of $\boldsymbol{\theta}$ are also vectors; i.e., angular velocity $\boldsymbol{\omega} = \frac{d\boldsymbol{\theta}}{dt}$ and angular acceleration $\boldsymbol{\alpha} = \frac{d\boldsymbol{\omega}}{dt} = \frac{d^2\boldsymbol{\theta}}{dt^2}$. They too are directed along the axis of rotation in accordance with the right-hand rule. Since fixed-axis rotation, however, is usually specified in terms of its projection upon a plane perpendicular to the axis of rotation, it is more convenient to denote θ, ω, and α as clockwise or counterclockwise quantities, letting the initial direction of rotation denote the positive sense of θ, ω, and α. This then agrees with the previously established convention for rectilinear motion; i.e., the initial direction of motion determines the positive sense of s, v, and a.

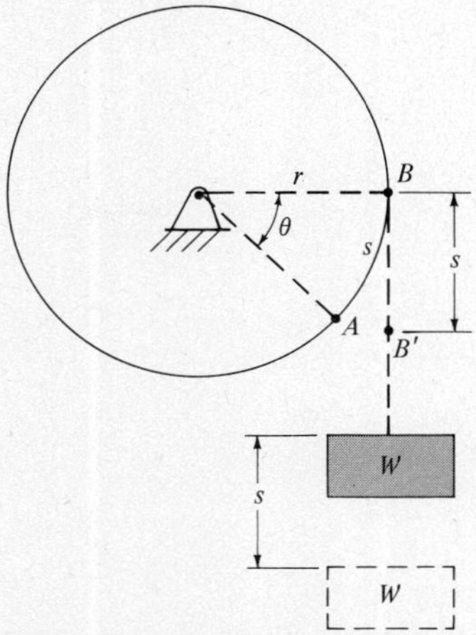

Figure 12-2.4 Relation between linear and angular displacement.

The correlation between linear and angular displacement is presented in Fig. 12-2.4 which shows a pulley free to rotate about an axle O under the action of a weight W suspended from a cord wound around the pulley. As the weight descends s ft, a point B on the rim of the pulley moves through an equal arc length to position A. The corresponding angular displacement θ of the pulley is obviously subtended by radii to points B and A. Evidently, the relation between the linear displacement of the weight and the angular displacement (in radians) of the pulley is given by

$$s = r\theta \tag{a}$$

Since the radius r is constant, differentiating Eq. (a) with respect to time gives

$$\frac{ds}{dt} = r\frac{d\theta}{dt} \quad \text{or} \quad v = r\omega \tag{b}$$

and a second differentiation of Eq. (b) gives

$$\frac{dv}{dt} = r\frac{d\omega}{dt} \quad \text{or} \quad a = r\alpha \tag{c}$$

The expression $\frac{dv}{dt}$ in Eq. (c) is the linear acceleration of the weight, but it also represents the tangential acceleration a_t of a point on the rim of the pulley.

Recalling Eq. (9-7.2), we may use $v = r\omega$ to write the corresponding normal acceleration of any point on the rim of the pulley in any of the following forms:

$$a_n = \frac{v^2}{r} = r\omega^2 = v\omega \tag{d}$$

Summarizing this discussion, we see that the differential kinematic equations for rectilinear motion and for rotation are completely analogous in form, differing only in the symbols used; namely,

Rectilinear Motion	*Rotation*
$v = \dfrac{ds}{dt}$	$\omega = \dfrac{d\theta}{dt}$
$a = \dfrac{dv}{dt} = \dfrac{d^2s}{dt^2}$	$\alpha = \dfrac{d\omega}{dt} = \dfrac{d^2\theta}{dt^2}$
$a\,ds = v\,dv$	$\alpha\,d\theta = \omega\,d\omega$

They can be transformed into each other by the relations

$$\begin{aligned} s &= r\theta \\ v &= r\omega \\ a_t &= r\alpha \\ a_n &= r\omega^2 \end{aligned} \tag{12-2.1}$$

Observe that a_t is tangent to the path in the sense that α rotates the radius r, while a_n is normal to the path and is always directed toward the center of rotation.

As an extension of the analogy between rectilinear motion and rotation, we observe that the procedures discussed previously in Section 9-3 for constant and variable acceleration will yield similar results in rotation that differ only in the symbols used. Thus, the equations of rotation with *constant angular acceleration* become

Rectilinear Motion	*(Related by)*	*Rotation*
$v = v_o + at$	$s = r\theta$	$\omega = \omega_o + \alpha t$
$s = v_o t + \frac{1}{2}at^2$	$v = r\omega$	$\theta = \omega_o t + \frac{1}{2}\alpha t^2$
$v^2 = v_o^2 + 2as$	$a = r\alpha$	$\omega^2 = \omega_o^2 + 2\alpha\theta$

ILLUSTRATIVE PROBLEMS

12-2.1. For the system of connected bodies shown in Fig. 12-2.5, the initial angular velocity of the compound pulley B is 6 rad per sec counterclockwise and weight D is decelerating at the constant rate of 4 fps². What distance will weight A travel before coming to rest?

Solution

To correlate the given data, we start by finding the kinematic relations among the bodies. Using Eq. (12-2.1) and denoting by E any point on the cord connecting B and C, we obtain

$$[s = r\theta] \qquad s_A = 3\theta_B, \qquad s_E = 2\theta_B = 3\theta_C, \qquad s_D = 1.5\theta_C$$

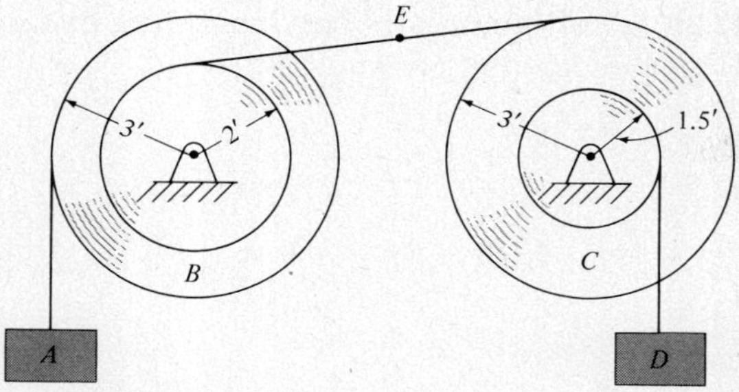

Figure 12-2.5

Combining these relations into one continuous equation, we have

$$s_A = 3\theta_B = \tfrac{9}{2}\theta_C = 3s_D \qquad (a)$$

The velocity and acceleration relations among the bodies are obtained by taking two successive time derivatives of Eq. (*a*). The results are equivalent to merely changing the symbols in Eq. (*a*) since $v = r\omega$ and $a_t = r\alpha$ have the same mathematical form as $s = r\theta$. Therefore, we obtain

$$v_A = 3\omega_B = \tfrac{9}{2}\omega_C = 3v_D \qquad (b)$$
$$a_A = 3\alpha_B = \tfrac{9}{2}\alpha_C = 3a_D \qquad (c)$$

Consider now the motion of A. From Eq. (*b*), $v_{o_A} = 3\omega_{o_B} = 3(6) = 18$ fps and, from Eq. (*c*), $a_A = 3a_D = 3(-4) = -12$ fps^2. Since the acceleration of *A* is constant, the distance it travels in coming to rest is

$$[v^2 = v_o{}^2 + 2as] \qquad 0 = (18)^2 - 2(12)s_A \qquad s_A = 13.5 \text{ ft} \quad \textbf{\textit{Ans.}}$$

If desired, the displacement of the other bodies can now be found from Eq. (*a*).

12-2.2. A body rotates according to the relation $\alpha = 3t^2 + 4$, angular displacement being measured in radians and time in seconds. If its initial angular velocity is 4 rad per sec and the initial angular displacement is zero, compute the values of ω and θ for the instant when $t = 3$ sec. Solve analytically and graphically.

Solution

Rewriting $\alpha = d\omega/dt$ as $d\omega = \alpha\, dt$ and integrating between the given limits, we have

$$\int_4^\omega d\omega = \int_0^t (3t^2 + 4)\, dt$$
$$\omega - 4 = t^3 + 4t$$
$$\omega = t^3 + 4t + 4 \qquad (a)$$

Applying $\omega = d\theta/dt$ in the form $d\theta = \omega\, dt$, substituting for ω its value from Eq. (*a*), and integrating gives

$$\int_0^\theta d\theta = \int_0^t (t^3 + 4t + 4)\, dt$$

$$\theta = \frac{t^4}{4} + 2t^2 + 4t \qquad (b)$$

Substituting $t = 3$ sec in Eqs. (*a*) and (*b*), we have

$$\omega = (3)^3 + 4 \times 3 + 4 = 43 \text{ rad per sec} \qquad \textbf{\textit{Ans.}}$$

$$\theta = \frac{(3)^4}{4} + 2 \times (3)^2 + 4 \times 3 = 50.25 \text{ rad} \qquad \textbf{\textit{Ans.}}$$

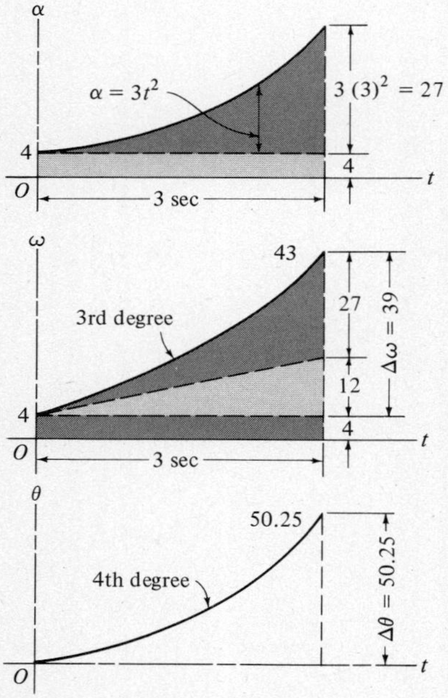

Figure 12-2.6 Motion curves.

Let us now check this solution by means of the motion curves shown in Fig. 12-2.6. Using an equation similar to Eq. (9-4.1), we find that the change in angular velocity is

$$[\Delta\omega = (\text{Area})_{\alpha-t}] \qquad \Delta\omega = 4 \times 3 + \tfrac{1}{3}(3)(27)$$
$$= 12 + 27 = 39 \text{ rad per sec}$$

Adding this value to $\omega_o = 4$ gives $\omega = 43$ rad per sec at $t = 3$ sec as before.

The area under the ω-t curve is subdivided into parts shaded to correspond to the similarly shaded subdivisions of the α-t curve. From an equation similar to Eq. (9-4.2), we then obtain

$$[\Delta\theta = (\text{Area})_{\omega-t}] \quad \Delta\theta = 4 \times 3 + \tfrac{1}{2} \times 3 \times 12 + \tfrac{1}{4} \times 3 \times 27$$
$$= 50.25 \text{ rad} \qquad \textbf{\textit{Check}}$$

If preferred, $\Delta\theta$ may be computed by applying an equation similar to Eq. (9-4.3) to the area under the α-t curve. This gives

$$[\Delta\theta = \omega_1(t_2 - t_1) + (\text{Area})_{\alpha-t}(\bar{t}_2)$$
$$\Delta\theta = 4 \times 3 + (4 \times 3)(\tfrac{1}{2} \times 3) + (\tfrac{1}{3} \times 3 \times 27)(\tfrac{1}{4} \times 3)$$
$$= 50.25 \text{ rad} \qquad \textbf{\textit{Check}}$$

which is identical term for term with the preceding computation although based on a different concept.

Finally, a comparison of these computations obtained from motion curves will be found to be identical term for term with those of the calculus solution.

PROBLEMS

12-2.3. A flywheel rotating at 3 rev per sec has its speed increased at the constant rate of 45 rad per min each second during an interval of

10 sec. Through how many revolutions does it rotate in this time?

67.5 rev **Ans.**

12-2.4. The rim of a 60-in. wheel on a brakeshoe testing machine has a speed of 60 mph when the brake is applied. It comes to rest after the rim has traveled a linear distance of 600 ft. What is the constant angular acceleration of the wheel and how many revolutions does it make after the brake is applied?

12-2.5. A gear is accelerated from rest to a speed of 1800 rpm and then immediately decelerated to a stop. If the total elapsed time is 12 sec, determine the total number of revolutions of the gear. Assume that both acceleration and deceleration are constant but not necessarily of the same magnitude.

12-2.6. When the angular velocity of a 4-ft diameter pulley is 3 rad per sec, the total acceleration of a point on its rim is 30 fps^2. Determine the angular acceleration of the pulley at this instant.

$\alpha = 12$ rad per sec^2 **Ans.**

12-2.7. Determine the horizontal and vertical components of the acceleration of point B on the rim of the flywheel shown in Fig. P-12-2.7. At the given position, $\omega = 4$ rad per sec and $\alpha = 12$ rad per sec^2, both clockwise.

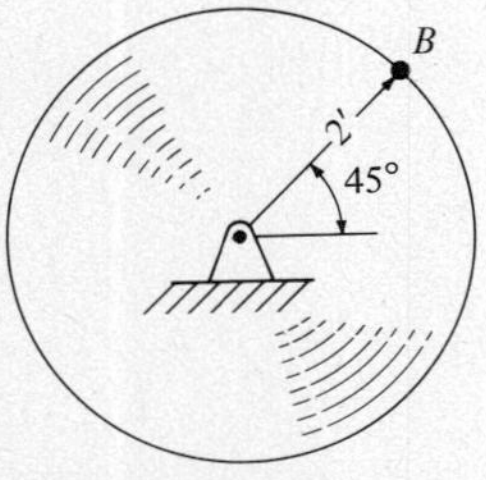

Figure P-12-2.7

12-2.8. Repeat the preceding problem if α is changed to 10 rad per sec^2 clockwise.

$a_h = 36.7$ fps^2 left; $a_v = 8.48$ fps^2 down **Ans.**

12-2.9. A pulley has a constant angular acceleration of 3 rad per sec^2. When the angular velocity is 2 rad per sec, the total acceleration of a point on the rim of the pulley is 10 fps^2. Compute the diameter of the pulley?

$d = 4$ ft **Ans.**

12-2.10. The step pulleys shown in Fig. P-12-2.10 are connected by a crossed belt. If the angular acceleration of C is 2 rad per sec^2, what time is required for A to travel 64 ft from rest? Through what distance will D move while A moves 100 ft?

$t = 4$ sec; $s_D = 75$ ft **Ans.**

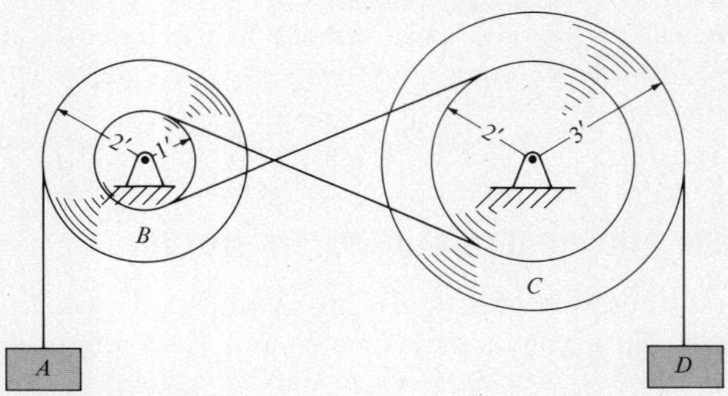

Figure P-12-2.10

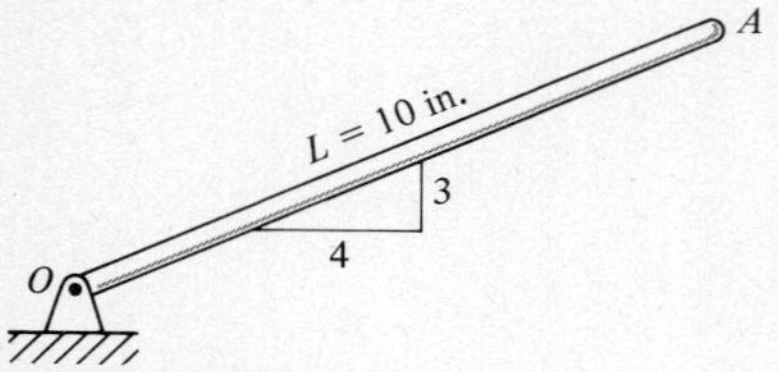

Figure P-12-2.12

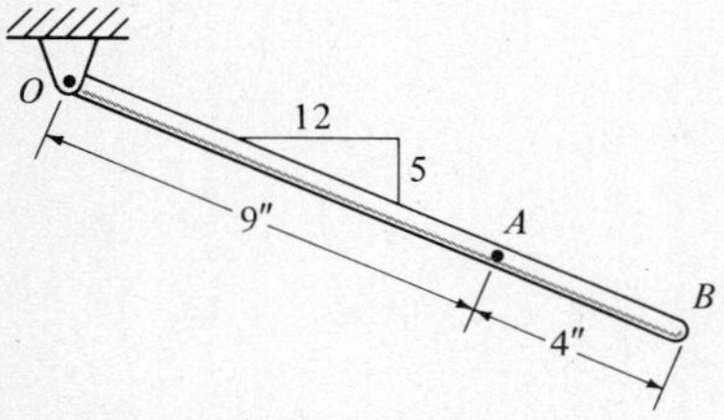

Figure P-12-2.13

12-2.11. Repeat Prob. 12-2.10 if the radii of pulley B are changed to 30 in. and 18 in.

12-2.12. At the position shown in Fig. P-12-2.12, the end A of the rod has a rightward component of velocity of 2 fps and an upward component of acceleration of 4 fps^2. Determine the angular acceleration of the rod at this position.

$\alpha = 18$ rad per sec^2 counterclockwise **Ans.**

12-2.13. The rod BO in Fig. P-12-2.13 rotates in a vertical plane about a horizontal axis at O. At the given position, end B has a downward vertical component of velocity of 6 fps and also a downward vertical component of acceleration of 9 fps^2. Compute the angular acceleration of the rod BO and the total acceleration of point A.

12-2.14. The rotation of a flywheel is governed by the equation $\omega = 4\sqrt{t}$; ω is in radians per second and t is in seconds. If $\theta = 2$ rad when $t = 1$ sec, determine the values of θ and α at the instant when $t = 3$ sec.

$\theta = 13.21$ rad; $\alpha = 1.154$ rad per sec^2 **Ans.**

12-2.15. The angular acceleration of a flywheel is given by $\alpha = 8 - t$ where α is in radians per second^2 and t is in seconds. If the angular velocity of the flywheel is 42 rad per sec at the end of 6 sec, determine its initial angular velocity and the number of revolutions made during the 6 sec.

$\omega_o = 12$ rad per sec; $\Delta\theta = 28.7$ rev **Ans.**

12-2.16. The angular acceleration of a flywheel increases uniformly from 3 rad per sec^2 to 9 rad per sec^2 in an interval of 6 sec. If its angular velocity was 12 rad per sec at the start of the interval, compute the number of revolutions made during the 6 sec.

12-2.17. Determine the number of revolutions through which a pulley will rotate from rest if its angular acceleration is increased uniformly from zero to 12 rad per sec^2 during 4 sec and then uniformly decreased to 4 rad per sec^2 during the next 3 sec.

$\theta = 23.2$ rev **Ans.**

12-2.18. The rotation of a flywheel is governed by the relation $\alpha = 10t - t^2$ where α is in radians per second^2 and t is in seconds. How many revolutions will the flywheel make, starting from rest, before it momentarily stops prior to reversing its direction? Solve both graphically and analytically.

$\Delta\theta = 224$ rev **Ans.**

12-3 DEFINITION AND ANALYSIS OF PLANE MOTION

Plane motion is that motion of a rigid body in which all particles in the body remain at a constant distance from a fixed reference plane. Consequently, all particles move in parallel planes and all particles on the same straight line perpendicular to the reference plane have identical values of displacement, velocity, and acceleration. The plane in which the mass center moves is defined as the *plane of motion*. Examples of bodies having plane motion are rolling wheels, the connecting rod of a reciprocating engine, or the links joining rotating elements of machines. In all these cases, the identifying characteristic of plane motion is that a reference line connecting any two points in the same plane of motion simultaneously undergoes an angular displacement and a linear displacement. Actually, a plane motion is a simultaneous combination of the motions of translation and rotation. The physical meaning of this characteristic will be amplified shortly.

We shall discuss three methods of relating the displacement, velocity, and acceleration of any two points in a body having plane motion. These methods are differing ways of solving the same problem; sometimes one or another will be simpler to apply. One method involves scalar calculus; another, vector calculus; and the third is their geometric application. An understanding of each method will reinforce comprehension of the others.

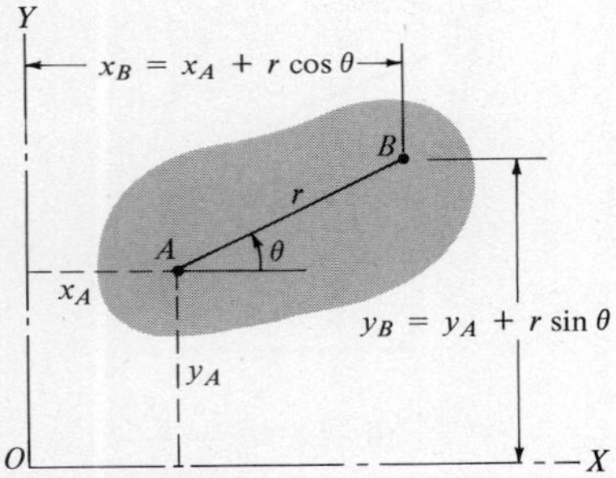

Figure 12-3.1

The most direct approach (but not necessarily the simplest to apply) is by means of scalar calculus. In Fig. 12-3.1 we represent a rigid body having plane motion by its projection upon the parallel fixed XY reference plane. Here A and B are the projections upon this plane of any two points in the body, r is the constant distance separating them, and θ is the angular coordinate of r measured from the X direction. Evidently the coordinates of B and A are related by

$$x_B = x_A + r\cos\theta \qquad y_B = y_A + r\sin\theta \tag{a}$$

Assuming that these coordinates, including θ but not r (which is a constant length), vary with time, we may find the relations between the components of velocity and acceleration by successive differentiation of Eq. (a). Thus, we obtain

$$\dot{x}_B = \dot{x}_A - r\dot{\theta}\sin\theta \qquad \dot{y}_B = \dot{y}_A + r\dot{\theta}\cos\theta \tag{b}$$

$$\ddot{x}_B = \ddot{x}_A - r\ddot{\theta}\sin\theta - r\dot{\theta}^2\cos\theta \qquad \ddot{y}_B = \ddot{y}_A + r\ddot{\theta}\cos\theta - r\dot{\theta}^2\sin\theta \tag{c}$$

By combining these scalar components in the order shown in parts (a) and (b) of Fig. 12-3.2, we can visualize the geometric relations among the velocity and acceleration vectors of B and A. From part (a), we see that $\mathbf{v}_{B/A}$ is the difference between $\mathbf{v}_B$ and $\mathbf{v}_A$. Its magnitude is

$$v_{B/A} = \sqrt{(r\dot{\theta}\sin\theta)^2 + (r\dot{\theta}\cos\theta)^2} = r\dot{\theta}\sqrt{\sin^2\theta + \cos^2\theta} = r\dot{\theta}$$

and its direction with the vertical is

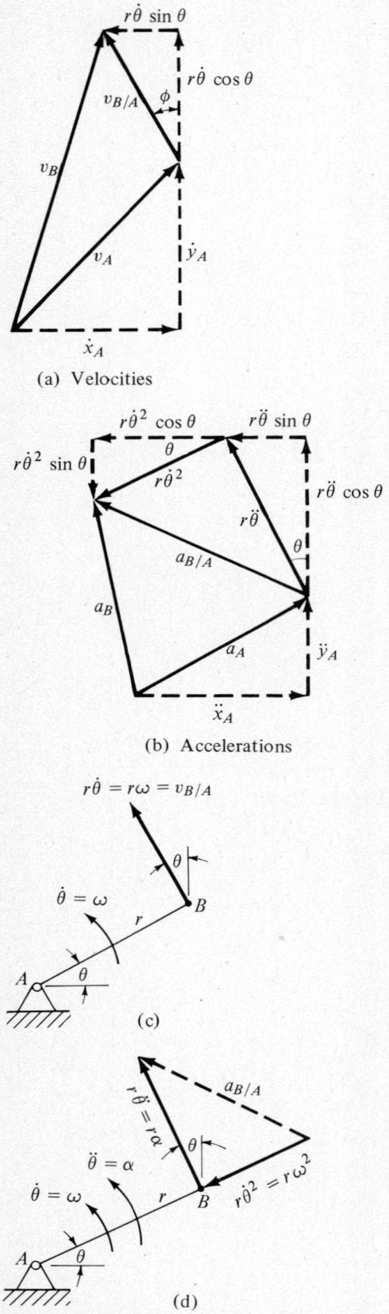

Figure 12-3.2 Interpretation of rectangular components of velocity and acceleration.

$$\tan\phi = \frac{r\dot{\theta}\sin\theta}{r\dot{\theta}\cos\theta} = \tan\theta$$

whence

$$\phi = \theta$$

A simple physical meaning of these results is obtained by comparing them with the velocity of B as found from a fixed-axis rotation of the body about A as represented in part (c). Since they are identical, we conclude that $\mathbf{v}_{B/A}$ is equivalent to the velocity of B as found from a fixed-axis rotation of B about A. For this reason, it is convenient to designate $\mathbf{v}_{B/A}$ as the relative velocity of B rotating about A.

Similarly, from a comparison of parts (b) and (d), we see that the difference $\mathbf{a}_{B/A}$ between the accelerations $\mathbf{a}_B$ and $\mathbf{a}_A$ is composed of terms which denote the normal and tangential components of acceleration of B rotating about A.

This discussion may be summarized by the equations[2]

$$v_B = v_A \;+\!\!\!\!\rightarrow (v_{B/A} = r\omega) \tag{12-3.1}$$

$$a_B = a_A \;+\!\!\!\!\rightarrow (a_{B/A} = r\omega^2 \;+\!\!\!\!\rightarrow r\alpha) \tag{12-3.2}$$

which define the instantaneous velocity and acceleration of B in terms of the velocity and acceleration of *any* reference point A plus vectorially the velocity and acceleration of a body point B assumed simultaneously rotating about a fixed axis through the reference point A. Thus, we confirm the statement made earlier that a plane motion is a simultaneous combination of a rigid-body rotation about *any* reference point plus the rigid-body translation corresponding to the motion of the reference point.

The conclusions just reached are part of Chasle's theorem which states that any motion of a rigid body may be resolved into the simultaneous combination of the translation of any reference point plus a rotation about that reference point.

Using a geometric approach to the application of Chasle's theorem, consider the rigid body represented in Fig. 12-3.3 by its projection upon a fixed reference plane coinciding with the plane of motion. Assume that the body changes its position from the full to the dashed outline during the time Δt. Assume the motion of two points A and B in the body to be as shown and to have the respective movements s_A and s_B.

[2] The use of parentheses in Eqs. (12-3.1) and (12-3.2) to enclose equivalent terms may be confusing at first, but they serve as a continual reminder that a relative motion, such as $a_{B/A}$, may be expressed in terms of its equivalent rotational components.

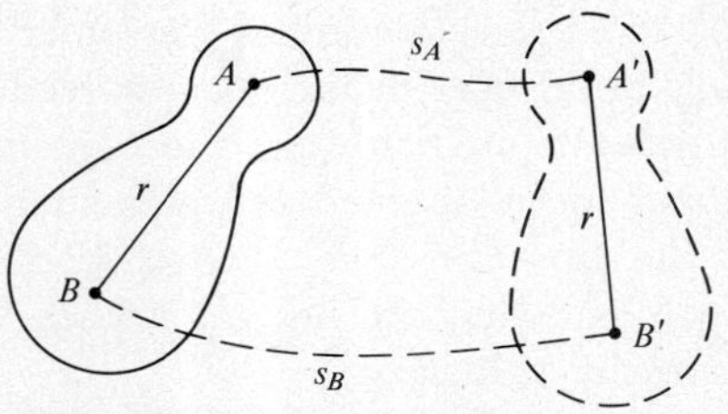

Figure 12-3.3 Plane motion.

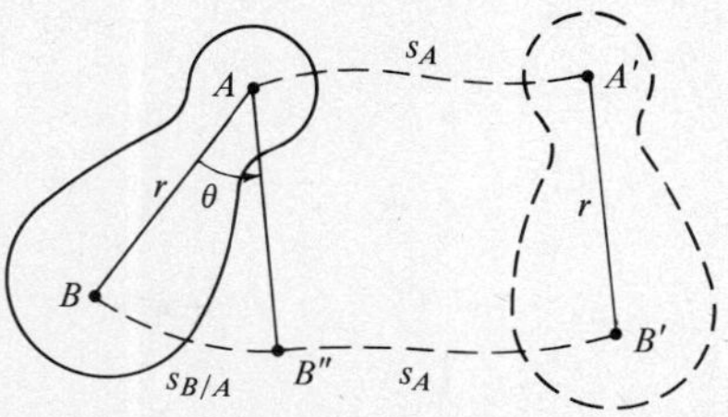

Figure 12-3.4 Plane motion resolved into a combined rotation about A and translation of A.

To determine a relation between the movements of points A, and B, assume that the body is rotated through an angle θ about an axis passing through A until line AB is parallel to line $A'B'$ (Fig. 12-3.4). The body may now be given the curvilinear translation defined by the motion of A, thereby causing the body to coincide with its final position. The motion of B thereby is the geometric sum of $s_{B/A}$ and s_A and may be expressed by the vector equation

$$s_B = s_A \nrightarrow (s_{B/A} = r\theta) \qquad (12\text{-}3.3)$$

The symbol $s_{B/A}$ is to read as the motion of B assumed rotating about A; it is equal to $r\theta$ where r is the distance between A and B.

In a finite time interval, the actual movement of point B will only approximately equal $s_A \nrightarrow s_{B/A}$ since this equation assumes that the translation and rotation of the body take place independently, whereas they actually occur simultaneously. For purposes of analysis, however, there is the valuable concept that plane motion is equivalent to a rigid-body rotation about *any* point plus the rigid-body translation of that point. *This assumption is exact when the time interval between any two successive positions approaches zero as a limit.* Then the actual motion is equivalent to successive infinitesimal displacements from one instantaneous position to the next. Each such infinitesimal displacement consists of the vector sum of the translation of an arbitrarily selected reference point plus a rotation about that reference point.

Differentiating Eq. (12-3.3) with respect to the time (i.e., assuming $\Delta t \to 0$), we have

$$v_B = v_A \nrightarrow (v_{B/A} = r\omega) \qquad (12\text{-}3.1)$$

The symbol $v_{B/A}$ is read as the velocity of B assumed to be rotating about A. It is equal to $r\omega$ where the angular velocity ω is defined as $d\theta/dt = \dot{\theta}$, i.e., the time rate of change of the body rotation θ, and r is the distance between A and B.

In like manner, differentiating the velocity relation with respect to time yields

$$a_B = a_A \nrightarrow (a_{B/A} = r\omega^2 \nrightarrow r\alpha) \qquad (12\text{-}3.2)$$

where the symbol $a_{B/A}$ is to be read as the acceleration of B assumed to be rotating about A. This acceleration may in turn be resolved into a normal component $r\omega^2$ directed from B toward the center of rotation A and a tangential component $r\alpha$ directed perpendicular to r in the sense that α moves the tip of r. (See Section 12-2, p. 428.) The angular acceleration α is defined by $d\omega/dt$, i.e., the time rate of change of the body angular velocity ω.

Thus we obtain the same equations reached by our first purely analytical development. It should now be clear that any motion of a rigid body is equivalent to the simultaneous translation of any reference point plus a rotation about that reference point. The use of parentheses to enclose equivalent terms should serve as a continual reminder that a relative motion, such as $a_{B/A}$, should be expressed in terms of its equivalent rotational components. It should be noted that the equations we have developed will also determine the relative motion between any two points whose absolute motion is known.

Finally, it is very important to understand that the rotational components of plane motion are properties of the body and are independent of the choice of a reference point. Our discussion of rotation on p. 426 proved that θ, ω, and α represent the angular properties of any line in the rotating body. Similarly, in plane motion, θ, ω, and α are defined in terms of the angular motion of any line in the body. It is easy to demonstrate that such a reference line has the same angular motion regardless of the choice of any arbitrarily selected reference point. For example, if the body in Fig. 12-3.5 were rotating about B instead of A as previously assumed in Fig. 12-3.4, its motion would be equivalent to a rotation about B, plus the translation of B. The reference line AB would then rotate about B to the position BA'' parallel to the final position $B'A'$, after which the body would be translated along the path $BB''B'$ to its final position. Obviously the angle θ defined by ABA'' is counterclockwise about B and equal to the angle BAB'' which defines the same counterclockwise rotation θ about A as a reference point. Hence, in plane motion, the values of θ, ω, and α refer to the body and have the same values and directional sense regardless of the reference point.

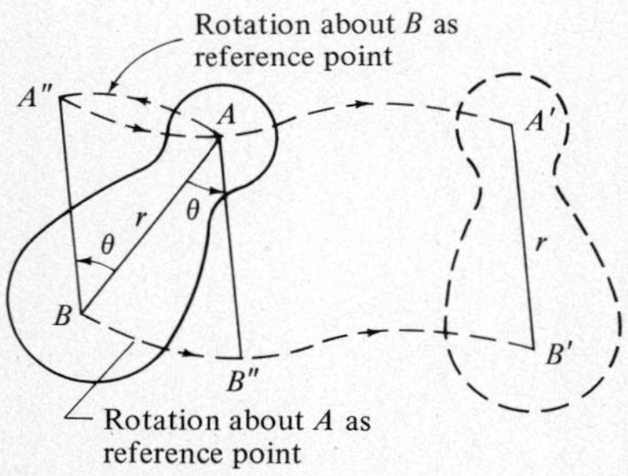

Figure 12-3.5 Angular motion is independent of reference point.

12-4 APPLICATION OF KINEMATIC EQUATIONS

As we shall see when we study Chapter 13 which relates applied forces to the motion of rigid bodies, it is necessary to determine the relation between the angular acceleration of a body and the linear acceleration of its center of gravity. This objective is achieved through the proper application of the kinematic equations. Since the acceleration equation involves the angular velocity ω of the body, it is usually necessary to first determine ω by a velocity analysis.

In the majority of cases involving plane motion, the linear motion A-A' in Fig. 12-3.3 of any point A in the body can be related to the angular motion only by careful application of the kinematic equations. However, in the special case of a disk or sphere rolling freely, (i.e., without slipping at the point of contact) along a plane as in Fig. 12-4.1, we can obtain a direct relation between the linear

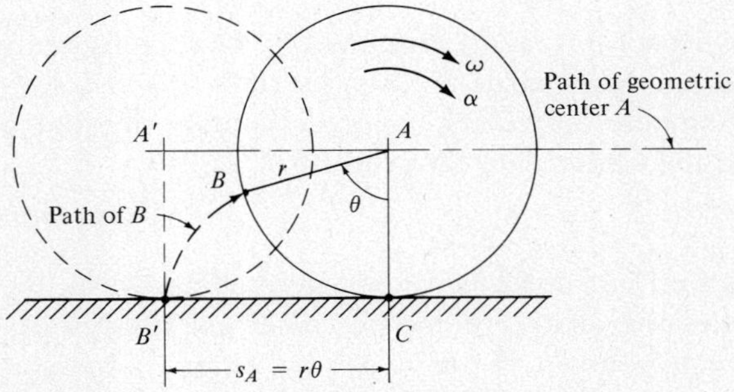

Figure 12-4.1 Free-rolling disk or sphere.

motion of its geometric center and its angular motion. In the position shown in Fig. 12-4.1, the disk has rolled to the right without slipping from a position with B originally coincident with B' i.e., from the dashed outline to the solid outline. This motion is equivalent to translating the disk through the distance $s_A = A'A$ and then rotating it about A through the angle θ. If there is no slipping, the distance $B'C$ equals the arc length $BC = r\theta$. But $B'C$ is equal to the linear motion of the geometric center A. Hence, we have

$$s_A = r\theta \tag{12-4.1}$$

and two successive time differentiations yield the following results which are true at any particular instant:

$$\frac{ds_A}{dt} = r\frac{d\theta}{dt} \quad \text{or} \quad v_A = r\,\omega \tag{12-4.2}$$

$$\frac{dv_A}{dt} = r\frac{d\omega}{dt} \quad \text{or} \quad a_A = r\,\alpha \tag{12-4.3}$$

Remember that these relations between the linear motion of the geometric center of a disk or sphere and its angular motion are correct only when the disk or sphere rolls without slipping. If slipping occurs, these relations are *not* valid. Note also that positive ω and α have the same direction as positive θ, i.e., clockwise for a rightward motion of the disk and counterclockwise for a leftward motion. These directions are illustrated in Fig. 12-4.1. Observe also that when the plane motion of the disk is resolved into a series of simultaneous infinitesimal translations and rotations, the actual path of B will be a cycloid.

ILLUSTRATIVE PROBLEMS

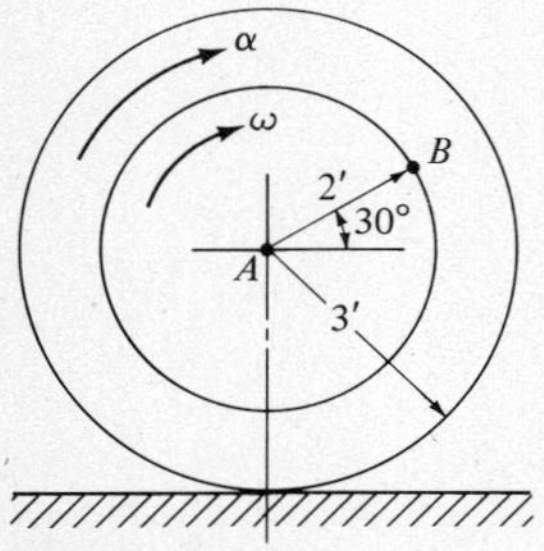

Figure 12-4.2

12-4.1. The wheel of 3-ft radius shown in Fig. 12-4.2 rolls freely to the right. At the given position, $\omega = 3$ rad/sec and $\alpha = 5$ rad/sec^2, both clockwise. Compute the velocity and acceleration of point B which is 2 ft from the center A of the wheel.

Solution

Applying the relations just developed for a free-rolling wheel, the velocity and acceleration of the geometric center are respectively $v_A = r\omega = (3)(3) = 9$ fps and $a_A = r\alpha = (3)(5) = 15$ fps^2, both rightward.

Since the motion of A is known, we use it as a reference point to resolve the plane motion of the wheel into a simultaneous rotation about A plus the translation of A. Then the velocity of B is determined as shown in part (a) of Fig. 12-4.3. The sketches below the velocity equation illustrate the magnitude and sense of each known term. Note that A is not actually stationary but is shown so only to emphasize the relative rotation of B about A.

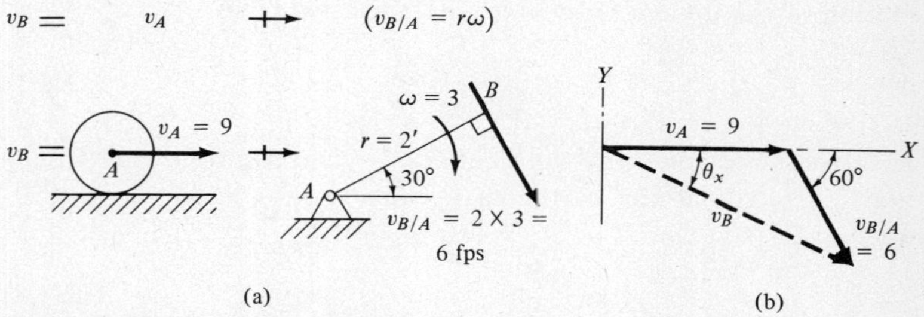

Figure 12-4.3 Interpretation of velocity equation.

If the terms in this vector equation are now plotted to scale as in part (b), the values of v_B and θ_x are scaled off as 13.1 fps and 23.4°, respectively. Analytically, the vector polygon can be used to compute these results either by the cosine law or by projection upon conveniently selected reference axes. Using this latter method, we find the components of v_B are

$$v_{B_x} = \Sigma v_x = 9 + 6 \cos 60° = 12 \text{ fps}$$
$$v_{B_y} = \Sigma v_y = -6 \sin 60° = -5.2 \text{ fps}$$

whence

$$v_B = \sqrt{(v_{B_x})^2 + (v_{B_y})^2} = 13.1 \text{ fps down to the right at } \theta_x = 23.4°$$

To determine the acceleration of B, use the acceleration equation shown in Fig. 12-4.4a. The sketches in part (a) show that the relative acceleration of B rotating about A consists of the normal component $r\omega^2 = 2(3)^2 = 18 \text{ fps}^2$ and the tangential component $r\alpha = 2(5) = 10 \text{ fps}^2$ each directed as shown. Plot the vector equation in tip-to-tail fashion to obtain the acceleration polygon shown in part (b) from which we scale the values $a_B = 18.2 \text{ fps}^2$ and $\theta_x = 76°$.

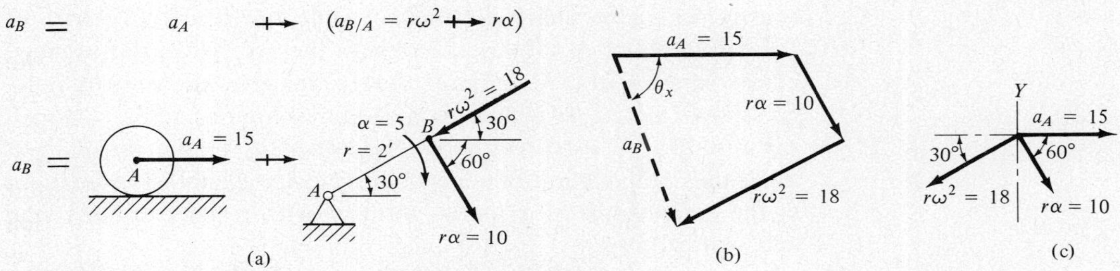

Figure 12-4.4 Interpretation of acceleration equation.

For an analytical solution, it is convenient to draw the vectors whose sum is a_B as acting at a common origin as shown in part (c) and apply the principle that the components of a resultant vector are the sums of the components of its parts. Then by taking summations with respect to the reference axes shown, we find the components of a_B are

$$\overset{+}{\rightarrow} a_{B_x} = \Sigma a_x = 15 + 10 \cos 60° - 18 \cos 30° = 4.4 \text{ fps}^2$$
$$+\uparrow a_{B_y} = \Sigma a_y = -10 \sin 60° - 18 \sin 30° = -17.7 \text{ fps}^2$$

whence

$$a_B = \sqrt{(a_{B_x})^2 + (a_{B_y})^2}$$
$$= 18.2 \text{ fps}^2 \text{ down to the right at } \theta_x = 76° \qquad \textbf{\textit{Ans.}}$$

12-4.2. A slender rod 10 in. long moves with its ends in contact with a horizontal floor and an inclined surface as shown in Fig. 12-4.5. At the given position, end A has a leftward velocity of 36 ips and a rightward acceleration of 7.5 ips². Determine the velocity and acceleration of end B.

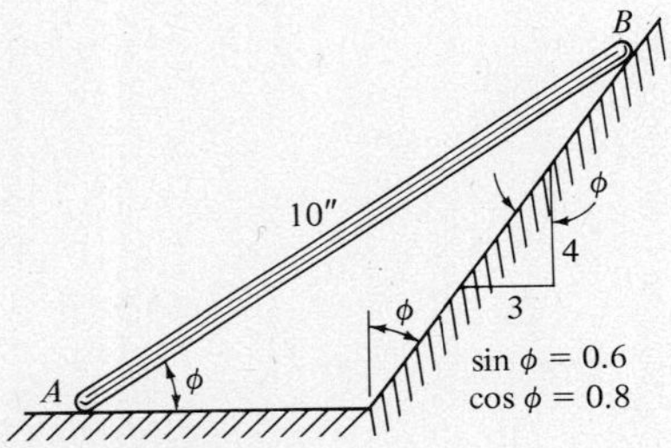

Figure 12-4.5

Solution

Since the motion of A is known, resolve the plane motion of the rod into a rotation about A plus the translation of A.

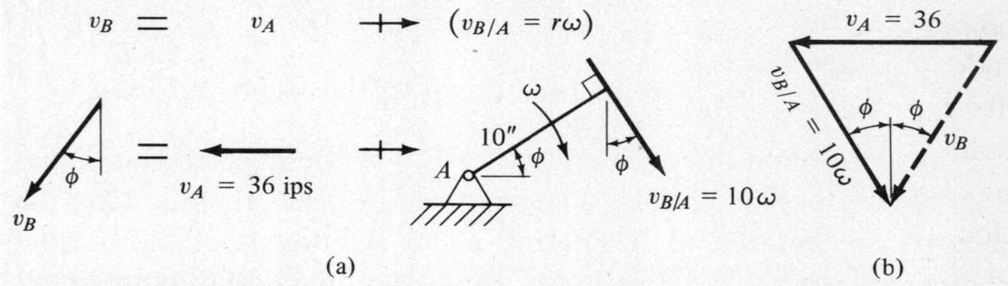

Figure 12-4.6 Interpretation of velocity equation.

To determine the velocity of B, use the vector equation shown in Fig. 12-4.6 in which the sketches below each velocity term specify its corresponding direction. The linear velocity of B is evidently constrained to be along the incline while the relative velocity of B rotating about A is perpendicular to the rod AB in the assumed clockwise sense of ω. Combining these terms gives the velocity polygon shown in part (b) where the tip-to-tail sum of v_A and $v_{B/A}$ equals the resultant v_B extending from the tail of v_A to the tip of $v_{B/A}$. Since the components of the resultant velocity equal the summations of the components of its parts, we have with respect to horizontal and vertical axes

$$[\overset{+}{\leftarrow} v_{B_h} = \Sigma v_h] \qquad v_B \sin\phi = 36 - 10\omega \sin\phi$$

$$[+\downarrow v_{B_v} = \Sigma v_v] \qquad v_B \cos\phi = 10\omega \cos\phi$$

from which, on substituting $\sin\phi = 0.6$ and $\cos\phi = 0.8$, we obtain

$$\omega = 3 \text{ rad/sec (clockwise)} \qquad \text{and} \qquad v_B = 30 \text{ ips} \qquad \textit{Ans.}$$

The positive value of ω confirms the assumption in Fig. 12-4.6a that B is rotating in a clockwise sense about A.

The velocity polygon also confirms the directions of v_B and $v_{B/A}$, and in addition is evidently an isosceles triangle in which $v_B = v_{B/A} = 10\omega$ by inspection.

To determine the acceleration of B, we use the acceleration equation shown in Fig. 12-4.7 in which the sketches below each term indicate the corresponding acceleration vector. Obviously, a_B is directed along the incline while a_A was given as 7.5 ips^2 rightward. The relative acceleration $a_{B/A}$ of B assumed to be rotating about A has the normal and tangential components shown. The normal

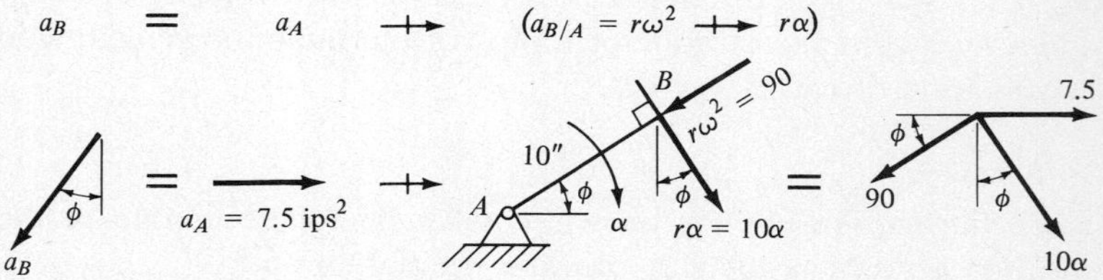

Figure 12-4.7 Interpretation of acceleration equation.

component is $r\omega^2$ and is directed from B toward A with the magnitude $10(3)^2 = 90$ ips^2. Observe that in order to evaluate $r\omega^2$ it was first necessary to determine ω from a velocity analysis. The tangential component $r\,\alpha$ is perpendicular to AB as shown with α assumed to be clockwise. Finally, the right-hand terms of the acceleration equation are shown acting at a common origin so that the components of the resultant acceleration a_B may be easily correlated with the summations of the components of its parts. Equating horizontal and vertical summations, we obtain

$$[\overset{+}{\leftarrow} a_{B_h} = \Sigma a_h] \qquad a_B \sin\phi = 90\cos\phi - 7.5 - 10\alpha\sin\phi$$

$$[+\downarrow a_{B_v} = \Sigma a_v] \qquad a_B \cos\phi = 90\sin\phi + 10\alpha\cos\phi$$

from which, on substituting $\sin\phi = 0.6$ and $\cos\phi = 0.8$, we obtain

$$\alpha = 2 \text{ rad/sec}^2 \text{ (clockwise)} \quad \text{and} \quad a_B = 87.5 \text{ ips}^2 \qquad \textit{Ans.}$$

The positive values of α and a_B confirm our assumptions of their directions.

Let us consider now two variations of the problem we have completed. The first variation is that of a slider-crank chain as shown in Fig. 12-4.8 in which the motion of B is controlled by the crank BC instead of the inclined plane. For ease of comparison, assume the 4-in. crank BC to have the angular position shown. Then the

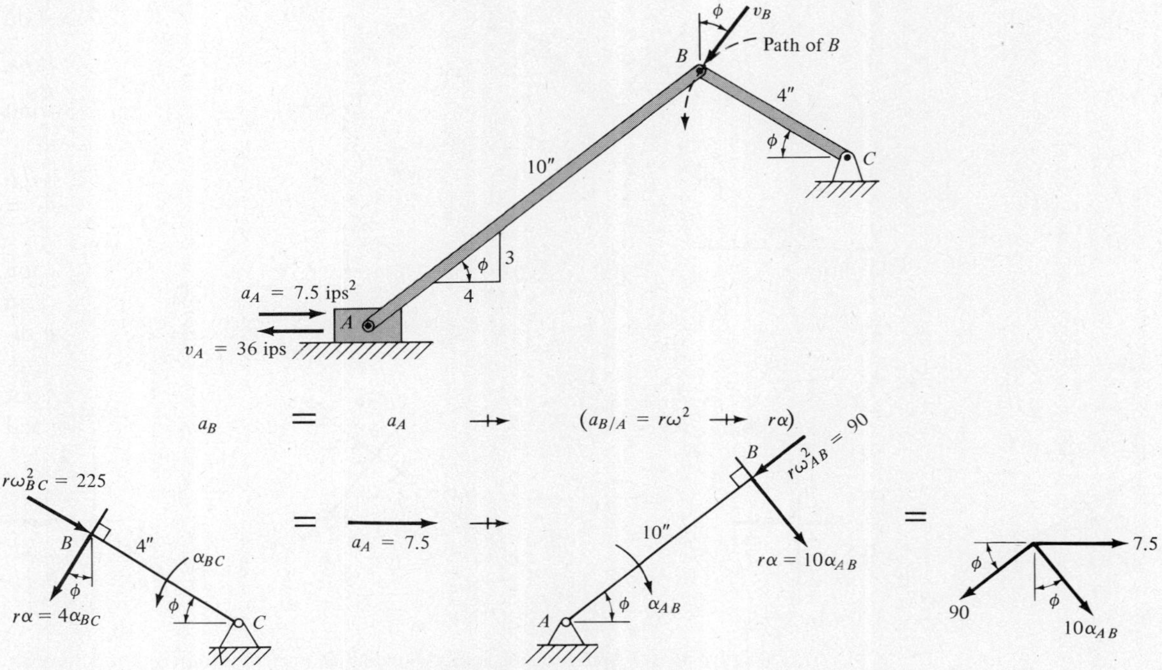

Figure 12-4.8 Effect on a_B caused by replacing incline by crank BC.

instantaneous velocity of B, acting perpendicular to BC, would be equivalent to the previous motion of sliding down the inclined plane. Hence, a velocity analysis of AB would result in $v_B = 30$ ips as before, but the angular velocity of crank BC would be

$$[v = r\omega] \qquad 30 = 4\omega_{BC} \qquad \omega_{BC} = 7.5 \text{ rad/sec}$$

The absolute acceleration of B, instead of being along the incline, now consists of a normal component $r\omega_{BC}^2 = (4)(7.5)^2 = 225$ ips^2 as well as a tangential component $r\alpha = 4\alpha_{BC}$ directed perpendicular to the crank BC. These components of a_B are shown below the term a_B in the acceleration equation shown in Fig. 12-4.8. The components of the acceleration equation become

$$[\overset{+}{\leftarrow} a_{B_h} = \Sigma a_h]$$
$$4\alpha_{BC} \sin\phi - 225\cos\phi = 90\cos\phi - 7.5 - 10\alpha_{AB}\sin\phi$$

$$[+\downarrow a_{B_v} = \Sigma a_v]$$
$$4\alpha_{BC}\cos\phi + 225\sin\phi = 90\sin\phi + 10\alpha_{AB}\cos\phi$$

from which, on substituting $\sin\phi = 0.6$ and $\cos\phi = 0.8$, we obtain

$$\alpha_{AB} = 26.4 \text{ rad/sec}^2 \quad \text{and} \quad \alpha_{BC} = 40.4 \text{ rad/sec}^2 \qquad \textbf{\textit{Ans.}}$$

As a second variation of the original problem, consider the

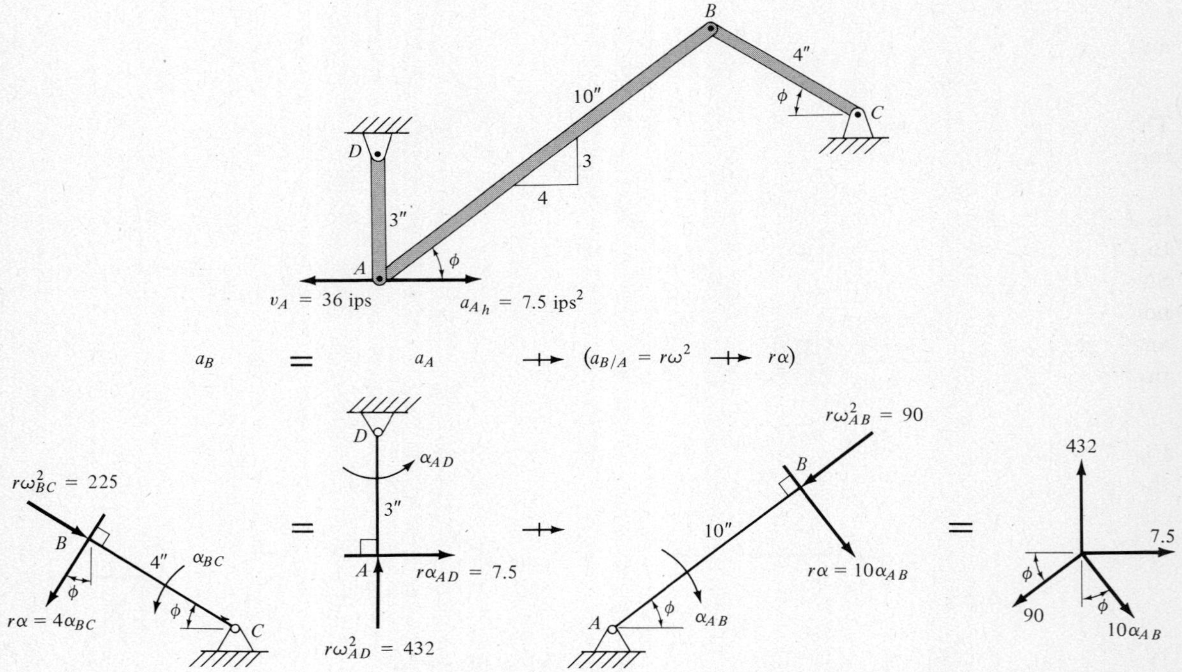

Figure 12-4.9 Acceleration analysis of AB as part of a four-link mechanism.

four-link mechanism shown in Fig. 12-4.9 in which the motions of B and A are controlled by the respective cranks BC and AD. If we let the cranks be in the positions shown and $v_A = 36$ ips leftward, the velocity analysis of rod AB is identical with the original problem resulting in $\omega_{AB} = 3$ rad/sec and $v_B = 30$ ips. When we apply the acceleration equation in this variation where both B and A now travel in circular paths, we must include the normal and tangential components of acceleration of both points B and A. These components are shown below the various terms of the acceleration equation in Fig. 12-4.9. Note that $r\omega_{AB}{}^2$ and $r\omega_{BC}{}^2$ for B are the same as in the slider-crank variation while the normal component of A rotating about D is $r\omega_{AD}{}^2 = \dfrac{v^2}{r} = \dfrac{(36)^2}{3} = 432 \text{ ips}^2$. Now the components of the acceleration equation are

$[\overset{+}{\leftarrow} a_{B_h} = \Sigma a_h]$

$$4\alpha_{BC} \sin\phi - 225 \cos\phi = 90 \cos\phi - 7.5 - 10a_{AB} \sin\phi$$

$[+\downarrow a_{B_v} = \Sigma a_v]$

$$4\alpha_{BC} \cos\phi + 225 \sin\phi = 90 \sin\phi - 432 + 10\alpha_{AB} \cos\phi$$

On substituting $\sin\phi = 0.6$ and $\cos\varphi = 0.8$ in these equations, we find

$$\alpha_{AB} = 53.2 \text{ rad/sec}^2 \circlearrowleft$$

and **Ans.**

$$\alpha_{BC} = -27.1 \text{ rad/sec}^2 = 27.1 \text{ rad/sec}^2 \circlearrowleft$$

The minus sign for α_{BC} indicates that it acts oppositely to the counterclockwise sense originally assumed.

Notice that the velocity analysis of AB produced the same results in all three variations of the original problem since we were careful to ensure that the velocity directions of A and B were unchanged. Similarly, except for the addition of the cranks which introduced normal components of acceleration in a_B and a_A, the acceleration analysis of AB in all three variations followed the same general procedure.

12-4.3. A link of length L is constrained to move as shown in Fig. 12-4.10. Its mass center G is at a distance b from end A. Use the scalar calculus approach to express the horizontal and vertical components of the acceleration of G in terms of the accelerations a_A and a_B of the ends of the link.

Solution

We first take successive time differentiations of the x coordinates of A and G as shown in the left margin on the next page.

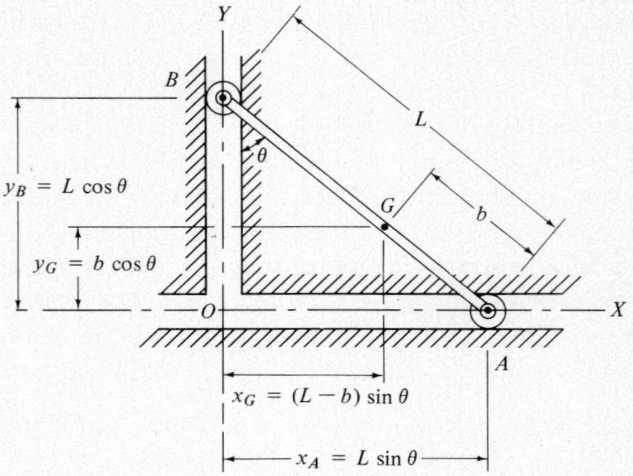

Figure 12-4.10 Analysis of motion by scalar calculus.

x components of A
$x_A = L \sin\theta$
$\dot{x}_A = L\dot{\theta} \cos\theta$

x components of G
$x_G = (L - b) \sin\theta$
$\dot{x}_G = (L - b)\dot{\theta} \cos\theta$

By eliminating the common term $\dot{\theta} \cos\theta$ between $\dot{x}_A$ and $\dot{x}_G$, we obtain

$$\dot{x}_G = \left(\frac{L - b}{L}\right)\dot{x}_A$$

whence another time differentiation gives

$$\ddot{x}_G = \left(\frac{L - b}{L}\right)\ddot{x}_A \quad \text{or} \quad a_{G_h} = \left(\frac{L - b}{L}\right)a_A \qquad \textit{Ans.}$$

Similarly, successive time differentiations of the y coordinates of B and G yield

y components of B	y components of G
$y_B = L \cos\theta$	$y_G = b \cos\theta$
$\dot{y}_B = -L\dot{\theta} \sin\theta$	$\dot{y}_G = -b\dot{\theta} \sin\theta$

Eliminating the common term $\dot{\theta} \sin\theta$ between $\dot{y}_B$ and $\dot{y}_G$ gives

$$\dot{y}_G = \frac{b}{L}\dot{y}_B$$

whence

$$\ddot{y}_G = \frac{b}{L}\ddot{y}_G \quad \text{or} \quad a_{G_v} = \frac{b}{L}a_B \qquad \textit{Ans.}$$

The geometric or relative motion approach of the previous problems gives results only for a particular position of a mechanism. On the other hand, the scalar calculus approach gives results for all positions of mechanisms which have repetitive cycles such as the slider-crank and four-link mechanisms. Such general results can then be adapted to computer analysis of a mechanism.

PROBLEMS

12-4.4. The compound wheel shown in Fig. P-12-4.4 rolls without slipping. At the given position, the velocity of A is 6 fps and the acceleration of A is 20 fps^2, both directed to the right. Compute the acceleration of points B and C.

$$a_B = 65.6\ \text{fps}^2;\ a_C = 36\ \text{fps}^2 \qquad \textbf{\textit{Ans.}}$$

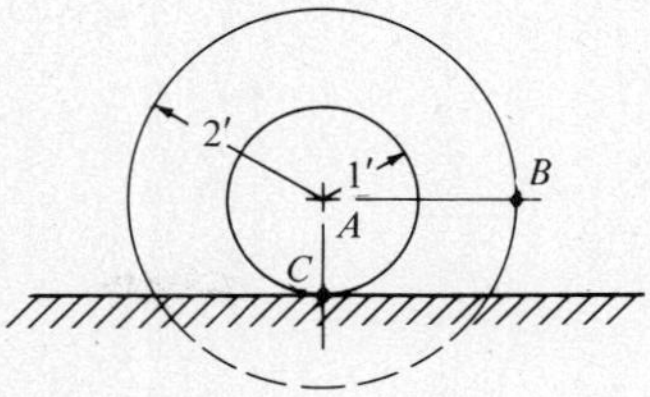

Figure P-12-4.4

12-4.5. The force P causes the unbalanced wheel shown in Fig. P-12-4.5 to have at the given instant $\omega = 3$ rad per sec and $\alpha = 6$ rad per sec^2, both clockwise. If the wheel does not slip, determine the acceleration of the mass center G.

$$a_G = 14.4\ \text{fps}^2 \qquad \textbf{\textit{Ans.}}$$

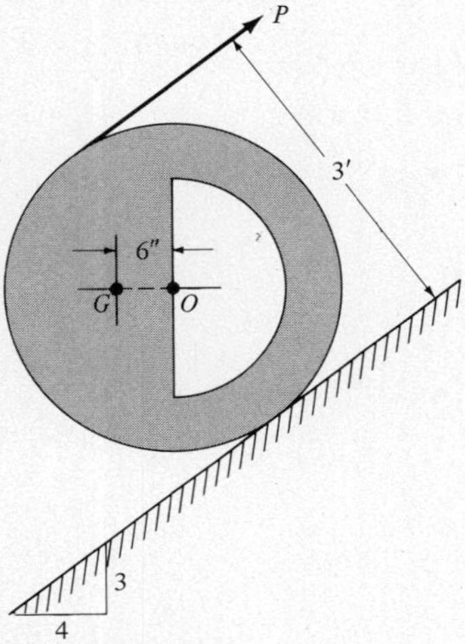

Figure P-12-4.5

12-4.6. As shown in Fig. P-12-4.6, a cord being unwrapped from the rim of a solid cylinder A of radius r passes over a fixed pulley P and back to the center of the cylinder. Determine the relation between the angular acceleration of the pulley and the cylinder.

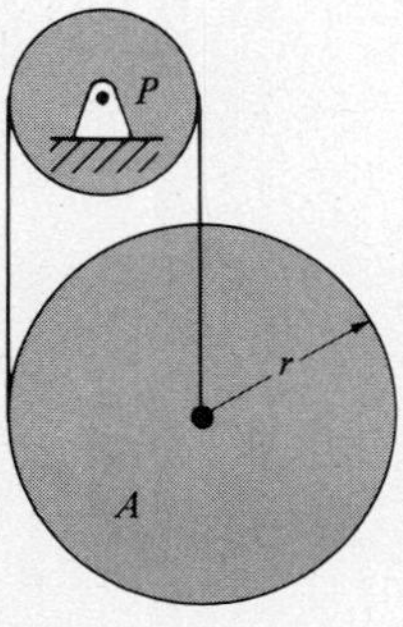

Figure P-12-4.6

12-4.7. Body B causes the compound drum in Fig. P-12-4.7 to roll without slipping up the incline. If the linear acceleration of B is 2 fps^2 down, compute the vertical acceleration of body A. Assume that the cord supporting A remains vertical.

$$a_{A_v} = 6\ \text{fps}^2 \qquad \textbf{\textit{Ans.}}$$

12-4.8. The compound wheel shown in Fig. P-12-4.8 rolls without slipping between the two parallel plates. At the given position, plate A has a rightward velocity and acceleration of 8 fps and 12 fps^2, respectively. At this same instant, the wheel has absolute values of $\omega = 4$ rad per sec and $\alpha = 10$ rad per sec^2, both clockwise. Compute the linear velocity and acceleration of plate B and of point D on the wheel.

Hint: First find v_G and a_G.

$$a_B = 8\ \text{fps}^2;\ a_D = 7.07\ \text{fps}^2 \qquad \textbf{\textit{Ans.}}$$

12-4.9. The bar AB of length $L = 10$ in. shown in Fig. P-12-4.9 moves with its ends in contact with the vertical wall and the horizontal floor.

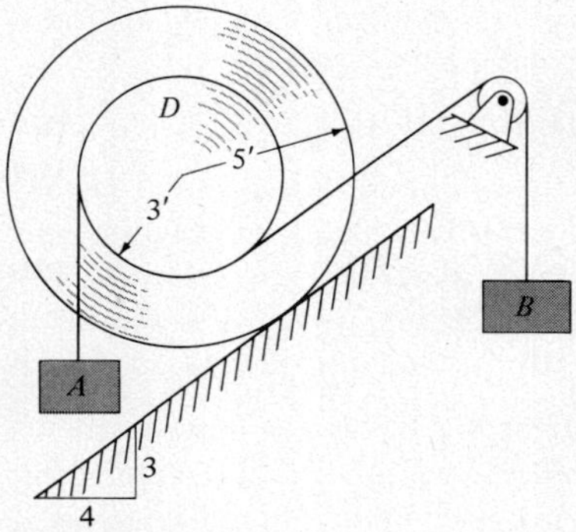

Figure P-12-4.7

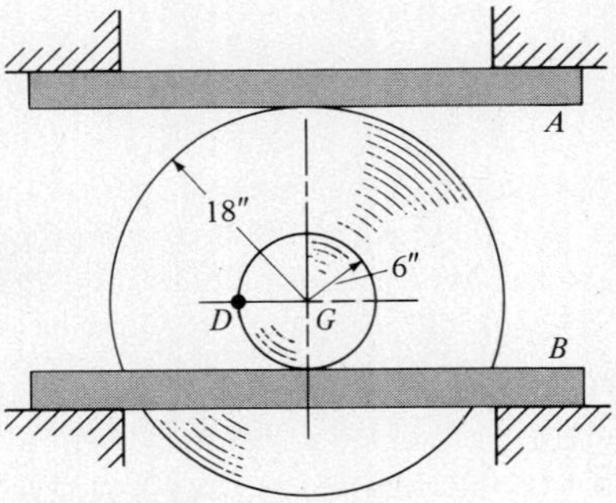

Figure P-12-4.8

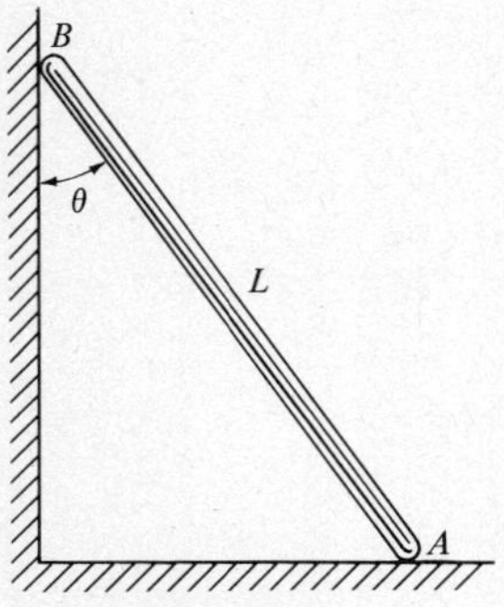

Figure P-12-4.9

Use the method of scalar calculus to find the velocity and acceleration of points A and B if, at the instant $\theta = 60°$, $\omega = 4$ rad per sec and $\alpha = 6$ rad per sec^2, both counterclockwise.

12-4.10. Referring to Fig. P-12-4.9 with $L = 10$ in., suppose that when $\tan\theta = \frac{3}{4}$, the velocity and acceleration of A is 24 ips rightward and 38 ips^2 leftward, respectively. Using scalar calculus, compute the acceleration of B.

$a_B = 84$ ips^2 down **_Ans._**

12-4.11. In Fig. P-12-4.9, assume that A is moving rightward with a constant velocity v_A. Use scalar calculus to find ω and α in terms of v_A and θ. What are these values expressed in terms of v_A and time t? Assume $t = 0$ when A is vertically below B.

12-4.12. The 6-ft rod in Fig. P-12-4.12 moves with its ends in contact with the floor and the inclined surface. At the given instant when $\omega = \sqrt{2}$ rad per sec and $\alpha = 3$ rad per sec^2 both clockwise, determine the velocity and acceleration of point B.

12-4.13. In Fig. P-12-4.12, assume that the velocity of A is 18 fps leftward and the acceleration of A is 20 fps^2 rightward. Compute the velocity and acceleration of point B.

$v_B = 18$ fps; $a_B = 42.4$ fps^2 **_Ans._**

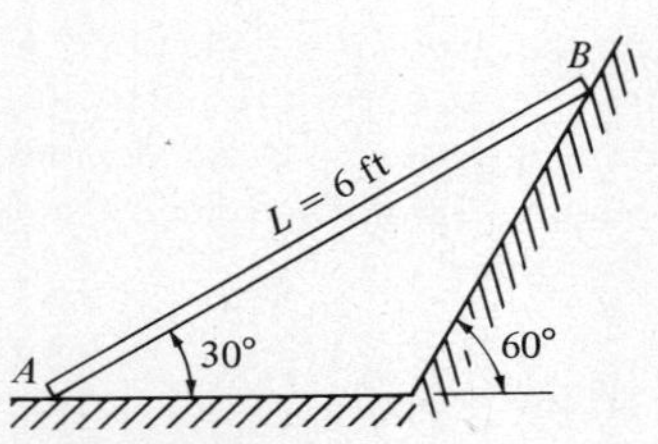

Figure P-12-4.12

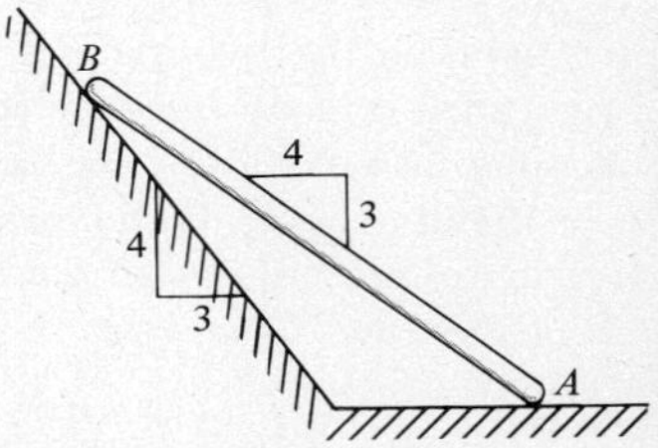

Figure P-12-4.14

12-4.14. A bar 3 ft long moves with its ends in contact with the inclined surface and horizontal floor as shown in Fig. P-12-4.14. At the given instant, $\omega = 2$ rad per sec and $\alpha = 3$ rad per sec^2, both counterclockwise. Compute the acceleration of the midpoint M of AB.

$a_A = 6.6$ fps^2; $a_B = 18.0$ fps^2; $a_M = 11.29$ fps^2 **_Ans._**

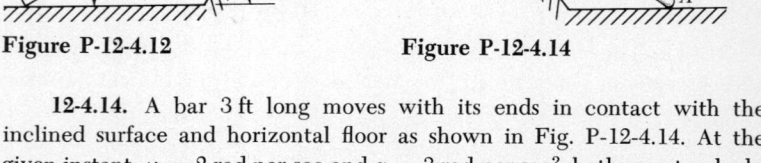

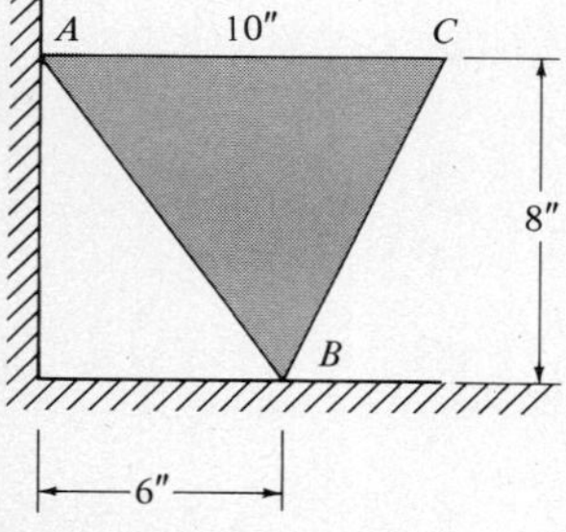

Figure P-12-4.16

12-4.15. In a rigid body having plane motion, let A and B be any two points in the plane of motion and let M be a point midway between A and B. Prove that the acceleration of M is one-half the vector sum of the accelerations of points A and B; i.e., $a_M = \frac{1}{2}(a_A \mathrel{+\!\!\!\!\!\rightarrow} a_B)$. Use this result to check the midpoint acceleration of bar AB in Prob. 12-4.14.

12-4.16. The triangular plate ABC in Fig. P-12-4.16 moves with its vertices A and B in contact with a vertical wall and a horizontal floor. If $\omega = \sqrt{5}$ rad per sec and $\alpha = 3$ rad per sec^2, both counterclockwise at the given position, compute the acceleration of points A, B, and C.

12-4.17. A plate *ABC* moves with its ends *A* and *B* on the horizontal and inclined guides shown in Fig. P-12-4.17. At the given position, $\omega = 4$ rad per sec and $\alpha = 5$ rad per sec², both clockwise. Find the accelerations of points *A*, *B*, and *C*.

$a_A = 2$ fps²; $a_B = 85$ fps; $a_C = 100$ fps² ***Ans.***

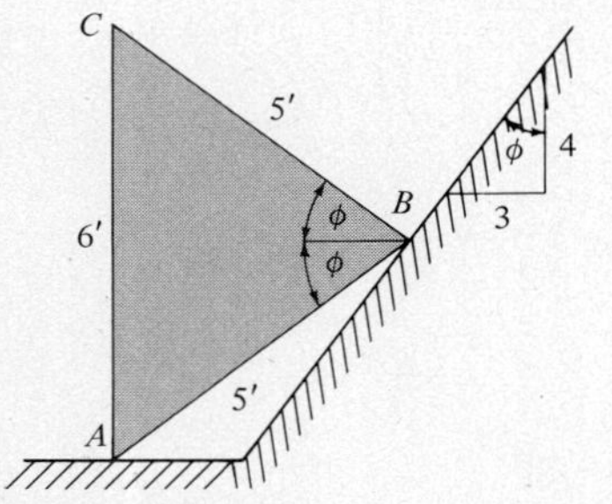

Figure P-12-4.17

12-4.18. The motion of the rigid link *ABC* in Fig. P-12-4.18 is governed by the control rods *AD* and *BE*. At the given instant when *AB* is horizontal, the angular velocity and acceleration of the link is $\omega = \sqrt{10}$ rad per sec and $\alpha = 5$ rad per sec², both clockwise. Determine the acceleration of point *C*.

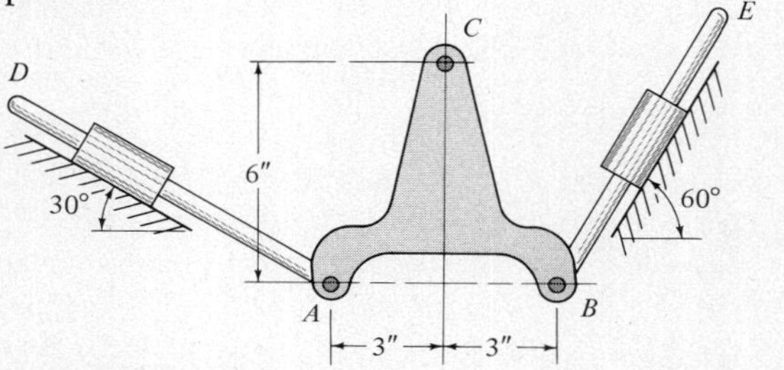

Figure P-12-4.18

12-4.19. The 6-ft diameter disk shown in Fig. P-12-4.19 rolls without slipping back and forth over a short distance. A 5-ft bar *BC* is pinned to the rim of the disk at *B* and its other end *C* drags on the ground. If the center of the disk has a constant rightward velocity $v_A = 12$ fps at the given position, compute v_C and a_C.

$v_C = 21$ fps; $a_C = 104.3$ fps² ***Ans.***

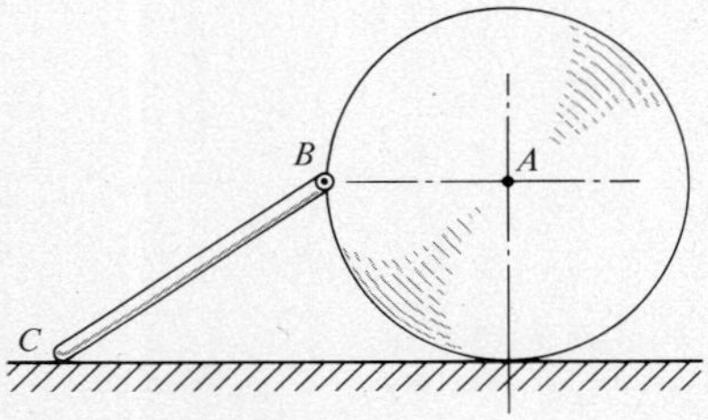

Figure P-12-4.19

12-4.20. In Fig. P-12-4.20, a vibratory motion of *B* in the vertical guide causes gear *A* to oscillate back and forth on the fixed rack. At the given position, $v_B = 26.4$ ips up and $a_B = 6.8$ ips² down. What is then the angular acceleration of link *AB*?

$\alpha_{AB} = 2$ rad per sec² ***Ans.***

12-4.21. In the offset slider-crank mechanism shown in Fig. P-12-4.21, the crank *OA* rotates at a constant clockwise speed of 10 rad per sec. Compute the acceleration of the midpoint of the connecting rod *AB* at the given position.

$a_M = 119.5$ fps² ***Ans.***

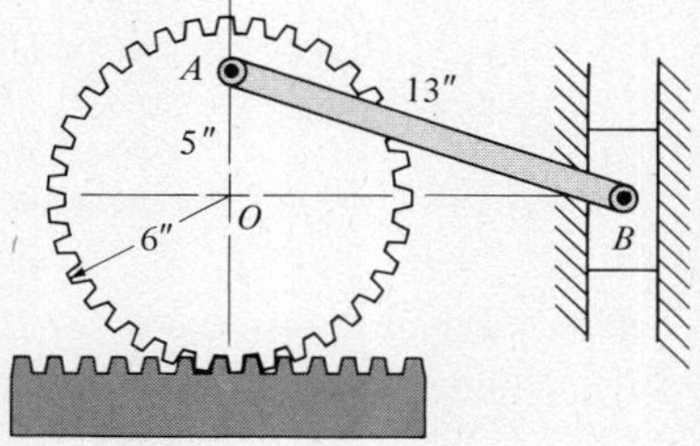

Figure P-12-4.20

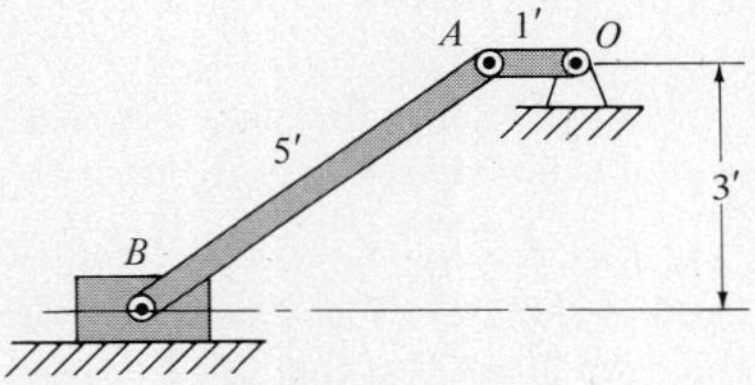

Figure P-12-4.21

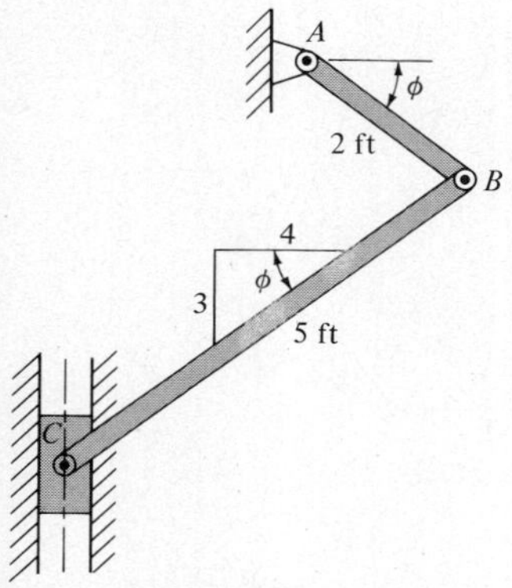

Figure P-12-4.22

12-4.22. When a slider-crank mechanism is in the position shown in Fig. P-12-4.22, the velocity and acceleration of slider C are $v_C = 16$ fps and $a_C = 2.8$ fps^2, both vertically downward. What is then the angular acceleration of crank AB?

$$\alpha_{AB} = 4 \text{ rad per sec}^2 \qquad \textit{Ans.}$$

12-4.23. At the instant shown in Fig. P-12-4.23, the driving crank AB of the four-link mechanism has a constant clockwise speed of 4 rad per sec. Find the angular acceleration of the link CD.

12-4.24. In Fig. P-12-4.23, find the angular acceleration of link CD if at the instant shown the driving crank AB has a clockwise angular velocity of 4 rad per sec and a counterclockwise angular acceleration of 2 rad per sec^2.

$$\alpha_{CD} = 37.5 \text{ rad per sec}^2 \qquad \textit{Ans.}$$

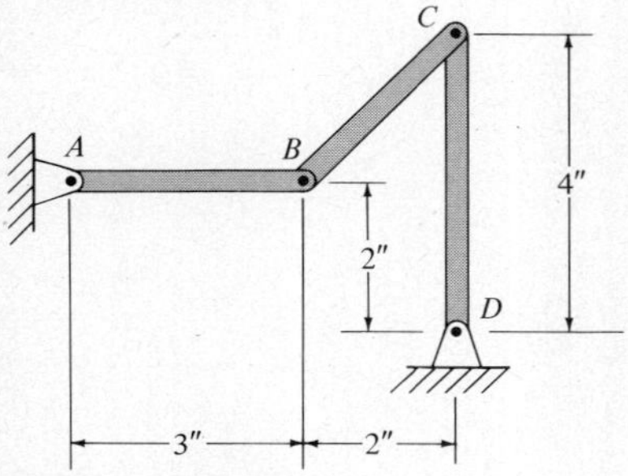

Figure P-12-4.23

12-4.25. A slider S is pinned at B to a collar C which is free to slide along AB as shown in Fig. P-12-4.25. If the slider S has a constant upward velocity v, use scalar calculus to find the angular velocity and angular acceleration of rod AB.

12-4.26. Use the coordinate system shown in Fig. P-12-4.26 to relate the velocity and acceleration of points A and B. Then apply these results to find the acceleration of B if $r = 5$ ft, $\tan\theta = \frac{3}{4}$, and A is moving down the incline with $v_A = 20$ fps and $a_A = 15$ fps^2.

$$a_B = -82 \text{ fps}^2 \qquad \textit{Ans.}$$

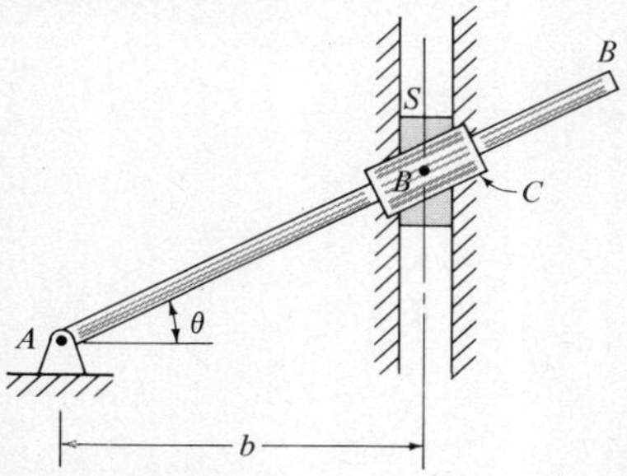

Figure P-12-4.25

12-4.27. Use the coordinate system shown in Fig. P-12-4.27 to relate the constant counterclockwise angular velocity ω of crank AO to the velocity and acceleration of B. Apply these results to the case when $\theta = 45°$, $\omega = 4$ rad per sec, $r = 3\sqrt{2}$ ft, and $L = 5$ ft.

$$v_B = -21 \text{ fps};\ a_B = -68.25 \text{ fps}^2 \qquad \textit{Ans.}$$

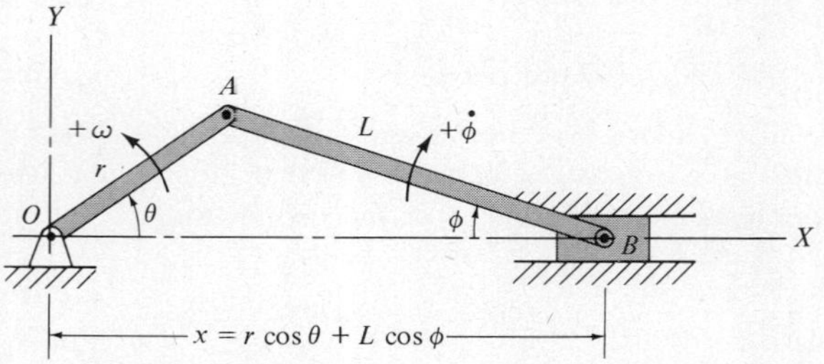

Figure P-12-4.27

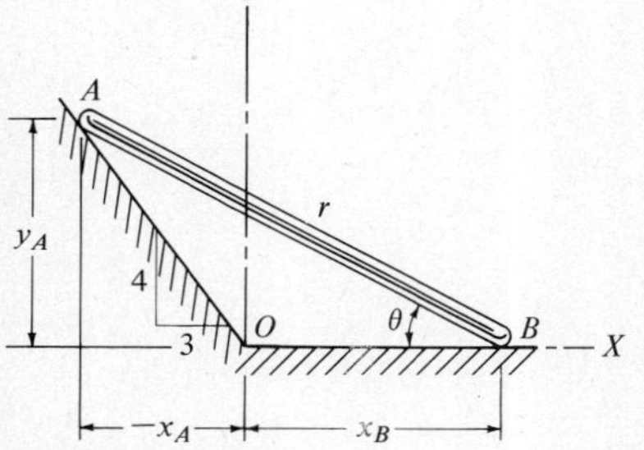

Figure P-12-4.26

12-4.28. Use the data specified in Prob. 12-4.27 to find the velocity and acceleration of B using the geometric approach.

12-4.29. An inversion of the slider-crank mechanism has the slider B fixed as shown in Fig. P-12-4.29. The rod OD slides back and forth through the fixed block B as the link AB oscillates about B and causes AO to describe plane motion. At the instant shown, the rod OD has a constant leftward

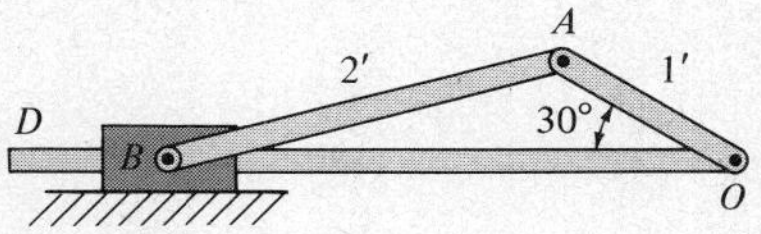

Figure P-12-4.29

velocity of 7.24 fps. Determine graphically using velocity and acceleration polygons the angular acceleration of link AO for this position.

$\alpha_{AO} = 160$ rad per sec^2 counterclockwise *Ans.*

12-5 INSTANT CENTER AND INSTANTANEOUS AXIS OF ROTATION

In the preceding sections, we considered the plane motion of a rigid body as equivalent to successive infinitesimal displacements from one instantaneous position to the next. Each such infinitesimal displacement consisted of the vector sum of the translation of an arbitrarily selected reference point plus a rotation about that reference point. If we could find a reference point that is momentarily at rest, however, each such successive displacement would consist only of a rotation about that reference point.

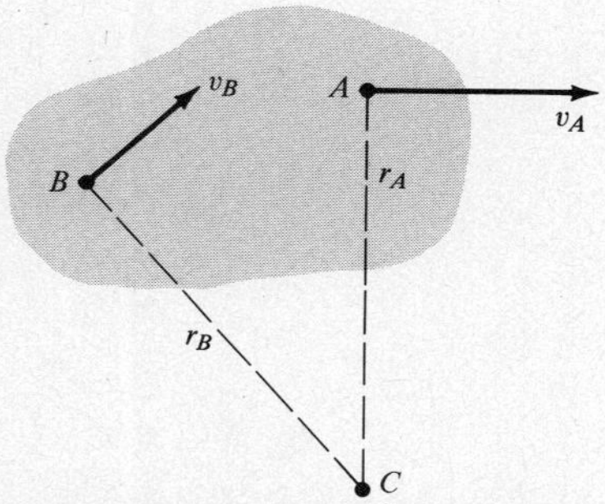

Figure 12-5.1 Graphic determination of instant center.

Consider now the body shown in Fig. 12-5.1 in which, at a particular instant, the velocities of points A and B have the given directions. If there is a center momentarily at rest about which the body rotates, it will be located at the intersection C of two lines drawn through A and B perpendicular to v_A and v_B respectively. The point C is on the instantaneous axis of rotation. This is the line, perpendicular to the plane of motion, which joins those points in the body which are instantaneously at rest and have zero velocity. The intersection of this instantaneous axis with the plane of motion is called the instantaneous center of rotation or, more briefly, the instant center.

The value of using the instant center to determine velocities is evident from a consideration of the kinematic equation for velocity. Thus, denoting the instant center by C and assuming the plane motion to be resolved into a rotation about C plus the translation of C, we can find the velocity of any point B from the equation $v_B = v_C \nrightarrow (v_{B/C} = r\omega)$. The velocity of C, however, must be zero by the definition of an instant center. Hence the kinematic equation for determining velocities reduces to $v_B = r\omega$, which is equivalent to the kinematic equation of rotation. During the instantaneous rotation, all points in the body have the same angular velocity about the instant center, and the direction of the instantaneous velocity of any point is perpendicular to the line joining the point with the instant center.

Returning to Fig. 12-5.1, if the velocity of one of the points, say v_A, is known, we can use the following equation to determine the angular velocity of the body and the linear velocity of any other point B in the body:

$$\omega = \frac{v_A}{r_A} = \frac{v_B}{r_B}$$

If the two points A and B have parallel but unequal velocities, as shown in Fig. 12-5.2a or (b), the instant center C lies on the common perpendicular to these velocities and is located by direct proportion.

The instant center can also be located if the linear velocity of one point and the angular velocity of the body are both known at any instant. Thus, if point A in Fig. 12-5.3 has a known rightward velocity v_A in a body with a known clockwise value of ω, the instant center C is located along the line passing through A perpendicular to v_A and lying a distance r below A as given by the relation

$$r = \frac{v_A}{\omega}$$

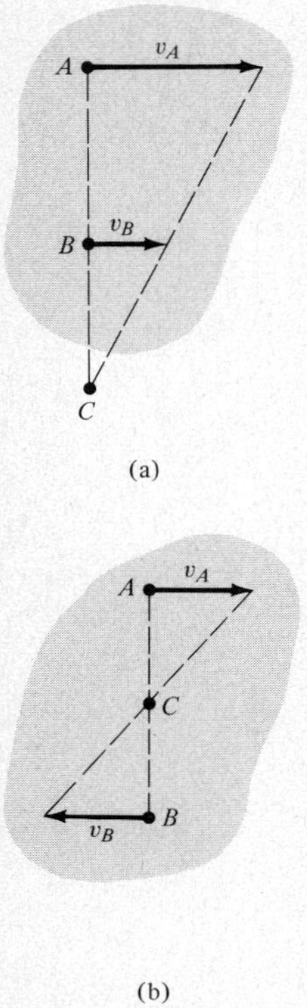

Figure 12-5.2 Determination of instant center when velocities are parallel, but unequal.

In the case of a wheel or sphere of radius r rolling freely along a plane, the velocity of the geometric center is expressed by $v_A = r\omega$, as shown in Section 12-4.1. Hence, we conclude that the instant center of a wheel is located at a distance r from the geometric center and, from the above discussion, that it is at the point of contact with the plane. There is visual evidence to confirm this, as for example the clear imprint left by a tire after it has run over a patch of wet pavement.

INSTANT CENTER OF ZERO ACCELERATION

Analogous to the instant center of zero velocity, there exists an instant center of zero acceleration.[3] This can be used to compute accelerations in plane motion as though the body were in pure rotation about the center of zero acceleration. The two instant centers of velocity and of acceleration, unfortunately, do not coincide except for the special case of a body starting from rest. A general method of locating the instant center of acceleration follows.

Consider the acceleration of a point B with respect to reference center C which instantaneously has a zero acceleration. The acceleration equation becomes

$$a_B = \overset{=0}{\cancel{a_C}} \mathrel{+\!\!\!\!\rightarrow} (a_{B/C} = r\omega^2 \mathrel{+\!\!\!\!\rightarrow} r\alpha)$$

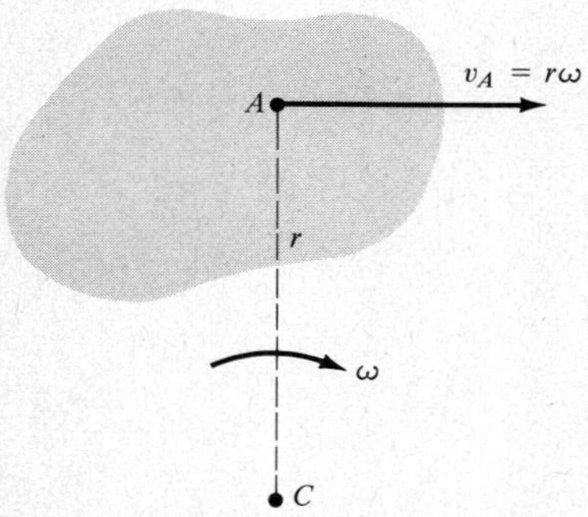

Figure 12-5.3 Analytical determination of instant center.

[3] Hereafter, the unmodified expression "instant center" will always mean the instant center of zero velocity. If there is any possibility of confusion, the explicit terms "instant center of velocity" or "instant center of acceleration" will be used.

so that with $a_C = 0$, the total acceleration a_B consists only of the relative acceleration $a_{B/C}$ of B rotating about C, which has the components $r\omega^2$ and $r\alpha$ acting as shown in Fig. 12-5.4. Clearly, C, the instant center of acceleration, lies on a line through B collinear with $r\omega^2$. The angle this line makes with a_B is defined by $\tan\phi = \alpha/\omega^2$. Obviously, both ω and α must be known before ϕ can be determined, but usually in most problems, prior knowledge of either or both ω and α is unknown. However, when this information is given, and the *direction* of the total acceleration of two points A and B in Fig. 12-5.5 is known, then the instant center of acceleration is located at the intersection of lines through A and B making the angle ϕ with a_A and a_B as is shown.

For the special case of a body starting from rest (i.e., $\omega = 0$), the angle $\phi = 90°$ and the instant center of acceleration will coincide with the instant center of velocity. Except in this special case, however, the instant center of acceleration is of limited value. It is usually easier to determine accelerations by using the fundamental methods described in Sections 12-4 and 12-7.

There is another important special case, that of a free-rolling disk, in which the tangential acceleration of certain points can be computed as though the disk were actually rotating about the instant center of zero velocity. These points lie on the line joining the geometric center of the disk with its instant center of zero velocity. Thus, consider the free-rolling disk in Fig. 12-5.6a whose geometric center has the acceleration $a_O = r\alpha$ as shown in Section 12-4 on page 437. If the plane motion is resolved into a rotation about O plus the translation of O, the acceleration of a point B is given by the equation

$$a_B = (a_O = r\alpha) \nrightarrow (a_{B/O} = b\omega^2 \nrightarrow b\alpha)$$

The graphic interpretation of this equation is shown in Fig. 12-5.6b. Obviously, the tangential acceleration of B (directed perpendicular to the line BOC) is given by

$$a_{B_t} = r\alpha + b\alpha = (r + b)\alpha = r_1\alpha$$

where r_1 is the distance from B to the instant center C.

If we disregard the normal component of acceleration, we can conclude from the preceding discussion that the motion of any point along the line joining the geometric center of a free-rolling disk with its instant center of velocity is the same as though the disk were actually rotating about an axis through this instant center. This concept gives us a convenient way to relate the plane motion of a free-rolling disk with the motion of a body connected to it. Several examples are given in the following illustrative problems.

Finally, let us show that the acceleration of the velocity instant

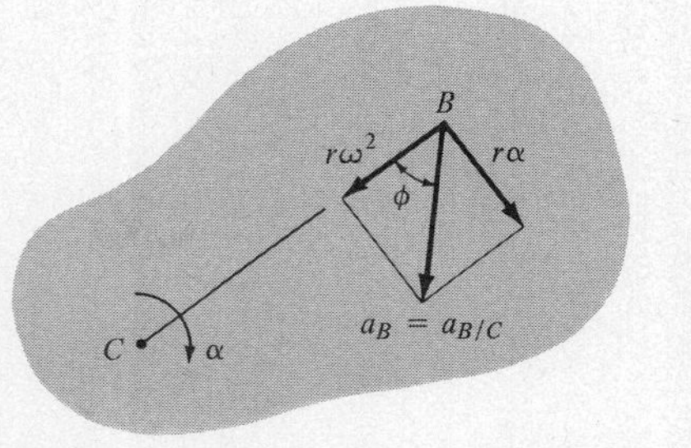

Figure 12-5.4 Acceleration of B with C as instant center of acceleration.

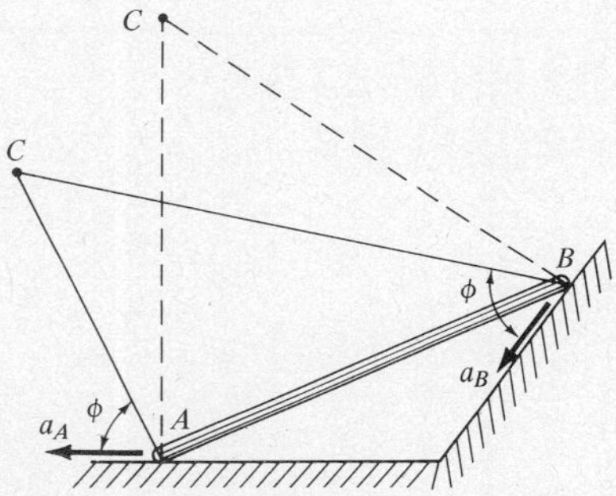

Figure 12-5.5 Location of instant center of acceleration C when $\phi = \tan^{-1}\alpha/\omega^2$ is known. When $\omega = 0$, instant center of acceleration is at C', and coincides instantaneously with the instant center of velocity.

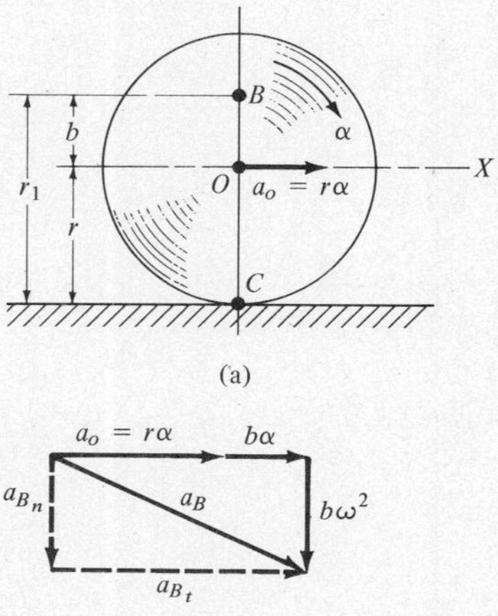

Figure 12-5.6 Acceleration of any point B on the line joining the geometric center O with the instant center C.

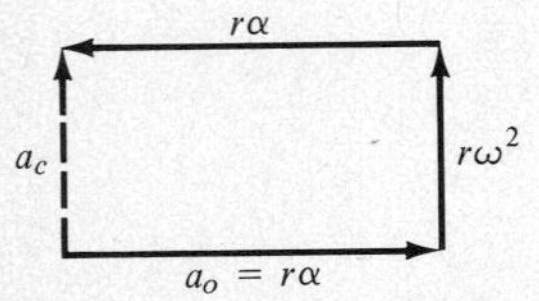

Figure 12-5.7 Acceleration of the instant center C.

center C of the disk is not zero. Figure 12-5.7 shows the graphic interpretation of the equation

$$a_C = (a_O = r\alpha) \nrightarrow (a_{C/O} = r\omega^2 \nrightarrow r\alpha)$$

from which it is evident that the acceleration of C is equal to $r\omega^2$ and is directed from C toward O. Observe that the spatial position of the instant center of velocity is never stationary but always moves with some acceleration. Were the instant center of velocity to have no acceleration, it would remain fixed in position and plane motion would reduce to rotation about a fixed axis.

ILLUSTRATIVE PROBLEMS

12-5.1. The compound wheel shown in Fig. 12-5.8a rolls without slipping between two parallel plates. The velocity of the upper plate A is $v_A = 12$ fps to the right and that of the lower plate B is $v_B = 6$ fps to the left. Find the angular velocity of the wheel and the linear velocity of point D on it using the instant center method.

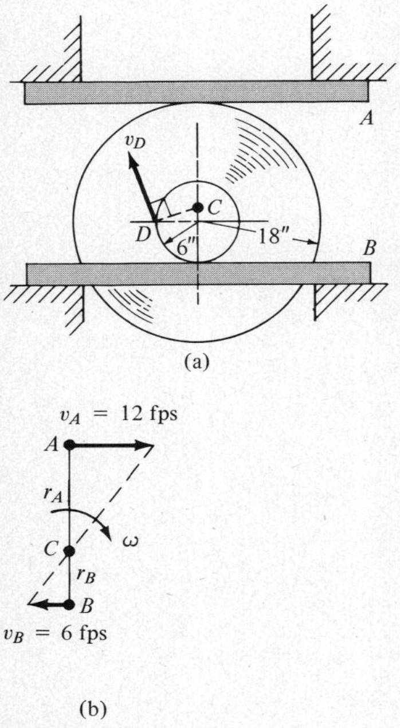

Figure 12-5.8

Solution

The points on the vertical centerline of the wheel which are in contact with the plates have the velocities shown in part (b) of the figure. The instant center C lies on the line AB which is the common perpendicular to v_A and v_B. Its location from points A and B is determined from

$$\left[\frac{v_A}{r_A} = \frac{v_B}{r_B}\right] \qquad \frac{12}{r_A} = \frac{6}{r_B} \quad \text{or} \quad r_A = 2r_B$$

which, substituted into $r_A + r_B = 24$ in. obtained from the dimensions of the wheel, gives $r_A = 16$ in. and $r_B = 8$ in. Therefore, the angular velocity is

$$\omega = \frac{v_B}{r_B} = \frac{6}{8/12} = 9 \text{ rad per sec clockwise} \qquad \textit{Ans.}$$

Returning to Fig. 12-5.8a, we find the velocity of D is perpendicular to the length CD drawn from the instant center and its magnitude is

$$[v = r\omega] \qquad v_D = \left(\sqrt{(2)^2 + (6)^2}\right)(9) = 57 \text{ ips} = 4.75 \text{ fps} \qquad \textit{Ans.}$$

It is useful to observe that the horizontal and vertical components of v_D are equal to the vertical and horizontal components of CD respectively multiplied by ω; i.e.,

$$v_{D_h} = 2(9) = 18 \text{ ips left} \quad \text{and} \quad v_{D_v} = 6(9) = 54 \text{ ips up}$$

12-5.2. The disks in Fig. 12-5.9 are free to roll on the inclined planes. Disk A is fastened to pulley P by cord Q, which is wrapped around the drum of the disk and around the smaller pulley. Disk B is similarly attached to the outer wheel of the pulley. The linear acceleration of the center of disk A is constant at 10 fps^2 down the incline. Determine the linear displacement, velocity, and acceleration of the center of disk B after 4 sec, starting from rest.

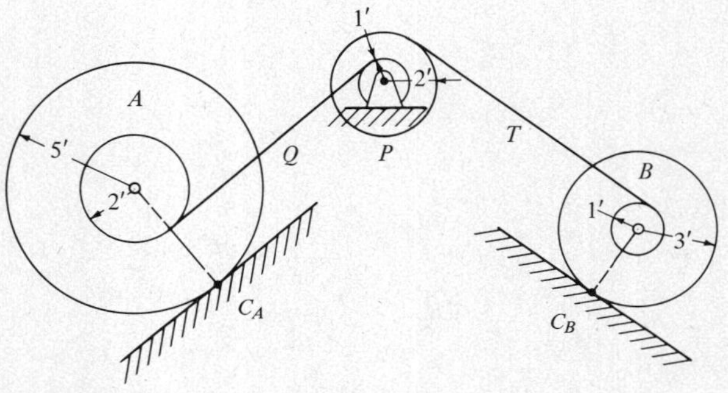

Figure 12-5.9

Solution

The motion of pulley P is related to the motions of disks A and B by cords Q and T. All the bodies can be considered as in a state of rotation—P about its axle, A about its instant center C_A, and B about its instant center C_B. Applying the relation between linear and angular displacement in rotation, we may express the linear displacements of points on cords Q and T in terms of the angular displacements of the bodies by

$[s = r\theta]$

$$s_Q = 3\theta_A = 1 \times \theta_P \qquad (a)$$
$$s_T = 2\theta_P = 4\theta_B \qquad (b)$$

From Eqs. (a) and (b) we obtain

$$3\theta_A = \theta_P = 2\theta_B \qquad (c)$$

Velocity and acceleration relations between the disks and the pulley have similar forms because $v = r\omega$ and $a_t = r\alpha$ have the same mathematical form as $s = r\theta$. Therefore we may also write

$$3\omega_A = \omega_P = 2\omega_B \qquad (d)$$
$$3\alpha_A = \alpha_P = 2\alpha_B \qquad (e)$$

Not all of the above discussion is really necessary to solve this problem; however, it will be found useful in applying the work-energy method in Section 14-8.

From the given data the angular acceleration of disk A can be expressed in terms of its linear acceleration by

$[\bar{a}_t = \bar{r}\alpha] \qquad 10 = 5\alpha_A \qquad \alpha_A = 2$ rad per sec^2

Applying the relation (*e*), we now compute the angular acceleration of disk *B*.

$[3\alpha_A = 2\alpha_B] \qquad 3 \times 2 = 2 \times \alpha_B \qquad \alpha_B = 3$ rad per sec^2

whence the acceleration of the geometric center of disk *B* is found from

$[\bar{a}_t = \bar{r}\alpha] \qquad \bar{a}_B = 3 \times 3 \qquad \bar{a}_B = 9$ fps^2

Since the motion of the center of disk *B* is rectilinear with constant acceleration $\bar{a}_B = 9$ fps^2, we may apply Eqs. (9-3.4) and (9-3.5) of Section 9-3 to obtain

$[v = v_o + at] \qquad \bar{v}_B = 0 + 9 \times 4 \qquad \bar{v}_B = 36$ fps ***Ans.***

$[s = v_o t + \frac{1}{2}at^2] \qquad \bar{s}_B = 0 + \frac{1}{2} \times 9 \times (4)^2 \qquad \bar{s}_B = 72$ ft ***Ans.***

In this problem as disk *A* rolls down the incline, its center travels the distance $\bar{s}_A = 5\theta_A$ while the cord *Q* travels $s_Q = 3\theta_A$. Hence for each radian of angular displacement of *A*, 2 ft of cord are unwound from the drum of *A*. Actually the unwinding of $2\theta_A$ ft of cord from *A* is the result of the rotation of the drum about the geometric center of *A*. Similarly we may show that for each radian of angular displacement of *B*, 1 ft of cord is unwound from *B* as it rolls up the incline. If the motion of the system were reversed with *A* rolling up and *B* rolling down their respective inclines, cords *Q* and *T* would wind up these amounts on *A* and *B* respectively. In either case, the geometric center of *A* would still move faster than cord *Q* while cord *T* would still move faster than the geometric center of *B*.

12-5.3. At a certain instant a four-link mechanism occupies the position shown in Fig. 12-5.10. The dimensions are as shown. The angular velocity of *AB* is $\omega_{AB} = 11$ rad per sec. Locate the instant center of *BC* and use it to determine the linear velocity of *C* and the angular velocity of *CD*.

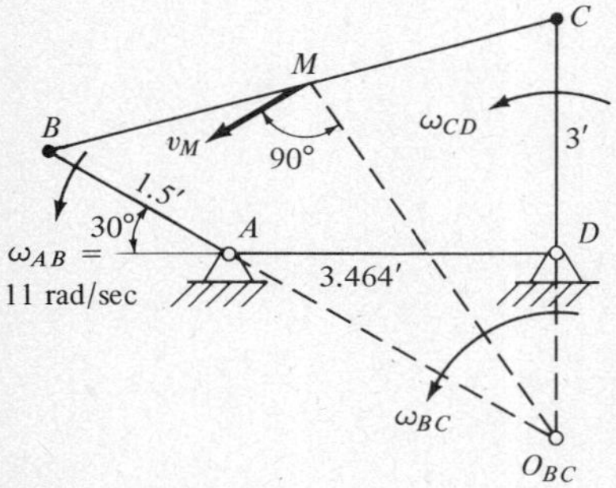

Figure 12-5.10

Solution

Points *B* and *C* are constrained to rotate about the fixed bearings *A* and *D* respectively. Since the velocity of *B* is therefore perpendicular to *AB* and the velocity of *C* is perpendicular to *CD*, the intersection of *AB* and *CD* prolonged determines the instant center O_{BC}. From the diagram, we have

$$DO = AD \tan 30° = 3.464 \times 0.577 = 2 \text{ ft}$$

$$AO = \frac{AD}{\cos 30°} = \frac{3.464}{0.866} = 4 \text{ ft}$$

These distances could have been obtained graphically by drawing the mechanism to scale. The graphic method is generally used in this type of problem.

Applying the principle that the velocity of any point is equivalent to rotation about the instant center, we have

$[v_B = r_{AB}\omega_{AB} = r_{BO}\omega_{BC}]$ $1.5 \times 11 = 5.5\omega_{BC}$ $\omega_{BC} = 3$ rad per sec

$[v_C = r_{CO}\omega_{BC}]$ $v_C = (3 + 2) \times 3 = 15$ fps ***Ans.***

Also

$[v_C = r_{CD}\omega_{CD}]$ $15 = 3 \times \omega_{CD}$ $\omega_{CD} = 5$ rad per sec ***Ans.***

The velocity of any other point M is found by scaling the distance MO from a scale diagram of the mechanism and applying $v_M = r_{MO}\omega_{BC}$. The direction of v_M is perpendicular to MO as shown.

PROBLEMS

12-5.4. Using the instant center method, resolve Prob. 12-4.11 for ω and then differentiate ω with respect to time to obtain α in terms of ω.

12-5.5. Rollers constrain the ends of rod AB to move along the guides shown in Fig. P-12-5.5. If $v_B = 10$ fps upward at the given position, find the velocity of A. Also sketch the position of AB at the instant end B has zero velocity.

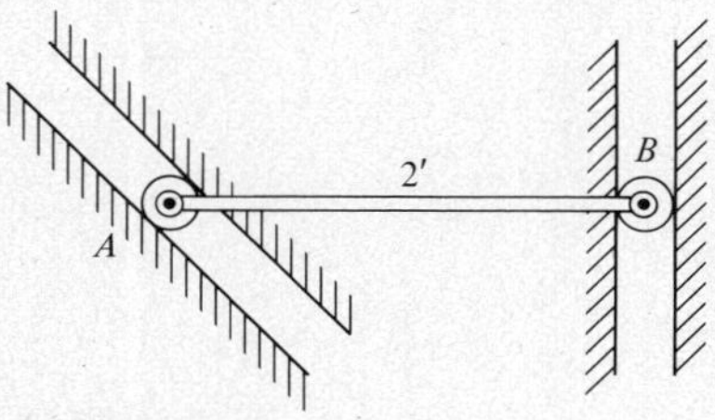

Figure P-12-5.5

12-5.6. The compound disk in Fig. P-12-5.6 rolls and slips on the horizontal floor. The linear velocity of the cord leaving A is 12 fps rightward and the disk rotates counterclockwise at 6 rad per sec. Use the instant center method to find the velocity of points O and B.

$$v_O = 6 \text{ fps right; } v_B = 12 \text{ fps left} \qquad \textbf{\textit{Ans.}}$$

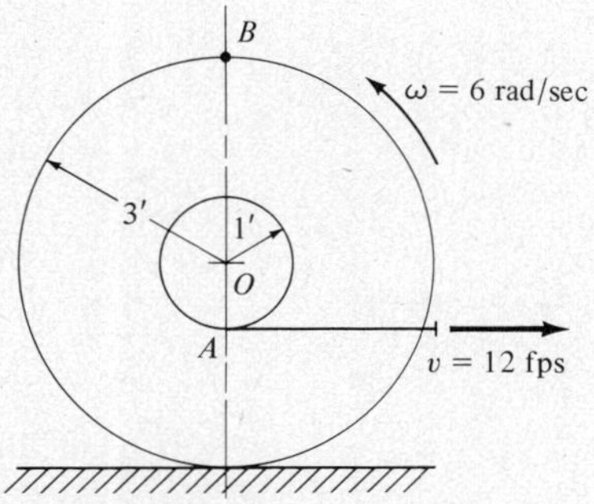

Figure P-12-5.6

12-5.7. If the center of the free-rolling disk in Fig. P-12-5.7 travels 9 ft to the right, how far does body A move?

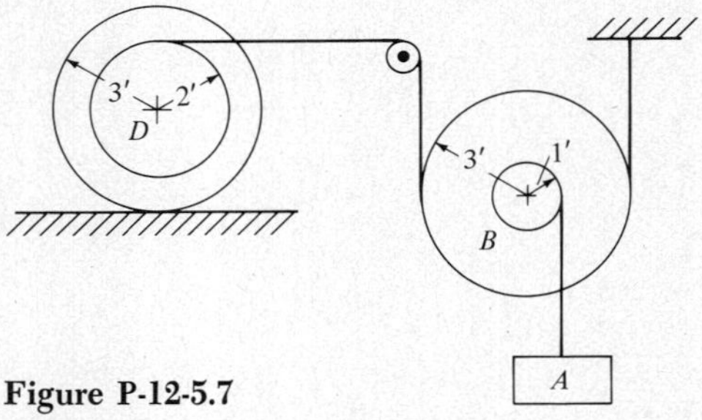

Figure P-12-5.7

12-5.8. Make the following changes in Illus. Prob. 12-5.2: The radii of disk A are changed to 3 ft and 5 ft; of disk B to 2 ft and 4 ft; of pulley

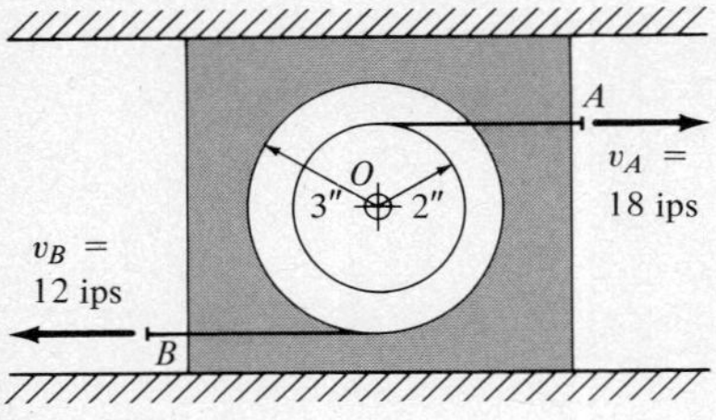

Figure P-12-5.9

P to 9 in. and 18 in. If the angular acceleration of disk A is 3 rad per sec^2 clockwise, determine the time for the center of disk B to travel 60 ft from rest.

$$t = 3.87 \text{ sec} \qquad \textbf{\textit{Ans.}}$$

12-5.9. The compound disk in Fig. P-12-5.9 is pinned to the sliding block at O. Cords attached to the disk have the constant velocities shown. Determine the velocity of the sliding block and the number of inches of cords A and B wound on or unwound from the disk each second.

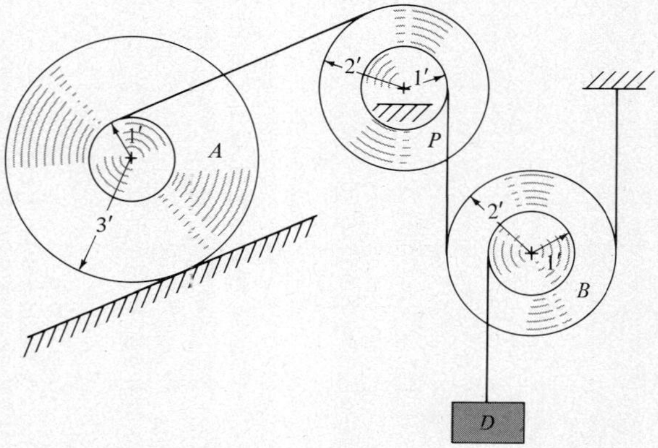

Figure P-12-5.10

12-5.10. In Fig. P-12-5.10, the initial downward velocity of body D is 12 ips and its acceleration is constant at 2 ips^2. Determine the linear displacement of the center of the free-rolling disk A after 2 sec.

56 in. ***Ans.***

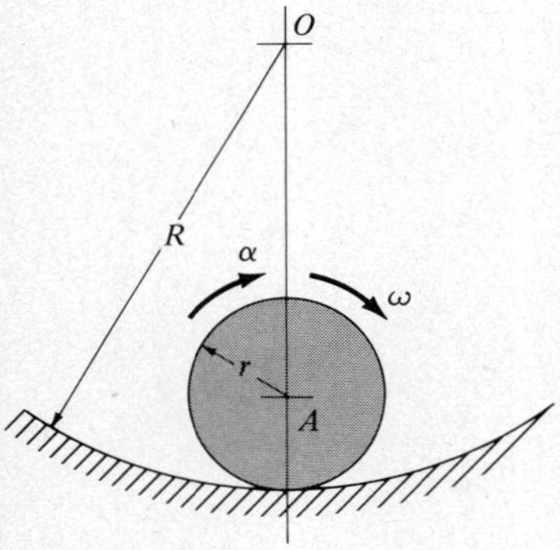

Figure P-12-5.11

12-5.11. The wheel in Fig. P-12-5.11 rolls freely on the circular arc. Show that $v_A = r\omega$ and $a_{A_t} = r\alpha$.

12-5.12. Find the acceleration of the instant center of the wheel rolling freely on the circular arc in Fig. P-12-5.11.

$$a = \left(\frac{Rr}{R - r}\right)\omega^2 = \frac{Rv_A{}^2}{(R - r)r} \qquad \textbf{\textit{Ans.}}$$

12-5.13. At the position shown in Fig. P-12-5.13, crank AB has a

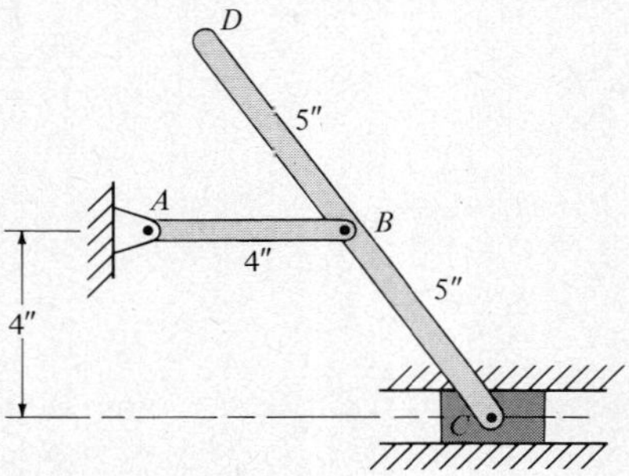

Figure P-12-5.13

clockwise angular velocity of 3 rad per sec. Use the instant center method to find the linear velocity of point D.

12-5.14. At the given position of the mechanism shown in Fig. P-12-5.14, the velocity of the center of the free-rolling wheel is $v_A = 30$ ips rightward. Determine the corresponding velocity of slider D.

$v_D = 16$ ips **Ans.**

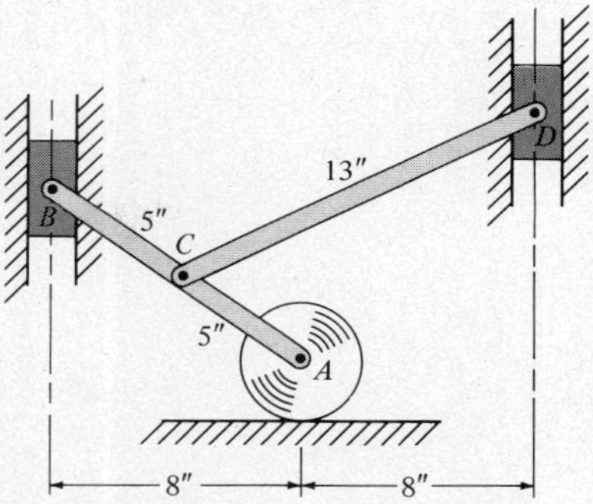

Figure P-12-5.14

12-5.15. The rigid-slab in Fig. P-12-5.15 moves in the XY plane with $v_{A_y} = 6$ ips, $v_{B_x} = 18$ ips, and $v_{C_y} = -12$ ips. Determine the coordinates of the instant center.

$x = 2$ in.; $y = -4$ in. **Ans.**

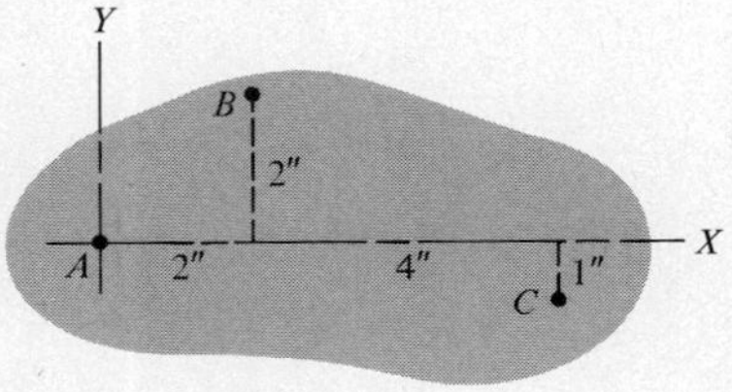

Figure P-12-5.15

12-5.16. Solve the preceding problem if $v_{A_x} = 6$ ips, $v_{B_y} = 2$ ips, and $v_{C_x} = 4$ ips.

12-5.17. When the slider-crank mechanism is in the position shown in Fig. P-12-5.17, the slider C has a downward velocity of 16 fps. Determine the velocity of the midpoint M of BC and the angular velocity of crank AB.

$v_M = 12.37$ fps; $\omega_{AB} = 5$ rad per sec **Ans.**

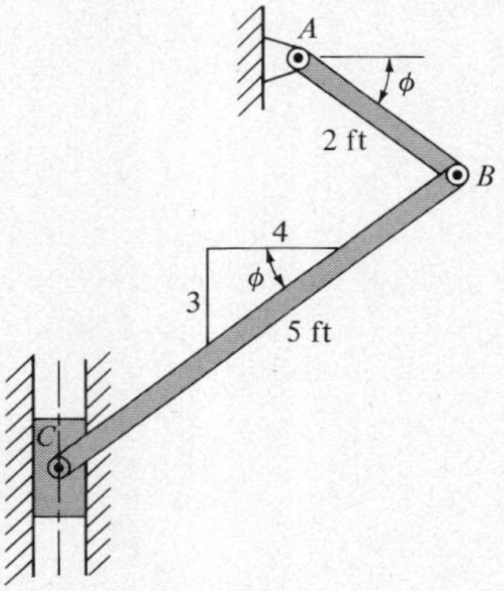

Figure P-12-5.17

12-5.18. At the instant shown in Fig. P-12-5.18, point B has an upward velocity of 22 ips. Use the instant center method to determine the velocity of point C and the angular velocity of crank AO.

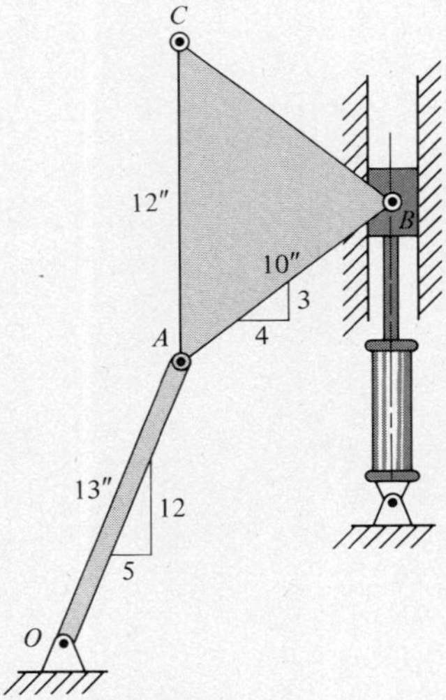

Figure P-12-5.18

12-5.19. In the four-link mechanism shown in Fig. P-12-5.19, the crank

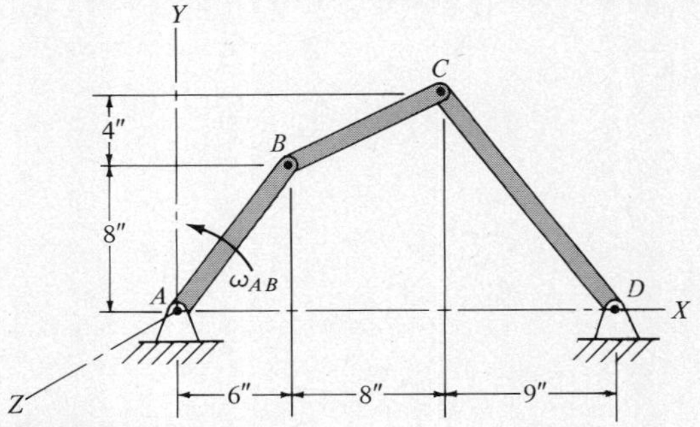

Figure P-12-5.19

AB has a constant angular speed of $\omega_{AB} = 3$ rad per sec. Use the instant center method to determine ω_{BC} and ω_{CD}. Compare your results with those in Illus. Prob. 12-7.2 on p. 464.

12-5.20. Rods AB and CD are pinned together at B as shown in Fig. P-12-5.20 and move in a vertical plane with the absolute angular velocities $\omega_{AB} = 4$ rad per sec and $\omega_{CD} = 3$ rad per sec. Determine the linear velocities of points C and D.

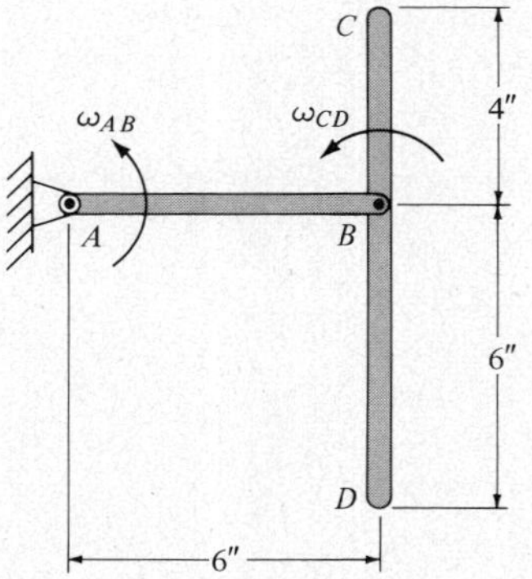

Figure P-12-5.20

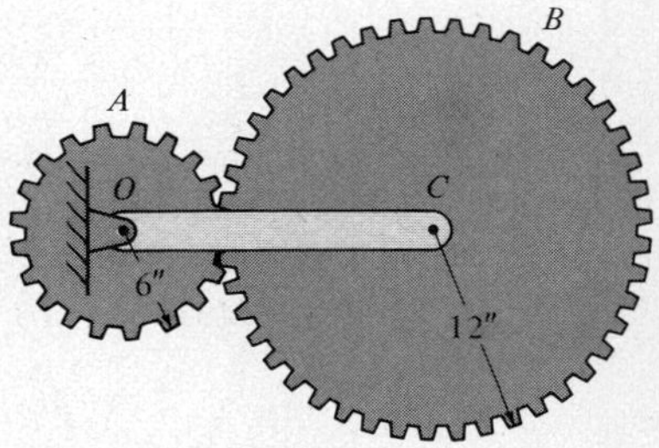

Figure P-12-5.21

12-5.21. Figure P-12-5.21 shows the pitch circles of two gears A and B connected by the arm OC. Gear A turns counterclockwise at 12 rad per sec while arm OC turns clockwise at 6 rad per sec, thereby causing gear B to roll around gear A. Locate the instant center of gear B and compute its angular velocity.

7.2 in. to the left of C; $\omega_B = 15$ rad per sec ***Ans.***

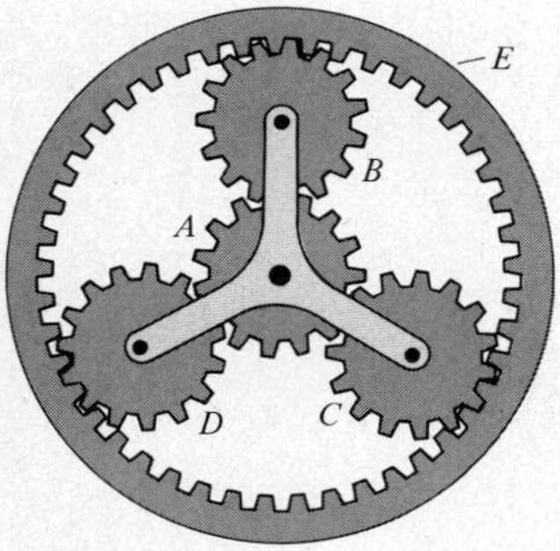

Figure P-12-5.23

12-5.22. Repeat Prob. 12-5.21 if gear A turns counterclockwise at 9 rad per sec while arm OC also turns counterclockwise at 1 rad per sec.

12-5.23. In the planetary gear system shown in Fig. P-12-5.23, the pitch circle radius of gears A, B, C, and D is r and that of gear E is $3r$. If the outer gear E is held stationary while gear A rotates clockwise at ω_A rad per sec, find the angular velocity of the spider connecting the planetary gears.

$\omega_{AB} = \frac{1}{4}\omega_A$ clockwise ***Ans.***

12-5.24. For the planetary gear system described in the preceding problem, assume that gear A is held fixed while the spider rotates clockwise at ω_{AB} rad per sec. What will then be the angular velocity of the ring gear E?

*12-6 THE OMEGA THEOREM

We consider now an application of the rotation of a rigid body to a concept known as the omega theorem. This theorem shows how to find the time derivative of any vector of constant magnitude, but whose direction changes as a result of rotating with the body in which it is fixed.

In Fig. 12-6.1 we show a body rotating about a fixed point O with an absolute angular velocity $\boldsymbol{\omega}$ coinciding with the axis of rotation. If the axis of rotation were fixed in space, any point P, located by the position vector $\mathbf{r}$ from O, would describe a circular path perpendicular to the axis of rotation. However, with only one point of a body fixed in space as in a spinning top, the axis of rotation will move or precess, and the path of P will be circular only for a differential time dt. During this time dt, the radius of this circular path about the instantaneous axis of rotation is $r \sin \phi$ and it sweeps through the angle $\omega\, dt$, thereby causing the displacement of P to be

$$d\mathbf{r} = (\omega\, dt)(r \sin \phi)\hat{\mathbf{e}}_t$$

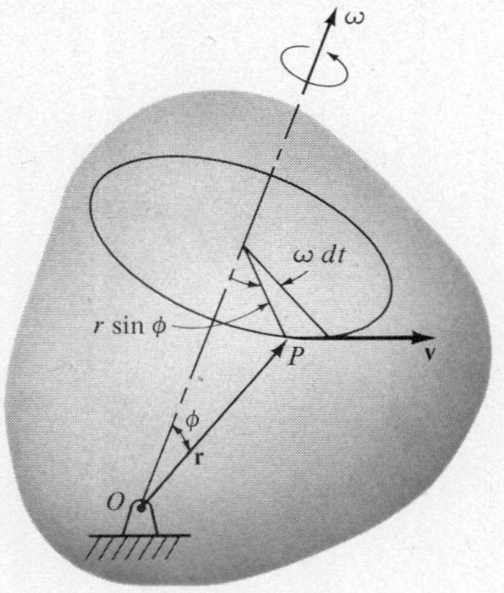

Figure 12-6.1

The velocity of P is therefore

$$\mathbf{v} = \frac{d\mathbf{r}}{dt} = (\omega r \sin \phi)\hat{\mathbf{e}}_t$$

It is evident that $\mathbf{v}$ is perpendicular to the plane formed by the vectors $\boldsymbol{\omega}$ and $\mathbf{r}$ so that the right side of this result represents the definition of the cross product $\boldsymbol{\omega} \times \mathbf{r}$. Hence, the velocity of any point in a rotating body is expressed by

$$\mathbf{v} = \dot{\mathbf{r}} = \boldsymbol{\omega} \times \mathbf{r} \tag{12-6.1}$$

where $\mathbf{r}$ may be the position vector to P from *any* origin on the axis of rotation regardless of whether the axis of rotation is fixed or moving. If the axis has a constant spatial orientation, $\boldsymbol{\omega}$ can change only in magnitude so that then $\boldsymbol{\alpha} = \dot{\boldsymbol{\omega}}$ will be collinear with $\boldsymbol{\omega}$. With a moving axis, however, $\boldsymbol{\omega}$ will always change direction in addition to possibly changing its magnitude which will cause $\boldsymbol{\alpha}$ to have a different direction from $\boldsymbol{\omega}$. This is analogous to the differing directions of $\mathbf{v}$ and $\mathbf{a}$ in curvilinear motion.

But more significantly, Eq. (12-6.1) represents the rate at which a constant length vector fixed in a rotating body changes its direction. Thus, consider a vector $\boldsymbol{\rho}$, fixed in a rotating body, joining any two particles A and B as in Fig. 12-6.2. The time derivative of $\boldsymbol{\rho}$ can be found by differentiating the relation $\boldsymbol{\rho} = \mathbf{r}_B - \mathbf{r}_A$ to obtain

$$\dot{\boldsymbol{\rho}} = \dot{\mathbf{r}}_B - \dot{\mathbf{r}}_A \tag{a}$$

But from Eq. (12-6.1), we have $\dot{\mathbf{r}}_B = \boldsymbol{\omega} \times \mathbf{r}_B$ and $\dot{\mathbf{r}}_A = \boldsymbol{\omega} \times \mathbf{r}_A$ which we substitute into Eq. (*a*) to obtain

$$\dot{\boldsymbol{\rho}} = \boldsymbol{\omega} \times \mathbf{r}_B - \boldsymbol{\omega} \times \mathbf{r}_A = \boldsymbol{\omega} \times (\mathbf{r}_B - \mathbf{r}_A) = \boldsymbol{\omega} \times \boldsymbol{\rho} \tag{b}$$

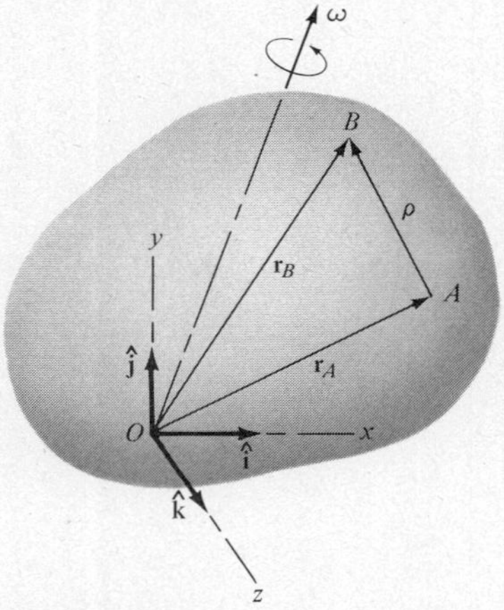

Figure 12-6.2

Thus we reach the important conclusion known as the omega theo-

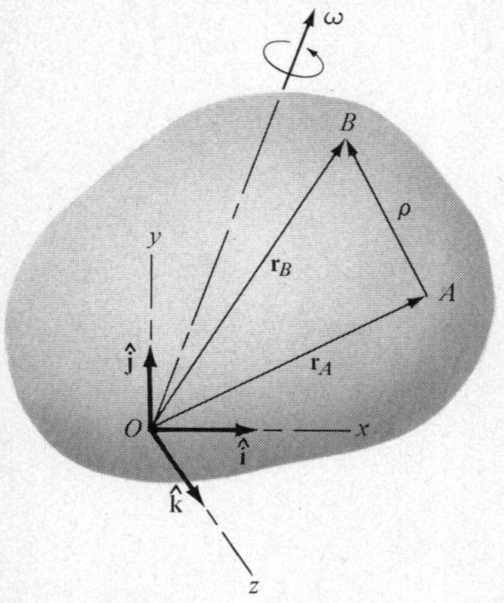

Figure 12-6.2

rem that the time derivative of a constant length vector fixed in a rotating body is simply the cross product of the angular velocity with the vector. We shall see later that the omega theorem can be extended to include the *directional* rate of change of any type of vector quantity.

Since the angular velocity $\boldsymbol{\omega}$ is the time rate of change of the angular displacement of *any* line in the body as was proved on p. 426, we may consider $\boldsymbol{\omega}$ to be a free vector. Consequently, we may define $\dot{\boldsymbol{\rho}}$ as the velocity of B relative to A, or

$$\mathbf{v}_{B/A} = \dot{\boldsymbol{\rho}} = \boldsymbol{\omega} \times \boldsymbol{\rho} \tag{12-6.2}$$

We can now apply the omega theorem to determine the time derivatives of the unit vectors in a coordinate system rotating with the body. In Fig. 12-6.2, we also show a reference frame xyz attached to the body and rotating with it at the angular velocity $\boldsymbol{\omega}$. Here the unit vectors $\hat{\mathbf{i}}$, $\hat{\mathbf{j}}$, and $\hat{\mathbf{k}}$, although of constant length, change their directions as a result of rotating with the body. Using the omega theorem, their time derivatives are

$$\begin{aligned} \dot{\hat{\mathbf{i}}} &= \boldsymbol{\omega} \times \hat{\mathbf{i}} \\ \dot{\hat{\mathbf{j}}} &= \boldsymbol{\omega} \times \hat{\mathbf{j}} \\ \dot{\hat{\mathbf{k}}} &= \boldsymbol{\omega} \times \hat{\mathbf{k}} \end{aligned} \tag{12-6.3}$$

and, in general, for any unit vector $\hat{\mathbf{e}}$ changing its direction at a rate $\boldsymbol{\omega}$,

$$\dot{\hat{\mathbf{e}}} = \boldsymbol{\omega} \times \hat{\mathbf{e}} \tag{12-6.4}$$

As an application of Eq. (12-6.4), consider the rates of change of the unit vectors $\hat{\mathbf{e}}_r$, $\hat{\mathbf{e}}_\theta$, in polar or cylindrical coordinates which we derived previously by other methods in Section 9-8. Using $\boldsymbol{\omega} = \theta\hat{\mathbf{k}}$ and the right-hand system of unit vectors $\hat{\mathbf{e}}_r$, $\hat{\mathbf{e}}_\theta$, $\hat{\mathbf{k}}$, we obtain at once

$$\dot{\hat{\mathbf{e}}}_r = \boldsymbol{\omega} \times \hat{\mathbf{e}}_r = \dot{\theta}\hat{\mathbf{k}} \times \hat{\mathbf{e}}_r = \dot{\theta}\hat{\mathbf{e}}_\theta$$

and

$$\dot{\hat{\mathbf{e}}}_\theta = \boldsymbol{\omega} \times \hat{\mathbf{e}}_\theta = \dot{\theta}\hat{\mathbf{k}} \times \hat{\mathbf{e}}_\theta = -\dot{\theta}\hat{\mathbf{e}}_r$$

Finally, let us re-examine the case of fixed-axis rotation using the omega theorem. The velocity of any point at a distance $\mathbf{r}$ from the axis of rotation is

$$\mathbf{v} = \dot{\mathbf{r}} = \boldsymbol{\omega} \times \mathbf{r} \tag{12-6.5}$$

and its acceleration is

$$\begin{aligned} \mathbf{a} = \dot{\mathbf{v}} &= \dot{\boldsymbol{\omega}} \times \mathbf{r} + \boldsymbol{\omega} \times (\dot{\mathbf{r}} = \mathbf{v}) \\ &= \boldsymbol{\alpha} \times \mathbf{r} + \boldsymbol{\omega} \times (\boldsymbol{\omega} \times \mathbf{r}) \end{aligned} \tag{12-6.6}$$

These vector equations are readily correlated with our previous discussion of rotation by referring to Fig. 12-6.3. There, the rotating body is represented by its projection upon a plane perpendicular to the axis of rotation. Noting that $\boldsymbol{\omega}$ and $\boldsymbol{\alpha}$ are perpendicular to $\mathbf{r}$, the definition of a cross product gives the magnitude of $\mathbf{v}$ as $v = |\boldsymbol{\omega} \times \mathbf{r}| = r\omega$, and the magnitude of the components of $\mathbf{a}$ as $a_t = |\boldsymbol{\alpha} \times \mathbf{r}| = r\alpha$ and $a_n = |\boldsymbol{\omega} \times \mathbf{v}| = r\omega^2$. Further, in agreement with Section 12-2, $\mathbf{v}$ is perpendicular to $\mathbf{r}$ in the sense that ω rotates r; while a_t is perpendicular to r in the sense of α, and a_n is directed inward toward the center of rotation. For plane motion problems, therefore, a convenient way of denoting the vector component of normal acceleration is the reduction

$$\mathbf{a}_n = \boldsymbol{\omega} \times (\boldsymbol{\omega} \times \mathbf{r}) = -\omega^2\mathbf{r} \tag{12-6.7}$$

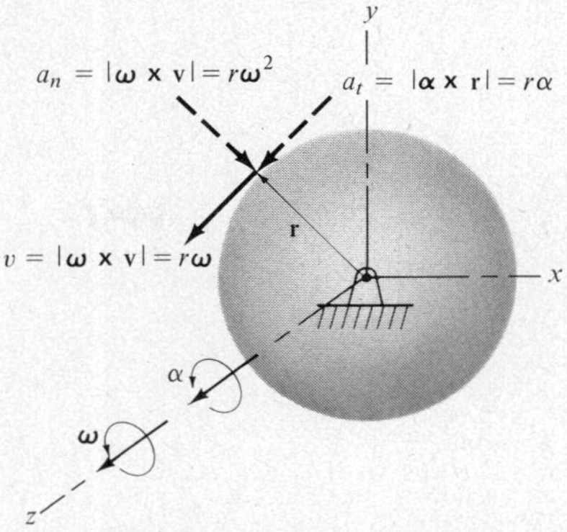

Figure 12-6.3 Interpretation of velocity and acceleration components.

which we may use whenever $\boldsymbol{\omega}$ and $\mathbf{r}$ are mutually perpendicular. (See Illus. Prob. 12-7.1.) Be warned, however, that this reduction does not apply to spatial motion where $\boldsymbol{\omega}$ and $\mathbf{r}$ generally are *not* perpendicular.

*12-7 PLANE MOTION ANALYSIS BY VECTOR ANALYSIS

We now present a third approach to plane motion analysis using vector calculus which is more direct and concise than the geometric approach, although not as physically meaningful. Its application requires strict adherence to the sign conventions of vector analysis instead of the visual approach of the previous sections. It is given here as an introduction to more complex spatial motions in Sections 12-8 and 12-9. The vector approach which follows is equally applicable to planar or spatial motion.

Consider the rigid body shown in Fig. 12-7.1. The body may be moving in any manner with respect to a fixed reference frame XYZ. Its instantaneous angular velocity and acceleration are $\boldsymbol{\omega}$ and $\boldsymbol{\alpha} = \dot{\boldsymbol{\omega}}$, both of which are free vectors. With respect to the fixed origin O, the position of any point B can be related to the position of any other point A by the relation

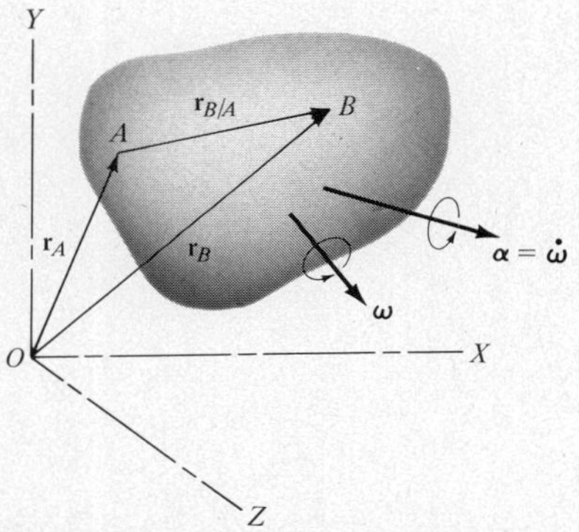

Figure 12-7.1 General spatial motion of a rigid body.

$$\mathbf{r}_B = \mathbf{r}_A + \mathbf{r}_{B/A} \tag{a}$$

where $\mathbf{r}_{B/A}$ is the position vector of B relative to A.

By differentiating this position equation with respect to time, we obtain (in dot notation) the following relation between the velocities of B and A:

$$\dot{\mathbf{r}}_B = \dot{\mathbf{r}}_A + \dot{\mathbf{r}}_{B/A} \tag{b}$$

Since $\mathbf{r}_{B/A}$ is a vector of constant length fixed in a rigid body having an angular velocity $\boldsymbol{\omega}$, the omega theorem (see Section 12-6) gives its directional time rate of change as $\dot{\mathbf{r}}_{B/A} = \boldsymbol{\omega} \times \mathbf{r}_{B/A}$. Then Eq. (*b*) becomes

$$\dot{\mathbf{r}}_B = \dot{\mathbf{r}}_A + \boldsymbol{\omega} \times \mathbf{r}_{B/A} \qquad (c)$$

The acceleration relation is found by differentiating Eq. (*c*) with respect to time which gives

$$\ddot{\mathbf{r}}_B = \ddot{\mathbf{r}}_A + \frac{d}{dt}(\boldsymbol{\omega} \times \mathbf{r}_{B/A}) \qquad (d)$$

Using Eq. (9-5.3) for differentiating cross products, the last term in Eq. (*d*) becomes

$$\frac{d}{dt}(\boldsymbol{\omega} \times \mathbf{r}_{B/A}) = \frac{d\boldsymbol{\omega}}{dt} \times \mathbf{r}_{B/A} + \boldsymbol{\omega} \times \frac{d\mathbf{r}_{B/A}}{dt}$$

Noting that $\boldsymbol{\alpha} = \dot{\boldsymbol{\omega}} = \dfrac{d\boldsymbol{\omega}}{dt}$ and, from the omega theorem, that $\dfrac{d\mathbf{r}_{B/A}}{dt} = \boldsymbol{\omega} \times \mathbf{r}_{B/A}$, we can rewrite Eq. (*d*) as

$$\ddot{\mathbf{r}}_B = \ddot{\mathbf{r}}_A + \boldsymbol{\alpha} \times \mathbf{r}_{B/A} + \boldsymbol{\omega} \times (\boldsymbol{\omega} \times \mathbf{r}_{B/A}) \qquad (e)$$

Summarizing, we have the following equivalent forms of Eqs. (*c*) and (*e*):

$$\mathbf{v}_B = \mathbf{v}_A + (\mathbf{v}_{B/A} = \boldsymbol{\omega} \times \mathbf{r}_{B/A}) \qquad (12\text{-}7.1)$$

$$\mathbf{a}_B = \mathbf{a}_A + [\mathbf{a}_{B/A} = \boldsymbol{\alpha} \times \mathbf{r}_{B/A} + \boldsymbol{\omega} \times (\boldsymbol{\omega} \times \mathbf{r}_{B/A})] \qquad (12\text{-}7.2)$$

In applying Eqs. (12-7.1) and (12-7.2) to plane motion, note that $\boldsymbol{\alpha} = \alpha\hat{\mathbf{k}}$ and $\boldsymbol{\omega} = \omega\hat{\mathbf{k}}$.

The application of Eqs. (12-7.1) and (12-7.2) is straightforward mathematically, but careful attention must be given to signs in expanding the cross products. It is useful to recall from Section 12-6 that for plane motion, the triple vector product $\boldsymbol{\omega} \times (\boldsymbol{\omega} \times \mathbf{r}_{B/A})$ reduces to $-\omega^2\mathbf{r}_{B/A}$. The application of the full vector approach, however, does obscure the physical significance of the terms. Recall the comparison in statics (Illus. Prob. 2-11.2) between the vector approach and the geometric approach. In plane motion, the geometric approach will generally be preferable, but the vector approach comes into its own when spatial motions are considered later in this and other chapters.

ILLUSTRATIVE PROBLEMS

12-7.1. Solve Illus. Prob. 12-4.2 using the method of vector analysis. For convenience, we restate the data: A slender rod 10 in. long moves with its ends in contact with a horizontal floor and an inclined plane as shown in Fig. 12-7.2. At the given position, end A has a leftward velocity of 36 ips and a rightward acceleration of 7.5 ips^2. Determine the velocity and acceleration of end B.

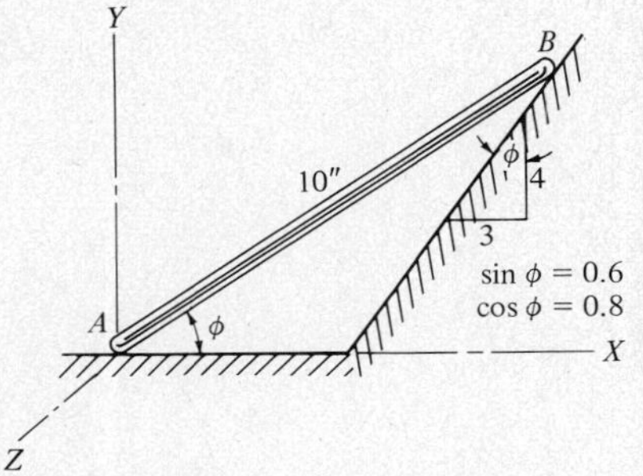

Figure 12-7.2

Solution

Expressing the given data in vector notation, we have $\mathbf{v}_A = -36\hat{\mathbf{i}}$ ips and $\mathbf{a}_A = 7.5\hat{\mathbf{i}}$ ips^2. Assuming $\boldsymbol{\omega}$ clockwise and $\mathbf{v}_B$ directed down the incline, we obtain $\boldsymbol{\omega} = -\omega\hat{\mathbf{k}}$ and $\mathbf{v}_B = -0.6v_B\hat{\mathbf{i}} - 0.8v_B\hat{\mathbf{j}}$. Applying the velocity equation, we obtain

$$[\mathbf{v}_B = \mathbf{v}_A + (v_{B/A} = \boldsymbol{\omega} \times \mathbf{r}_{B/A})]$$

$$-0.6v_B\hat{\mathbf{i}} - 0.8v_B\hat{\mathbf{j}} = -36\hat{\mathbf{i}} - \omega\hat{\mathbf{k}} \times (8\hat{\mathbf{i}} + 6\hat{\mathbf{j}})$$
$$= -36\hat{\mathbf{i}} + 8\omega\hat{\mathbf{j}} - 6\omega\hat{\mathbf{i}}$$

which, on equating the coefficients of the $\hat{\mathbf{i}}$ and $\hat{\mathbf{j}}$ terms (actually equivalent to horizontal and vertical summations), reduces to

[$\hat{\mathbf{i}}$ terms] $\quad -0.6v_B = -36 + 6\omega$

[$\hat{\mathbf{j}}$ terms] $\quad -0.8v_B = \quad - 8\omega$

which are solved to yield

$$\omega = +3 \text{ rad/sec} \quad \text{and} \quad v_B = 30 \text{ ips}$$

Observe the complete correspondence of these equations with those of the geometric analysis. The positive results confirm our assumptions. If we had completely ignored our intuitive assumptions and assumed positive values for $\boldsymbol{\omega}$ and $\mathbf{v}_B$; i.e., respectively counterclockwise $\boldsymbol{\omega}$ and upward motion of $\mathbf{v}_B$, we would have obtained the same numerical answers but with negative signs.

Assuming $\mathbf{a}_B$ to be directed down the incline, the acceleration equation becomes

$$[\mathbf{a}_B = \mathbf{a}_A + (\mathbf{a}_{B/A} = \boldsymbol{\alpha} \times \mathbf{r}_{B/A} - \omega^2\mathbf{r}_{B/A})]$$

$$-0.6a_B\hat{\mathbf{i}} - 0.8a_B\hat{\mathbf{j}} = 7.5\hat{\mathbf{i}} + \alpha\hat{\mathbf{k}} \times (8\hat{\mathbf{i}} + 6\hat{\mathbf{j}}) - (3)^2(8\hat{\mathbf{i}} + 6\hat{\mathbf{j}})$$
$$= 7.5\hat{\mathbf{i}} + 8\alpha\hat{\mathbf{j}} - 6\alpha\hat{\mathbf{i}} - 72\hat{\mathbf{i}} - 54\hat{\mathbf{j}}$$

which, on equating the coefficients of the $\hat{\mathbf{i}}$ and $\hat{\mathbf{j}}$ terms, reduces to

[$\hat{\mathbf{i}}$ terms] $\quad -0.6a_B = 7.5 - 72 - 6\alpha$

[$\hat{\mathbf{j}}$ terms] $\quad -0.8a_B = \quad - 54 + 8\alpha$

These equations are identical to those obtained previously from horizontal and vertical summations using the geometric method and yield $\alpha = -2$ rad/sec^2 and $a_B = 87.5$ ips^2. Here the negative sign for α denotes a clockwise sense which agrees with our previous solution.

12-7.2. In the four-link mechanism shown in Fig. 12-7.3, the crank AB has a constant counterclockwise angular velocity $\omega_{AB} =$ 3 rad per sec. Find the acceleration of point C.

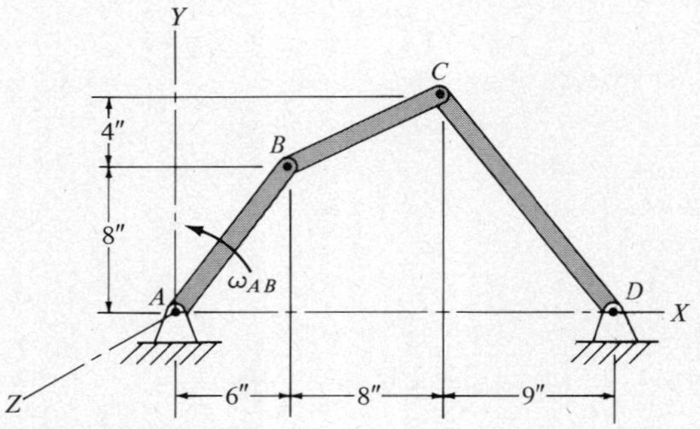

Figure 12-7.3

Solution

As was discussed in Section 9-2, different ground observers at A and D see the same motion of C. Hence, with respect to the ground the time derivatives of $\mathbf{r}_{DC}$ and $\mathbf{r}_C = \mathbf{r}_{AB} + \mathbf{r}_{BC}$ are identical. Each of these position vectors is a constant length vector fixed in a rotating body (i.e., link) so that their time derivatives, using the omega theorem, gives the following velocity equation:

$$\begin{aligned} \mathbf{v}_C &= \mathbf{v}_B + \mathbf{v}_{C/B} \\ \boldsymbol{\omega}_{CD} \times \mathbf{r}_{DC} &= \boldsymbol{\omega}_{AB} \times \mathbf{r}_{AB} + \boldsymbol{\omega}_{BC} \times \mathbf{r}_{BC} \end{aligned} \qquad (a)$$

Noting that each angular velocity vector is perpendicular to the XY plane, we evaluate each cross product separately as follows:

$$\boldsymbol{\omega}_{CD} \times \mathbf{r}_{DC} = \omega_{CD}\hat{\mathbf{k}} \times (-9\hat{\mathbf{i}} + 12\hat{\mathbf{j}}) = -9\omega_{CD}\hat{\mathbf{j}} - 12\omega_{CD}\hat{\mathbf{i}}$$
$$\boldsymbol{\omega}_{AB} \times \mathbf{r}_{AB} = 3\hat{\mathbf{k}} \times (6\hat{\mathbf{i}} + 8\hat{\mathbf{j}}) = 18\hat{\mathbf{j}} - 24\hat{\mathbf{i}}$$
$$\boldsymbol{\omega}_{BC} \times \mathbf{r}_{BC} = \omega_{BC}\hat{\mathbf{k}} \times (8\hat{\mathbf{i}} + 4\hat{\mathbf{j}}) = 8\omega_{BC}\hat{\mathbf{j}} - 4\omega_{BC}\hat{\mathbf{i}}$$

Substituting these results in Eq. (a) and equating the coefficients of the $\hat{\mathbf{i}}$ and $\hat{\mathbf{j}}$ terms, we get

$$[\hat{\mathbf{i}} \text{ terms}] \qquad -12\omega_{CD} = -24 - 4\omega_{BC}$$
$$[\hat{\mathbf{j}} \text{ terms}] \qquad -9\omega_{CD} = 18 + 8\omega_{BC}$$

On solving these equations, we obtain

$$\omega_{BC} = -3.27 \text{ rad/sec} \quad \text{and} \quad \omega_{CD} = 0.909 \text{ rad/sec}$$

We now have the angular velocities necessary to apply in the following acceleration equation:

$$\mathbf{a}_C = \mathbf{a}_B + \mathbf{a}_{C/B} \qquad (b)$$

Each term of Eq. (*b*) is evaluated separately as follows. Note that the angular acceleration vectors are perpendicular to the XY plane and that $\boldsymbol{\alpha}_{AB} = 0$ because AB has a constant angular velocity.

$$\begin{aligned} \mathbf{a}_C &= \boldsymbol{\alpha}_{CD} \times \mathbf{r}_{DC} - \omega_{CD}{}^2\mathbf{r}_{DC} \\ &= \alpha_{CD}\hat{\mathbf{k}} \times (-9\hat{\mathbf{i}} + 12\hat{\mathbf{j}}) - (0.909)^2(-9\hat{\mathbf{i}} + 12\hat{\mathbf{j}}) \\ &= -9\alpha_{CD}\hat{\mathbf{j}} - 12\alpha_{CD}\hat{\mathbf{i}} + 7.44\hat{\mathbf{i}} - 9.92\hat{\mathbf{j}} \end{aligned} \qquad (c)$$

$$\begin{aligned} \mathbf{a}_B &= \boldsymbol{\alpha}_{AB} \times \mathbf{r}_{AB} - \omega_{AB}{}^2\mathbf{r}_{AB} \\ &= 0 - (3)^2(6\hat{\mathbf{i}} + 8\hat{\mathbf{j}}) = -54\hat{\mathbf{i}} - 72\hat{\mathbf{j}} \end{aligned} \qquad (d)$$

$$\begin{aligned} \mathbf{a}_{C/B} &= \boldsymbol{\alpha}_{BC} \times \mathbf{r}_{BC} - \omega_{BC}{}^2\mathbf{r}_{BC} \\ &= \alpha_{BC}\hat{\mathbf{k}} \times (8\hat{\mathbf{i}} + 4\hat{\mathbf{j}}) - (3.27)^2(8\hat{\mathbf{i}} + 4\hat{\mathbf{j}}) \\ &= 8\alpha_{BC}\hat{\mathbf{j}} - 4\alpha_{BC}\hat{\mathbf{i}} - 85.4\hat{\mathbf{i}} - 42.7\hat{\mathbf{j}} \end{aligned} \qquad (e)$$

Substituting these results in Eq. (*b*) and equating the coefficients of the $\hat{\mathbf{i}}$ and $\hat{\mathbf{j}}$ terms, we obtain

$$\begin{aligned} -12\alpha_{CD} &= -147 - 4\alpha_{BC} \\ -9\alpha_{CD} &= -105 + 8\alpha_{BC} \end{aligned}$$

whose solution yields

$$\alpha_{CD} = 12.09 \text{ rad/sec}^2 \circlearrowleft \quad \text{and} \quad \alpha_{BC} = -0.477 \text{ rad/sec}^2 \circlearrowright$$

The absolute sense of each angular acceleration is indicated by the appended curved arrow.

Finally, the acceleration of C is found by substituting α_{CD} in Eq. (*c*) to give

$$\begin{aligned} \mathbf{a}_C &= -9(12.09)\hat{\mathbf{j}} - 12(12.09)\hat{\mathbf{i}} + 7.44\hat{\mathbf{i}} - 9.92\hat{\mathbf{j}} \\ &= -137.6\hat{\mathbf{i}} - 118.7\hat{\mathbf{j}} \text{ ips}^2 \end{aligned} \qquad \textbf{\textit{Ans.}}$$

PROBLEMS

Most of the following problems are repeats of previous problems solved geometrically, but here are to be solved by vector analysis. If time permits, a comparison of solutions by both methods will be very worthwhile.

12-7.3. In Fig. P-12-7.3, rod OA is rotating in a vertical plane about a horizontal axis at O. At the given instant, point A has an upward component of velocity of 40 ips and a leftward component of acceleration of 100 ips². Find the angular acceleration of rod OA.

$$\boldsymbol{\alpha} = -100\hat{\mathbf{k}} \text{ rad per sec}^2 \qquad \textbf{\textit{Ans.}}$$

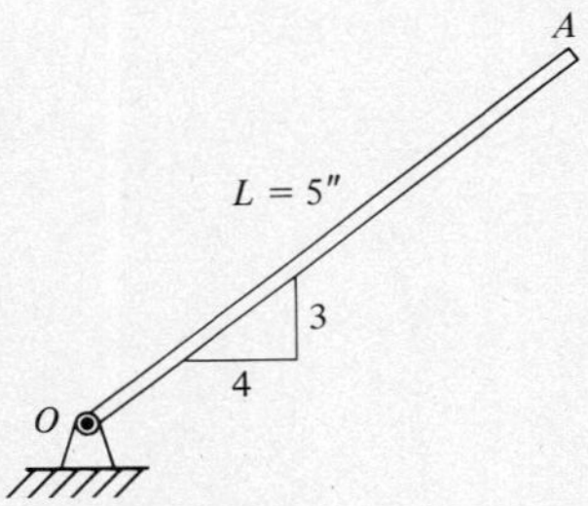

Figure P-12-7.3

12-7.4. Find the angular acceleration of the rod in the preceding problem if the data are changed as follows: The length $L = 10$ in. and point A has a rightward component of velocity of 24 ips and an upward component of acceleration of 48 ips².

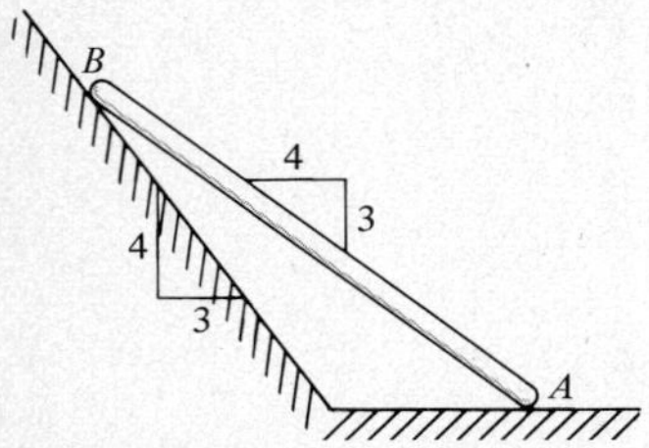

Figure P-12-7.5

12-7.5. A bar 3 ft long moves with its ends in contact with the inclined surface and the horizontal floor as shown in Fig. P-12-7.5. At the given instant, $\boldsymbol{\omega} = 2\hat{\mathbf{k}}$ rad per sec and $\boldsymbol{\alpha} = 3\hat{\mathbf{k}}$ rad per sec^2. Compute the acceleration of the midpoint M of AB. (Repeat of Prob. 12-4.14.)

12-7.6. In a rigid body having any type of motion, let A and B be any two particles in the body and let M be a particle midway between A and B. Prove that the acceleration of M is one-half the vector sum of the accelerations of points A and B.

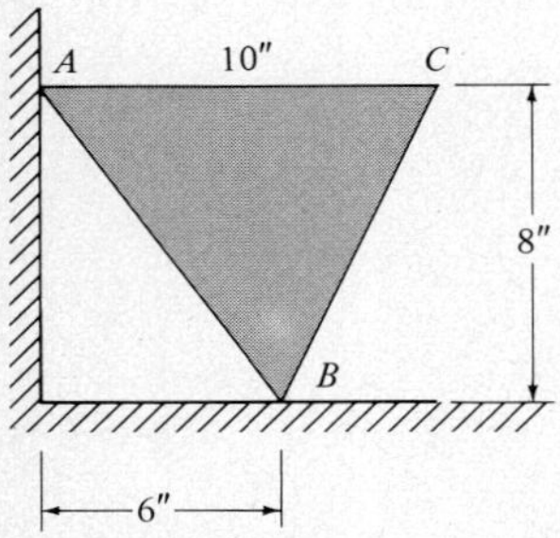

Figure P-12-7.7

12-7.7. The triangular plate ABC in Fig. P-12-7.7 moves with its vertices A and B in contact with a vertical wall and a horizontal floor. If $\boldsymbol{\omega} = \sqrt{5}\hat{\mathbf{k}}$ rad per sec and $\boldsymbol{\alpha} = 3\hat{\mathbf{k}}$ rad per sec^2, at the given instant, compute the accelerations of points A, B, and C. (Repeat of Prob. 12-4.16.)

$$\mathbf{a}_A = -58\hat{\mathbf{j}} \text{ fps}^2;\ \mathbf{a}_B = -6\hat{\mathbf{i}} \text{ fps}^2;\ \mathbf{a}_C = -50\hat{\mathbf{i}} - 28\hat{\mathbf{j}} \text{ fps}^2 \qquad \textit{Ans.}$$

12-7.8. A plate ABC moves with its ends A and B on the horizontal and inclined guides shown in Fig. P-12-7.8. At the given position $\boldsymbol{\omega} = -4\hat{\mathbf{k}}$ rad per sec and $\boldsymbol{\alpha} = -5\hat{\mathbf{k}}$ rad per sec^2. Find the accelerations of points A, B, and C. (Repeat of Prob. 12-4.17.)

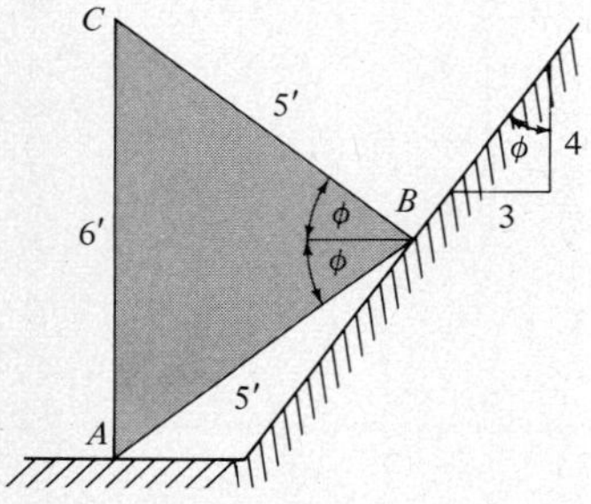

Figure P-12-7.8

12-7.9. The motion of the rigid link ABC in Fig. P-12-7.9 is governed by the control rods AD and BE. At the given instant when AB is horizontal, $\boldsymbol{\omega} = -\sqrt{10}\hat{\mathbf{k}}$ rad per sec and $\boldsymbol{\alpha} = -5\hat{\mathbf{k}}$ rad per sec^2. Determine the acceleration of point C. (Repeat of Prob. 12-4.18.)

$$\mathbf{a}_C = 32\hat{\mathbf{i}} - 93.5\hat{\mathbf{j}} \text{ ips}^2 \qquad \textit{Ans.}$$

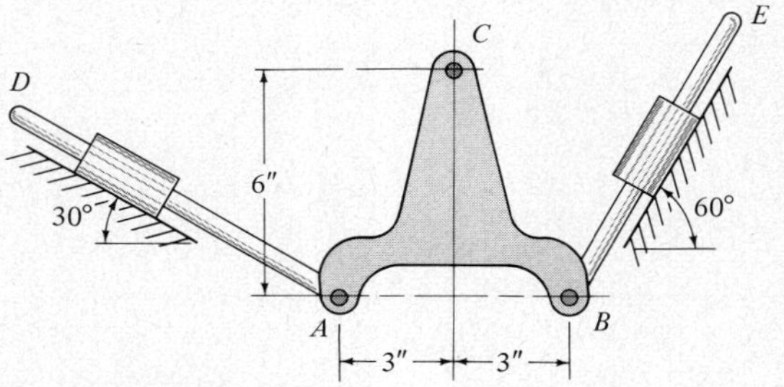

Figure P-12-7.9

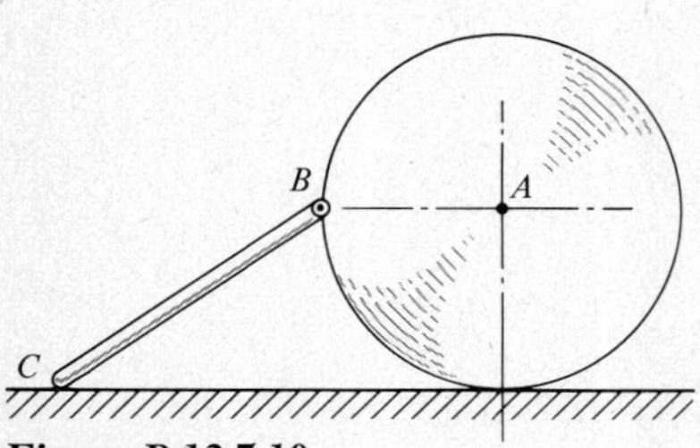

Figure P-12-7.10

12-7.10. The 6-ft diameter disk in Fig. P-12-7.10 rolls back and forth over a short distance without slipping. A 5-ft bar BC is pinned to the rim of the disk at B and its other end C drags on the ground. If the center of the disk has a constant velocity $\mathbf{v}_A = 12\hat{\mathbf{i}}$ fps at the given position, compute $\mathbf{v}_C$ and $\mathbf{a}_C$. (Repeat of Prob. 12-4.19.)

12-7.11. In the offset slider-crank mechanism shown in Fig. P-12-7.11, the crank OA rotates at $\boldsymbol{\omega}_{OA} = -10\hat{\mathbf{k}}$ rad per sec. Compute the acceleration of the midpoint M of the connecting rod AB at the given position. (Repeat of Prob. 12-4.21.)

12-7.12. When a slider-crank mechanism is in the position shown in Fig. P-12-7.12, the velocity and acceleration of slider C are $\mathbf{v}_C = -16\hat{\mathbf{j}}$ fps and $\mathbf{a}_C = -2.8\hat{\mathbf{j}}$ fps^2. What is then the angular acceleration of crank AB? (Repeat of Prob. 12-4.22.)

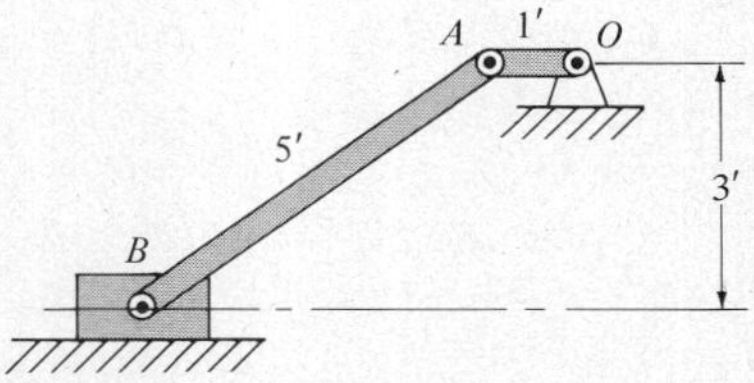

Figure P-12-7.11

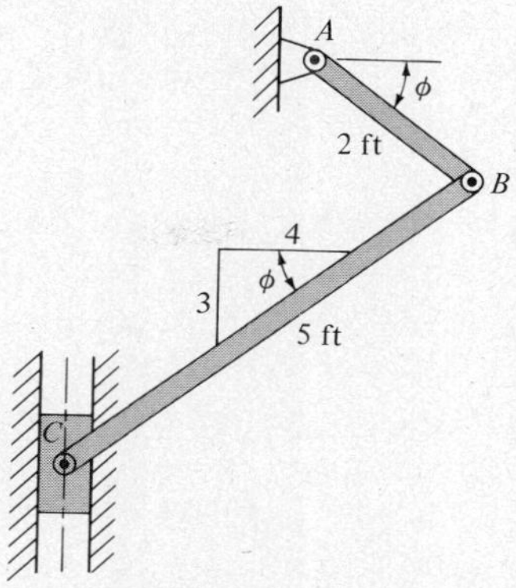

Figure P-12-7.12

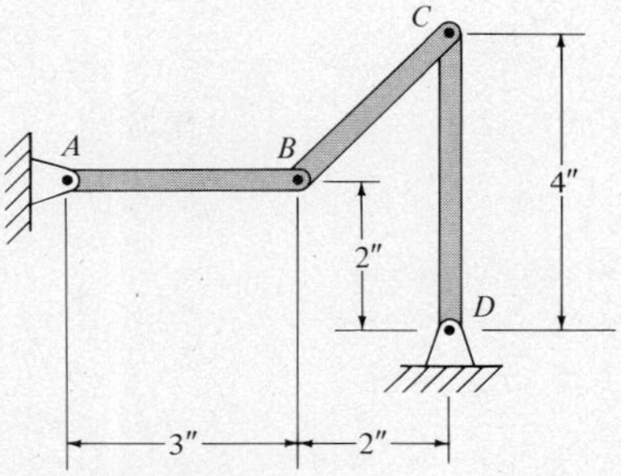

Figure P-12-7.13

12-7.13. At the instant shown in Fig. P-12-7.13, the driving crank AB of the four-link mechanism has a constant clockwise speed of 4 rad per sec. Find the angular acceleration of the link CD. (Repeat of Prob. 12-4.23.)

$$\boldsymbol{\alpha}_{DC} = 39\hat{\mathbf{k}} \text{ rad per sec}^2 \qquad \textit{Ans.}$$

12-7.14. The motion of the rigid plate in Fig. P-12-7.14 is controlled by the crank OA and the slider B. Find the velocity of point C if at the given instant $\mathbf{v}_B = 22\hat{\mathbf{j}}$ ips.

$\mathbf{v}_C = -24\hat{\mathbf{i}} - 10\hat{\mathbf{j}}$ ips; $\omega_{OA} = -2\hat{\mathbf{k}}$ rad per sec; $\boldsymbol{\omega} = 4\hat{\mathbf{k}}$ rad per sec *Ans.*

12-7.15. Using the results of the velocity analysis in the preceding problem, determine the acceleration of point C if at the given instant $\mathbf{a}_B = -139.6\hat{\mathbf{j}}$ ips².

$$\mathbf{a}_C = 56\hat{\mathbf{i}} - 332\hat{\mathbf{j}} \text{ ips}^2 \qquad \textit{Ans.}$$

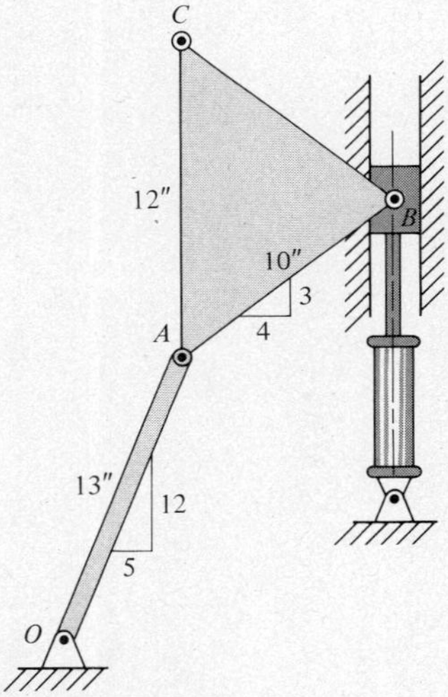

Figure P-12-7.14

* 12-8 ABSOLUTE SPATIAL MOTION

In Section 12-7 we discussed the relations between any two points of a rigid body moving in any manner with respect to a fixed reference frame XYZ. This general motion is equivalent to combining the motion of any point A of the body with a rigid-body rotation about this point. Then the velocity and acceleration of any other point B in the body is determined by the following equations:

$$\mathbf{v}_B = \mathbf{v}_A + (\mathbf{v}_{B/A} = \boldsymbol{\omega} \times \mathbf{r}_{B/A}) \tag{12-7.1}$$

$$\mathbf{a}_B = \mathbf{a}_A + [\mathbf{a}_{B/A} = \boldsymbol{\alpha} \times \mathbf{r}_{B/A} + \boldsymbol{\omega} \times (\boldsymbol{\omega} \times \mathbf{r}_{B/A})] \tag{12-7.2}$$

where the angular velocity $\boldsymbol{\omega}$ and angular acceleration $\boldsymbol{\alpha} = \dot{\boldsymbol{\omega}}$ of the body are referred to a fixed reference frame and thus are known as absolute quantities.

The application of these equations to plane motion was simplified by the fact that $\boldsymbol{\omega}$ and $\boldsymbol{\alpha} = \dot{\boldsymbol{\omega}}$ were collinear free vectors, both directed perpendicular to the plane of motion. When these equations

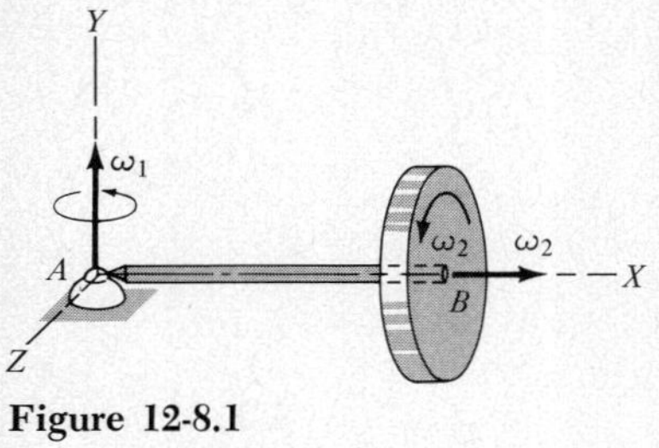

Figure 12-8.1

are applied to spatial motion, however, $\boldsymbol{\omega}$ and $\dot{\boldsymbol{\omega}}$ are not generally collinear and their absolute values are not as obvious. For example, consider the model of a gyroscope in Fig. 12-8.1 whose rotor is revolving at a constant rate $\boldsymbol{\omega}_2$ about its axis AB which is simultaneously precessing about the fixed vertical axis at the constant rate $\boldsymbol{\omega}_1$. The combination of these separate motions determines the resultant absolute angular velocity of the gyroscope to be $\boldsymbol{\omega} = \boldsymbol{\omega}_1 + \boldsymbol{\omega}_2$.

Although both $\boldsymbol{\omega}_1$ and $\boldsymbol{\omega}_2$ are of constant magnitude, we cannot jump to the conclusion that the angular acceleration of the gyroscope is zero. Instead, observe that although $\boldsymbol{\omega}_1$ has a constant direction, $\boldsymbol{\omega}_2$ is changing its direction due to the precession $\boldsymbol{\omega}_1$. Indeed, $\boldsymbol{\omega}_2$ may be considered as a vector fixed in a body which is changing its direction at the rate ω_1. Applying the omega theorem, we obtain

$$\boldsymbol{\omega} = \boldsymbol{\omega}_1 + \boldsymbol{\omega}_2 \tag{a}$$

$$\dot{\boldsymbol{\omega}} = \overset{=0}{\dot{\boldsymbol{\omega}}_1} + \overset{=0}{\dot{\boldsymbol{\omega}}_2} + \boldsymbol{\omega}_1 \times \boldsymbol{\omega}_2 = \boldsymbol{\omega}_1 \times \boldsymbol{\omega}_2 \tag{b}$$

A more complex example is that in Fig. 12-8.2 where the rod AB is rotating at $\boldsymbol{\omega}_2$ about a horizontal axis fixed in a platform which

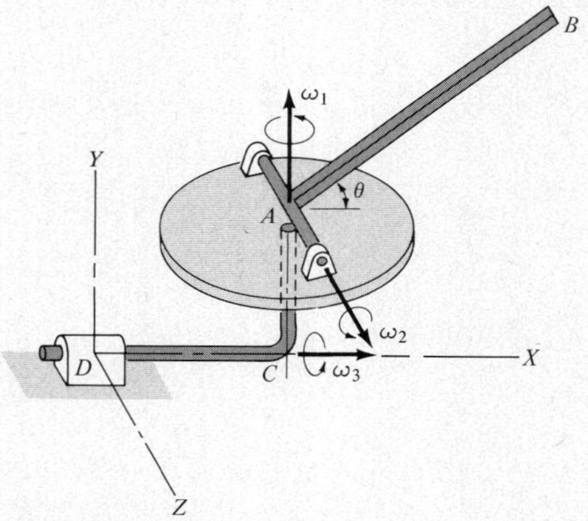

Figure 12-8.2

is rotating at $\boldsymbol{\omega}_1$ about AC of the bent bar ACD. If the bar ACD is also rotating at $\boldsymbol{\omega}_3$ about the axis DC, the absolute angular velocity of AB is

$$\boldsymbol{\omega}_{AB} = \boldsymbol{\omega}_3 + \boldsymbol{\omega}_1 + \boldsymbol{\omega}_2 \tag{c}$$

and its absolute angular acceleration is

$$\dot{\boldsymbol{\omega}}_{AB} = \dot{\boldsymbol{\omega}}_3 + \dot{\boldsymbol{\omega}}_1 + \boldsymbol{\omega}_3 \times \boldsymbol{\omega}_1 + \dot{\boldsymbol{\omega}}_2 + (\boldsymbol{\omega}_3 + \boldsymbol{\omega}_1) \times \boldsymbol{\omega}_2 \tag{d}$$

Notice the application of the omega theorem to determine how $\boldsymbol{\omega}_3$ affects the rate of change of the direction of $\boldsymbol{\omega}_1$, while the directional rate of change of $\boldsymbol{\omega}_2$ is caused by the sum of $\boldsymbol{\omega}_3$ and $\boldsymbol{\omega}_1$.

To illustrate the application of Eqs. (*c*) and (*d*) to finding the absolute velocity and acceleration of point B in rod AB, we would select A as a reference point and add the motion of B considered rotating about A at the rates $\boldsymbol{\omega}_{AB}$ and $\dot{\boldsymbol{\omega}}_{AB}$. The motion of A being that of a point in the bar ACD which has fixed-axis rotation about DC, we obtain

$$\mathbf{r}_B = \mathbf{r}_{DA} + \mathbf{r}_{AB} \tag{e}$$

$$\begin{aligned}\mathbf{v}_B &= \mathbf{v}_A + \mathbf{v}_{B/A} \\ &= \boldsymbol{\omega}_3 \times \mathbf{r}_{DA} + \boldsymbol{\omega}_{AB} \times \mathbf{r}_{AB}\end{aligned} \tag{f}$$

$$\begin{aligned}\mathbf{a}_B &= \mathbf{a}_A + \mathbf{a}_{B/A} \\ &= (\dot{\boldsymbol{\omega}}_3 \times \mathbf{r}_{DA} + \boldsymbol{\omega}_3 \times \mathbf{v}_A) + (\dot{\boldsymbol{\omega}}_{AB} \times \mathbf{r}_{AB} + \boldsymbol{\omega}_{AB} \times \mathbf{v}_{B/A})\end{aligned} \tag{g}$$

Specific details of obtaining numerical results are given in the illustrative problems. The procedure is straightforward except for careful determination of the absolute values of $\boldsymbol{\omega}_{AB}$ and $\dot{\boldsymbol{\omega}}_{AB}$. A relative motion analysis which uses a rotating and translating reference frame will be explained in the next section. Both absolute and relative motion analysis should be mastered as sometimes one or the other will be more direct.

ILLUSTRATIVE PROBLEMS

12-8.1. A gyroscope consists of a wheel of 2-ft radius mounted in a yoke as shown in Fig. 12-8.3. The wheel spins about its horizontal axis at a constant rate of 4 rad per sec while the yoke rotates about a fixed vertical axis at a constant rate of 5 rad per sec. Find the velocity and acceleration of points B and E on the wheel at the given instant.

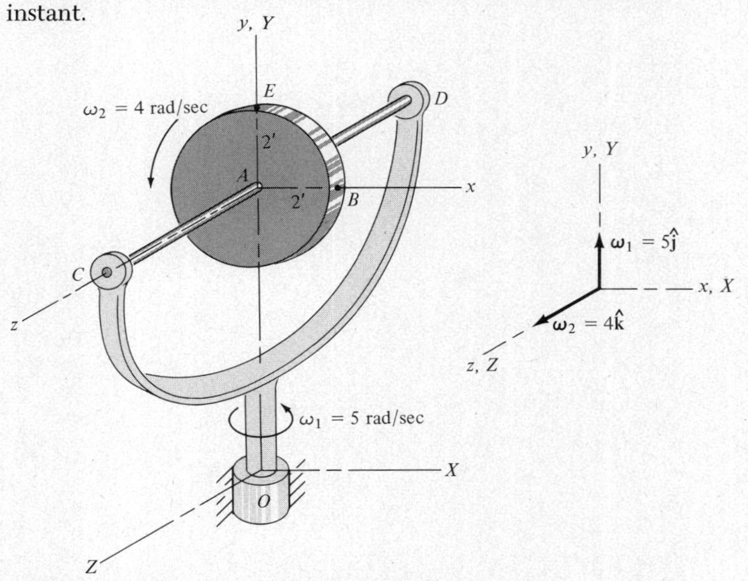

Figure 12-8.3 Kinematic analysis of gyroscope.

Solution

The wheel is in rigid-body rotation about the fixed point A. To evaluate formally the angular velocity and angular acceleration of the wheel, select xyz axes fixed in and rotating with the wheel whose directions instantaneously coincide with those of the stationary axes XYZ. Then the unit vectors $\hat{\mathbf{i}}$, $\hat{\mathbf{j}}$, $\hat{\mathbf{k}}$ of axes xyz instantaneously have the same direction as unit vectors $\hat{\mathbf{I}}$, $\hat{\mathbf{J}}$, $\hat{\mathbf{K}}$ of axes XYZ. Note that $\hat{\mathbf{j}}$ has a fixed direction, but $\hat{\mathbf{i}}$ and $\hat{\mathbf{k}}$ rotate at the rate $\boldsymbol{\omega}_1$.

The absolute angular velocity of the wheel and the attached xyz axes is

$$\boldsymbol{\omega} = \boldsymbol{\omega}_1 + \boldsymbol{\omega}_2 = (5\hat{\mathbf{j}} + 4\hat{\mathbf{k}}) \text{ rad per sec}$$

and their absolute angular acceleration is

$$\begin{aligned}\boldsymbol{\alpha} = \dot{\boldsymbol{\omega}} &= \frac{d}{dt}(5\hat{\mathbf{j}} + 4\hat{\mathbf{k}}) = 5\dot{\hat{\mathbf{j}}} + 4\dot{\hat{\mathbf{k}}} \\ &= 0 + 4(\boldsymbol{\omega} \times \hat{\mathbf{k}}) = 4(5\hat{\mathbf{j}} + 4\hat{\mathbf{k}}) \times \hat{\mathbf{k}} \\ &= 20\hat{\mathbf{i}} \text{ rad per sec}^2\end{aligned}$$

Note the use of the omega theorem to evaluate $\dot{\hat{\mathbf{k}}}$.

As a check on this fundamental approach, we can also find $\dot{\boldsymbol{\omega}}$ directly from the omega theorem by noting that the direction of $\boldsymbol{\omega}_1$ is fixed but the direction of $\boldsymbol{\omega}_2$ changes as axes xyz attached to the wheel simultaneously rotate about CD and OY. Thus, by treating $\boldsymbol{\omega}_2$ as a vector fixed in a body rotating at $\boldsymbol{\omega}_1$, we obtain again

$$\boldsymbol{\alpha} = \dot{\boldsymbol{\omega}} = \boldsymbol{\omega}_1 \times \boldsymbol{\omega}_2 = 5\hat{\mathbf{j}} \times 4\hat{\mathbf{k}} = 20\hat{\mathbf{i}} \text{ rad per sec}^2$$

As discussed in Section 12-6, recall that the velocity and acceleration of a point located by a position vector $\mathbf{r}$ in a rotating frame (or body) is given by $\mathbf{v} = \boldsymbol{\omega} \times \mathbf{r}$ and $\mathbf{a} = \dot{\boldsymbol{\omega}} \times \mathbf{r} + \boldsymbol{\omega} \times (\boldsymbol{\omega} \times \mathbf{r}) = \dot{\boldsymbol{\omega}} \times \mathbf{r} + \boldsymbol{\omega} \times \mathbf{v}$. Now that we know $\dot{\boldsymbol{\omega}}$ and $\boldsymbol{\omega}$, we apply these equations as follows to determine the velocity and acceleration of points B and E.

Velocity and Acceleration of Point B: $\mathbf{r} = 2\hat{\mathbf{i}}$ ft

$$\mathbf{v}_B = \boldsymbol{\omega} \times \mathbf{r} = (5\hat{\mathbf{j}} + 4\hat{\mathbf{k}}) \times (2\hat{\mathbf{i}}) = (-10\hat{\mathbf{k}} + 8\hat{\mathbf{j}}) \text{ fps} \qquad \textit{Ans.}$$

$$\begin{aligned}\mathbf{a}_B &= \dot{\boldsymbol{\omega}} \times \mathbf{r} + \boldsymbol{\omega} \times \mathbf{v} \\ &= 20\hat{\mathbf{i}} \times 2\hat{\mathbf{i}} + (5\hat{\mathbf{j}} + 4\hat{\mathbf{k}}) \times (-10\hat{\mathbf{k}} + 8\hat{\mathbf{j}}) \\ &= 0 + (-50\hat{\mathbf{i}}) + (-32\hat{\mathbf{i}}) = -82\hat{\mathbf{i}} \text{ fps}^2 \qquad \textit{Ans.}\end{aligned}$$

Velocity and Acceleration of Point E: $\mathbf{r} = 2\hat{\mathbf{j}}$ ft

$$\mathbf{v}_E = \boldsymbol{\omega} \times \mathbf{r} = (5\hat{\mathbf{j}} + 4\hat{\mathbf{k}}) \times 2\hat{\mathbf{j}} = -8\hat{\mathbf{i}} \text{ fps} \qquad \textit{Ans.}$$

$$\begin{aligned}\mathbf{a}_E &= \dot{\boldsymbol{\omega}} \times \mathbf{r} + \boldsymbol{\omega} \times \mathbf{v} \\ &= 20\hat{\mathbf{i}} \times 2\hat{\mathbf{j}} + (5\hat{\mathbf{j}} + 4\hat{\mathbf{k}}) \times (-8\hat{\mathbf{i}}) \\ &= 40\hat{\mathbf{k}} + 40\hat{\mathbf{k}} - 32\hat{\mathbf{j}} = (80\hat{\mathbf{k}} - 32\hat{\mathbf{j}}) \text{ fps}^2 \qquad \textit{Ans.}\end{aligned}$$

12-8.2. The front view of a mechanism like that in Fig. 12-8.2 is shown here with pertinent dimensions as Fig. 12-8.4. At the given instant when $\tan\theta = \frac{3}{4}$, the 5-ft bar AB is rotating about an axis at A with the rates $\omega_2 = 2$ rad per sec and $\dot{\omega}_2 = -3$ rad per sec^2. Simultaneously, the platform to which this axis is attached is spinning about AC at the rates $\omega_1 = 3$ rad per sec and $\dot{\omega}_1 = 1$ rad per sec^2 while the bar ACD supporting the platform is rotating about the fixed axis DC at the rates $\omega_3 = -4$ rad per sec and $\dot{\omega}_3 = -2$ rad per sec^2. Determine the velocity and acceleration of the end B of bar AB.

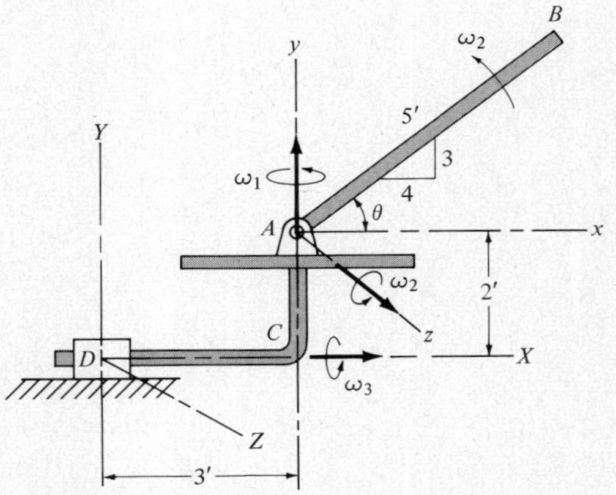

Figure 12-8.4

Solution

Here we illustrate the numerical details of combining various motions. The motion of any point in bar AB is due to the absolute rotation of the bar about A plus the motion of A. Select xyz axes attached to bar AB whose origin is at A and which are instantaneously parallel to the fixed axes XYZ. With respect to these axes, the given rates of rotation are

$$\boldsymbol{\omega}_1 = 3\hat{\mathbf{j}} \qquad \boldsymbol{\omega}_2 = 2\hat{\mathbf{k}} \qquad \boldsymbol{\omega}_3 = -4\hat{\mathbf{i}}$$
$$\dot{\boldsymbol{\omega}}_1 = \hat{\mathbf{j}} \qquad \dot{\boldsymbol{\omega}}_2 = -3\hat{\mathbf{k}} \qquad \dot{\boldsymbol{\omega}}_3 = -2\hat{\mathbf{i}}$$

Then as discussed on p. 468 the absolute angular velocity and acceleration of AB and the attached xyz axes are

$$\boldsymbol{\omega}_{AB} = \boldsymbol{\omega}_3 + \boldsymbol{\omega}_1 + \boldsymbol{\omega}_2 = -4\hat{\mathbf{i}} + 3\hat{\mathbf{j}} + 2\hat{\mathbf{k}} \text{ rad/sec}$$

and

$$\begin{aligned}\dot{\boldsymbol{\omega}}_{AB} &= \dot{\boldsymbol{\omega}}_3 + \dot{\boldsymbol{\omega}}_1 + \boldsymbol{\omega}_3 \times \boldsymbol{\omega}_1 + \dot{\boldsymbol{\omega}}_2 + (\boldsymbol{\omega}_3 + \boldsymbol{\omega}_1) \times \boldsymbol{\omega}_2 \\ &= -2\hat{\mathbf{i}} + \hat{\mathbf{j}} + (-4\hat{\mathbf{i}} \times 3\hat{\mathbf{j}}) - 3\hat{\mathbf{k}} + (-4\hat{\mathbf{i}} + 3\hat{\mathbf{j}}) \times 2\hat{\mathbf{k}} \\ &= -2\hat{\mathbf{i}} + \hat{\mathbf{j}} - 12\hat{\mathbf{k}} - 3\hat{\mathbf{k}} + 8\hat{\mathbf{j}} + 6\hat{\mathbf{i}} \\ &= 4\hat{\mathbf{i}} + 9\hat{\mathbf{j}} - 15\hat{\mathbf{k}} \text{ rad/sec}^2\end{aligned}$$

Now that $\boldsymbol{\omega}_{AB}$ and $\dot{\boldsymbol{\omega}}_{AB}$ are known, we can combine the motion of B as a point in AB rotating about A with the motion of A. Note that the motion of A is determined by the fixed-axis rotation of ACD. At the given instant, the absolute position vector of B is

$$\mathbf{r}_B = (\mathbf{r}_{DA} = 3\hat{\mathbf{i}} + 2\hat{\mathbf{j}}) + (\mathbf{r}_{AB} = 4\hat{\mathbf{i}} + 3\hat{\mathbf{j}})$$

and the corresponding velocity of B is

$$\mathbf{v}_B = (\mathbf{v}_A = \boldsymbol{\omega}_3 \times \mathbf{r}_{DA}) + (\mathbf{v}_{B/A} = \boldsymbol{\omega}_{AB} \times \mathbf{r}_{AB})$$

Evaluate each cross product separately as follows:

$$\mathbf{v}_A = \boldsymbol{\omega}_B \times \mathbf{r}_{DA} = -4\hat{\mathbf{i}} \times (3\hat{\mathbf{i}} + 2\hat{\mathbf{j}}) = -8\hat{\mathbf{k}}$$

$$\mathbf{v}_{B/A} = \boldsymbol{\omega}_{AB} \times \mathbf{r}_{AB} = (-4\hat{\mathbf{i}} + 3\hat{\mathbf{j}} + 2\hat{\mathbf{k}}) \times (4\hat{\mathbf{i}} + 3\hat{\mathbf{j}})$$

$$= \begin{vmatrix} -4 & 3 & 2 \\ 4 & 3 & 0 \\ \hat{\mathbf{i}} & \hat{\mathbf{j}} & \hat{\mathbf{k}} \end{vmatrix} = -6\hat{\mathbf{i}} + 8\hat{\mathbf{j}} - 24\hat{\mathbf{k}}$$

whence

$$\mathbf{v}_B = \mathbf{v}_A + \mathbf{v}_{B/A} = (-6\hat{\mathbf{i}} + 8\hat{\mathbf{j}} - 32\hat{\mathbf{k}}) \text{ fps}$$

The acceleration of B is

$$\mathbf{a}_B = (\mathbf{a}_A = \dot{\boldsymbol{\omega}}_3 \times \mathbf{r}_{DA} + \boldsymbol{\omega}_3 \times \mathbf{v}_A) + (\mathbf{a}_{B/A} = \dot{\boldsymbol{\omega}}_{AB} \times \mathbf{r}_{AB} + \boldsymbol{\omega}_{AB} \times \mathbf{v}_{BA})$$

Using the previously obtained values of $\mathbf{v}_A$, $\mathbf{v}_{B/A}$, and $\dot{\boldsymbol{\omega}}_{AB}$, we evaluate each cross product as follows:

$$\dot{\boldsymbol{\omega}}_3 \times \mathbf{r}_{DA} = -2\hat{\mathbf{i}} \times (3\hat{\mathbf{i}} + 2\hat{\mathbf{j}}) = -4\hat{\mathbf{k}}$$

$$\boldsymbol{\omega}_3 \times \mathbf{v}_A = -4\hat{\mathbf{i}} \times (-8\hat{\mathbf{k}}) = -32\hat{\mathbf{j}}$$

$$\dot{\boldsymbol{\omega}}_{AB} \times \mathbf{r}_{AB} = (4\hat{\mathbf{i}} + 9\hat{\mathbf{j}} - 15\hat{\mathbf{k}}) \times (4\hat{\mathbf{i}} + 3\hat{\mathbf{j}})$$

$$= \begin{vmatrix} 4 & 9 & -15 \\ 4 & 3 & 0 \\ \hat{\mathbf{i}} & \hat{\mathbf{j}} & \hat{\mathbf{k}} \end{vmatrix} = 45\hat{\mathbf{i}} - 60\hat{\mathbf{j}} - 24\hat{\mathbf{k}}$$

$$\boldsymbol{\omega}_{AB} \times \mathbf{v}_{B/A} = (-4\hat{\mathbf{i}} + 3\hat{\mathbf{j}} + 2\hat{\mathbf{k}}) \times (-6\hat{\mathbf{i}} + 8\hat{\mathbf{j}} - 24\hat{\mathbf{k}})$$

$$= \begin{vmatrix} -4 & 3 & 2 \\ -6 & 8 & -24 \\ \hat{\mathbf{i}} & \hat{\mathbf{j}} & \hat{\mathbf{k}} \end{vmatrix} = -88\hat{\mathbf{i}} - 108\hat{\mathbf{j}} - 14\hat{\mathbf{k}}$$

whence on collating the coefficients of the unit vectors, we finally obtain

$$\mathbf{a}_B = (-43\hat{\mathbf{i}} - 200\hat{\mathbf{j}} - 42\hat{\mathbf{k}}) \text{ fps}^2 \qquad \textit{Ans.}$$

PROBLEMS

12-8.3. At a given instant, the angular velocity of a body is $\boldsymbol{\omega} = 2\hat{\mathbf{i}} - 3\hat{\mathbf{j}} + 4\hat{\mathbf{k}}$ rad per sec and the linear velocity of a point O in the body is $\mathbf{v}_O = 12\hat{\mathbf{i}} - 8\hat{\mathbf{j}}$ fps. Measured from O, the position vectors of two other points in the body are $\mathbf{r}_A = 2\hat{\mathbf{i}} + 3\hat{\mathbf{j}}$ and $\mathbf{r}_B = 5\hat{\mathbf{i}} + 3\hat{\mathbf{j}} + 4\hat{\mathbf{k}}$ ft. Find the velocity of point B in two ways: (a) as $\mathbf{v}_B = \mathbf{v}_A + \mathbf{v}_{B/A}$ and (b) as $\mathbf{v}_B = \mathbf{v}_O + \mathbf{v}_{B/O}$.

12-8.4. In addition to the data of the preceding problem, assume that $\boldsymbol{\alpha} = \dot{\boldsymbol{\omega}} = 3\hat{\mathbf{i}} + 1\hat{\mathbf{j}} - 2\hat{\mathbf{k}}$ rad per sec^2 and the acceleration of point O is $\mathbf{a}_O = 12\hat{\mathbf{i}} + 36\hat{\mathbf{j}} - 17\hat{\mathbf{k}}$ fps^2. Find the acceleration of point A, and of B relative to A. Is it correct to add these results to determine the acceleration of B?

$$\mathbf{a}_A = 50\hat{\mathbf{i}} - 40\hat{\mathbf{j}} - 30\hat{\mathbf{k}} \text{ fps}^2;\ \mathbf{a}_{B/A} = -39\hat{\mathbf{i}} - 124\hat{\mathbf{j}} - 61\hat{\mathbf{k}} \text{ fps}^2 \quad \textit{Ans.}$$

12-8.5. At the instant shown in Fig. P-12-8.5, the top is spinning with a constant absolute angular velocity $\boldsymbol{\omega} = 20\hat{\mathbf{i}} + 10\hat{\mathbf{j}}$ rad per sec. Determine the rate at which the top is spinning about its geometric axis OA and use this result to find the velocity of point B relative to point A. Both A and B are in the XY plane at the given instant. Observe that this method is more direct than taking the difference between $\mathbf{v}_B$ and $\mathbf{v}_A$.

$$v_{B/A} = 40 \text{ ips} \quad \textit{Ans.}$$

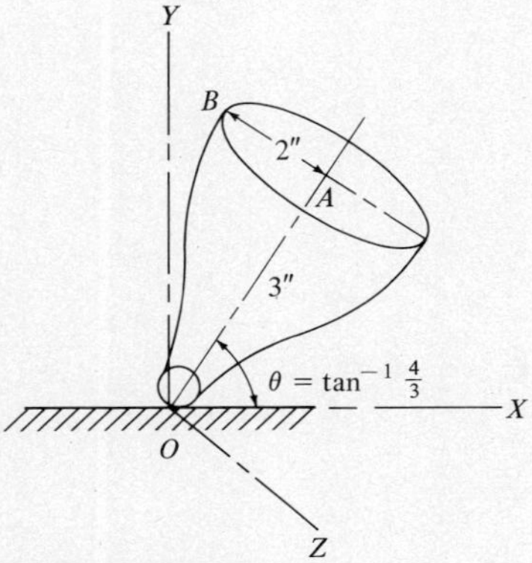

Figure P-12-8.5

12-8.6. Determine the absolute angular acceleration of the spin axis OA in the previous problem.

$$\boldsymbol{\alpha} = -200\hat{\mathbf{k}} \text{ rad per sec}^2 \quad \textit{Ans.}$$

12-8.7. In Fig. P-12-8.7, a disk rotates about a vertical axis at A at $\omega_2 = 10$ rad per sec relative to the wheel carrying the disk. Meanwhile, the wheel rotates about a vertical axis at O at $\omega_1 = 5$ rad per sec. Both ω_1 and ω_2 are constant. Find the velocity and acceleration of particles B and C on the rim of the disk at the given instant.

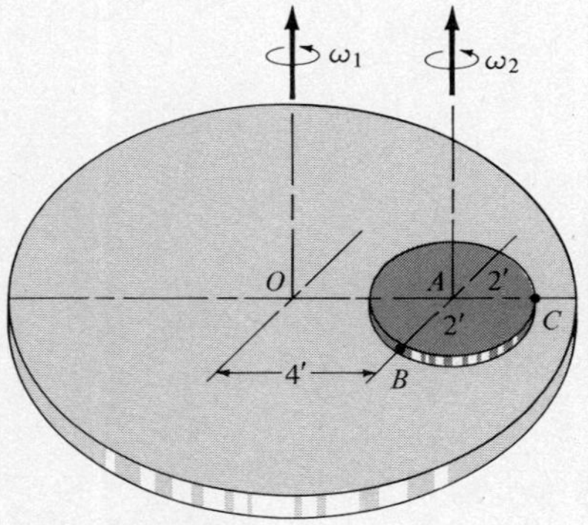

Figure P-12-8.7

12-8.8. In Fig. P-12-8.8, a disk rotates at the constant rate of

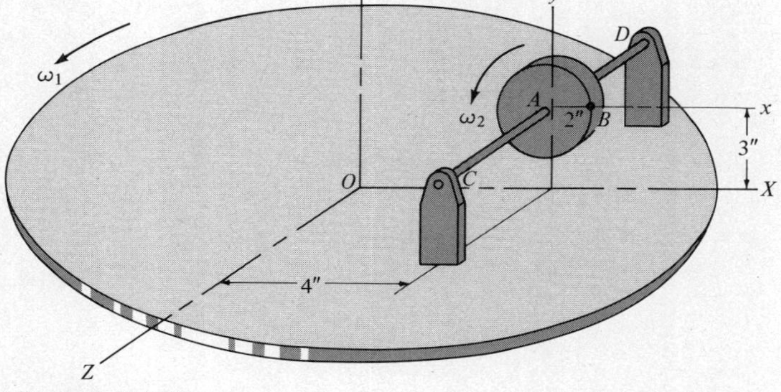

Figure P-12-8.8

$\omega_2 = 4$ rad per sec about the axle CD mounted on a wheel which is itself rotating at a constant rate of $\omega_1 = 5$ rad per sec about a vertical axis at O. Compute the acceleration of point B on the rim of the disk at the given instant.

$$\mathbf{a}_B = -182\hat{\mathbf{i}} \text{ ips}^2 \qquad \textit{Ans.}$$

12-8.9. The tail rotor controls the spin of the fuselage of the helicopter in Fig. P-12-8.9 about its vertical axis. At the given instant, the fuselage is rotating at a constant rate of $\omega_1 = 5$ rad per sec counterclockwise when

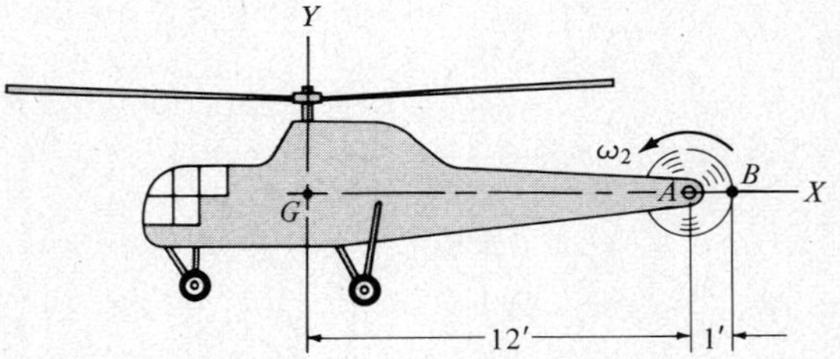

Figure P-12-8.9

viewed from above. Relative to the fuselage, the tail rotor is spinning at a constant rate of $\omega_2 = 20$ rad per sec. Determine the acceleration of point B. Assume that the helicopter is hovering in space so that the gravity center G is stationary.

$$\mathbf{a}_B = -725\hat{\mathbf{i}} \text{ fps}^2 \qquad \textit{Ans.}$$

12-8.10. The center A of the wheel in Fig. P-12-8.10 has a constant velocity of v fps. The wheel rolls without slipping on a horizontal surface as it also rotates about the shaft axis AB. Determine the absolute angular

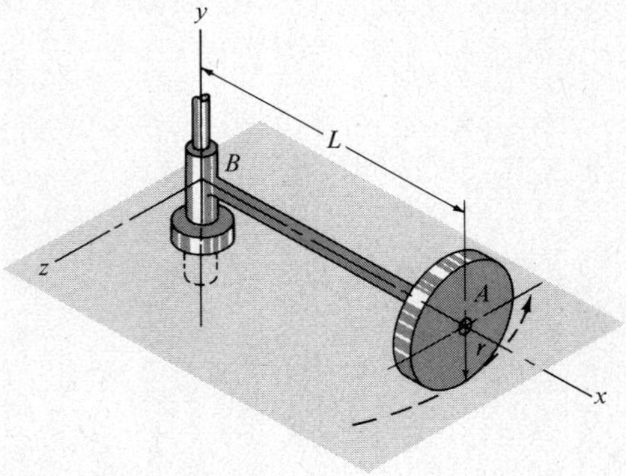

Figure P-12-8.10

acceleration of the wheel. Apply this result to find the angular acceleration of a tire of 28-in. outside diameter on an auto moving at 30 mph on a horizontal turn of 300-ft radius.

$$\boldsymbol{\alpha} = \frac{v^2}{Lr}\hat{\mathbf{k}} \qquad \textit{Ans.}$$

12-8.11. In Fig. P-12-8.11, a solid cone of base radius r and height h rolls on a horizontal floor without slipping. The center line OA rotates

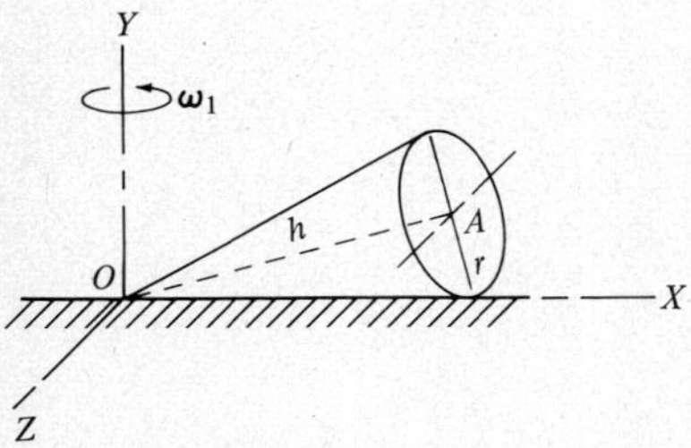

Figure P-12-8.11

about the fixed Y axis at ω_1 rad per sec. Find the absolute angular velocity and acceleration of the cone.

12-8.12. As the bevel gear A rolls on the fixed gear B in Fig. P-12-8.12, it also rotates about the axle CD of length h which is hinged to and rotates with the vertical shaft DE. The pitch circle radii of gears A and B are respectively a and b. Assuming that DE rotates at a constant rate of ω_1 rad per sec, determine the spin rate ω_2 of gear A about its axle and the absolute angular acceleration of gear A.

$$\omega_2 = \frac{b}{a}\omega_1; \ \boldsymbol{\alpha} = \frac{b}{a}\omega_1^2 \sin \gamma \hat{\mathbf{k}} \qquad \textit{Ans.}$$

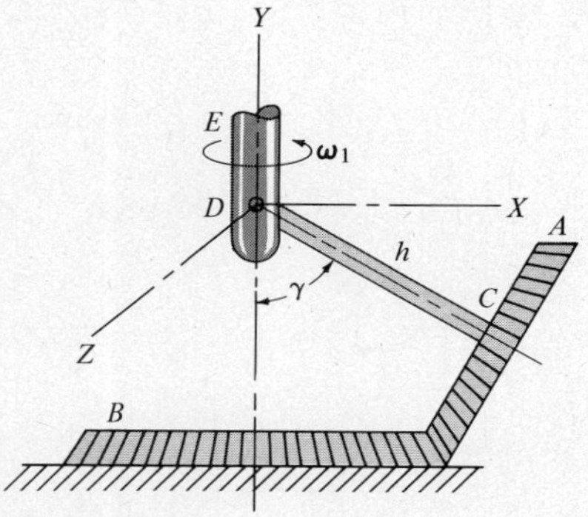

Figure P-12-8.12

12-8.13. At the instant shown in Fig. P-12-8.13, the plate ABC is rotating at the constant rate of 2 rad per sec about the edge AB which moves in a vertical plane. At the same instant, A has a leftward velocity of 8 fps and a rightward acceleration of 10 fps². Find the absolute velocity and acceleration of C.

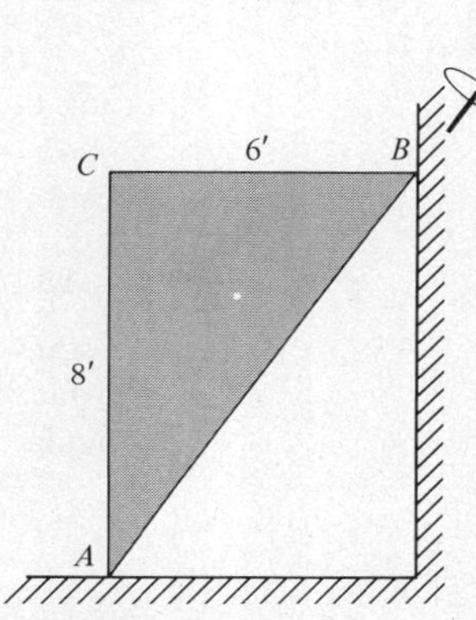

Figure P-12-8.13

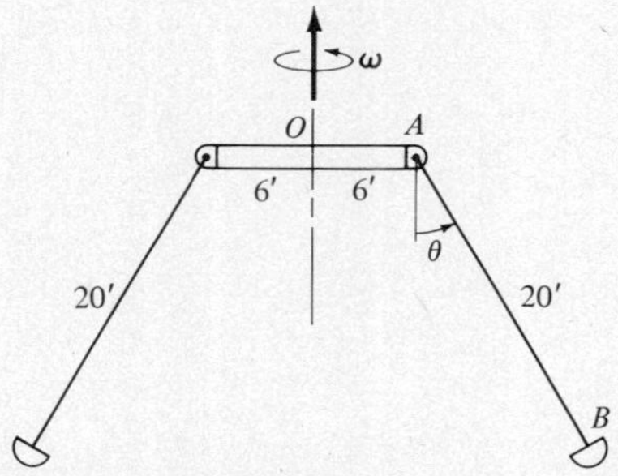

Figure P-12-8.14

12-8.14. In an airplane swing in an amusement park, the arm AB in Fig. P-12-8.14 is swinging outward at $\dot{\theta} = 2$ rad per sec and $\ddot{\theta} = 6$ rad per sec² while the main vertical mast is rotating at a constant speed $\omega = 3$ rad per sec. Compute the absolute acceleration of B when $\theta = 30°$.

$$a_B = 257 \text{ fps}^2 \qquad \textit{Ans.}$$

12-8.15. A wheel of 2-ft radius rotates about the axis OA at $\omega_2 = 6$ rad per sec as the platform carrying the wheel rotates about the fixed Y axis at $\omega_1 = 4$ rad per sec. Find the absolute velocity and acceleration of point B at the top of the wheel at the given position in Fig. P-12-8.15.

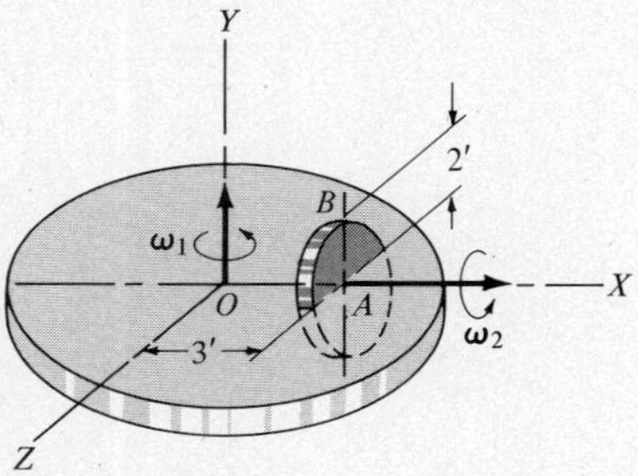

Figure P-12-8.15

12-8.16. A gyroscope consists of a 2-in. radius wheel rotating at $\omega_2 = 40$ rad per sec about axis OA in Fig. P-12-8.16. Simultaneously, axis OA is precessing (i.e., rotating) about a fixed vertical Y axis at $\omega_1 = 2$ rad per sec. Find the absolute velocity and acceleration of point B which is at the top of the wheel in the given position. Assume the angle ϕ does not change; i.e., axis OA stays at the same inclination with the horizontal as it rotates about the Y axis.

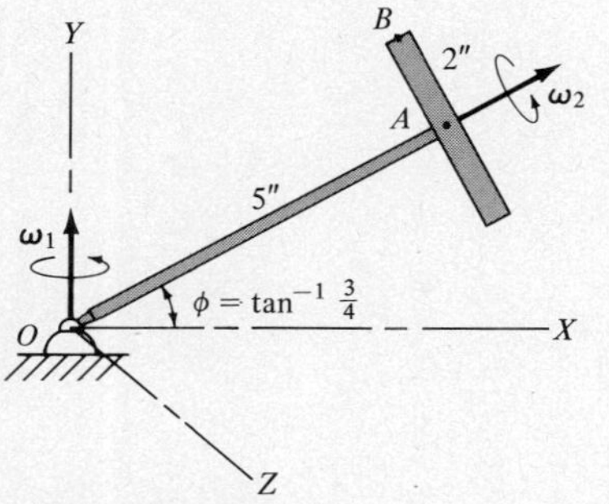

Figure P-12-8.16

12-8.17. Solve the preceding problem if, in addition to the given data, the angle ϕ of axis OA is increasing counterclockwise at $\omega_3 = 1$ rad per sec.

$$\mathbf{v}_B = -4.6\hat{\mathbf{i}} + 2.8\hat{\mathbf{j}} + 74.4\hat{\mathbf{k}} \text{ ips}$$
$$\mathbf{a}_B = 2226\hat{\mathbf{i}} - 2564.6\hat{\mathbf{j}} + 18.4\hat{\mathbf{k}} \text{ ips}^2$$
Ans.

12-8.18. A flight simulator for the training of astronauts is shown schematically in Fig. P-12-8.18. The cockpit containing the astronaut may rotate about the horizontal axis of the main arm OB while the seat to which the astronaut is strapped may rotate inside the cockpit about an

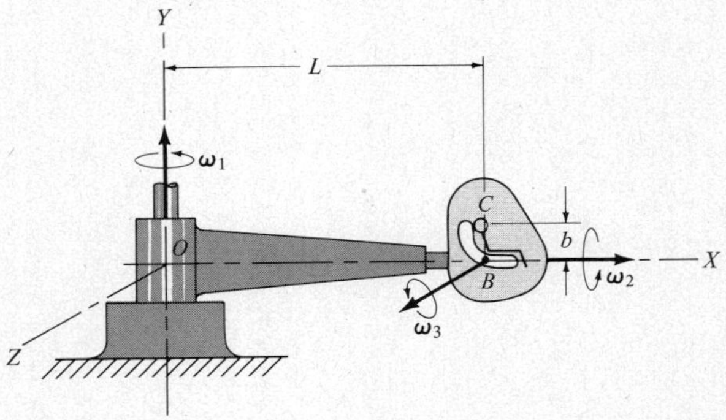

Figure P-12-8.18

axis at B. For the given position, find the velocity and acceleration of the astronaut's head at C if the main arm, cockpit, and seat rotate at the following constant rates: main arm $= \omega_1$; cockpit $= \omega_2$ relative to main arm; seat $= \omega_3$ relative to cockpit.

$$\mathbf{v}_C = -b\omega_3\hat{\mathbf{i}} + (b\omega_2 - L\omega_1)\hat{\mathbf{k}}$$
$$\mathbf{a}_C = (2b\omega_1\omega_2 - L\omega^2)\hat{\mathbf{i}} - b(\omega_2^2 + \omega_3^2)\hat{\mathbf{j}} + 2b\omega_1\omega_3\hat{\mathbf{k}}$$
Ans.

*12-9 RELATIVE SPATIAL MOTION. ROTATING REFERENCE FRAMES

So far we have limited the discussion of kinematics (both plane and spatial) to rigid bodies or systems of connected rigid bodies, but situations exist in which a body moves within another body. While this situation can be handled by absolute motion analysis, it is much easier to do it by a more general approach known as relative motion analysis which uses a reference frame which both rotates and translates. As we shall see, relative motion analysis is a completely general method applicable not only to kinematic analysis but also to finding the time derivative of any vector which changes *both its direction and its magnitude*. In effect, we shall develop an extension of the omega theorem which applied only to vectors of constant magnitude.

Consider the situation in Fig. 12-9.1 where a particle B moves along a curved path fixed in a body as the body rotates with an angular velocity $\boldsymbol{\omega}$. Let the position vector of B be denoted by $\boldsymbol{\rho}$ from an origin A of a reference frame fixed in and rotating with the rigid body. With respect to this frame

$$\boldsymbol{\rho} = x\hat{\mathbf{i}} + y\hat{\mathbf{j}} + z\hat{\mathbf{k}}$$

and

$$\mathbf{v}_{B/A} = \dot{\boldsymbol{\rho}} = \dot{x}\hat{\mathbf{i}} + \dot{y}\hat{\mathbf{j}} + \dot{z}\hat{\mathbf{k}} + x\dot{\hat{\mathbf{i}}} + y\dot{\hat{\mathbf{j}}} + z\dot{\hat{\mathbf{k}}} \qquad (a)$$

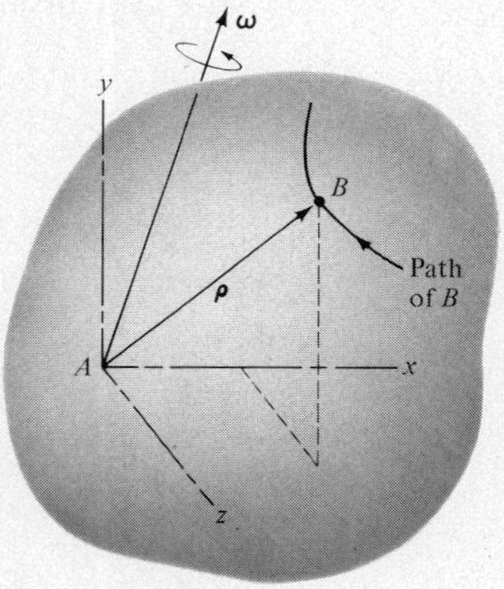

Figure 12-9.1

The first three right-hand terms represent the velocity of B as if the frame were not rotating or what would be seen by an observer attached to and rotating with the frame. We designate this as velocity relative to the frame (or, more briefly, as frame velocity) and denote it by $\dot{\boldsymbol{\rho}}_r = \dfrac{\delta\boldsymbol{\rho}}{\delta t}$ where the subscript r means relative to the frame. The symbol δ is used to indicate differentiation with respect to time as seen by an observer attached to and rotating with the frame. The last three right-hand terms of Eq. (a) exist because the unit vectors of the frame change their directions as the frame rotates with the rigid body. Noting from the omega theorem that $\dot{\hat{\mathbf{i}}} = \boldsymbol{\omega} \times \hat{\mathbf{i}}$, etc., their sum is

$$\begin{aligned} x\dot{\hat{\mathbf{i}}} + y\dot{\hat{\mathbf{j}}} + z\dot{\hat{\mathbf{k}}} &= x(\boldsymbol{\omega} \times \hat{\mathbf{i}}) + y(\boldsymbol{\omega} \times \hat{\mathbf{j}}) + z(\boldsymbol{\omega} \times \hat{\mathbf{k}}) \\ &= \boldsymbol{\omega} \times (x\hat{\mathbf{i}}) + \boldsymbol{\omega} \times (y\hat{\mathbf{j}}) + \boldsymbol{\omega} \times (z\hat{\mathbf{k}}) = \boldsymbol{\omega} \times (x\hat{\mathbf{i}} + y\hat{\mathbf{j}} + z\hat{\mathbf{k}}) \\ &= \boldsymbol{\omega} \times \boldsymbol{\rho} \end{aligned}$$

Thus we find the total velocity of B, moving along a path in a rotating body, to be

$$\mathbf{v}_{B/A} = \dot{\boldsymbol{\rho}} = \frac{d\boldsymbol{\rho}}{dt} = \frac{\delta\boldsymbol{\rho}}{\delta t} + \boldsymbol{\omega} \times \boldsymbol{\rho} = \dot{\boldsymbol{\rho}}_r + \boldsymbol{\omega} \times \boldsymbol{\rho} \qquad (12\text{-}9.1)$$

The absolute velocity of B would be found by adding to this result the absolute velocity of A.

Equation (12-9.1) may be interpreted physically as the sum of two separate effects occurring simultaneously: (1) the motion of B with respect to the frame computed as though the frame were not rotating, and (2) the motion of B due to the rotation of the frame. Figure 12-9.2 attempts to visualize these separate simultaneous motions. Assuming first that $\boldsymbol{\rho}$ is of constant length and fixed in the body, in a differential time dt, the frame rotation causes the displacement $(\boldsymbol{\omega}\, dt) \times \boldsymbol{\rho}$ while the relative motion of $\boldsymbol{\rho}$ due to its swing and stretch in the frame itself during the *equal* time δt is $\delta\boldsymbol{\rho}_r$. The total displacement of B is then

$$d\boldsymbol{\rho} = \delta\boldsymbol{\rho}_r + (\boldsymbol{\omega}\, dt) \times \boldsymbol{\rho}$$

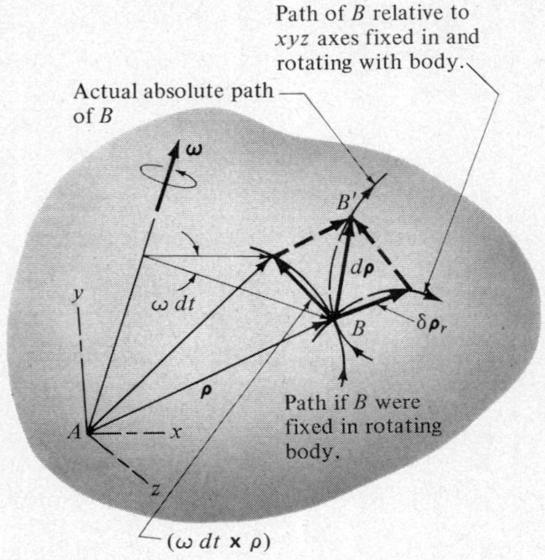

Figure 12-9.2

whence dividing by the elapsed time $dt = \delta t$, we get

$$\frac{d\boldsymbol{\rho}}{dt} = \frac{\delta \boldsymbol{\rho}_r}{\delta t} + \boldsymbol{\omega} \times \boldsymbol{\rho}$$

or

$$\mathbf{v}_{B/A} = \dot{\boldsymbol{\rho}} = \dot{\boldsymbol{\rho}}_r + \boldsymbol{\omega} \times \boldsymbol{\rho}$$

as before.

Equation (12-9.1) may be interpreted as the special form of a more general equation expressing the absolute time derivative of *any* vector **A** in terms of its rate of change with respect to a rotating coordinate system, namely,[4]

$$\frac{d\mathbf{A}}{dt} = \frac{\delta \mathbf{A}}{\delta t} + \boldsymbol{\omega} \times \mathbf{A} \tag{12-9.2}$$

We are now ready, with the aid of Eq. (12-9.2), to develop the relative motion analysis of kinematics. This approach will be very useful in situations where a relative motion occurs within a moving

[4] Note that if the absolute system is called system S and the rotating coordinate system is called system B, we can write Eq. (12-9.2) as

$$\left(\frac{d\mathbf{A}}{dt}\right)_S = \left(\frac{\delta \mathbf{A}}{\delta t}\right)_B + \boldsymbol{\omega}_{B/S} \times \mathbf{A}$$

where $\boldsymbol{\omega}_{B/S}$ is the rotational rate of system B as viewed from system S. On the other hand, we could also write Eq. (12-9.2) as

$$\left(\frac{d\mathbf{A}}{dt}\right)_B = \left(\frac{\delta \mathbf{A}}{\delta t}\right)_S + \boldsymbol{\omega}_{S/B} \times \mathbf{A}$$

since $\boldsymbol{\omega}_{S/B} = \boldsymbol{\omega}_{B/S}$ and **A** is the same when viewed from either system.

body or when two moving bodies are connected by a sliding link. It will also simplify analyses involving several relative angular motions between connected bodies. As we shall see, the simplification is accomplished by attaching to one body a reference frame with respect to which another body has either a rectilinear or a circular motion.

For example, consider finding the velocity and acceleration of a particle B moving along a path fixed in a rigid body whose outline is shown in Fig. 12-9.3. Select a body point A whose motion is known (or can easily be found) as the origin of xyz axes fixed in the rigid body and rotating with it at the absolute rates $\boldsymbol{\omega}$ and $\boldsymbol{\alpha}$ with respect to the stationary XYZ axes (also called inertial axes or a Newtonian frame). In general spatial motion, the free vectors $\boldsymbol{\omega}$ and $\boldsymbol{\alpha}$ may have any direction and are not collinear except when the direction of $\boldsymbol{\omega}$ does not change.

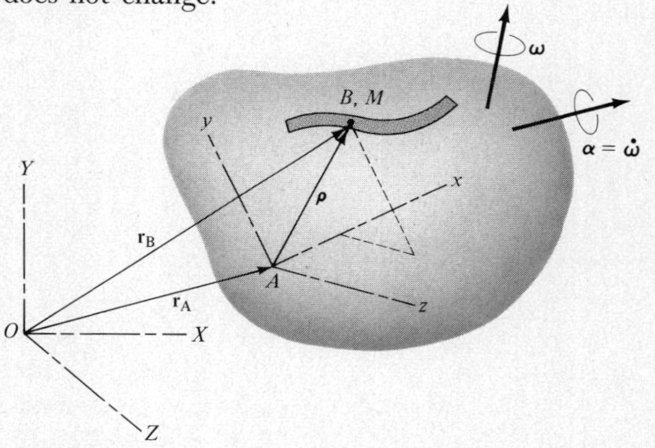

Figure 12-9.3

Let $\boldsymbol{\rho}$ be the variable length position vector of B relative to A and let $\mathbf{r}_B$ and $\mathbf{r}_A$ be the absolute position vectors of B and A. Then we have

$$\mathbf{r}_B = \mathbf{r}_A + \boldsymbol{\rho}$$

which is differentiated with respect to time to give the following velocity relation:

$$\left[\mathbf{v} = \frac{d\mathbf{r}}{dt} = \dot{\mathbf{r}}\right] \qquad \mathbf{v}_B = \dot{\mathbf{r}}_B = \dot{\mathbf{r}}_A + \dot{\boldsymbol{\rho}}$$

Since we are considering the motion of $\boldsymbol{\rho}$ relative to a rotating reference frame xyz, the absolute time derivative $\dot{\boldsymbol{\rho}}$ is found from Eq. (12-9.2) or its special equivalent Eq. (12-9.1). Using the latter equation, we obtain

$$\mathbf{v}_B = \dot{\mathbf{r}}_B = \dot{\mathbf{r}}_A + \boldsymbol{\omega} \times \boldsymbol{\rho} + \dot{\boldsymbol{\rho}}_r \qquad (12\text{-}9.3)$$

Differentiating this velocity relation with respect to time gives the following acceleration equation:

$$\mathbf{a}_B = \ddot{\mathbf{r}}_B = \ddot{\mathbf{r}}_A + \frac{d\boldsymbol{\omega}}{dt} \times \boldsymbol{\rho} + \boldsymbol{\omega} \times \frac{d\boldsymbol{\rho}}{dt} + \frac{d\dot{\boldsymbol{\rho}}_r}{dt} \tag{b}$$

Since $\boldsymbol{\rho}$ and $\dot{\boldsymbol{\rho}}_r$ are variable length vectors carried by the rotating reference frame, we must use the general result Eq. (12-9.2) to find their absolute time derivatives. Thus, we obtain

$$\frac{d\boldsymbol{\rho}}{dt} = \boldsymbol{\omega} \times \boldsymbol{\rho} + \frac{\delta\boldsymbol{\rho}}{\delta t} = \boldsymbol{\omega} \times \boldsymbol{\rho} + \dot{\boldsymbol{\rho}}_r \tag{c}$$

and

$$\frac{d}{dt}(\dot{\boldsymbol{\rho}}_r) = \boldsymbol{\omega} \times \dot{\boldsymbol{\rho}}_r + \frac{\delta\dot{\boldsymbol{\rho}}_r}{\delta t} = \boldsymbol{\omega} \times \dot{\boldsymbol{\rho}}_r + \ddot{\boldsymbol{\rho}}_r \tag{d}$$

Substituting these results back into Eq. (*b*) then gives

$$\begin{aligned}\mathbf{a}_B = \ddot{\mathbf{r}}_B &= \ddot{\mathbf{r}}_A + \dot{\boldsymbol{\omega}} \times \boldsymbol{\rho} + \boldsymbol{\omega} \times (\boldsymbol{\omega} \times \boldsymbol{\rho} + \dot{\boldsymbol{\rho}}_r) + \boldsymbol{\omega} \times \dot{\boldsymbol{\rho}}_r + \ddot{\boldsymbol{\rho}}_r \\ &= \ddot{r}_A + \dot{\boldsymbol{\omega}} \times \boldsymbol{\rho} + \boldsymbol{\omega} \times (\boldsymbol{\omega} \times \boldsymbol{\rho}) + \boldsymbol{\omega} \times \dot{\boldsymbol{\rho}}_r + \boldsymbol{\omega} \times \dot{\boldsymbol{\rho}}_r + \ddot{\boldsymbol{\rho}}_r\end{aligned}$$

which becomes finally

$$\mathbf{a}_B = \mathbf{a}_A + \boldsymbol{\alpha} \times \boldsymbol{\rho} + \boldsymbol{\omega} \times (\boldsymbol{\omega} \times \boldsymbol{\rho}) + \ddot{\boldsymbol{\rho}}_r + 2\boldsymbol{\omega} \times \dot{\boldsymbol{\rho}}_r \tag{12-9.4}$$

Both Eqs. (12-9.3) and (12-9.4) are cumbersome to remember, but they can be reinterpreted in a simple physical manner if we imagine a point M fixed in the rigid body which is instantaneously coincident with B. With this concept, the motion of M is then that of a particle in a rigid body rotating about A which we have discussed at length in previous sections. For this rigid-body rotation of M (in which $\overrightarrow{AM} = \boldsymbol{\rho}$ is a vector of constant length), the velocity of M relative to A is

$$\mathbf{v}_{M/A} = \dot{\boldsymbol{\rho}} = \boldsymbol{\omega} \times \boldsymbol{\rho} \tag{e}$$

and the acceleration of M relative to A is

$$\mathbf{a}_{M/A} = \dot{\boldsymbol{\omega}} \times \boldsymbol{\rho} + \boldsymbol{\omega} \times \dot{\boldsymbol{\rho}} = \boldsymbol{\alpha} \times \boldsymbol{\rho} + \boldsymbol{\omega} \times (\boldsymbol{\omega} \times \boldsymbol{\rho}) \tag{f}$$

On comparing Eqs. (*e*) and (*f*) with their equivalent terms in Eqs. (12-9.3) and (12-9.4), the latter equations can be interpreted as follows:

$$\mathbf{v}_B = \dot{\mathbf{r}}_A + \underbrace{\boldsymbol{\omega} \times \boldsymbol{\rho}} + \dot{\boldsymbol{\rho}}_r$$

Absolute velocity of A ($\dot{\mathbf{r}}_A$); Velocity of M due to rigid-body rotation about A ($\boldsymbol{\omega} \times \boldsymbol{\rho}$); Velocity of B relative to path fixed in xyz reference frame ($\dot{\boldsymbol{\rho}}_r$)

or

$$\mathbf{v}_B = \mathbf{v}_A + (\mathbf{v}_{M/A} = \boldsymbol{\omega} \times \boldsymbol{\rho}) + \mathbf{v}_r \tag{12-9.5}$$

Also,

$$\underbrace{\mathbf{a}_B}_{\text{Absolute acceleration of } A} = \mathbf{a}_A + \underbrace{\boldsymbol{\alpha} \times \boldsymbol{\rho} + \boldsymbol{\omega} \times (\boldsymbol{\omega} \times \boldsymbol{\rho})}_{\text{Acceleration of } M \text{ due to rigid-body rotation about } A} + \underbrace{\ddot{\boldsymbol{\rho}}_r}_{\text{Acceleration of } B \text{ relative to path fixed in } xyz \text{ reference frame}} + \underbrace{2\boldsymbol{\omega} \times \dot{\boldsymbol{\rho}}_r}_{\text{Coriolis' acceleration}}$$

or

$$\mathbf{a}_B = \mathbf{a}_A + (a_{M/A} = \boldsymbol{\alpha} \times \boldsymbol{\rho} + \boldsymbol{\omega} \times \mathbf{v}_{M/A}) + \ddot{\boldsymbol{\rho}}_r + (\mathbf{a}_c = 2\boldsymbol{\omega} \times \mathbf{v}_r) \tag{12-9.6}$$

To summarize, the motion of any point B is the vector sum of the translational motion of any convenient base point A plus the rigid-body rotation about A of a point M instantaneously coincident with B plus the relative motion of B with respect to a path fixed in the body. In addition, whenever B is moving relative to a path in the body, the acceleration equation must also include the Coriolis component of acceleration[5] indicated by $\mathbf{a}_c$ in Eq. (12-9.6).

Note that since the path along which B moves is stationary in the xyz frame, the relative velocity and acceleration of B may be determined by any of the methods developed in Chapter 9 to determine the motion of a particle in an inertial frame; i.e., in terms of rectangular components relative to the moving and rotating xyz axes, or in terms of components tangent and normal to its path in the body, or in terms of cylindrical components described by $\boldsymbol{\rho}$ with respect to the xyz axes. Remember that cylindrical components are equivalent to radial and transverse components in one plane—say the xy plane—and an axial component perpendicular to this plane—say the z axis. Generally, try to select a moving reference frame in which B will have either a linear motion or that of a pure rotation. As a general guide when several angular motions are involved, let one of them represent the relative motion $\boldsymbol{\omega}_r$ of the path to the reference frame and combine the others to form the absolute motion $\boldsymbol{\omega}$ of the frame.

The rest of this section is included only to show the generality of the relative motion approach. We show how all possible accelerations heretofore considered can be deduced by merely changing some of the conditions under which the general acceleration equation [Eq. (12-9.4)] was derived. Some of the possible variations follow:

1. *Translation of a Rigid Body*. Here the rigid body moves in such a manner that the line joining any two points A and B remains fixed in length and direction. This is equivalent to having $\boldsymbol{\omega} = \boldsymbol{\alpha} = 0$

[5] Named after the French engineer and scientist, G. C. Coriolis (1792–1842) who first called attention to this component of acceleration.

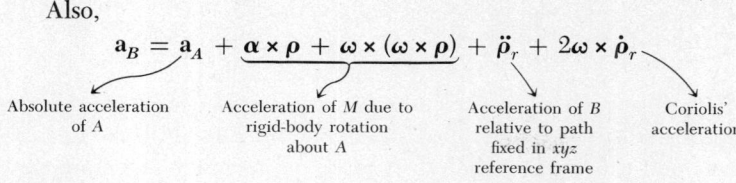

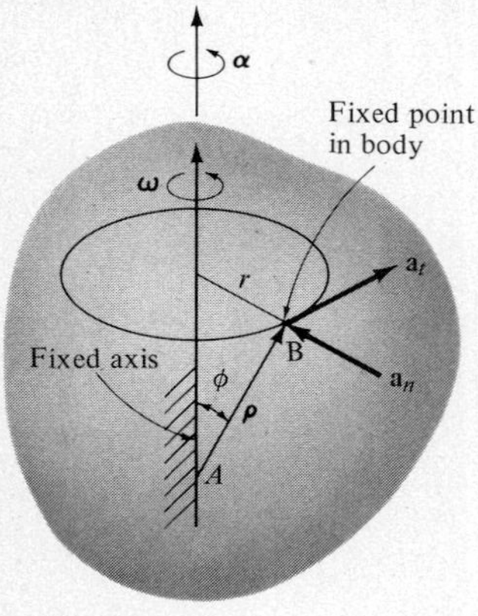

Figure 12-9.4

and $\dot{\boldsymbol{\rho}}_r = \ddot{\boldsymbol{\rho}}_r = 0$. Then Eq. (12-9.4) becomes

$$\mathbf{a}_B = \mathbf{a}_A$$

which means that all points in a translating rigid body have the same acceleration. The acceleration of the body is caused by the variation in $\mathbf{r}_A$ and may be computed in terms of rectangular components, or normal and tangential components, or cylindrical components. Frequently, the motion is parallel to the XY plane so that cylindrical components reduce to the case of radial and transverse components.

2. *Rotation of a Rigid Body About a Fixed Axis.* This is equivalent to fixing point A in space and keeping the position vector $\boldsymbol{\rho}$ from A to any other point B fixed in length and at a constant inclination ϕ with the axis of rotation. Then B describes a circle centered on the axis of rotation the radius of which is $r = \rho \sin \phi$ as shown in Fig. 12-9.4. Here $\mathbf{a}_A = 0$ and $\dot{\boldsymbol{\rho}}_r = \ddot{\boldsymbol{\rho}}_r = 0$ and Eq. (12-9.4) reduces to

$$\mathbf{a}_{B/A} = \boldsymbol{\alpha} \times \boldsymbol{\rho} + \boldsymbol{\omega} \times (\boldsymbol{\omega} \times \boldsymbol{\rho})$$

or

$$a_{B/A} = r\alpha \;{+\!\!\!\!\rightarrow}\; r\omega^2 \tag{g}$$

3. *Plane Motion of a Rigid Body Moving Parallel to the XY Plane.* Here point A moves while point B describes a circle of radius r centered on a Z axis through A. However, $\boldsymbol{\rho}$ is fixed in length so that $\dot{\boldsymbol{\rho}}_r = \ddot{\boldsymbol{\rho}}_r = 0$ and Eq. (12-9.4) becomes

$$\mathbf{a}_B = \mathbf{a}_A + \boldsymbol{\alpha} \times \boldsymbol{\rho} + \boldsymbol{\omega} \times (\boldsymbol{\omega} \times \boldsymbol{\rho})$$

or

$$a_B = a_A \;{+\!\!\!\!\rightarrow}\; (a_{B/A} = r\alpha \;{+\!\!\!\!\rightarrow}\; r\omega^2) \tag{h}$$

Observe that the alternate form of vector notation used in Eqs. (g) and (h) indicates only the magnitudes of the quantities to be combined. Their directions depend on the physical meanings of these quantities; for example, that the normal component $r\omega^2$ is always directed toward the center of rotation of the path while the tangential component $r\alpha$ is tangent to the path in the sense determined by α.

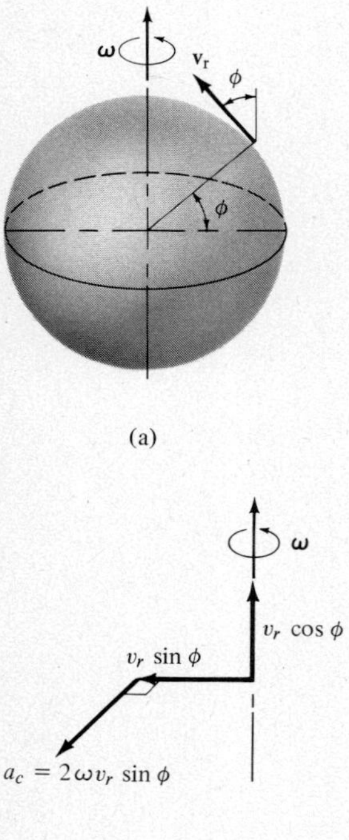

(a)

(b)

Figure 12-9.5

4. *Coriolis' Acceleration.* To clarify the application of the Coriolis acceleration (hereafter denoted by a_c), consider a particle moving along a meridian of the earth as shown in Fig. 12-9.5. Since the earth rotates about its axis, we have here the case of a particle moving along a rotating path. In effect we are referring the motion of the particle to an inertial origin at the center of the earth.

According to Eq. (12-9.6) the Coriolis acceleration is given by $\mathbf{a}_c = 2\boldsymbol{\omega} \times \mathbf{v}_r$. From the definition of a cross product, its magnitude is $a_c = 2\omega v_r \sin \phi$ and it is directed perpendicular to the plane of $\boldsymbol{\omega}$ and $\mathbf{v}_r$ in the direction a right-hand screw would advance as $\boldsymbol{\omega}$ is turned toward $\mathbf{v}_r$. Looking now at (b) of Fig. 12-9.5, we see that at the latitude ϕ, the velocity $\mathbf{v}_r$ has the components $v_r \sin\phi$ and $v_r \cos\phi$. The component $v_r \sin\phi$ is perpendicular to $\boldsymbol{\omega}$ and causes $a_c = 2\omega v_r \sin\phi$ while the other component $v_r \cos\phi$ is parallel to $\boldsymbol{\omega}$ and its direction is unchanged by $\boldsymbol{\omega}$. In other words, the Coriolis acceleration is caused only by that component of $\mathbf{v}_r$ that is perpendicular to $\boldsymbol{\omega}$ and its value is twice the rate at which the tip of that velocity component is rotated by $\boldsymbol{\omega}$. Its direction is determined by the sense of $\boldsymbol{\omega}$ as is shown in Fig. 12-9.5b.

ILLUSTRATIVE PROBLEMS

12-9.1. The 6-in. crank OA of an oscillating quick-return mechanism is rotating counterclockwise at a constant rate of 10 rad/sec. When the mechanism is in the position shown in Fig. 12-9.6, determine the angular acceleration of arm BD.

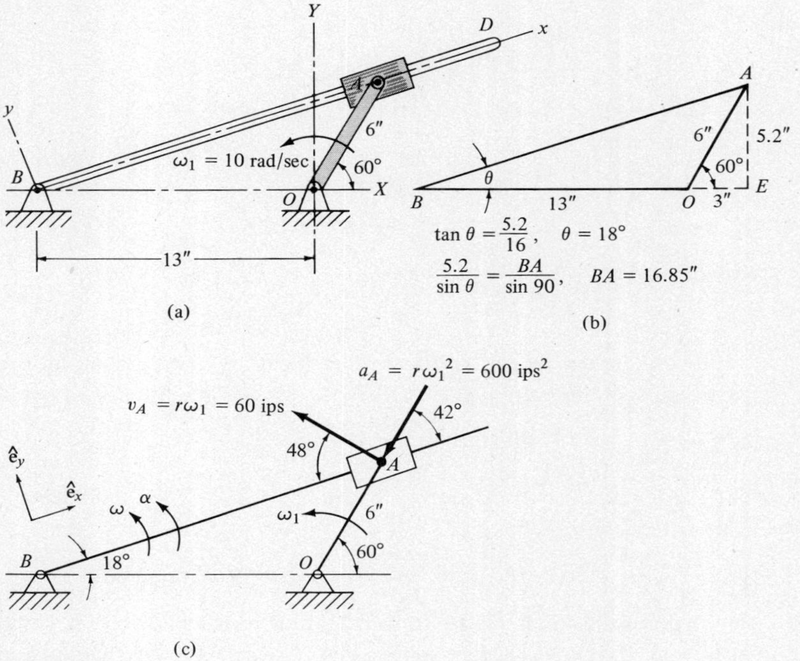

Figure 12-9.6

Solution

Here we present an alternate solution to Illus. Prob. 9-8.2, this time using a rotating xyz reference frame attached to BD at B. Then A has a rectilinear motion in this frame. Simultaneously as part of the crank OA, point A has pure rotation about O which causes the absolute values of v_A and a_A shown in part (c). From the preliminary geometry indicated in part (b), we find $\theta = 18°$ and $BA = 16.85$ in. Expressing v_A and a_A in terms of unit vectors $\hat{\mathbf{e}}_x$ and $\hat{\mathbf{e}}_y$ of the rotating xyz frame,[6] we have

$$\mathbf{v}_A = -60 \cos 48° \hat{\mathbf{e}}_x + 60 \sin 48° \hat{\mathbf{e}}_y \qquad (a)$$

$$\mathbf{a}_A = -600 \cos 42° \hat{\mathbf{e}}_x - 600 \sin 42° \hat{\mathbf{e}}_y \qquad (b)$$

When we consider the motion of A in the xyz frame where $BA = |\boldsymbol{\rho}| = 16.85$ in., and M is a point in BD instantaneously coincident with A, Eqs. (12-9.5) and (12-9.6) give

$$\begin{aligned} \mathbf{v}_A &= \mathbf{v}_B + (\mathbf{v}_{M/B} = \boldsymbol{\omega} \times \boldsymbol{\rho}) + \mathbf{v}_r \\ &= 0 + 16.85\omega \hat{\mathbf{e}}_y + v_r \hat{\mathbf{e}}_x \qquad (c) \end{aligned}$$

and

$$\begin{aligned} \mathbf{a}_A &= \mathbf{a}_B + (a_{M/B} = \boldsymbol{\alpha} \times \boldsymbol{\rho} + \boldsymbol{\omega} \times \mathbf{v}_{M/B}) + \mathbf{a}_r + (\mathbf{a}_c = 2\boldsymbol{\omega} \times \mathbf{v}_r) \\ &= 0 + 16.85\alpha \hat{\mathbf{e}}_y - 16.85\omega^2 \hat{\mathbf{e}}_x + a_r \hat{\mathbf{e}}_x + 2\omega v_r \hat{\mathbf{e}}_y \qquad (d) \end{aligned}$$

On equating the coefficients of the unit vectors in Eqs. (a) and (c) for $\mathbf{v}_A$, we get

[$\hat{\mathbf{e}}_x$ terms] $\qquad -60 \cos 48° = v_r$

[$\hat{\mathbf{e}}_y$ terms] $\qquad 60 \sin 48° = 16.85\,\omega$

which yield

$$v_r = -40.1 \text{ ips} \quad \text{and} \quad \omega = 2.64 \text{ rad/sec} \circlearrowleft$$

These values of v_r and ω are now used when equating the coefficients of the unit vectors in Eqs. (b) and (d) for $\mathbf{a}_A$ to yield

[$\hat{\mathbf{e}}_x$ terms] $\qquad -600 \cos 42° = -16.85(2.64)^2 + a_r$

[$\hat{\mathbf{e}}_y$ terms] $\qquad -600 \sin 42° = 16.85\alpha + 2(2.64)(-40.1)$

from which we obtain

$$a_r = -369 \text{ ips}^2 \quad \text{and} \quad \alpha = -11.2 = 11.2 \text{ rad/sec}^2 \circlearrowright$$

A comparison of this solution with that of Illus. Prob. 9-8.4 will show that they are exactly equivalent. Notice that the radial and transverse components of the motion of A in BD are automatically included here.

12-9.2. As shown in Fig. 12-9.7, a particle P is moving at a constant speed v around a circular tube carried by the square plat-

[6] When the axes of the rotating xyz frame have different directions from those of the fixed XYZ frame, we suggest using $\hat{\mathbf{e}}_x$, $\hat{\mathbf{e}}_y$, and $\hat{\mathbf{e}}_z$ as unit vectors for the rotating frame to avoid confusion with $\hat{\mathbf{i}}$, $\hat{\mathbf{j}}$, and $\hat{\mathbf{k}}$ of the fixed frame.

form. The platform is rotating about the axis AB at a constant rate of ω rad/sec. Determine the velocity and acceleration of P at the given instant.

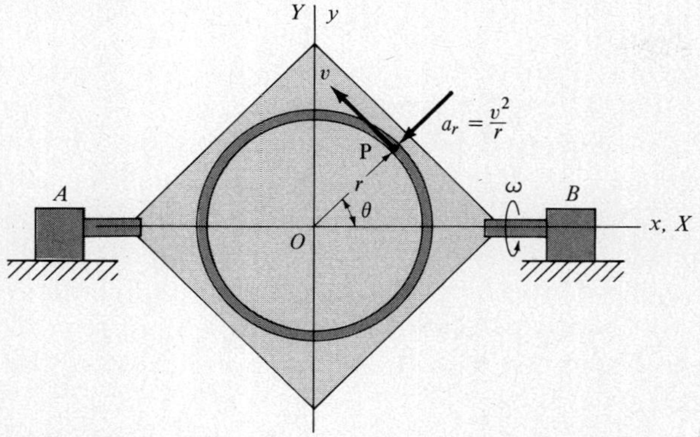

Figure 12-9.7

Solution

Attach the reference frame xyz to rotate with the plate at the rate $\boldsymbol{\omega} = \omega\hat{\mathbf{i}}$, and note that $\dot{\boldsymbol{\omega}} = 0$. With respect to this frame, P is rotating about O with a relative velocity v and a relative acceleration $a_r = v^2/r$.

Applying Eq. (12-9.5), we have

$$\begin{aligned}\mathbf{v}_P &= \mathbf{v}_O + (\mathbf{v}_{P/O} = \boldsymbol{\omega} \times \boldsymbol{\rho}) + v_r \\ &= 0 + \omega\hat{\mathbf{i}} \times (r\cos\theta\hat{\mathbf{i}} + r\sin\theta\hat{\mathbf{j}}) - v\sin\theta\hat{\mathbf{i}} + v\cos\theta\hat{\mathbf{j}} \\ &= r\omega\sin\theta\hat{\mathbf{k}} - v\sin\theta\hat{\mathbf{i}} + v\cos\theta\hat{\mathbf{j}} \qquad \textit{Ans.}\end{aligned}$$

The acceleration is now found by using Eq. (12-9.6):

$$\begin{aligned}\mathbf{a}_P &= \mathbf{a}_O + (\mathbf{a}_{P/O} = \dot{\boldsymbol{\omega}} \times \boldsymbol{\rho} + \boldsymbol{\omega} \times v_{P/O}) + \mathbf{a}_r + (\mathbf{a}_c = 2\boldsymbol{\omega} \times v_r) \\ &= 0 + [0 + \omega\hat{\mathbf{i}} \times (r\omega\sin\theta\hat{\mathbf{k}})] + \frac{v^2}{r}(-\cos\theta\hat{\mathbf{i}} - \sin\theta\hat{\mathbf{j}}) \\ &\quad + 2\omega\hat{\mathbf{i}} \times (-v\sin\theta\hat{\mathbf{i}} + v\cos\theta\hat{\mathbf{j}}) \\ &= -\frac{v^2}{r}\cos\theta\hat{\mathbf{i}} - \left(r\omega^2\sin\theta + \frac{v^2}{r}\sin\theta\right)\hat{\mathbf{j}} + 2\omega v\cos\theta\hat{\mathbf{k}} \qquad \textit{Ans.}\end{aligned}$$

Alternate Solution

If the xyz reference frame were chosen to rotate with P in the plane of the plate, its angular velocity would be

$$\boldsymbol{\Omega} = \omega\hat{\mathbf{i}} + \frac{v}{r}\hat{\mathbf{k}}$$

and its angular acceleration would be

$$\dot{\boldsymbol{\Omega}} = \omega\hat{\mathbf{i}} \times \frac{v}{r}\hat{\mathbf{k}} = -\frac{\omega v}{r}\hat{\mathbf{j}}$$

Note that P has no relative motion with respect to this frame; actually we revert to an absolute motion analysis. However, show that applying Eqs. (12-9.5) and (12-9.6) literally gives the answers obtained above.

12-9.3. Solve Illus. Prob. 12-8.2 using relative motion analysis. For convenience Fig. 12-8.4 is repeated here as Fig. 12-9.8. The pertinent data are: At the given instant when $\tan\theta = \frac{3}{4}$, the 5-ft bar AB is rotating about an axis at A with the rates $\omega_2 = 2$ rad/sec and $\dot{\omega}_2 = -3$ rad/sec^2. Simultaneously, the platform to which this axis is attached is spinning about AC at the rates $\omega_1 = 3$ rad/sec and $\dot{\omega}_1 = 1$ rad/sec^2 while the bent bar ACD supporting the platform is rotating about the fixed axis DC at the rates $\omega_3 = -4$ rad/sec and $\dot{\omega}_3 = -2$ rad/sec^2. Determine the velocity and acceleration of the end B of bar AB.

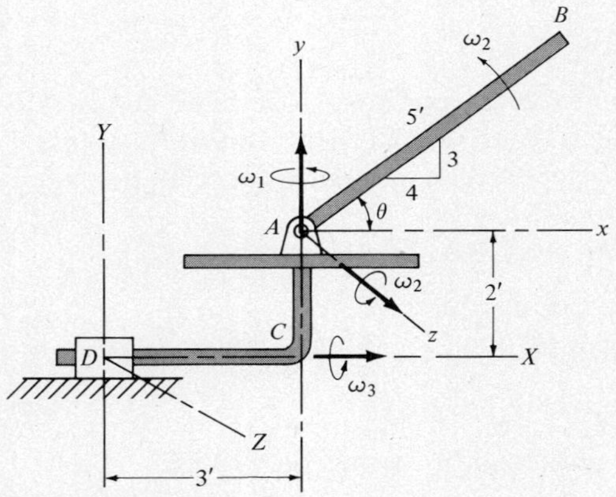

Figure 12-9.8

Solution

When several sets of angular motions are involved, let one of them be used to define motion relative to the moving reference frame and combine the others to determine the absolute angular properties of the frame. Thus, select reference axes xyz attached to the platform at A and which are instantaneously parallel to the fixed axes XYZ. With respect to these axes, the given rates of rotation are

$$\boldsymbol{\omega}_1 = 3\hat{\mathbf{j}} \qquad \boldsymbol{\omega}_2 = 2\hat{\mathbf{k}} \qquad \boldsymbol{\omega}_3 = -4\hat{\mathbf{i}}$$
$$\dot{\boldsymbol{\omega}}_1 = \hat{\mathbf{j}} \qquad \dot{\boldsymbol{\omega}}_2 = -3\hat{\mathbf{k}} \qquad \dot{\boldsymbol{\omega}}_3 = -2\hat{\mathbf{i}}$$

With respect to the xyz reference frame, the bar AB is rotating instantaneously at the relative rates $\boldsymbol{\omega}_r = 2\hat{\mathbf{k}}$ rad/sec and $\dot{\boldsymbol{\omega}}_r = -3\hat{\mathbf{k}}$ rad/sec^2. The absolute rates for the xyz axes are

$$\boldsymbol{\omega} = \boldsymbol{\omega}_3 + \boldsymbol{\omega}_1 = -4\hat{\mathbf{i}} + 3\hat{\mathbf{j}} \text{ rad/sec}$$

and

$$\begin{aligned}\dot{\boldsymbol{\omega}} &= \dot{\boldsymbol{\omega}}_3 + \dot{\boldsymbol{\omega}}_1 + \boldsymbol{\omega}_3 \times \boldsymbol{\omega}_1 \\ &= -2\hat{\mathbf{i}} + \hat{\mathbf{j}} + (-4\hat{\mathbf{i}}) \times 3\hat{\mathbf{j}} = -2\hat{\mathbf{i}} + \hat{\mathbf{j}} - 12\hat{\mathbf{k}} \text{ rad/sec}^2\end{aligned}$$

Velocity Analysis

Let M be a point fixed in xyz which coincides with the moving point B. For subsequent use, observe that

$$\mathbf{r}_{DA} = 3\hat{\mathbf{i}} + 2\hat{\mathbf{j}} \quad \text{and} \quad \boldsymbol{\rho}_{AB} = 4\hat{\mathbf{i}} + 3\hat{\mathbf{j}}$$

where we use $\mathbf{r}$ to denote an absolute position vector whereas $\boldsymbol{\rho}$ is used to denote a position vector in the moving reference frame. Applying Eq. (12-9.5), we have

$$\mathbf{v}_B = \mathbf{v}_A + (\mathbf{v}_{M/A} = \boldsymbol{\omega} \times \boldsymbol{\rho}) + (\mathbf{v}_r = \boldsymbol{\omega}_r \times \boldsymbol{\rho})$$

where

$$\begin{aligned}\mathbf{v}_A &= \boldsymbol{\omega}_3 \times \mathbf{r}_{DA} = -4\hat{\mathbf{i}} \times (3\hat{\mathbf{i}} + 2\hat{\mathbf{j}}) = -8\hat{\mathbf{k}} \\ \mathbf{v}_{M/A} &= \boldsymbol{\omega} \times \boldsymbol{\rho} = (-4\hat{\mathbf{i}} + 3\hat{\mathbf{j}}) \times (4\hat{\mathbf{i}} + 3\hat{\mathbf{j}}) = -24\hat{\mathbf{k}} \\ \mathbf{v}_r &= \boldsymbol{\omega}_r \times \boldsymbol{\rho} = 2\hat{\mathbf{k}} \times (4\hat{\mathbf{i}} + 3\hat{\mathbf{j}}) = 8\hat{\mathbf{j}} - 6\hat{\mathbf{i}}\end{aligned}$$

Summing up:

$$\mathbf{v}_B = -6\hat{\mathbf{i}} + 8\hat{\mathbf{j}} - 32\hat{\mathbf{k}} \text{ fps} \qquad \textit{Ans.}$$

Acceleration Analysis

Applying Eq. (12-9.6), we obtain

$$\mathbf{a}_B = \mathbf{a}_A + (\mathbf{a}_{M/A} = \dot{\boldsymbol{\omega}} \times \boldsymbol{\rho} + \boldsymbol{\omega} \times \mathbf{v}_{M/A}) + \mathbf{a}_r + (\mathbf{a}_c = 2\boldsymbol{\omega} \times v_r)$$

where

$$\begin{aligned}\mathbf{a}_A &= \dot{\boldsymbol{\omega}}_3 \times \mathbf{r}_{DA} + \boldsymbol{\omega}_3 \times \mathbf{v}_A \\ &= -2\hat{\mathbf{i}} \times (3\hat{\mathbf{i}} + 2\hat{\mathbf{j}}) + (-4\hat{\mathbf{i}}) \times (-8\hat{\mathbf{k}}) = -4\hat{\mathbf{k}} - 32\hat{\mathbf{j}}\end{aligned}$$

To compute $\mathbf{a}_{M/A}$, it is computationally easier to express the cross products as the following determinants, using the known values of $\dot{\boldsymbol{\omega}}$, $\boldsymbol{\rho}$, $\boldsymbol{\omega}$, and $\mathbf{v}_{M/A}$.

$$\mathbf{a}_{M/A} = \begin{vmatrix} -2 & 1 & -12 \\ 4 & 3 & 0 \\ \hat{\mathbf{i}} & \hat{\mathbf{j}} & \hat{\mathbf{k}} \end{vmatrix} + \begin{vmatrix} -4 & 3 & 0 \\ 0 & 0 & -24 \\ \hat{\mathbf{i}} & \hat{\mathbf{j}} & \hat{\mathbf{k}} \end{vmatrix} = -36\hat{\mathbf{i}} - 144\hat{\mathbf{j}} - 10\hat{\mathbf{k}}$$

$$\begin{aligned}\mathbf{a}_r &= \dot{\boldsymbol{\omega}}_r \times \boldsymbol{\rho} + \boldsymbol{\omega}_r \times \mathbf{v}_r \\ &= -3\hat{\mathbf{k}} \times (4\hat{\mathbf{i}} + 3\hat{\mathbf{j}}) + 2\hat{\mathbf{k}} \times (8\hat{\mathbf{j}} - 6\hat{\mathbf{i}}) = -7\hat{\mathbf{i}} - 24\hat{\mathbf{j}} \\ \mathbf{a}_c &= 2\boldsymbol{\omega} \times \mathbf{v}_r = 2(-4\hat{\mathbf{i}} + 3\hat{\mathbf{j}}) \times (6\hat{\mathbf{i}} + 8\hat{\mathbf{j}}) = -28\hat{\mathbf{k}}\end{aligned}$$

Summing up:

$$\mathbf{a}_B = -43\hat{\mathbf{i}} - 200\hat{\mathbf{j}} - 42\hat{\mathbf{k}} \text{ fps}^2 \qquad \textit{Ans.}$$

As expected, these answers verify those of Illus. Prob. 12-8.2. The effort involved in both problems is about equal, but more care was required in evaluating and applying $\boldsymbol{\omega}$ and $\dot{\boldsymbol{\omega}}$ in Illus. Prob. 12-8.2.

Alternate Solution

Referring again to Fig. 12-9.8 but using another approach, attach the moving reference frame xyz at A, but this time to rotate with the bar ACD with the absolute angular properties

$$\boldsymbol{\omega} = \boldsymbol{\omega}_3 = -4\hat{\mathbf{i}} \text{ rad/sec} \quad \text{and} \quad \dot{\boldsymbol{\omega}} = \dot{\boldsymbol{\omega}}_3 = -2\hat{\mathbf{i}} \text{ rad/sec}^2$$

The relative angular motion of the platform and bar AB with respect to the moving reference frame now is

$$\boldsymbol{\omega}_r = \boldsymbol{\omega}_1 + \boldsymbol{\omega}_2 = 3\hat{\mathbf{j}} + 2\hat{\mathbf{k}} \text{ rad/sec}$$

$$\dot{\boldsymbol{\omega}}_r = \dot{\boldsymbol{\omega}}_1 + \dot{\boldsymbol{\omega}}_2 + \boldsymbol{\omega}_1 \times \boldsymbol{\omega}_2 = 6\hat{\mathbf{i}} + \hat{\mathbf{j}} - 3\hat{\mathbf{k}} \text{ rad/sec}^2$$

Velocity Analysis

The motion of B relative to xyz is now more complex than in the preceding solution, but Eqs. (12-9.5) and (12-9.6) are so general that they automatically compensate for this. As before, let M be a point in xyz that instantaneously coincides with B. The velocity of B is

$$\mathbf{v}_B = \mathbf{v}_A + \mathbf{v}_{M/A} + \mathbf{v}_r$$

where

$$\mathbf{v}_A = \boldsymbol{\omega}_3 \times \mathbf{r}_{DA} = -4\hat{\mathbf{i}} \times (3\hat{\mathbf{i}} + 2\hat{\mathbf{j}}) = -8\hat{\mathbf{k}}$$

$$\mathbf{v}_{M/A} = \boldsymbol{\omega} \times \boldsymbol{\rho} = -4\hat{\mathbf{i}} \times (4\hat{\mathbf{i}} + 3\hat{\mathbf{j}}) = -12\hat{\mathbf{k}}$$

$$\mathbf{v}_r = \boldsymbol{\omega}_r \times \boldsymbol{\rho} = \begin{vmatrix} 0 & 3 & 2 \\ 4 & 3 & 0 \\ \hat{\mathbf{i}} & \hat{\mathbf{j}} & \hat{\mathbf{k}} \end{vmatrix} = -6\hat{\mathbf{i}} + 8\hat{\mathbf{j}} - 12\hat{\mathbf{k}}$$

Summing up:

$$\mathbf{v}_B = -6\hat{\mathbf{i}} + 8\hat{\mathbf{j}} - 32\hat{\mathbf{k}} \text{ fps} \qquad \textit{Ans.}$$

Acceleration Analysis

The acceleration of B is given by

$$\mathbf{a}_B = \mathbf{a}_A + \mathbf{a}_{M/A} + \mathbf{a}_r + \mathbf{a}_c$$

where

$$\mathbf{a}_A = \dot{\boldsymbol{\omega}}_3 \times \mathbf{r}_{DA} + \boldsymbol{\omega}_3 \times \mathbf{v}_A$$

$$= -2\hat{\mathbf{i}} \times (3\hat{\mathbf{i}} + 2\hat{\mathbf{j}}) + (-4\hat{\mathbf{i}}) \times (-8\hat{\mathbf{k}}) = -4\hat{\mathbf{k}} - 32\hat{\mathbf{j}}$$

$$\mathbf{a}_{M/A} = \dot{\boldsymbol{\omega}} \times \boldsymbol{\rho} + \boldsymbol{\omega} \times \mathbf{v}_{M/A}$$
$$= -2\hat{\mathbf{i}} \times (4\hat{\mathbf{i}} + 3\hat{\mathbf{j}}) + (-4\hat{\mathbf{i}}) \times (-12\hat{\mathbf{k}}) = -6\hat{\mathbf{k}} - 48\hat{\mathbf{j}}$$

To compute $\mathbf{a}_r$, it is simpler to express the following cross products as determinants, using the previously determined values of $\dot{\boldsymbol{\omega}}_r$, $\boldsymbol{\rho}$, $\boldsymbol{\omega}_r$, and $\mathbf{v}_r$.

$$\mathbf{a}_r = \dot{\boldsymbol{\omega}}_r \times \boldsymbol{\rho} + \boldsymbol{\omega}_r \times \mathbf{v}_r$$
$$= \begin{vmatrix} 6 & 1 & -3 \\ 4 & 3 & 0 \\ \hat{\mathbf{i}} & \hat{\mathbf{j}} & \hat{\mathbf{k}} \end{vmatrix} + \begin{vmatrix} 0 & 3 & 2 \\ -6 & 8 & -12 \\ \hat{\mathbf{i}} & \hat{\mathbf{j}} & \hat{\mathbf{k}} \end{vmatrix} = -43\hat{\mathbf{i}} - 24\hat{\mathbf{j}} + 32\hat{\mathbf{k}}$$
$$\mathbf{a}_c = 2\boldsymbol{\omega} \times \mathbf{v}_r = -8\hat{\mathbf{i}} \times (-6\hat{\mathbf{i}} + 8\hat{\mathbf{j}} - 12\hat{\mathbf{k}}) = -64\hat{\mathbf{k}} - 96\hat{\mathbf{j}}$$

Summing up:

$$\mathbf{a}_B = \hat{\mathbf{i}}(-43) + \hat{\mathbf{j}}(-32 - 48 - 24 - 96) + \hat{\mathbf{k}}(-4 - 6 + 32 - 64)$$
$$= -43\hat{\mathbf{i}} - 200\hat{\mathbf{j}} - 42\hat{\mathbf{k}} \text{ fps}^2 \qquad \textit{Ans.}$$

12-9.4. At the instant shown in Fig. 12-9.9, the connected system of bodies A and B is moving relative to the boom CD at $v = 13$ fps and $a = 6.5$ fps^2. Simultaneously the boom is rising by rotating about C at the constant rate of $\omega_2 = 0.4$ rad/sec while the platform carrying the boom is rotating about the fixed Y axis at the constant rate of $\omega_1 = 0.5$ rad/sec. Determine the absolute velocity and acceleration of body A at the given instant when $\tan\theta = \frac{5}{12}$.

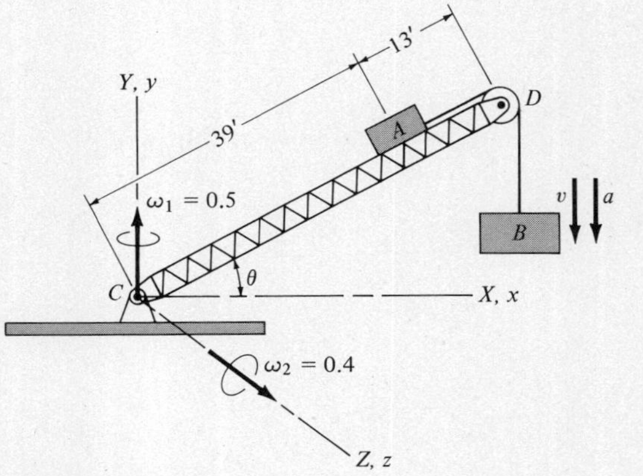

Figure 12-9.9

Preliminary

To reinforce understanding of the concepts involved, review again Fig. 12-9.3. Then you will see that if we select reference axes xyz

at C fixed in and moving with the boom, we must combine the rigid-body rotation of a coincident point M fixed in xyz axes with the rectilinear motion of A as it moves along a track in the boom. In effect, we combine an absolute motion analysis of M with the relative motion of A with respect to M. Alternatively, we may select reference axes xyz at C fixed in and attached to the platform which simplifies the absolute motion analysis of the coincident point M, but requires the relative motion of A in xyz to be described in terms of radial and transverse components. In either case, it should be understood that the absolute motion of A is the combination of the rigid-body motion of a coincident point M fixed in the xyz reference frame plus the motion of A relative to the reference frame.

Solution

Here we use the first approach and leave the second approach to be solved as Prob. 12-9.20 in the exercises. Attach a reference frame xyz at C fixed in and having the motion of the boom. Relative to this frame, A has rectilinear motion with $v_r = 13$ fps and $a_r = 6.5$ fps^2. To facilitate computation, let the directions of the xyz axes be instantaneously parallel to the inertial XYZ axes.

The absolute angular properties of the reference frame are

$$\boldsymbol{\omega} = \boldsymbol{\omega}_1 + \boldsymbol{\omega}_2 = 0.5\hat{\mathbf{j}} + 0.4\hat{\mathbf{k}} \text{ rad/sec}$$

$$\dot{\boldsymbol{\omega}} = \dot{\boldsymbol{\omega}}_1 + \dot{\boldsymbol{\omega}}_2 + \boldsymbol{\omega}_1 \times \boldsymbol{\omega}_2 = 0.5\hat{\mathbf{j}} \times 0.4\hat{\mathbf{k}} = 2\hat{\mathbf{i}} \text{ rad/sec}^2$$

Note that although $\dot{\boldsymbol{\omega}}_1$ and $\dot{\boldsymbol{\omega}}_2$ are both zero, the omega theorem gives the xyz frame an angular acceleration as a result of $\boldsymbol{\omega}_2$ changing its direction at the rate $\boldsymbol{\omega}_1$. Also note that the absolute position vector of M is $\mathbf{r}_{CM} = 36i + 15j$ ft.

Velocity Analysis

Since we are combining the absolute rigid-body rotation of M with the relative motion of A in xyz, Eq. (12-9.5) becomes

$$\mathbf{v}_A = \mathbf{v}_C + \mathbf{v}_{M/C} + v_r$$

where

$$\mathbf{v}_C = 0$$

$$\mathbf{v}_{M/C} = \boldsymbol{\omega} \times \mathbf{r}_{CM} = \begin{vmatrix} 0 & 0.5 & 0.4 \\ 36 & 15 & 0 \\ \hat{\mathbf{i}} & \hat{\mathbf{j}} & \hat{\mathbf{k}} \end{vmatrix} = -6\hat{\mathbf{i}} + 14.4\hat{\mathbf{j}} - 18\hat{\mathbf{k}}$$

$$\mathbf{v}_r = v_r(\cos\theta\,\hat{\mathbf{i}} + \sin\theta\,\hat{\mathbf{j}}) = 13(\tfrac{12}{13}\hat{\mathbf{i}} + \tfrac{5}{13}\hat{\mathbf{j}}) = 12\hat{\mathbf{i}} + 5\hat{\mathbf{k}}$$

Summing up terms, we get

$$\mathbf{v}_A = 6\hat{\mathbf{i}} + 19.4\hat{\mathbf{j}} - 18\hat{\mathbf{k}} \text{ fps} \qquad \textit{Ans.}$$

Acceleration Analysis

Applying Eq. (12-9.6) the acceleration of A is expressed by

$$\mathbf{a}_A = \mathbf{a}_C + \mathbf{a}_{M/C} + \mathbf{a}_r + \mathbf{a}_C$$

where

$$\mathbf{a}_C = 0$$

$$\mathbf{a}_{M/C} = \dot{\boldsymbol{\omega}} \times \mathbf{r}_{CM} + \boldsymbol{\omega} \times \mathbf{v}_{M/C}$$

$$= 0.2\hat{\mathbf{i}} \times (36\hat{\mathbf{i}} + 15\hat{\mathbf{j}}) + \begin{vmatrix} 0 & 0.5 & 0.4 \\ -6 & 14.4 & -18 \\ \hat{\mathbf{i}} & \hat{\mathbf{j}} & \hat{\mathbf{k}} \end{vmatrix}$$

$$= -14.76\hat{\mathbf{i}} - 2.4\hat{\mathbf{j}} + 6\hat{\mathbf{k}}$$

$$\mathbf{a}_r = 6.5(\tfrac{12}{13}\hat{\mathbf{i}} + \tfrac{5}{13}\hat{\mathbf{j}}) = 6\hat{\mathbf{i}} + 2.5\hat{\mathbf{j}}$$

$$\mathbf{a}_C = 2\boldsymbol{\omega} \times \mathbf{v}_r = 2\begin{vmatrix} 0 & 0.5 & 0.4 \\ 12 & 5 & 0 \\ \hat{\mathbf{i}} & \hat{\mathbf{j}} & \hat{\mathbf{k}} \end{vmatrix} = -4\hat{\mathbf{i}} + 9.6\hat{\mathbf{j}} - 12\hat{\mathbf{k}}$$

Summing up gives

$$\mathbf{a}_A = -12.76\hat{\mathbf{i}} + 9.7\hat{\mathbf{j}} - 6\hat{\mathbf{k}}\ \text{fps}^2 \qquad \textit{Ans.}$$

Looking ahead to the dynamics of spatial systems, the complexity of determining absolute accelerations indicates how difficult it may be to apply Newton's law $\mathbf{F} = m\mathbf{a}$ which is valid only in an inertial system. Thus, to apply Newton's law to each of the bodies in the connected system here of A and B would require the absolute acceleration of each body. Fortunately, other methods are available, such as those developed by Lagrange and Hamilton.

PROBLEMS

12-9.5. In Fig. P-12-9.5, a disk rotates about a vertical axis at A at $\omega_2 = 10$ rad/sec relative to the wheel carrying the disk. Meanwhile, the wheel rotates about a vertical axis at O at $\omega_1 = 5$ rad/sec. Both ω_1 and ω_2 are constant. Find the velocity and acceleration of particles B and C on the rim of the disk at the given instant.

$$\mathbf{a}_B = -100\hat{\mathbf{i}} - 450\hat{\mathbf{k}}\ \text{fps}^2;\ \mathbf{a}_C = -550\hat{\mathbf{i}}\ \text{fps}^2 \qquad \textit{Ans.}$$

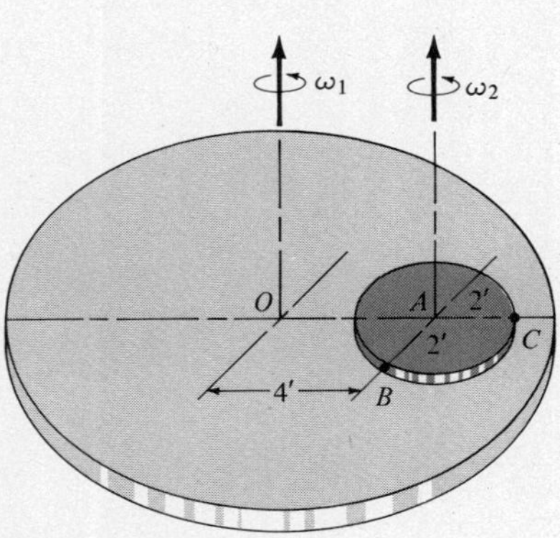

Figure P-12-9.5

12-9.6. In the top and front views of the assembly shown in Fig. P-12-9.6, the arm OA is rotating counterclockwise at a constant rate of 3 rad/sec. The arm carries a wheel at A which rotates relative to the arm at a constant clockwise rate of 5 rad/sec. In the wheel is a slot along which a particle B moves outward from A at a speed of 6 fps relative to the

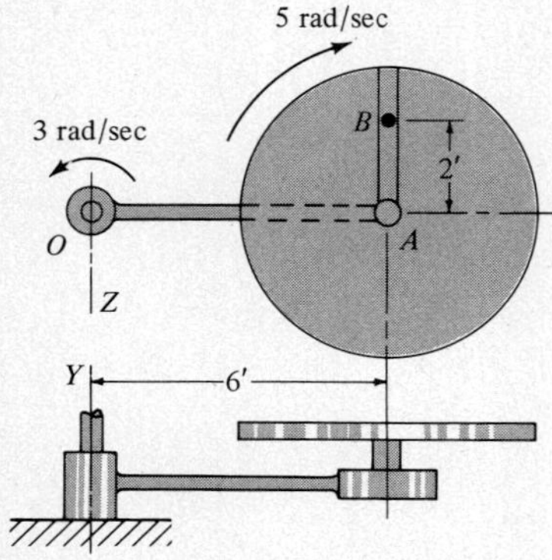

Figure P-12-9.6

wheel and is decelerating at 10 fps^2. At the given instant, B is 2 ft from A. Find the absolute acceleration of B at this instant.

$$\mathbf{a}_B = -30\hat{\mathbf{i}} + 18\hat{\mathbf{k}}\text{ fps}^2;\ a_B = 35\text{ fps}^2 \qquad \textbf{\textit{Ans.}}$$

12-9.7. In Fig. P-12-9.7, a particle P moves in a circular tube of 2-ft radius at a constant speed of 6 fps relative to the tube. If the tube is rotating about a horizontal axis at A with a counterclockwise velocity of 3 rad/sec, find its angular acceleration α that will cause the absolute acceleration of P to be horizontal at the position shown.

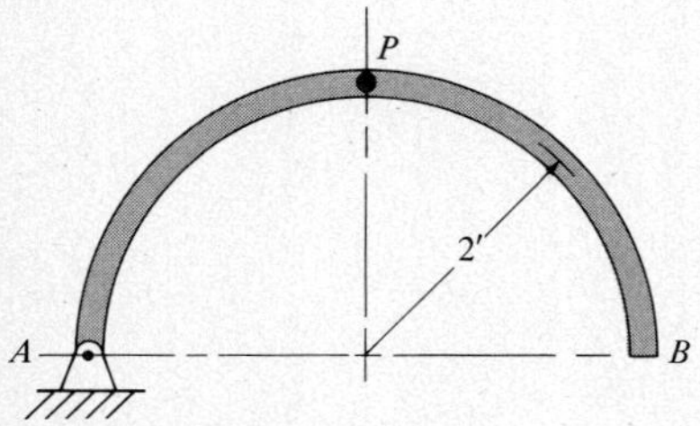

Figure P-12-9.7

12-9.8. At the instant shown in Fig. P-12-9.8, a particle P is moving at a constant speed of 10 ips outward along BC in the bent tube ABC. The tube is welded to a wheel which rotates at a constant counterclockwise rate of 3 rad/sec. Find the velocity and acceleration of P.

$$\mathbf{a}_P = -126\hat{\mathbf{i}} + 21\hat{\mathbf{j}}\text{ ips}^2 \qquad \textbf{\textit{Ans.}}$$

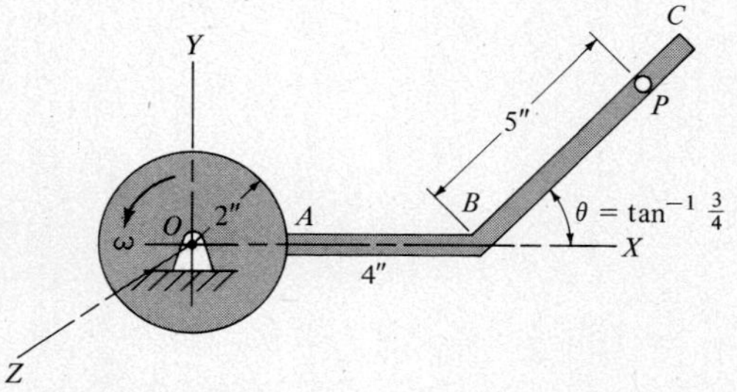

Figure P-12-9.8

12-9.9. Solve the preceding problem if the segment BC is circular as shown in Fig. P-12-9.9 and the position of particle P is defined by $\tan\theta = \frac{4}{3}$.

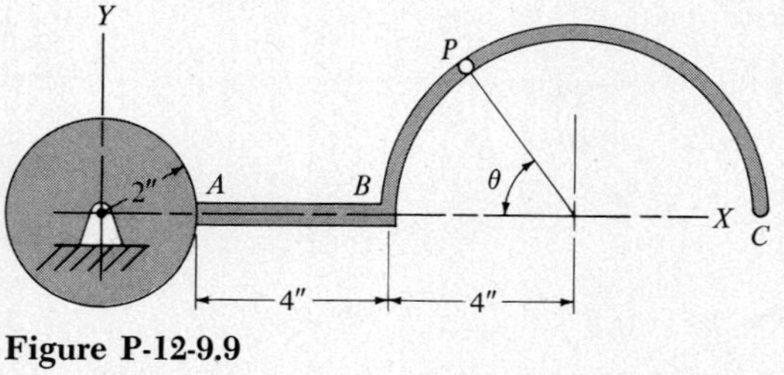

Figure P-12-9.9

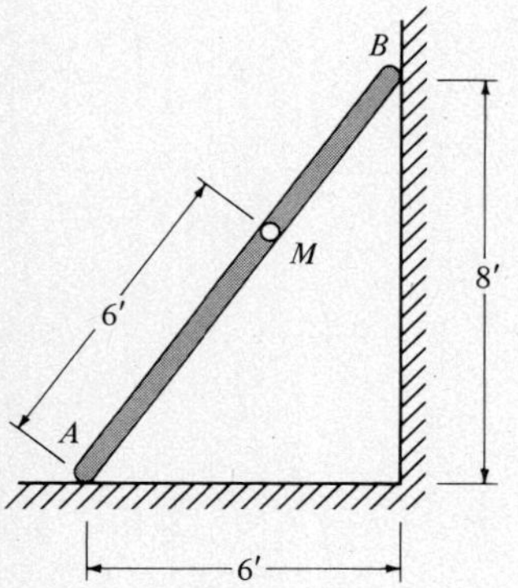

Figure P-12-9.10

12-9.10. A marble M rolls down a hollow tube 10 ft long. At the instant shown in Fig. P-12-9.10, the marble has a velocity of 6 fps and an acceleration of 3 fps^2 both relative to the tube and directed toward A. At the same instant, the end A of the tube has a leftward velocity of 8 fps and a rightward acceleration of 10 fps^2. Find the absolute acceleration of the marble at this instant.

Hint: Attach a reference frame at A with its x axis along the tube and its y axis perpendicular to the tube.

$$\mathbf{a}_M = -3\hat{\mathbf{i}} + 7\hat{\mathbf{j}}\text{ fps}^2;\ \mathbf{a}_M = 7.62\text{ fps}^2 \text{ at } \theta_h = 13.7^\circ \qquad \textbf{\textit{Ans.}}$$

12-9.11. Gimbal rings are used to mount the wheel in Fig. P-12-9.11 so that it may have a constant absolute angular velocity $\boldsymbol{\omega}$ about one or more of the fixed XYZ axes. A 4-in. pendulum AB swings at a constant relative rate of $\omega_r = 2.5$ rad/sec about an axis at A which is perpendicular to the plane of the wheel. Determine the absolute acceleration of the pendulum bob B at the given position (a) if $\boldsymbol{\omega} = 4\hat{\mathbf{k}}$ rad/sec and (b) if $\boldsymbol{\omega} = 4\hat{\mathbf{j}}$ rad/sec.

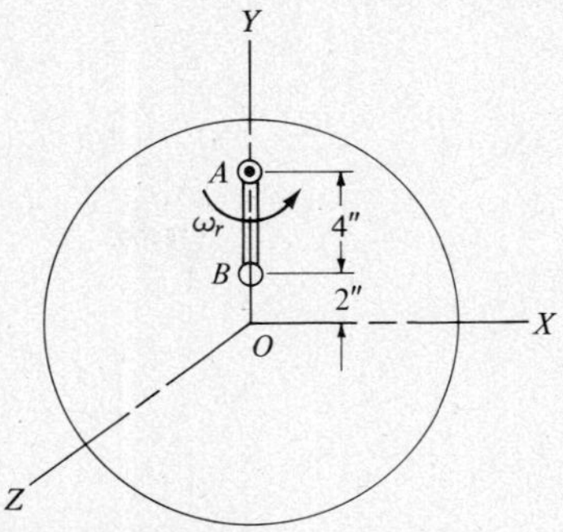

Figure P-12-9.11

12-9.12. Solve the preceding problem if $\boldsymbol{\omega} = 2\hat{\mathbf{i}} + 3\hat{\mathbf{j}} + 4\hat{\mathbf{k}}$ rad/sec.

$$\mathbf{a}_B = 65\hat{\mathbf{j}} - 60\hat{\mathbf{k}}\ \text{fps}^2 \qquad \textbf{\textit{Ans.}}$$

12-9.13. The driving crank AB of the oscillating arm quick-return mechanism shown in Fig. P-12-9.13 rotates at a constant clockwise rate of 11.2 rad/sec. Compute the angular acceleration of the arm CD at the instant when crank AB is horizontal as shown.

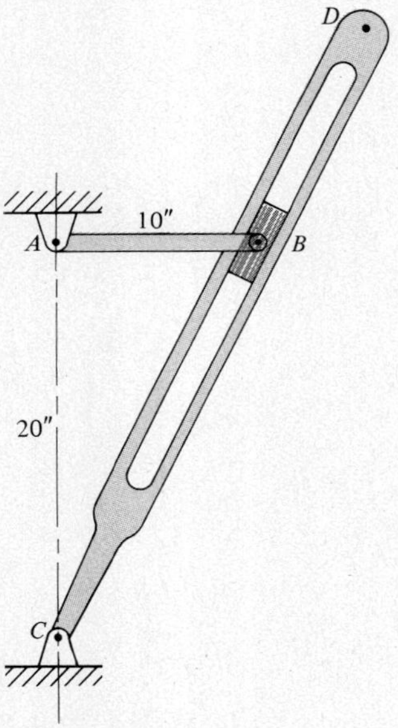

Figure P-12-9.13

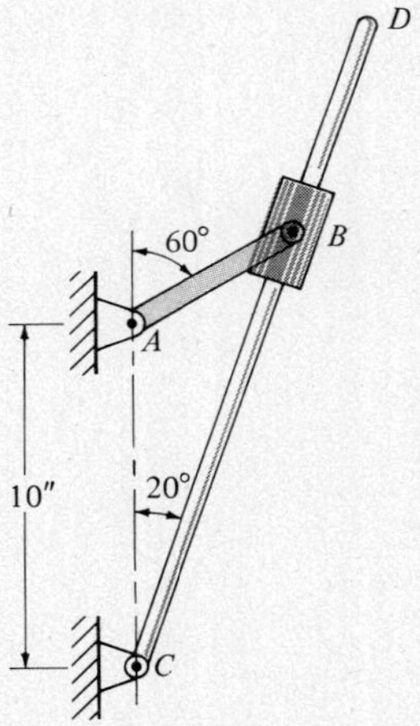

Figure P-12-9.14

12-9.14. The elements of a quick-return mechanism have the given configuration at the instant shown in Fig. P-12-9.14. Find the angular acceleration of arm CD if the driving crank AB rotates at a constant clockwise rate of 5 rad/sec.

$$\alpha_{CD} = 2.51\ \text{rad/sec}^2\ \text{counterclockwise} \qquad \textbf{\textit{Ans.}}$$

12-9.15. In Fig. P-12-9.15, a disk rotates at the constant rate of $\omega_2 =$ 4 rad/sec about the axle CD mounted on a wheel which is itself rotating at a constant rate of $\omega_1 = 5$ rad/sec about a vertical axis at O. Compute the acceleration of point B on the rim of the disk at the given instant.

12-9.16. A wheel rotates about the longitudinal axis of shaft AB at a constant rate $\omega_2 = 10$ rad/sec while the shaft AB rotates about the vertical

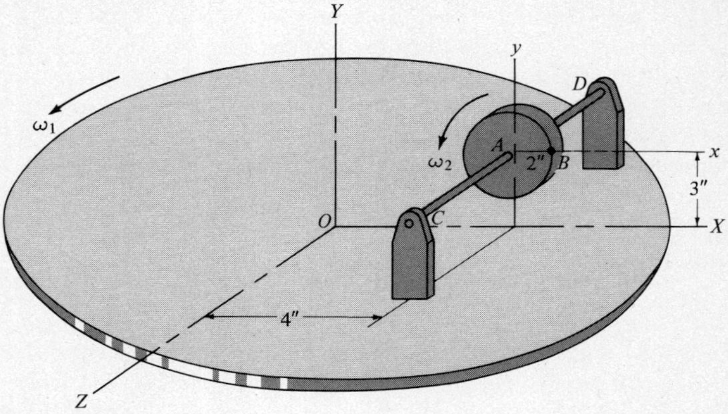

Figure P-12-9.15

Y axis with the values of ω_1 and $\dot{\omega}_1$ shown in Fig. P-12-9.16. The wheel has a radial slot in it along which, at the given instant, a particle P is moving outward from B at 10 fps and accelerating at 4 fps^2. Compute the absolute velocity and acceleration of P.

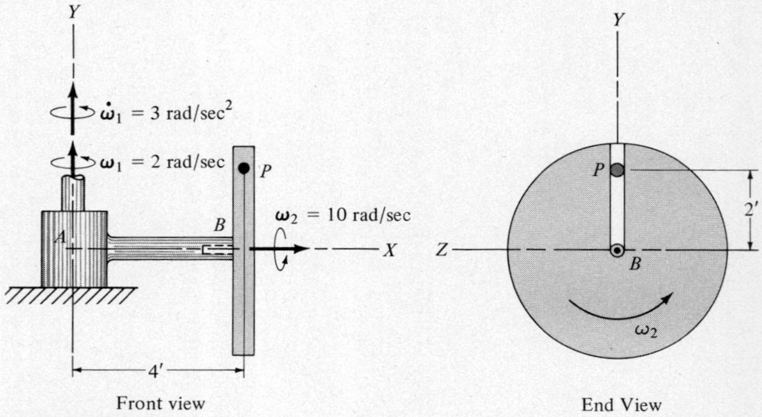

Figure P-12-9.16

Hint: Select a rotating reference frame relative to which particle P has rectilinear motion.

$$\mathbf{a}_P = -64\hat{\mathbf{i}} - 196\hat{\mathbf{j}} + 188\hat{\mathbf{k}} \text{ fps}^2 \qquad \textit{Ans.}$$

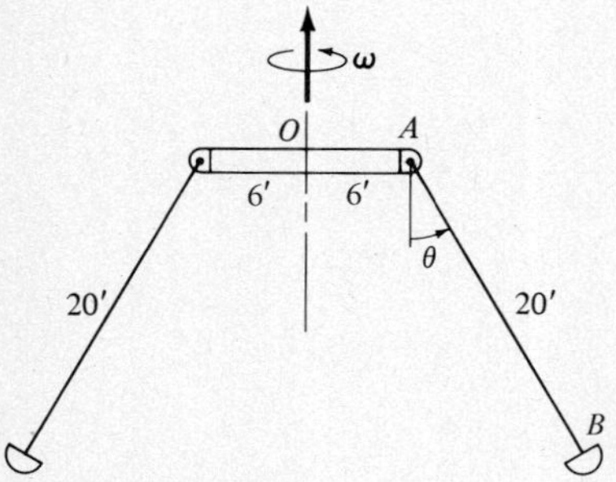

Figure P-12-9.17

12-9.17. In an airplane swing in an amusement park, the arm AB in Fig. P-12-9.17 is swinging outward about A at $\dot{\theta} = 2$ rad/sec and $\ddot{\theta} = 6$ rad/sec^2 while the main vertical mast is rotating at a constant speed $\omega = 3$ rad/sec. Compute the absolute acceleration of B when $\theta = 30°$.

$$\mathbf{a}_B = -80\hat{\mathbf{i}} + 129.3\hat{\mathbf{j}} - 207\hat{\mathbf{k}} \text{ fps}^2 \qquad \textit{Ans.}$$

12-9.18. A gyroscope consists of a 2-in. radius wheel rotating at $\omega_2 = 40$ rad/sec about axis OA in Fig. P-12-9.18. Simultaneously, axis OA is processing (i.e., rotating) about a fixed vertical Y axis at $\omega_1 = 2$ rad/sec. Find the absolute velocity and acceleration of point B which is at the top of the wheel in the given position. Assume the angle ϕ does not change; i.e., axis OA stays at the same inclination with the horizontal as it rotates about the Y axis.

$$\mathbf{v}_B = 74.4k \text{ fps}; \ \mathbf{a}_B = 2228.8i - 2560j \text{ fps}^2 \qquad \textit{Ans.}$$

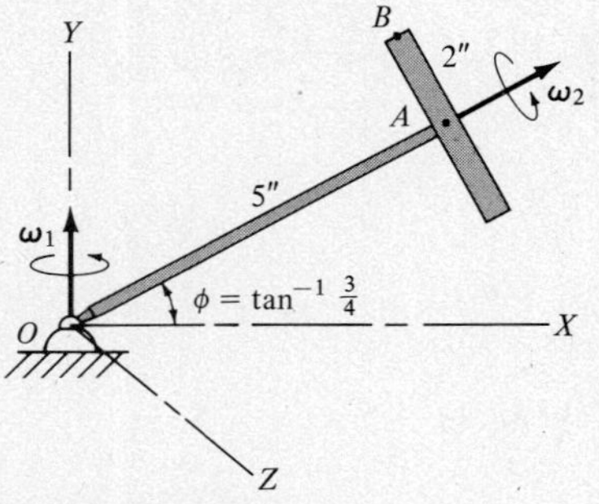

Figure P-12-9.18

12-9.19. Solve the preceding problem if, in addition to the given data, the angle ϕ of axis OA is increasing counterclockwise at $\omega_3 = 1$ rad/sec.

12-9.20. Solve Illus. Prob. 12-9.4 by using reference axes xyz at C fixed in and attached to the platform.

12-9.21. Find the absolute acceleration of body B in Illus. Prob. 12-9.4.

Hint: Refer the motion of B to nonrotating translating reference axes attached to point D.

12-9.22. The semicircular link of the mechanism shown in Fig. P-12-9.22 is rotating at a constant clockwise rate of 5 rad/sec. Determine the angular acceleration of link AB at the given instant.

$$\alpha_{AB} = 5.63 \text{ rad/sec}^2 \text{ counterclockwise} \qquad \textit{Ans.}$$

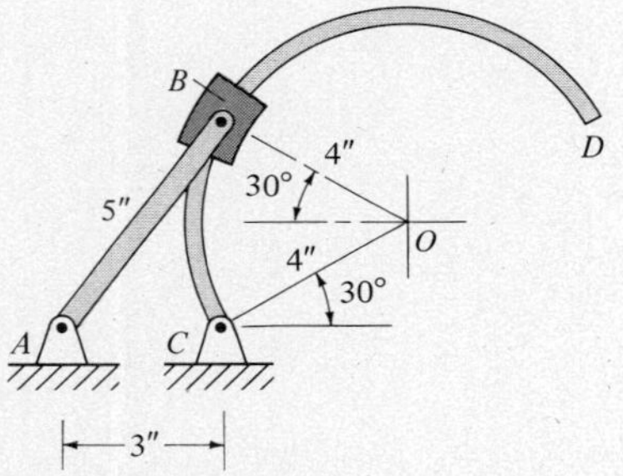

Figure P-12-9.22

12-9.23. A flight simulator for the training of astronauts is shown schematically in Fig. P-12-9.23. The cockpit containing the astronaut may rotate about the horizontal axis of the main arm OB while the seat to which the astronaut is strapped may rotate inside the cockpit about an axis at B. For the given position, find the velocity and acceleration of the astronaut's head at C if the main arm, cockpit, and seat rotate at the following constant rates: main arm $= \boldsymbol{\omega}_1$; cockpit $= \boldsymbol{\omega}_2$ relative to main arm; seat $= \boldsymbol{\omega}_3$ relative to cockpit.

Hint: Select a reference frame about which the astronaut is in circular motion.

See Prob. 12-8.18 for results. ***Ans.***

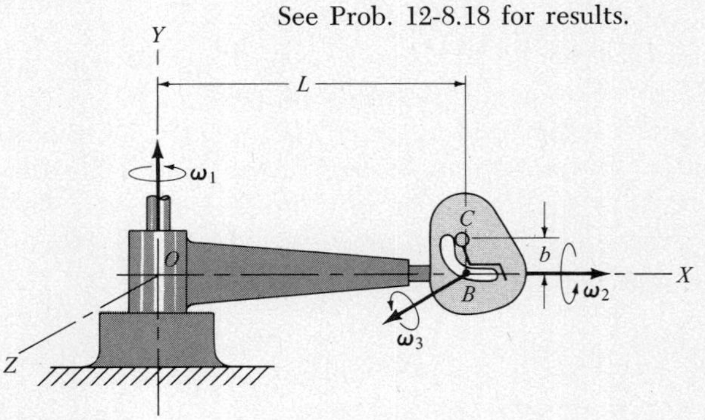

Figure P-12-9.23

SUMMARY

Fixed-axis rotation is that motion of a rigid body in which the particles move in circular paths with their centers on a fixed straight line called the axis of rotation. Angular displacement is measured in radians by the angular distance swept through by the projection of any line in the body upon a plane perpendicular to the axis of rotation. The kinematic characteristic of rotation is that all lines in the body have identical values of angular displacement, angular velocity, and angular acceleration. The linear values of displacement, velocity, and acceleration, however, vary with the distance of a particle from the axis of rotation.

The kinematic differential equations for rotation are tabulated below for comparison with similar equations for rectilinear motion.

Rectilinear Motion	*Rotation*
$v = \dfrac{ds}{dt}$	$\omega = \dfrac{d\theta}{dt}$
$a = \dfrac{dv}{dt} = \dfrac{d^2s}{dt^2}$	$\alpha = \dfrac{d\omega}{dt} = \dfrac{d^2\theta}{dt^2}$
$a\,ds = v\,dv$	$\alpha\,d\theta = \omega\,d\omega$

Because these equations differ only in the symbols used, all the methods used for the kinematics of rectilinear motion can be applied directly to the kinematics of rotation. Furthermore, rectilinear motion and rotation are related by

$$\begin{aligned} s &= r\theta \\ v &= r\omega \\ a_t &= r\alpha \\ a_n &= r\omega^2 = \frac{v^2}{r} = v\omega \end{aligned} \tag{12-2.1}$$

Plane motion (Section 12-3) is the motion of a rigid body in which all particles move in parallel planes. The fundamental concept of plane motion is that it may be considered equivalent to an instantaneous combination of rotation about *any* reference point in the body plus geometrically the translation of the reference point. The geometric approach to plane motion uses the following equations which are true at any particular instant:

$$s_B = s_A \mathbin{+\!\!\!\!\rightarrow} (s_{B/A} = r\theta) \tag{12-3.3}$$

$$v_B = v_A \mathbin{+\!\!\!\!\rightarrow} (v_{B/A} = r\omega) \tag{12-3.1}$$

$$a_B = a_A \mathbin{+\!\!\!\!\rightarrow} (a_{B/A} = r\omega^2 \mathbin{+\!\!\!\!\rightarrow} r\alpha) \tag{12-3.2}$$

In these equations, A is the reference point and B is any other point in the body. Note that values of θ, ω, and α are properties of the body and are independent of the choice of the reference point. The application of these equations is explained in Section 12-4.

For free-rolling disks or spheres (Section 12-4), relations between the translation of the geometric center of the body and its rotational motion are given by

$$s_A = r\theta; \qquad v_A = r\omega; \qquad a_A = r\alpha$$

in which A represents the geometric center of the body and r is the distance from A to the surface on which the body rolls.

The instantaneous axis of rotation in plane motion is the line joining points in the body having zero velocity. The instant center (denoted by C) is located at the intersection of this line with the plane of motion. It is used to determine the velocity of any point in the body from the concept that the body is instantaneously rotating about the instant center, but cannot be used similarly to determine acceleration components except in special cases (see p. 451).

There also exists an instant center of zero acceleration. It can be used to determine the acceleration components of any point in the body. However, the instant center of acceleration is usually so difficult to determine that it is rarely used. Only in the special case of a body in plane motion starting from rest does the instant center of zero acceleration coincide with the instant center of zero velocity.

In Section 12-6, the omega theorem is developed. It states that the time derivative of a vector fixed in a rotating body is simply the cross product of the angular velocity $\boldsymbol{\omega}$ of the body with the vector. Using it, the velocity of any point in a rotating body located by a position vector $\mathbf{r}$ emanating from any point on the axis of rotation is given by

$$\mathbf{v} = \dot{\mathbf{r}} = \boldsymbol{\omega} \times \mathbf{r} \tag{12-6.1}$$

and the velocity of a point B with respect to any other point A in the rotating body is

$$\mathbf{v}_{B/A} = \dot{\boldsymbol{\rho}} = \boldsymbol{\omega} \times \boldsymbol{\rho} \tag{12-6.2}$$

where $\boldsymbol{\rho}$ is the relative position vector from A to B.

Similarly, the omega theorem will determine the time derivatives of unit vectors in an xyz coordinate system rotating with an angular velocity $\boldsymbol{\omega}$. We obtain

$$\dot{\hat{\mathbf{i}}} = \boldsymbol{\omega} \times \hat{\mathbf{i}}; \qquad \dot{\hat{\mathbf{j}}} = \boldsymbol{\omega} \times \hat{\mathbf{j}}; \qquad \dot{\hat{\mathbf{k}}} = \boldsymbol{\omega} \times \hat{\mathbf{k}} \tag{12-6.3}$$

and, in general, for any unit vector $\hat{\mathbf{e}}$ changing its direction at a rate $\boldsymbol{\omega}$, we have

$$\dot{\hat{\mathbf{e}}} = \boldsymbol{\omega} \times \hat{\mathbf{e}} \tag{12-6.4}$$

The general omega theorem developed in Section 12-9 gives the time derivative of a general vector **A** whose magnitude changes with respect to a coordinate system which is itself rotating with an angular velocity $\boldsymbol{\omega}$. Denoting the time rate of change of **A** *relative* to the rotating coordinate system by $\frac{\delta \mathbf{A}}{\delta t}$, the *absolute* time rate of change of **A** with respect to inertial XYZ axes is

$$\frac{d\mathbf{A}}{dt} = \frac{\delta \mathbf{A}}{\delta t} + \boldsymbol{\omega} \times \mathbf{A} \tag{12-9.2}$$

The vector calculus approach to plane motion analysis (Section 12-7) may be used in place of the geometric approach. It also is an introduction to spatial kinematics. The geometric equations (12-3.1) and (12-3.2) are now written in the following form:

$$\mathbf{v}_B = \mathbf{v}_A + (\mathbf{v}_{B/A} = \boldsymbol{\omega} \times \mathbf{r}_{B/A}) \tag{12-7.1}$$
$$\mathbf{a}_B = \mathbf{a}_A + [\mathbf{a}_{B/A} = \boldsymbol{\alpha} \times \mathbf{r}_{B/A} + \boldsymbol{\omega} \times (\boldsymbol{\omega} \times \mathbf{r}_{B/A})] \tag{12-7.2}$$

It is useful to observe that in plane motion, the triple vector product $\boldsymbol{\omega} \times (\boldsymbol{\omega} \times \mathbf{r}_{B/A})$ in Eq. (12-7.2) reduces to $-\omega^2 \mathbf{r}_{B/A}$.

For absolute spatial motion, we also use Eqs. (12-7.1) and (12-7.2). However, the absolute values of $\boldsymbol{\omega}$ and $\boldsymbol{\alpha} = \dot{\boldsymbol{\omega}}$ are not as obvious as they are for plane motion, but usually are the result of combining several separate values. Refer to Section 12-8 for details.

The use of rotating reference frames in relative spatial motion (Section 12-9) is very useful in situations where a relative motion occurs within a moving body. Relative spatial motion analysis is based on the general omega theorem [Eq. (12-9.2)]. The velocity and acceleration of any point B moving relative to body-fixed reference axes with an origin at a body point A whose velocity and acceleration are known is expressed by the following equations:

$$\mathbf{v}_B = \mathbf{v}_A + (\mathbf{v}_{M/A} = \boldsymbol{\omega} \times \boldsymbol{\rho}) + \mathbf{v}_r \tag{12-9.5}$$

$$\mathbf{a}_B = \mathbf{a}_A + (\mathbf{a}_{M/A} = \boldsymbol{\alpha} \times \boldsymbol{\rho} + \boldsymbol{\omega} \times \mathbf{v}_{M/A}) + \mathbf{a}_r + (\mathbf{a}_C = 2\boldsymbol{\omega} \times \mathbf{v}_r) \tag{12-9.6}$$

These equations express the motion of B as the vector sum of the translational motion of any convenient reference point A plus the rigid-body rotation about A of a body point M instantaneously coincident with B plus the relative motion of B with respect to a path fixed in the body. In addition, the acceleration equation must also include the Coriolis component of acceleration.

13

KINETICS OF RIGID BODIES

13-1 INTRODUCTION

By now we have covered so many topics in such detail that you may have forgotten the overall preview of dynamics that was presented in Section 9-1. *Please read this section again*. It will enable you to see the "forest" of dynamics rather than the trees.

In this chapter we shall consider only the plane motion of rigid bodies that are symmetrical with respect to the plane in which the mass center moves. Since most engineering applications are of this type, they are separated here from the motion of unsymmetrical bodies and general spatial kinetics. The latter are discussed in later chapters and involve complications which, although important, are not essential to the plane motion of symmetrical bodies.

Only two basic equations are involved: $\mathbf{R} = m\bar{\mathbf{a}} = (W/g)\bar{\mathbf{a}}$ and $\Sigma\mathbf{M} = \dot{\mathbf{H}}$. These were derived in Sections 10-5 and 10-6. The first relates the resultant of applied forces to the acceleration of the mass center while the second relates the moment of applied forces to the angular motion. In addition, these equations are coupled by the correlation between linear and angular motion as presented in the preceding chapter on kinematics.

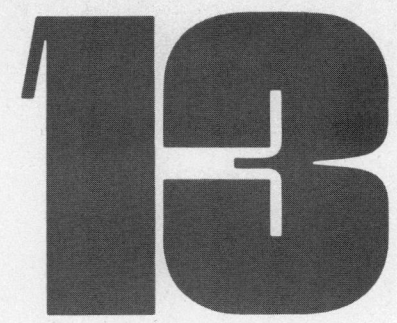

13-2 EQUATIONS OF PLANE MOTION

Consider a symmetrical rigid body constrained to move in the XY plane. Because of this constraint, all forces affecting the motion can be reduced to a resultant force $\mathbf{R}$ acting through the mass center G, and a resultant couple $\Sigma \mathbf{M}_G$ representing the moment sum of the applied forces about a Z axis through the mass center perpendicular to the plane of motion.

We know from the principle of the motion of the mass center that the relation between $\mathbf{R}$ and the linear acceleration of the mass center is expressed by $\mathbf{R} = (W/g)\bar{\mathbf{a}}$. The resultant couple $\Sigma \mathbf{M}_G$ causes an angular acceleration $\boldsymbol{\alpha}$, and the kinematics of constraints determine the relation between $\bar{\mathbf{a}}$ and $\boldsymbol{\alpha}$. What we do not yet know is the explicit relation between $\Sigma \mathbf{M}_G$ and $\boldsymbol{\alpha}$. We shall soon show that this is $\Sigma \mathbf{M}_G = I_G \boldsymbol{\alpha}$ by using several approaches which complement one another.

But first, let us review the possible effects of the resultant force-couple system upon the motion of a body starting from rest. If the resultant is a single force which passes through the mass center, we have learned (see Section 11-2) that a motion of translation ensues: rectilinear translation if the direction of the resultant force is constant, or curvilinear translation if the direction of the resultant force varies. A force system whose resultant is a couple will cause the body on which it acts to have a centroidal rotation even if there is no fixed axis through the mass center. Conversely, if motion occurs about a fixed centroidal axis, the resultant must be a couple. When the resultant is both a force and a couple, plane motion ensues, consisting of both a linear motion of the mass center and an angular acceleration. The angular acceleration is a property of the body and is independent of the actual or assumed axis of rotation (see Sections 12-2 and 12-3). The important cases of plane motion are noncentroidal rotation (i.e., rotation about a fixed axis which does not pass through the mass center), rolling bodies, and general plane motion. In all these cases, kinematic analyses must be used to relate the linear acceleration $\bar{\mathbf{a}}$ of the mass center with the angular acceleration $\boldsymbol{\alpha}$ of the body.

Let us now derive the relation between the moment sum of the applied forces and the angular acceleration of the body, and then later discuss its application to each of the various motions just described. To obtain this relation from the basic equation $\Sigma \mathbf{M} = \dot{\mathbf{H}}$, we must first determine the angular momentum $\mathbf{H}$ for the rigid-body motion.[1]

Consider in Fig. 13-2.1 the outline of a symmetrical rigid body constrained to move in the XY plane through its mass center. Because

[1] Later in Chapter 16, we shall develop a general expression for $\mathbf{H}$ applicable to any body having any motion.

of the plane motion restriction, the angular velocity and acceleration vectors of the body are perpendicular to the plane of motion and are expressed by $\boldsymbol{\omega} = \omega\hat{\mathbf{k}}$ and $\boldsymbol{\alpha} = \dot{\boldsymbol{\omega}} = \alpha\hat{\mathbf{k}}$. Fix xyz axes in the body as shown so that the z axis is always perpendicular to the XY plane of motion while the xy axes rotate with the body at the angular rates $\boldsymbol{\omega}$ and $\boldsymbol{\alpha}$. Let the origin of these axes be the mass center G, although it may also be a center A which is fixed in space, or else one whose acceleration is directed through the mass center. Recall from Section 10-6 that with any of these conditions, the moment equation has the form $\Sigma\mathbf{M} = \dot{\mathbf{H}}$.

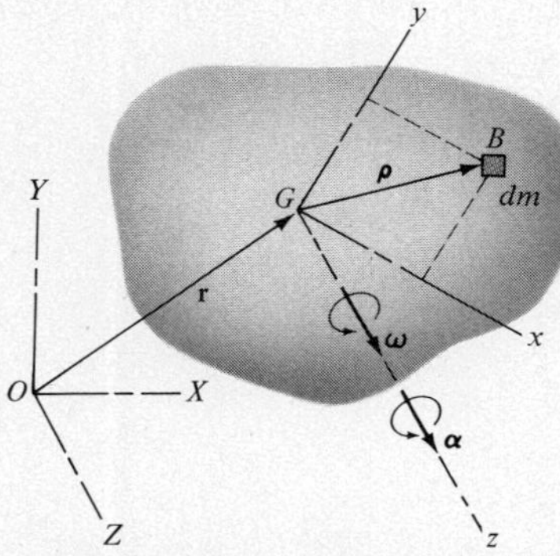

Figure 13-2.1 G and the xy axes move in the XY plane of motion.

Because of symmetry, the mass center of any pair of symmetrically placed elements lies in the XY plane. Denote their mass by dm and their location at B by the relative position vector from G having the constant length $\boldsymbol{\rho} = x\hat{\mathbf{i}} + y\hat{\mathbf{j}}$. The moment of their relative momentum about G is then $\boldsymbol{\rho} \times \mathbf{v}_{B/G}\, dm = \boldsymbol{\rho} \times (\boldsymbol{\omega} \times \boldsymbol{\rho})\, dm$ and the total relative moment of momentum of the body (or angular momentum) is then

$$\mathbf{H}_G = \int \boldsymbol{\rho} \times (\boldsymbol{\omega} \times \boldsymbol{\rho})\, dm \qquad (a)$$

To evaluate the vector triple product of the integrand, we can use the identity $\mathbf{a} \times (\mathbf{b} \times \mathbf{c}) = (\mathbf{a} \cdot \mathbf{c})\mathbf{b} - (\mathbf{a} \cdot \mathbf{b})\mathbf{c}$ [Eq. (2-8.6)] to yield $\rho^2\boldsymbol{\omega}$ directly[2] or proceed fundamentally as follows:

$$\boldsymbol{\omega} \times \boldsymbol{\rho} = \omega\hat{\mathbf{k}} \times (x\hat{\mathbf{i}} + y\hat{\mathbf{j}}) = \omega x\hat{\mathbf{j}} - \omega y\hat{\mathbf{i}}$$

whence

$$\begin{aligned}\boldsymbol{\rho} \times (\boldsymbol{\omega} \times \boldsymbol{\rho}) &= (x\hat{\mathbf{i}} + y\hat{\mathbf{j}}) \times (\omega x\hat{\mathbf{j}} - \omega y\hat{\mathbf{i}}) \\ &= (x^2 + y^2)\omega\hat{\mathbf{k}} = \rho^2\boldsymbol{\omega}\end{aligned}$$

On substituting this evaluation of the integrand into Eq. (a) and placing $\boldsymbol{\omega}$ outside the integral since it is a property of the body and thus independent of the position of B, we obtain

$$\mathbf{H}_G = \boldsymbol{\omega} \int \rho^2\, dm = I_G\boldsymbol{\omega} \qquad (b)$$

Note that the inertia integral is defined with respect to a z axis through G fixed in and moving with the body and hence is a constant quantity. Since $\boldsymbol{\omega}$ is always perpendicular to the plane of motion, there is no change in its direction, and the absolute time derivative of $\mathbf{H}_G$ is

$$\dot{\mathbf{H}}_G = I_G\dot{\boldsymbol{\omega}} = I_G\boldsymbol{\alpha} \qquad (c)$$

Thus, the application of $\Sigma\mathbf{M} = \dot{\mathbf{H}}$ results in $\Sigma\mathbf{M}_G = I_G\boldsymbol{\alpha}$ in which each of the vectors is perpendicular to the XY plane. The equivalent scalar forms of this equation and of $\mathbf{R} = m\bar{\mathbf{a}}$ are

$$\Sigma X = \frac{W}{g}\bar{a}_x; \qquad \Sigma Y = \frac{W}{g}\bar{a}_y; \qquad \Sigma\bar{M} = \bar{I}\alpha \qquad (13\text{-}2.1)$$

[2] Thus, $\boldsymbol{\rho} \times (\boldsymbol{\omega} \times \boldsymbol{\rho}) = (\boldsymbol{\rho} \cdot \boldsymbol{\rho})\boldsymbol{\omega} - (\boldsymbol{\rho} \cdot \boldsymbol{\omega})\boldsymbol{\rho} = \rho^2\boldsymbol{\omega}$ since $\boldsymbol{\rho} \cdot \boldsymbol{\omega} = 0$ because $\boldsymbol{\rho}$ and $\boldsymbol{\omega}$ are mutually perpendicular.

where the subscript G has been replaced by bar superscripts as is customary when referring to centroidal axes. These equations are valid for any plane motion of a rigid body. Their physical interpretation is shown in Fig. 13-2.2 which is the plane motion equivalent of Fig. 10-6.2 on page 391.

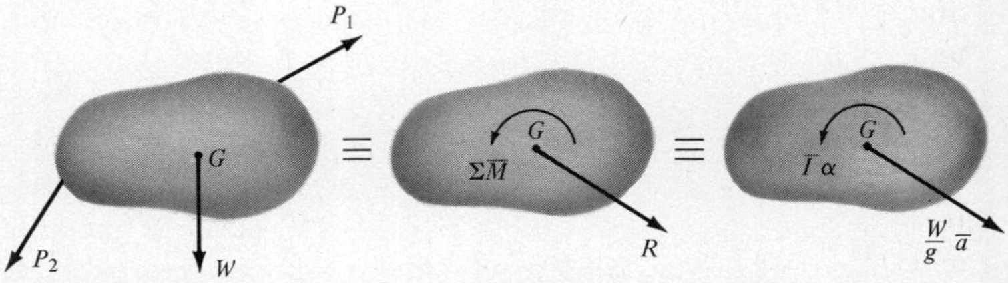

Figure 13-2.2 Equivalence of applied forces to their dynamic effects.

Observe that these three independent scalar equations are the coplanar kinetic equivalents of the three independent scalar equations of coplanar statics. Later in Chapter 16, we shall expand $\mathbf{R} = m\bar{\mathbf{a}}$ and $\Sigma\mathbf{M} = \dot{\mathbf{H}}$ to obtain six independent scalar equations for spatial motion. It is recommended that you clearly identify the number of unknowns in any problem to ensure that there are a corresponding number of independent equations available to determine them. In this regard, the suggestions made in Chapters 3 and 4 may be useful.

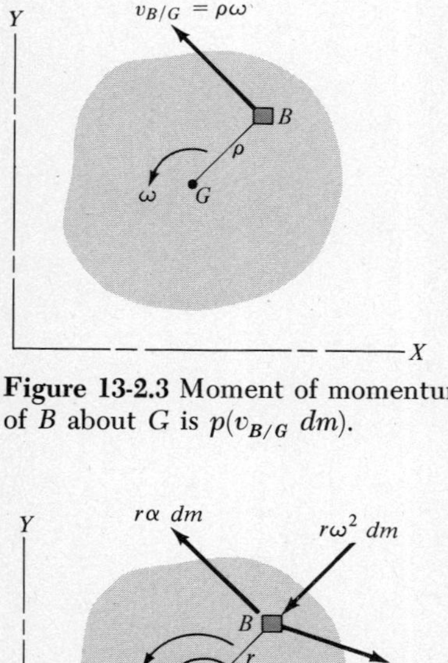

Figure 13-2.3 Moment of momentum of B about G is $p(v_{B/G}\ dm)$.

Before proceeding with the application of these equations, it may be informative to consider other methods of obtaining $\Sigma\bar{M} = \bar{I}\alpha$. A direct evaluation of the *magnitude* of $\mathbf{H}_G$ may be found, without using vector analysis in this plane motion, by summing up the moments of relative momentum $v_{B/G}\ dm$ of all typical particles such as B of mass dm in Fig. 13-2.3. Clearly, the relative velocity $v_{B/G}$ of B about G is $v_{B/G} = \rho\omega$ whence the total relative angular momentum is

$$H_G = \int \rho(v_{B/G}\ dm) = \omega \int \rho^2\ dm = \bar{I}\omega$$

and, from differentiating this expression for H_G, we obtain again

$$[\Sigma M = \dot{H}] \qquad \Sigma\bar{M} = \bar{I}\dot{\omega} = \bar{I}\alpha$$

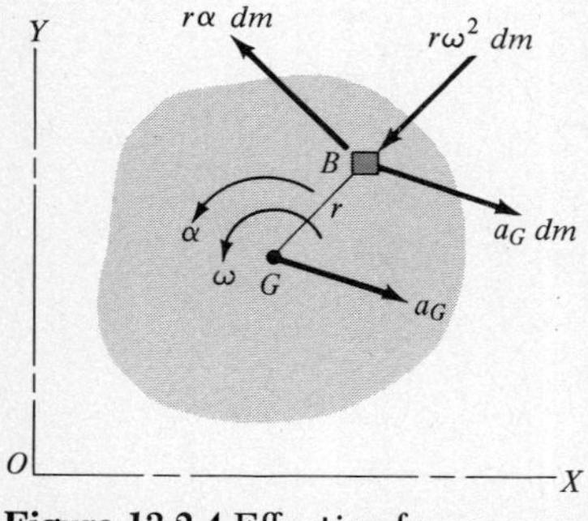

Figure 13-2.4 Effective forces on particle B.

Another fundamental way of obtaining this result is to equate the moments of the applied forces about the mass center G to the moment sum about G of the effective forces on each particle. Thus, a typical particle at B is subjected to the effective force system shown in Fig. 13-2.4 found by multiplying each term of $a_B = a_G \mathrel{+\!\!\!\!\rightarrow} (a_{B/G} = r\omega^2 \mathrel{+\!\!\!\!\rightarrow} r\alpha)$ by dm. Summing their moments about G eliminates all normal effective forces like $r\omega^2\ dm$ which pass through G and also the effects of the uniformly distributed parallel force system of $a_G\ dm$ acting on all particles, since their resultant is $a_G \int dm = ma_G$ which passes through G. We are left with only the moment sum $\int r(r\alpha\, dm)$ of the tangential effective forces, resulting once more in

$$\Sigma \bar{M} = \alpha \int r^2\, dm = \bar{I}\alpha$$

Finally, since $\Sigma \mathbf{M} = \dot{\mathbf{H}}$ is also valid about a center A which is fixed in space, it should be observed that if such a point were substituted for G in the preceding derivations, the corresponding moment equation becomes

$$\Sigma M_A = I_A \alpha \qquad (13\text{-}2.2)$$

where I_A is the mass moment of inertia about A. This equation is particularly useful in fixed-axis rotation since it automatically eliminates the usually unknown axis reaction at A from the moment equation.

13-3 FIXED-AXIS ROTATION

A common approach to all problems in plane motion consists of the following method: (1) Determine the kinematic relation between the acceleration $\bar{a}$ of the mass center and the angular acceleration α of the body. (2) Apply Eq. (13-2.1) diagramatically by equating a free-body diagram of the body to a diagram of the equivalent dynamic effects. (3) An alternative to item (2) is to use a single diagram showing both the applied forces and the *reversed* dynamic effects, thereby creating dynamic equilibrium. The principal advantage of dynamic equilibrium is that only one diagram need be drawn, and no special sign conventions need be observed since force and moment summations are zero with respect to any axis or moment center. We now illustrate how this common approach is applied to fixed-axis rotation.

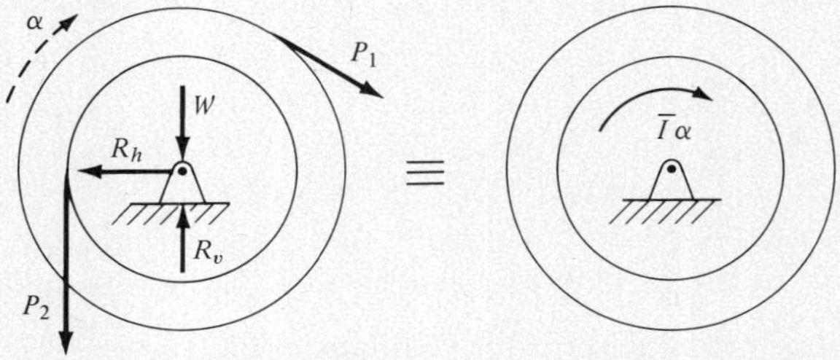

Figure 13-3.1 Applied forces and equivalent dynamic effect in centroidal rotation.

CENTROIDAL ROTATION

One of the commonest types of fixed-axis rotation is centroidal rotation in which the axis of rotation passes through the mass center.

Since $\bar{a} = 0$, the equivalent diagrams of applied forces and dynamic effects are as shown in Fig. 13-3.1 from which we deduce the following equations:

$$\Sigma X = 0; \qquad \Sigma Y = 0; \qquad \Sigma \bar{M} = \bar{I}\alpha \tag{13-3.1}$$

The bar sign is used to indicate that both $\Sigma \bar{M}$ and $\bar{I}$ are taken with respect to a centroidal axis of rotation.

Since both ΣX and ΣY equal zero for a centroidal rotation, the resultant of the impressed forces always is a couple of magnitude $\Sigma \bar{M}$. The converse of this observation is that if the force system applied to a body reduces to a couple, the body will undergo a centroidal rotation. The resultant couple will create centroidal rotation even if there is no fixed axis of rotation as we mentioned initially in Section 9-1.

NONCENTROIDAL ROTATION

Consider now the rigid body of weight W in Fig. 13-3.2 which is constrained to rotate about a noncentroidal axle at A. The action of the applied forces W and P, as well as the components of the bearing reaction at A, give the body the instantaneous values of the angular velocity ω and angular acceleration α as shown. The mass center G moves in a circle of radius $\bar{r}$ from the axis of rotation at A and has an acceleration $\bar{a}$ whose components are $\bar{a}_n = \bar{r}\omega^2$ directed toward A and $\bar{a}_t = \bar{r}\alpha$ directed perpendicular to $\bar{r}$ in the sense of α. The corresponding components of the resultant force $R = \left(\frac{W}{g}\right)\bar{a}$ therefore act as shown and, together with $\bar{I}\alpha$ in the sense of α, compose a dynamic system equivalent to the impressed forces.

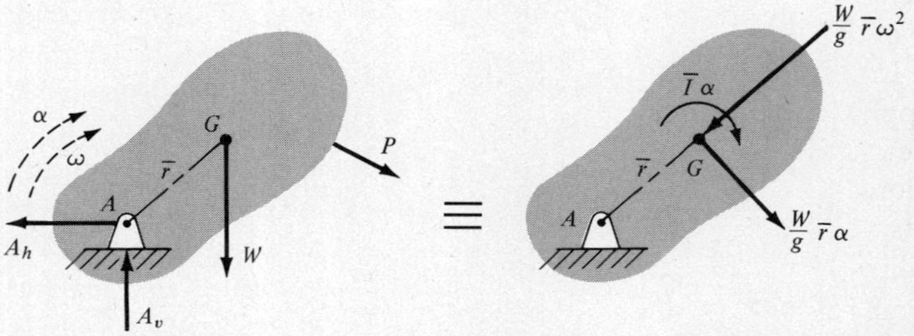

Figure 13-3.2 Equivalence of applied forces and dynamic effects in noncentroidal rotation.

Equating the moments of impressed forces about the axis of rotation at A (thereby eliminating the generally unknown components of the axle reaction) to the moment sum of the effective forces gives

$$\Sigma M_A = \bar{I}\alpha = \frac{W}{g}\bar{r}\alpha(\bar{r}) = \left(\bar{I} + \frac{W}{g}\bar{r}^2\right)\alpha$$

in which we recognize that the term in parentheses means that the centroidal moment of inertia is transferred from G through the distance $\bar{r}$ to the axle at A. Consequently, we obtain

$$\Sigma M_A = I_A\alpha$$

which corresponds with Eq. (13-2.2) as obtained on p. 503 directly from $\Sigma \mathbf{M} = \dot{\mathbf{H}}$.

In summary, the general equations of noncentroidal rotation of a symmetrical body are

$$\Sigma X = \frac{W}{g}\bar{r}\omega^2; \qquad \Sigma Y = \frac{W}{g}\bar{r}\alpha$$
$$\Sigma \bar{M} = \bar{I}\alpha \quad \text{or} \quad \Sigma M_A = I_A\alpha \tag{13-3.2}$$

where the X and Y axes through the mass center are positive in the directions of $\bar{a}_n$ and $\bar{a}_t$, respectively. The direction of $\bar{a}_n$ will always be toward the center of rotation; that of $\bar{a}_t$ is determined from the sense of α. The sense of $\Sigma\bar{M}$ and ΣM_A is positive in the initial direction of rotation.

In the practical application of Eqs. (13-3.2), it is usually best to apply $\Sigma M_A = I_A\alpha$ directly to a free-body diagram of the impressed forces, thereby eliminating the unknown axle reaction. Force summations can then be obtained from equating the free-body diagram to its equivalent dynamic effects, or else from creating dynamic equilibrium by adding the reversed dynamic effects to the free-body diagram. Both procedures are discussed in the illustrative problems.

ILLUSTRATIVE PROBLEMS

13-3.1. The pulley assembly shown in Fig. 13-3.3 weighs 150 lb and has a centroidal radius of gyration of 2 ft. The blocks are attached to the assembly by cords wrapped around the pulleys. Determine the acceleration of each body and the tension in each cord.

Solution

For any connected system, apply the procedure outlined in Section 11-3 on p. 404. Thus, after drawing a FBD of each body and determining the direction of motion, we next determine the kinematic relations among the bodies, and finally apply the kinetic equation appropriate to the motion of each body.

The moments of the weights about the center of rotation O give

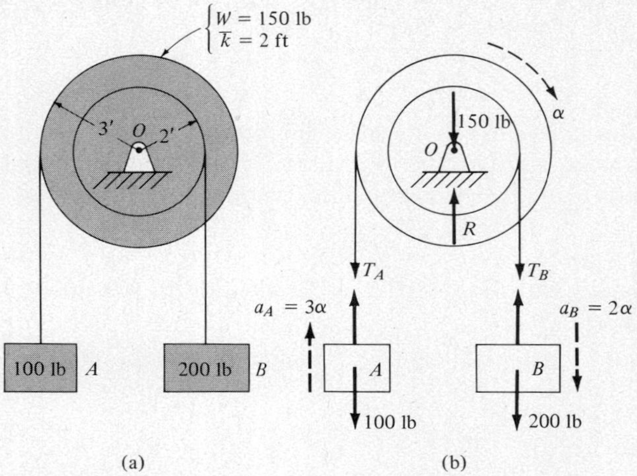

Figure 13-3.3

an unbalanced moment in a clockwise sense. Body B therefore moves down while body A rises. The FBD of each part of the system, showing the direction of motion of each body, can now be drawn as in part (b). In accordance with Section 12-2, the linear accelerations of A and B are each expressed in terms of the angular acceleration of the pulley by $a_t = r\alpha$, or in this case by $a_A = 3\alpha$ and $a_B = 2\alpha$ which we draw as dashed vectors adjacent to each body. Their directions also indicate the positive sense of force and moment summations.

The kinetic equations for the translating bodies and the rotating pulley are respectively $\Sigma X = \left(\frac{W}{g}\right) a$ and $\Sigma \bar{M} = \bar{I}\alpha$. We recall that the moment of inertia is given by $\bar{I} = \left(\frac{W}{g}\right) \bar{k}^2$ and the application of these equations gives

$$\text{For } B: \quad 200 - T_B = \frac{200}{g} a_B = \frac{200}{g}(2\alpha) \qquad (a)$$

$$\text{For } A: \quad T_A - 100 = \frac{100}{g} a_A = \frac{100}{g}(3\alpha) \qquad (b)$$

$$\text{For pulley:} \quad 2T_B - 3T_A = \frac{150}{g}(2)^2 = \frac{600}{g}\alpha \qquad (c)$$

Notice that we have retained the gravitational acceleration g in symbolic form rather than using its numerical equivalent of 32.2 fps^2. As a result, the above equations contain the common term $\frac{\alpha}{g}$ which is readily found by multiplying Eq. (a) by 2 and Eq. (b)

by 3, and then adding Eqs. (*a*), (*b*), and (*c*) which thereby eliminates the tensions. We obtain

$$\frac{\alpha}{g} = \frac{100}{2300}$$

from which, using g = 32.2,

$$\alpha = 1.4 \text{ rad/sec}^2; \quad a_A = 3\alpha = 4.2 \text{ fps}^2; \quad a_B = 2\alpha = 2.8 \text{ fps}^2 \qquad \textbf{\textit{Ans.}}$$

By substituting the known value of $\frac{\alpha}{g}$ in Eqs. (*a*) and (*b*), we then find

$$T_A = 113 \text{ lb} \quad \text{and} \quad T_B = 182.6 \text{ lb} \qquad \textbf{\textit{Ans.}}$$

The reaction on the pulley axle is obtained from the vertical summation

$$[\Sigma V = 0] \qquad R - 150 - T_A - T_B = 0 \qquad R = 445.6 \text{ lb} \quad \textbf{\textit{Ans.}}$$

13-3.2. Suspended from a horizontal axis at A in Fig. 13-3.4 is a 6-ft rod weighing 20 lb to which is welded a 2-ft diameter cylinder weighing 30 lb. A horizontal force $P = 40$ lb is applied to the assembly at a distance *d* from A. (a) Find the horizontal component A_h of the bearing reaction at A when $d = 2$ ft. (b) Find the distance *d* at which *P* should act so that A_h will be zero. This position of *P* is called the *center of percussion*.

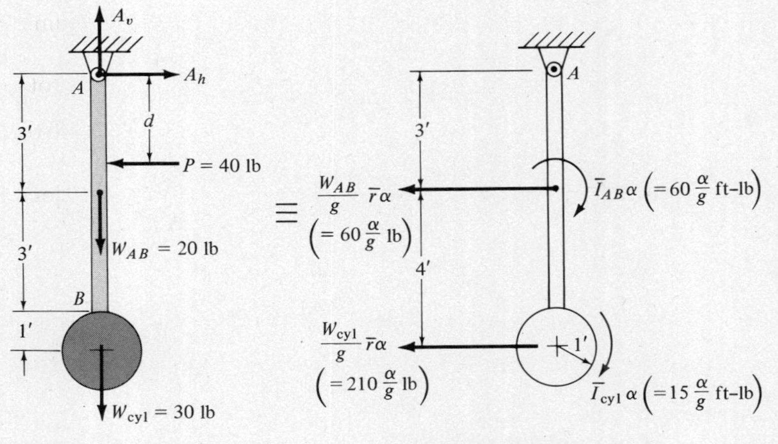

(a) Free-body diagram (b) Dynamic effects

Figure 13-3.4

Solution: Part (a)

The FBD of impressed forces in (a) is equivalent to the dynamic effects in (b). For each part of the assembly, these consist of the effective force $\left(\frac{W}{g}\right)\bar{a}$ (having only a component $\left(\frac{W}{g}\right)\bar{r}\alpha$ since $\omega = 0$) and a couple $\bar{I}\alpha$. Their values are

$$\text{Rod:}\quad \frac{W}{g}\bar{r}\alpha = \frac{20}{g}(3)\alpha = 60\frac{\alpha}{g}\ \text{lb};\qquad \bar{I}\alpha = \frac{1}{12}\frac{W}{g}L^2\alpha = \frac{20}{12\,g}(6)^2\alpha = 60\frac{\alpha}{g}\ \text{ft-lb}$$

$$\text{Cyl.:}\quad \frac{W}{g}\bar{r}\alpha = \frac{30}{g}(7)\alpha = 210\frac{\alpha}{g}\ \text{lb};\qquad \bar{I}\alpha = \frac{1}{2}\frac{W}{g}r^2\alpha = \frac{30}{2\,g}(1)^2\alpha = 15\frac{\alpha}{g}\ \text{ft-lb}$$

To eliminate the unknown components of the bearing reaction, we equate moments about A of the impressed forces and the equivalent dynamic effects. Noting that the moment of the effective couples are independent of any moment center, we obtain

$$[\circlearrowleft + (\Sigma M_A)_{\mathbf{FBD}} = (\Sigma M_A)_{\text{dyn}}]$$

$$40(2) = 60\frac{\alpha}{g} + \left(60\frac{\alpha}{g}\right)(3) + 15\frac{\alpha}{g} + \left(210\frac{\alpha}{g}\right)(7) = 1725\frac{\alpha}{g}$$

whence

$$\frac{\alpha}{g} = \frac{80}{1725}$$

Equating horizontal force summations for each system, we get

$$[\overset{+}{\leftarrow} (\Sigma F_h)_{\mathbf{FBD}} = (\Sigma F_h)_{\text{dyn}}]$$

$$40 - A_h = 60\frac{\alpha}{g} + 210\frac{\alpha}{g} = 270\left(\frac{\alpha}{g} = \frac{80}{1725}\right) \approx 12.5$$

$$A_h = 27.5\ \text{lb}$$ ***Ans.***

A few comments are pertinent. First, numerical simplicity is achieved by solving for $\frac{\alpha}{g}$ from which α can be found if desired. Second, if only α were to be found, we could have applied $\Sigma M_A = I_A\alpha$ directly to the FBD in (a), but to determine I_A requires transferring each centroidal moment of inertia to A as follows:

$$[I = \Sigma(\bar{I} + M\,d^2)]\qquad I_A = \frac{60}{g} + \frac{20}{g}(3)^2 + \frac{15}{g} + \frac{30}{g}(7)^2 = \frac{1725}{g}\ \text{ft-lb-sec}^2$$

whence

$$[\Sigma M_A = I_A\alpha] \qquad 40(2) = \frac{1725}{g}\alpha \qquad \text{or} \qquad \frac{\alpha}{g} = \frac{80}{1725}$$

as before. Notice that the previous solution automatically included the transfer terms for moments of inertia.

Part (b)

Here both the location of P and the value of $\frac{\alpha}{g}$ are unknown, but $A_h = 0$. Hence, equating force summations (with $A_h = 0$), we get

$$[\overset{+}{\leftarrow}(\Sigma F_h)_{\mathbf{FBD}}(\Sigma F_h)_{\text{dyn}}] \qquad P = 60\frac{\alpha}{g} + 210\frac{\alpha}{g}, \qquad \frac{\alpha}{g} = \frac{P}{270}$$

With this value of $\frac{\alpha}{g}$, the moment summations obtained previously gives

$$[\circlearrowleft + (\Sigma M_A)_{\mathbf{FBD}} = (\Sigma M_A)_{\text{dyn}}] \qquad P\,d = 1725\frac{\alpha}{g} = 1725\left(\frac{P}{270}\right)$$

$$d = \frac{1725}{270} = 6.39 \text{ ft} \qquad \textbf{\textit{Ans.}}$$

Note that the value of d is independent of the value of P. In other words, the center of percussion is a unique property of the body which depends only on the distribution of its mass.

13-3.3. A turntable rotating in a horizontal plane about a vertical axis O carries a bent bar weighing 16.1 lb/ft pinned to it at A and forced to rotate with it by a smooth peg at C. At the instant shown in Fig. 13-3.5, $\omega = 4$ rad/sec and $\alpha = 6$ rad/sec² both clockwise. Determine the forces acting on the bar at A and C.

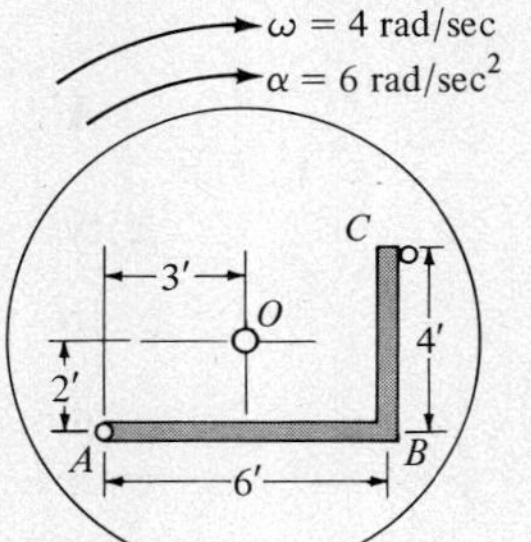

Figure 13-3.5

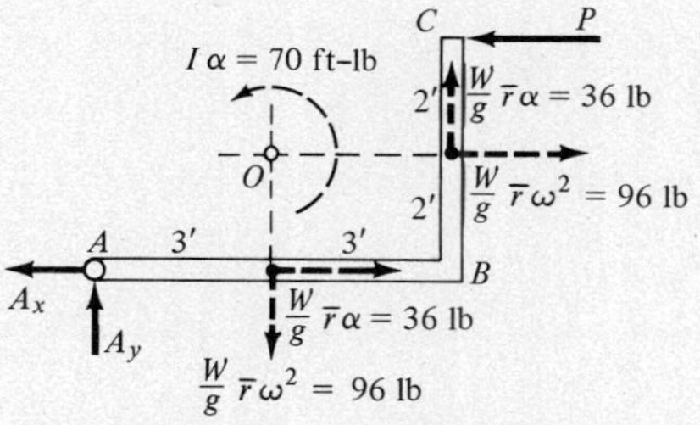

Figure 13-3.6 Dynamic equilibrium.

Solution

Instead of drawing two separate diagrams of impressed forces and equivalent dynamic effects, consider here the method of dynamic

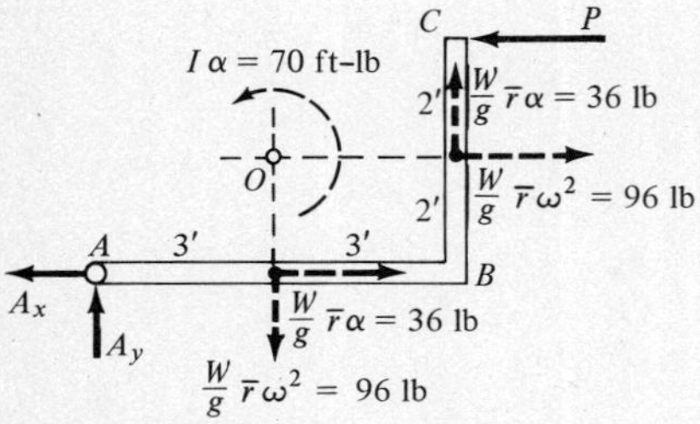

Figure 13-3.6

equilibrium which permits a free choice of axes and moment centers. Rather than locating the mass center of the entire bent bar, it is more convenient to apply the inertia components acting as shown in Fig. 13-3.6 at the mass center of each segment. Each of these inertia components act respectively opposite to $\bar{a}_n$ and $\bar{a}_t$ of each segment.

The values of the inertia components are

For AB:

$$\frac{W}{g}\bar{r}\omega^2 = \left(\frac{16.1 \times 6}{32.2}\right)(2)(4)^2 = 96 \text{ lb}$$

$$\frac{W}{g}\bar{r}\alpha = \left(\frac{16.1 \times 6}{32.2}\right)(2)(6) = 36 \text{ lb}$$

For BC:

$$\frac{W}{g}\bar{r}\omega^2 = \left(\frac{16.1 \times 4}{32.2}\right)(3)(4)^2 = 96 \text{ lb}$$

$$\frac{W}{g}\bar{r}\alpha = \left(\frac{16.1 \times 4}{32.2}\right)(3)(6) = 36 \text{ lb}$$

The resultant inertia couple $\bar{I}\alpha$ is the sum of the inertia couples acting opposite to α on each segment. Its value is

$$\bar{I}\alpha = \sum\left(\frac{1}{12}\frac{W}{g}L^2\right)\alpha = \left[\frac{1}{12}\left(\frac{96.6}{32.2}\right)(6^2)\right](6) + \left[\frac{1}{12}\left(\frac{64.4}{32.2}\right)(4)^2\right](6)$$

or

$$\bar{I}\alpha = 70 \text{ ft-lb}$$

We now apply the equations of dynamic equilibrium. The value of P is determined from a moment summation about A.

$$[+\circlearrowleft \Sigma M_A = 0] \qquad 4P + 36(6) - 96(2) - 96(3) + 70 = 0$$
$$P = 48.5 \text{ lb} \qquad \textit{Ans.}$$

Using force summations directed along the perpendicular components of the reaction at A, we obtain

$$[\overset{+}{\leftarrow}\Sigma X = 0] \qquad A_x + (P = 48.5) - 96 - 36 = 0 \qquad A_x = 87.5 \text{ lb}$$
$$[+\uparrow \Sigma Y = 0] \qquad A_y - 96 + 36 = 0 \qquad A_y = 60 \text{ lb}$$

from which the total reaction at A is found to be

$$A = \sqrt{(87.5)^2 + (60)^2} = 106.2 \text{ lb} \qquad \textit{Ans.}$$

PROBLEMS

13-3.4. A weight of 96.6 lb is suspended from a cord wrapped around a solid cylinder of 3-ft radius weighing 322 lb. The cylinder rotates about a horizontal axle through its mass center. Compute the total bearing reaction on this axle.

13-3.5. What torque applied to the cylinder of the previous problem will raise the weight with an acceleration of 12 fps^2? What will be the total bearing reaction?

$M = 578$ ft-lb; $R = 454.6$ lb ***Ans.***

13-3.6. During the operation of a punch press, its flywheel decelerates uniformly from 600 rpm to 400 rpm in 1 sec. The rim of the flywheel weighs 1288 lb, its inside and outside diameters are 56 in. and 60 in., and it is attached to its hub by 6 spokes. What average shearing force is developed between the rim and each spoke during the 1-sec interval?

$F = 350$ lb ***Ans.***

13-3.7. The train of gears in Fig. P-13-3.7 is represented by their pitch circles. Their centroidal moments of inertia, in units of in.-lb-sec^2, are $\bar{I}_A = 1.5$, $\bar{I}_B = 4.0$, and $\bar{I}_C = 1.2$. Determine the angular acceleration of gear A if the difference in torques applied at A and removed at C is 2 in.-lb.

Hint: Create dynamic equilibrium by applying inertia couples to each gear.

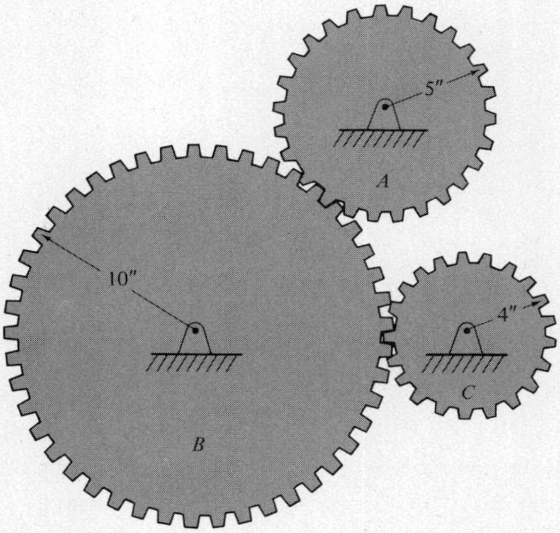

Figure P-13-3.7

13-3.8. A 3220-lb flywheel is mounted at the midpoint of a horizontal shaft 6 ft long. The mass center of the flywheel is 0.01 in. from the centerline of the shaft. When the flywheel is rotating at a constant speed of 1800 rpm, determine the maximum and minimum values of the bearing reactions at each end of the shaft.

max $R = 3090$ lb; min $R = 130$ lb ***Ans.***

13-3.9. A uniform bar, 10 ft long and weighing 1.61 lb/ft, is rotating in a horizontal plane about a vertical axis at one end. If the angular velocity is constant at 4 rad/sec, compute the axial force on a transverse section at its midpoint.

13-3.10. A 6-ft rod weighing 100 lb is rigidly fastened to a 200-lb cylinder as shown in Fig. P-13-3.10. Find the linear acceleration of the 300-lb block at the given position.

$a = 6.90 \text{ fps}^2$ ***Ans.***

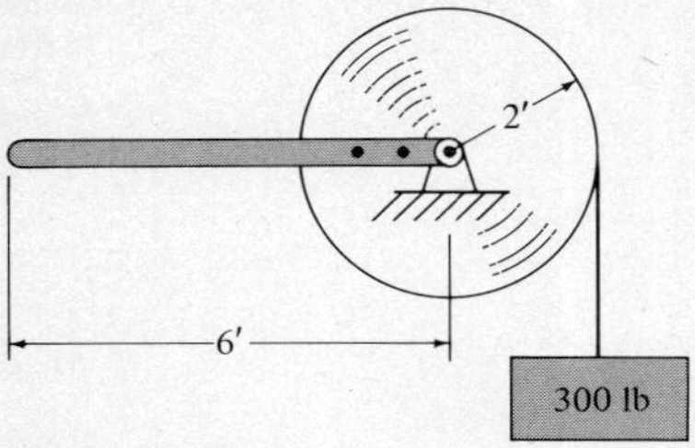

Figure P-13-3.10

13-3.11. If the coefficient of kinetic friction is 0.25 under each block in Fig. P-13-3.11, compute the total reaction at the axle of the compound cylinder.

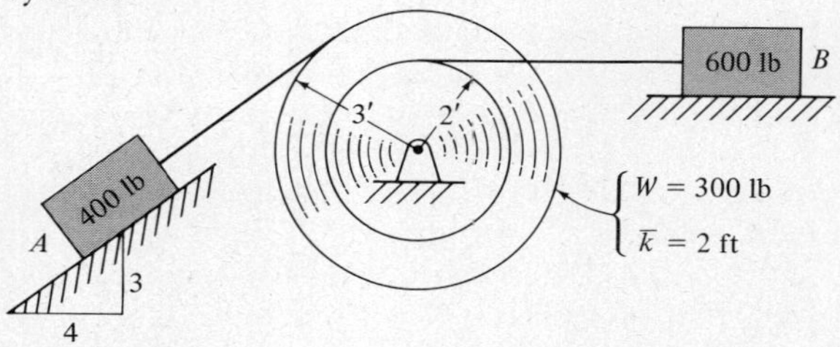

Figure P-13-3.11

13-3.12. For the system shown in Fig. P-13-3.12, determine the minimum and maximum weight of body C that will keep it at rest. For each incline, the coefficient of friction is 0.25.

min $W_C = 130$ lb; max $W_C = 260$ lb ***Ans.***

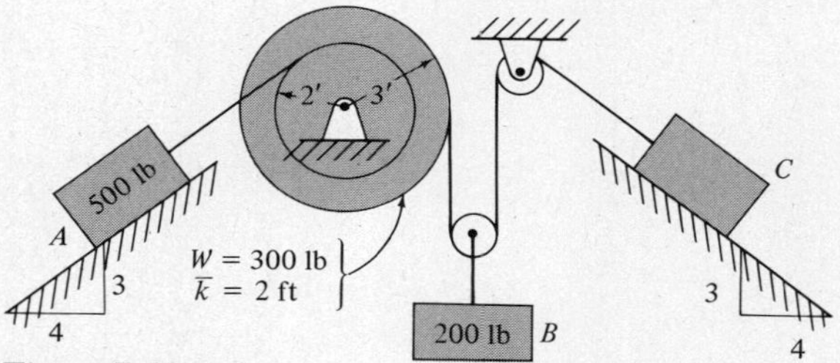

Figure P-13-3.12

13-3.13. Determine the maximum weight of body B in Fig. P-13-3.13 that will permit the homogeneous 50-lb block A to slide without tipping over.

$W_B = 42.9$ lb ***Ans.***

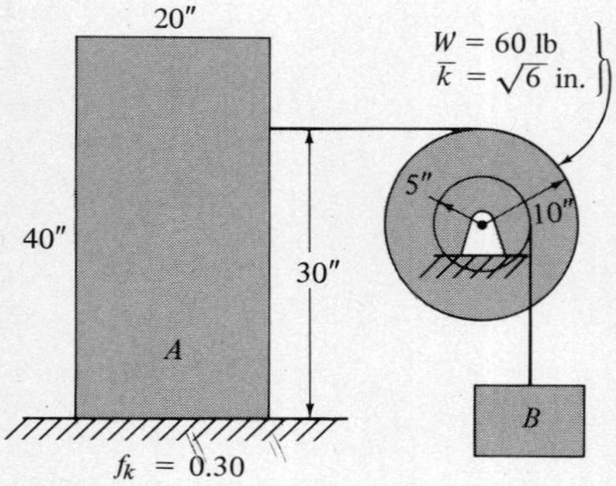

Figure P-13-3.13

13-3.14. In the system shown in Fig. P-13-3.14, body A is dropping at 40 fps when a brake force $P = 80$ lb is applied. Determine the time for A to come to rest. Neglect the thickness of the brake block.

$t = 11.05$ sec ***Ans.***

13-3.15. In the previous problem, assume the maximum strength of the cord supporting body A is 700 lb and of that joining drums B and C is 800 lb. If too large a brake force is applied, one of these cords will rupture. Which one will it be and at what brake force P?

$P = 393$ lb ***Ans.***

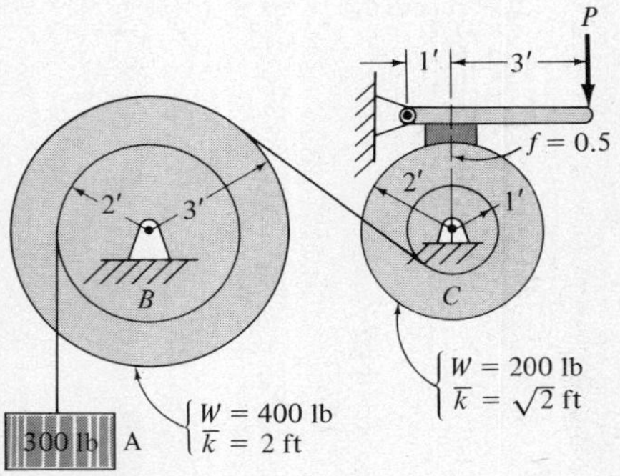

Figure P-13-3.14

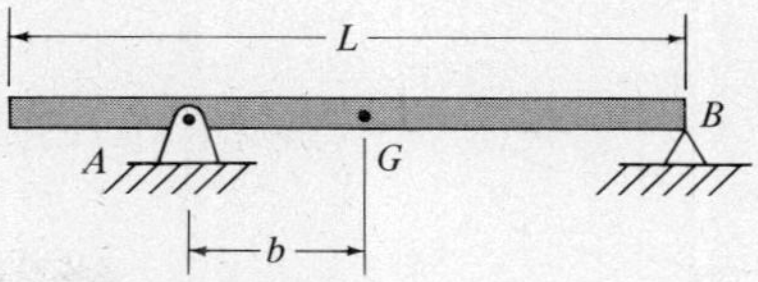

Figure P-13-3.16

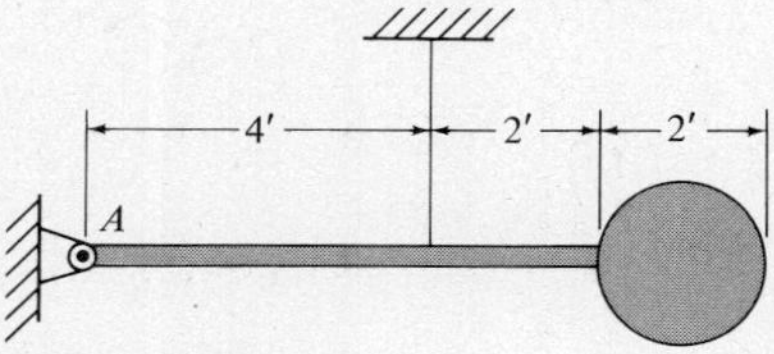

Figure P-13-3.17

13-3.16. As shown in Fig. P-13-3.16, a slender bar of weight W and length L is supported at one end and at a distance b from its mass center G. Determine the reaction at A the instant after the support at B is suddenly removed.

$$R_A = \frac{WL^2}{L^2 + 12b^2} \qquad \textit{Ans.}$$

13-3.17. A cylinder weighing 200 lb is welded to the end of a 100-lb bar. The assembly is supported by a horizontal axis at A and by a vertical cable as shown in Fig. P-13-3.17. Compute the reaction at A an instant after cutting the cable.

13-3.18. At the instant the assembly of the previous problem has rotated through an angle $\theta = \tan^{-1}\frac{3}{4}$ from the horizontal, its angular velocity is known to be $\omega = \sqrt{0.184g}$ rad/sec. What is then the magnitude of the axle reaction at A?

$$A = 494 \text{ lb} \qquad \textit{Ans.}$$

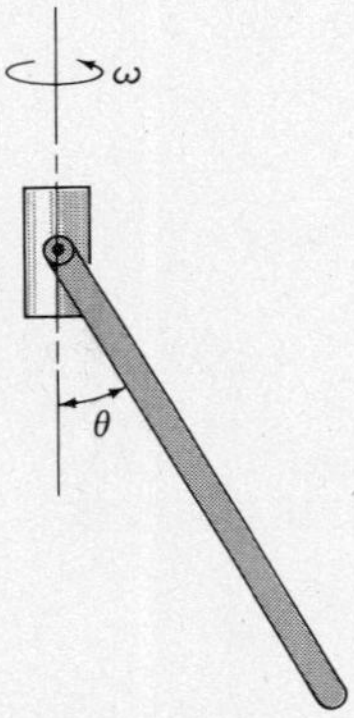

Figure P-13-3.19

13-3.19. A uniform slender rod of length L is hinged to a frame rotating about a vertical axis as in Fig. P-13-3.19. Show that the angle between the rod and the axis is defined by $\cos\theta = 3g/2L\omega^2$.

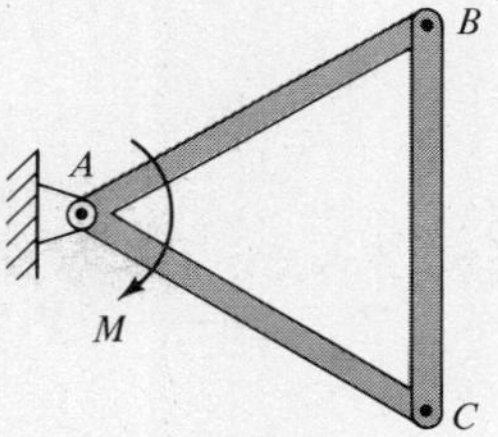

Figure P-13-3.20

13-3.20. Three bars, each 2 ft long and weighing 9.66 lb, are pinned together to form the equilateral frame shown in Fig. P-13-3.20. They rotate in a horizontal plane about a vertical axis at A. What torque M is required to cause a constant angular acceleration of 12 rad/sec²? What is the reaction at A when the frame has rotated through $\frac{2}{3}$ rad starting from rest?

$$M = 21.6 \text{ ft-lb};\ A = 20.8 \text{ lb} \qquad \textit{Ans.}$$

13-3.21. Two bars are welded together as shown in Fig. P-13-3.21. The assembly rotates about a horizontal axis at A. Segment AB weighs 60 lb and segment CD weighs 30 lb. At the given position it is known that $\omega = \sqrt{0.3g}$ rad/sec. Determine the horizontal and vertical components of the axle reaction and the moment transmitted at B by AB upon CD.

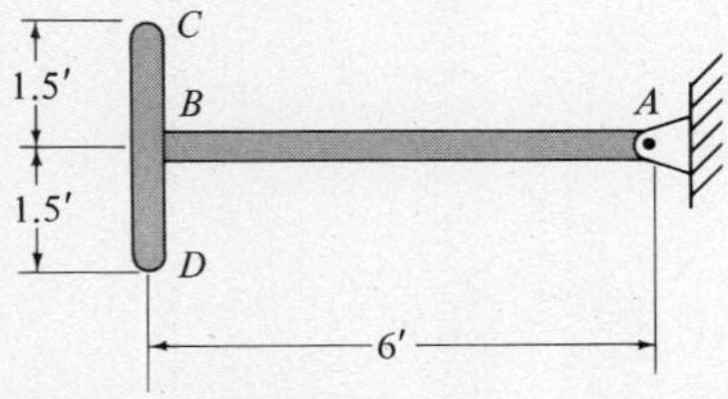

Figure P-13-3.21

13-3.22. A turntable, rotating in a horizontal plane about a vertical axis O, carries a bent bar pinned to it at A and forced to rotate with it by a smooth peg at C. Each segment of the bar weighs 9.66 lb. At the

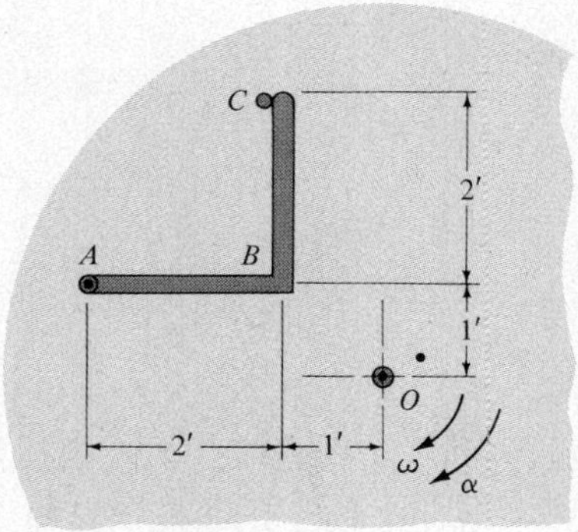

Figure P-13-3.22

instant shown in Fig. P-13-3.22, $\omega = \sqrt{20}$ rad/sec and $\alpha = 10$ rad/sec^2 both clockwise. Determine the forces acting at A and C.

$A = 14.22$ lb; $C = 16$ lb **Ans.**

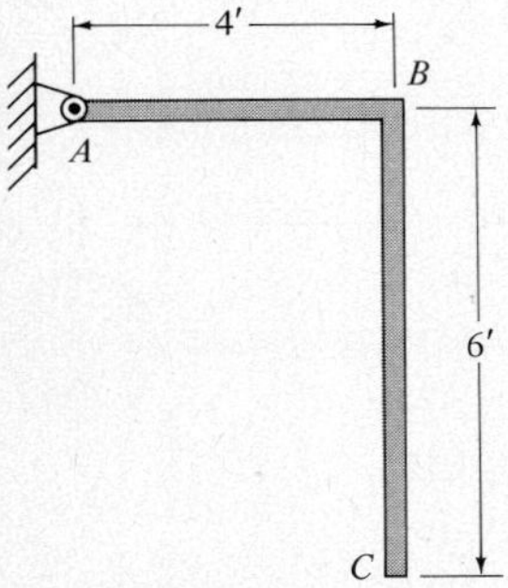

Figure P-13-3.23

13-3.23. The bent bar shown in Fig. P-13-3.23 weighs 10 lb/ft and is free to rotate in a vertical plane about a horizontal axis at A. Determine the moment at B exerted by AB on BC an instant after the bar is released from rest at the given position.

$M_B = 121.6$ ft-lb **Ans.**

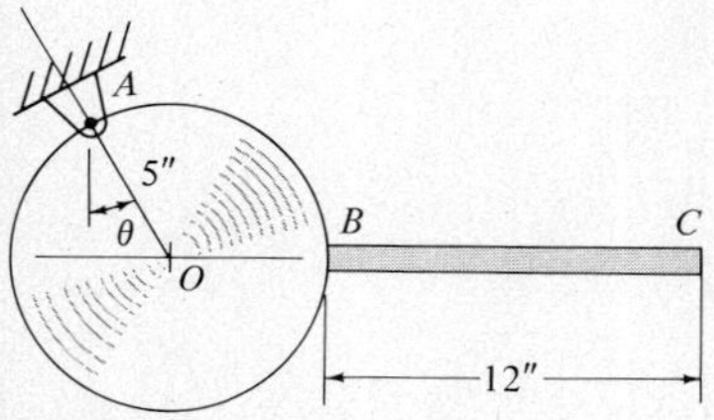

Figure P-13-3.24

13-3.24. A 12-in. bar weighing 3 lb is welded to the rim of an 80-lb cylinder as shown in Fig. P-13-3.24. Find the moment exerted on the weld at B when the system is released from rest at the given position where $\tan\theta = \frac{3}{4}$.

$M_B = 217$ in.-lb **Ans.**

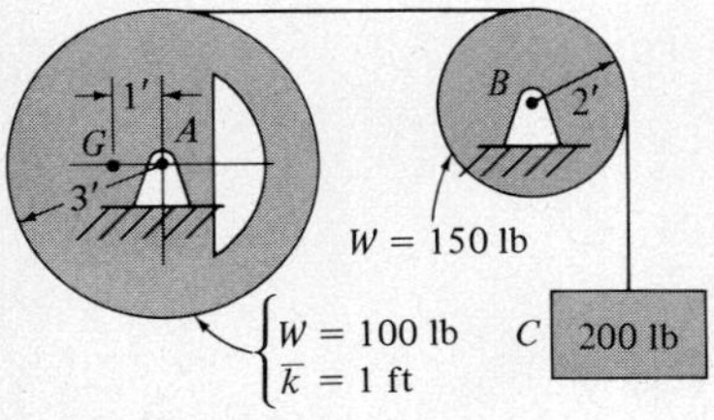

Figure P-13-3.25

13-3.25. In the system shown in Fig. P-13-3.25, A is an unbalanced wheel whose gravity center is at G, and B is a solid cylinder. Find the components of each axle reaction if $\omega_A = \sqrt{g}$ rad/sec at the instant shown.

$A_h = 51.2$ lb; $A_v = 118.7$ lb; $B_h = 40.8$ lb; $B_v = 237.8$ lb **Ans.**

13-3.26. In the system shown in Fig. P-13-3.26, a 50-lb bar 6 ft long is attached to drum B which has a clockwise angular velocity $\omega = \sqrt{g/3}$ rad/sec at the instant when $\tan\theta = \frac{3}{4}$. Find the axle reactions on drums B and C. Under D, $F = 0.225$.

$B = 194.4$ lb; $C = 98.5$ lb **Ans.**

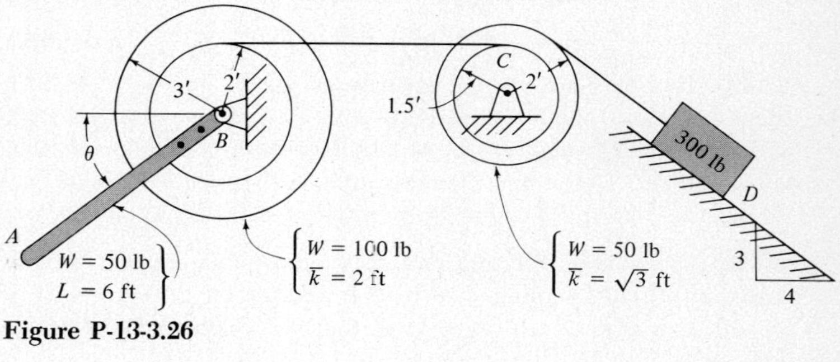

Figure P-13-3.26

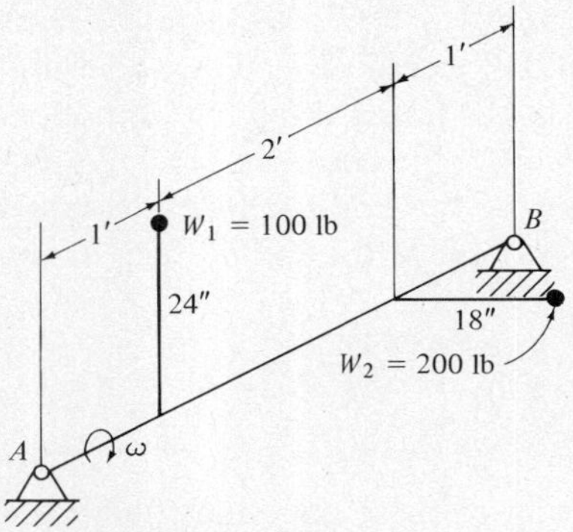

Figure P-13-3.27

13-3.27. Two eccentric weights $W_1 = 100$ lb and $W_2 = 200$ lb are fastened to the rotating horizontal shaft shown in Fig. P-13-3.27. Compute the values of balance weights concentrated 1 ft from the shaft and rotating in vertical planes through A and B that will balance the dynamic effects of W_1 and W_2. What are the angular positions of the balance weights measured from the plane containing W_1 and axis AB?

$W_A = 168$ lb, $\theta_A = 26.6°$; $W_B = 231$ lb, $\theta_B = 77.5°$ ***Ans.***

13-4 ROLLING BODIES

Let us consider next the application of Eq. (13-2.1) to rolling bodies, postponing the discussion of general plane motion to the next section. The case of the *homogeneous* rolling wheel is simplified by the fact that its mass center has a rectilinear motion parallel to the flat surface on which it rolls. Then, as shown in Fig. 13-4.1, the resultant effective forces consist of a single force $\left(\frac{W}{g}\right)\bar{a}$ parallel to the surface acting through the mass center G, and the couple $\bar{I}\alpha$. The

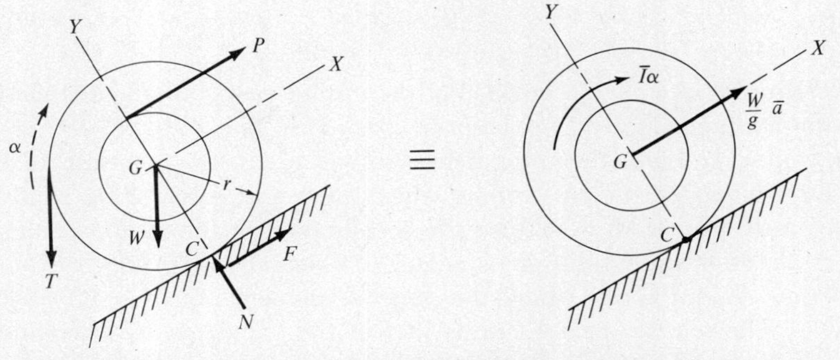

Figure 13-4.1 Equivalence of applied forces and resultant effective forces. With a free-rolling wheel, $\bar{a} = r\alpha$ and $\frac{\omega}{g}\bar{a}$ becomes $\frac{\omega}{g}r\alpha$.

best reference axes to choose are those which pass through the mass center with the X axis directed parallel to the surface on which the body rolls and considered positive in the initial direction of motion. The components of $\bar{a}$ are then given by $\bar{a}_x = \bar{a}$ and $\bar{a}_y = 0$. Applying the principle that the resultant of the applied forces is equal to that of the effective forces, the motion of the rolling body is determined by

$$\Sigma X = \frac{W}{g}\bar{a}; \qquad \Sigma Y = 0; \qquad \Sigma \bar{M} = \bar{I}\alpha \tag{13-4.1}$$

Before discussing the application of these equations, there are two important special cases to consider. The first is that in which the wheel rolls freely without slipping. It can do so only if sufficient frictional resistance acts at the instant center C to hold that point instantaneously at rest. The value of this static frictional resistance is generally unknown since it may have any value ranging from zero up to the limiting static friction force $f_s N$. It is therefore convenient to eliminate this unknown static frictional force by taking moments about the instant center C. Thus, by equating the moment sum about C of the applied forces (i.e., ΣM_C) to the moment sum of the effective forces, we obtain

$$\Sigma M_C = \bar{I}_C\alpha + \left(\frac{W}{g}\bar{a}\right)r \tag{a}$$

When free-rolling exists, then $\bar{a} = r\alpha$, so that Eq. (*a*) becomes

$$\Sigma M_C = \bar{I}\alpha + \left(\frac{W}{g}r\alpha\right)r = \left(\bar{I} + \frac{W}{g}r^2\right)\alpha \tag{b}$$

in which we recognize that the sum in parentheses is an application of the transfer formula and represents the moment of inertia I_C about the instant center C. Consequently, for free-rolling wheels, we obtain

$$\Sigma M_C = I_C\alpha \tag{13-4.2}$$

If slipping occurs, the position of the instant center is unknown and the relation $\Sigma M_C = I_C\alpha$ cannot be used. However, Eqs. (13-4.1) are always valid, whether or not free rolling exists. Observe also that for a homogeneous free-rolling wheel, we could obtain $\Sigma M_C = I_C\alpha$ directly from $\Sigma \mathbf{M} = \dot{\mathbf{H}}$ since the acceleration of the instant center is directed through the mass center (see Section 12-5) which is another condition for which the general moment equation is valid.

The second special case is that of the unbalanced free-rolling wheel or cylinder in Fig. 13-4.2 whose mass center G does *not* coincide with the geometric center O. The mass center in this case does *not* have a rectilinear motion and consequently Eq. (13-4.2) *does*

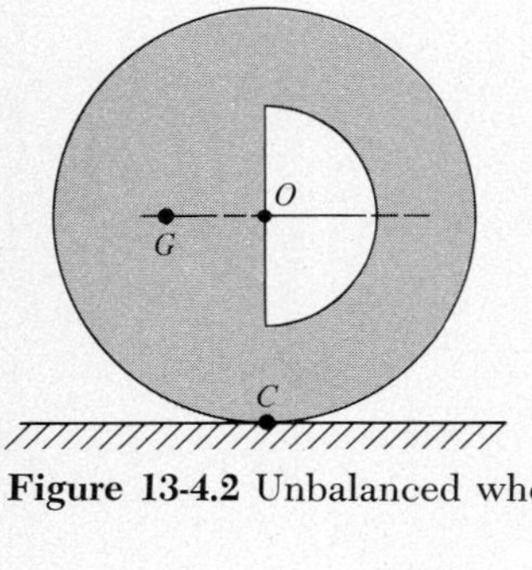

Figure 13-4.2 Unbalanced wheel.

not apply. Nevertheless, $\Sigma M_C = I_C\alpha$ is still valid at the two instants when G crosses the line containing the geometric center O and the instant center C because at such positions the acceleration of the instant center is directed toward the mass center.

ILLUSTRATIVE PROBLEMS

13-4.1. The solid cylinder in Fig. 13-4.3 is 4 ft in diameter and weighs 600 lb. It is acted upon by an upward force of 100 lb applied by a cord wrapped around it. Find the coefficient of friction required to prevent slipping.

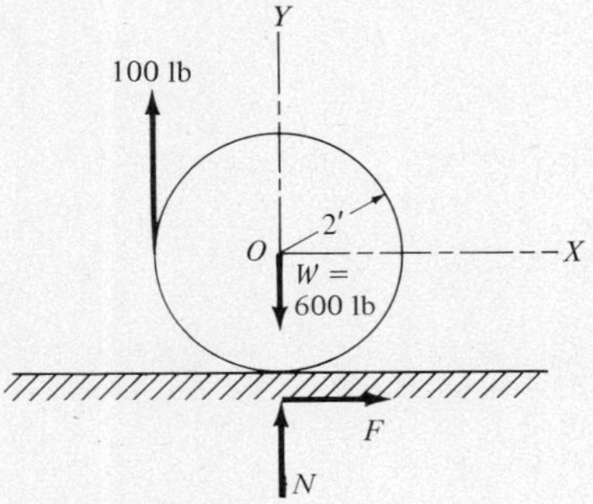

Figure 13-4.3

Solution: Part (a)

Since free rolling of the cylinder is implied in the statement of the problem, the instant center C is at the point of contact with the ground, and Eq. (13-4.2) may be used to obtain α. The centroidal mass moment of inertia of the cylinder is

$$\bar{I} = \frac{1}{2}\frac{W}{g}r^2 = \frac{1}{2}\left(\frac{600}{g}\right)(2)^2 = \frac{1200}{g}\text{ ft-lb-sec}^2$$

and the transfer formula gives the mass moment of inertia about the instant center as

$$[I = \bar{I} + m\,d^2] \qquad I_C = \frac{1200}{g} + \left(\frac{600}{g}\right)(2)^2 = \frac{3600}{g}\text{ ft-lb-sec}^2$$

We now have

$$[\circlearrowleft + \Sigma M_C = I_C\alpha] \qquad 100(2) = \frac{3600}{g}\alpha \qquad \frac{\alpha}{g} = \frac{1}{18}$$

whence, applying the general relations of Eqs. (13-4.1), we have

$$[\circlearrowleft + \Sigma\bar{M} = \bar{I}\alpha] \quad 100(2) - 2F = \frac{1200}{g}\alpha = 1200\left(\frac{1}{18}\right) \qquad F = 66.7\text{ lb}$$

$$[+\!\downarrow \Sigma Y = 0] \qquad N + 100 - 600 = 0 \qquad N = 500\text{ lb}$$

Applying the friction relationship gives

$$\left[f_s = \frac{F}{N}\right] \qquad f_s = \frac{66.7}{500} = 0.1334 \qquad \textit{Ans.}$$

Part (b)

If the coefficient of static friction were specified as 0.20, what would be the friction force? The available friction force would be found from the relation $F = f_s N = 0.2(500) = 100$ lb. But not all of this is required to prevent slipping; only 66.7 lb would be acting as before. This illustrates that the actual static friction force cannot be found from the relation defining the available static friction force.

Part (c)

If $f_s = 0.12$ and $f_k = 0.11$, what would happen? What would be the values of $\bar{a}$ and α? Under these conditions, the available static friction of $0.12(500) = 60$ lb would be less than required to prevent slipping. Therefore, slipping would occur at the ground and a kinetic friction force $F_k = f_k N = 0.11(500) = 55$ lb would act. $\Sigma M_C = I_C \alpha$ cannot then be used because it applies only to free-rolling bodies. Therefore, we must use the general relations of Eqs. (13-4.1) to obtain

$$[\circlearrowleft + \Sigma \bar{M} = \bar{I}\alpha] \qquad 100(2) - 55(2) = \frac{1200}{g}\alpha \qquad \alpha = 2.42 \text{ rad/sec}^2$$

$$\left[\overset{+}{\rightarrow} \Sigma X = \frac{W}{g}\bar{a}\right] \qquad 55 = \frac{600}{g}\bar{a} \qquad \bar{a} = 2.95 \text{ fps}^2$$

Note that when slipping occurs, the relation between $\bar{a}$ and α for free rolling is not valid; i.e.,

$$[\bar{a} \neq r\alpha] \qquad 2.95 \neq 2(2.42)$$

13-4.2. In the system shown in Fig. 13-4.4, the floating pulley D is supported by a cord wound around the drum and a second cord wound around its outer radius. The second cord, after passing over a frictionless pulley of negligible weight, is wound around the drum of disk A. Determine the angular accelerations and the tensions T and P if both bodies roll without slipping.

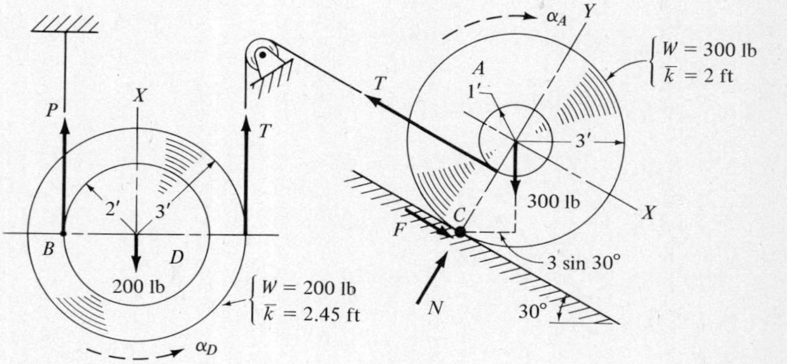

Figure 13-4.4

Solution

To determine whether motion occurs and in which direction, temporarily assume disk A to be at rest. Setting a moment summation about A's instant center C equal to zero, we find $2T = 300(3 \sin 30°)$ from which a preliminary value of T is 225 lb. Consider now disk D whose instant center is at B. If $\Sigma M_B = 0$, there will be equilibrium; but if $\Sigma M_B \neq 0$, the disk D will move in the direction of the unbalanced moment. Thus $\Sigma M_B = 225(5) - 200(2)$ indicates an unbalanced counterclockwise moment acting on disk D. It therefore moves up, climbing along cord P which is wound up on its drum. Simultaneously, cord T, moving up, is unwound from disk D. The absolute motion of T is also *down* the incline, its point of attachment with disk A moving slower than the mass center of A, so that T also unwinds from A as disk A rolls down the incline. These directions of motion specify the positive directions of moments and forces.

The kinematic relations between disks A and D are found as described in Illus. Prob. 12-5.2 on p. 453. The bodies may be assumed to be instantaneously rotating about their instant centers B and C so that the motion of the cord which connects them is given by

$$[s = r\theta] \qquad s_T = 5\theta_D = 2\theta_A$$

whence successive differentiation yields the following relation between their accelerations:

$$5\alpha_D = 2\alpha_A \tag{a}$$

The centroidal mass moments of inertia are

$$[\bar{I} = m\bar{k}^2] \qquad \text{For } A\text{:} \quad \bar{I} = \frac{300}{g}(2)^2 = \frac{1200}{g} \text{ ft-lb-sec}^2$$

$$\text{For } D\text{:} \quad \bar{I} = \frac{200}{g}(2.45)^2 = \frac{1200}{g} \text{ ft-lb-sec}^2$$

whence, using the transfer formula, the values with respect to the instant centers are

$$[I = \bar{I} + md^2] \qquad \text{For } A\text{:} \quad I_C = \frac{1200}{g} + \frac{300}{g}(3)^2 = \frac{3900}{g} \text{ ft-lb-sec}^2$$

$$\text{For } D\text{:} \quad I_B = \frac{1200}{g} + \frac{200}{g}(2)^2 = \frac{2000}{g} \text{ ft-lb-sec}^2$$

We are now ready to consider the kinetics of each disk. The friction force F acting on disk A is a static friction holding point C momentarily at rest. It is directed so as to resist rotation about the mass center. Since a static friction may vary from zero to its maximum available value, F will be unknown even though f_s may be specified. Only when slipping impends will F be determined by f_sN. However, this unknown friction force can be easily eliminated by taking advantage of the condition of free rolling and applying $\Sigma M_C = I_C\alpha$ which permits a moment summation about the instant center C through which F acts. Similarly, the unknown tension P on disk B is eliminated by a moment summation about its instant center B. Taking moments positive in the initial directions of motion, we obtain

$$[\Sigma M_C = I_C\alpha] \quad 300(3 \sin 30°) - 2T = \frac{3900}{g}\alpha_A \tag{b}$$

$$[\Sigma M_B = I_B\alpha] \quad 5T - 200(2) = \frac{2000}{g}\alpha_D = \frac{2000}{g}\left(\frac{2}{5}\alpha_A\right) \tag{c}$$

The tension T is easily eliminated from these equations by multiplying Eq. (*c*) by $\frac{2}{5}$ and adding them, whence

$$\frac{\alpha_A}{g} = \frac{290}{4220}$$

and

$$\alpha_A = 2.21 \text{ rad/sec}^2; \quad \alpha_D = \tfrac{2}{5}\alpha_A = 0.885 \text{ rad/sec}^2 \qquad \textit{Ans.}$$

after which either value of α may be substituted in (*b*) or (*c*) to determine

$$T = 91 \text{ lb} \qquad \textit{Ans.}$$

To determine P, we apply the general relations of Eq. (13-4.1) to disk D, using both ΣM and ΣX as a mutual check. We obtain

$$[+\circlearrowleft \Sigma \bar{M} = \bar{I}\alpha] \quad 91(3) - 2P = \frac{1200}{32.2}(0.885)$$

$$P = 120 \text{ lb} \qquad \textit{Ans.}$$

$$\left[+\uparrow \Sigma X = \frac{W}{g}\bar{a}\right] \quad 91 + P - 200 = \frac{200}{32.2}(2 \times 0.885)$$

$$P = 120 \text{ lb} \qquad \textit{Check}$$

PROBLEMS

13-4.3. A solid cylinder and a homogeneous sphere, each of weight W and radius r, roll without slipping down a plane inclined at $\theta°$ with the horizontal. For each body, determine the minimum coefficient of friction to prevent slipping and the acceleration of the mass center.

For cylinder: $f = \frac{1}{3}\tan\theta$; $\bar{a} = \frac{2}{3}g \sin\theta$ *Ans.*

13-4.4. In Fig. P-13-4.4, at what height h above the table should a billiard ball of radius r be stroked so that it will roll without slipping on a frictionless surface?

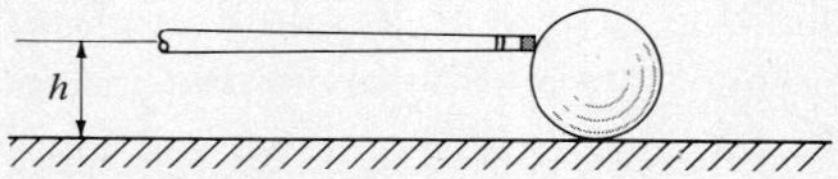

Figure P-13-4.4

13-4.5. A spherical shell and a circular hoop of the same weight and mean radius are side by side on an incline. If they are simultaneously released from rest, will the shell or the hoop reach the bottom of the incline first, or do they arrive together?

13-4.6. The body shown in Fig. P-13-4.6 consists of two solid cylinders each weighing 100 lb. They are separated by a hub of negligible weight. Find the acceleration of the mass center, the friction force acting, the angular acceleration, and the direction of motion.

$\bar{a} = 9.66$ fps^2; $F = 40$ lb; $\alpha = 0.717$ rad/sec^2 **Ans.**

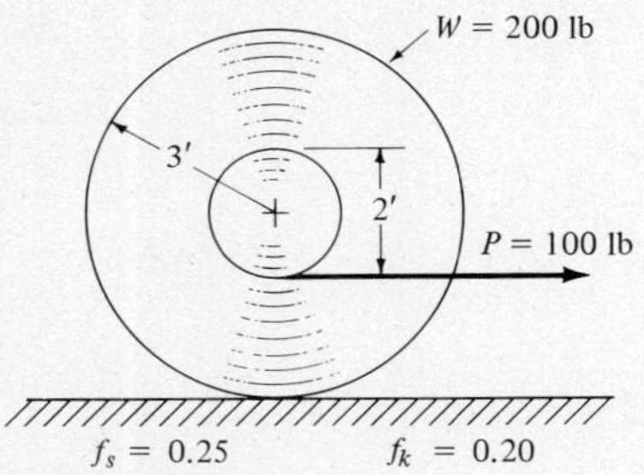

Figure P-13-4.6

13-4.7. The compound disk and drum shown in Fig. P-13-4.7 is acted upon by a force $P = 100$ lb which always remains horizontal. Assuming free rolling, determine $\bar{a}$ and the required friction force.

$\bar{a} = 7.08$ fps^2; $F = 84$ lb **Ans.**

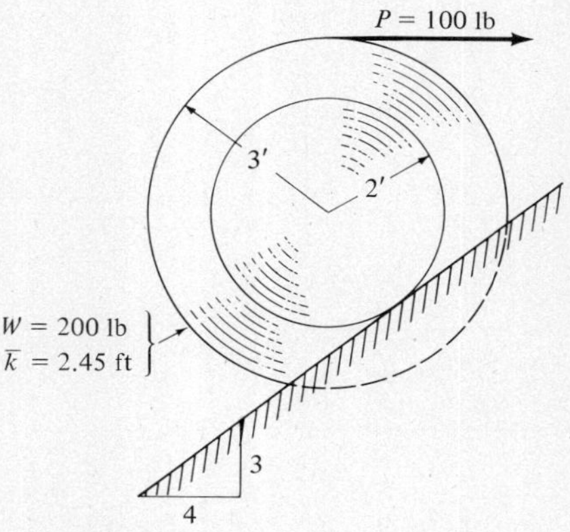

Figure P-13-4.7

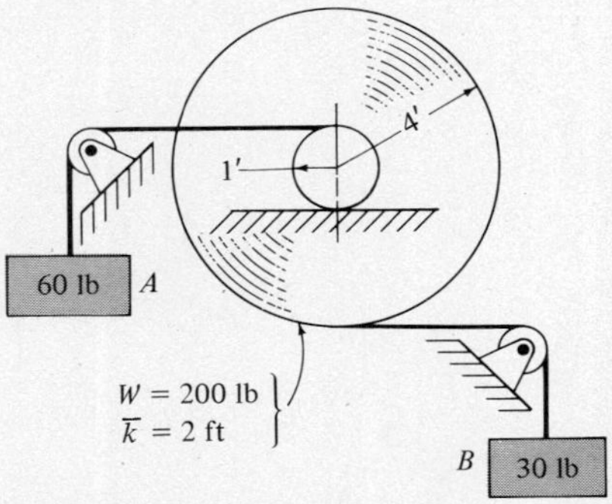

Figure P-13-4.9

13-4.8. A bowling ball is sent down an alley with an initial velocity v_o and without any spin. Assuming the coefficient of friction is f, how far does it skid before pure rolling occurs.

$$s = \frac{12v_o^2}{49\,fg} \qquad \textbf{Ans.}$$

13-4.9. If the compound disk shown in Fig. P-13-4.9 rolls without slipping, determine the acceleration of its mass center. Assume the fixed pulleys to be frictionless and of negligible weight.

13-4.10. The 80-lb plank shown in Fig. P-13-4.10 rests on two cylindrical rollers of radius r each weighing 30 lb. Determine the force P that will accelerate the plank up the incline at 6 fps^2. Assume no slipping occurs either at the plank or at the incline.

$P = 56.7$ lb **Ans.**

13-4.11. In the preceding problem, compute the acceleration of the plank after the force P is removed.

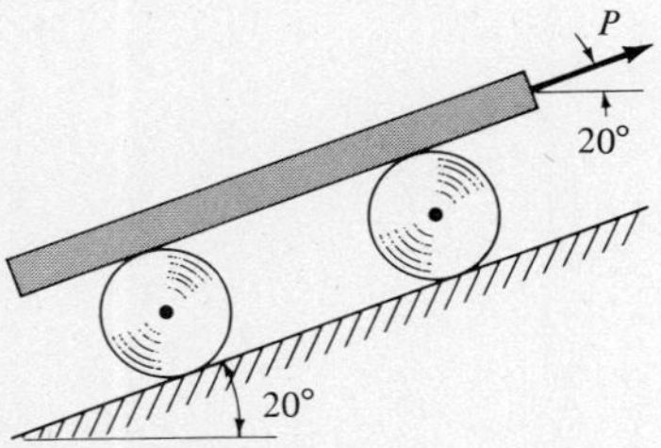

Figure P-13-4.10

13-4.12. Determine the tension in the cord supporting the 50-lb block shown in Fig. P-13-4.12.

$$T = 42 \text{ lb} \qquad \textbf{\textit{Ans.}}$$

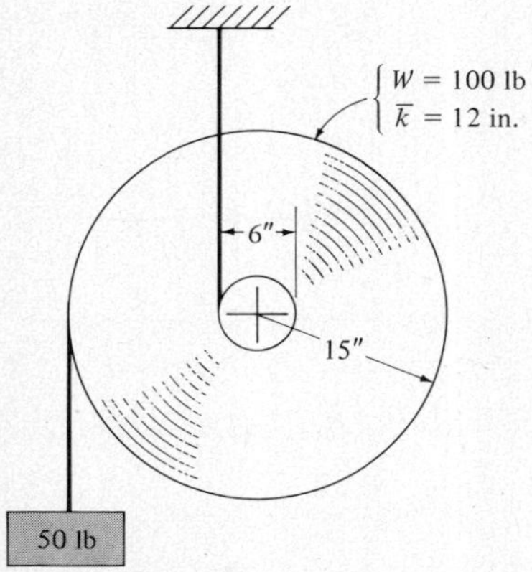

Figure P-13-4.12

13-4.13. Determine the tensions in the cords supporting the floating pulley shown in Fig. P-13-4.13.

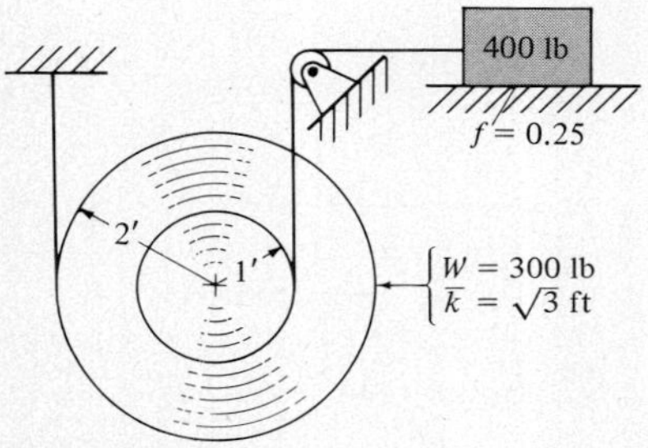

Figure P-13-4.13

13-4.14. The 100-lb weight causes the compound disk shown in Fig. P-13-4.14 to roll and slip on the horizontal floor. If the coefficient of friction between the disk and the floor is 0.40, determine the acceleration $\bar{a}$ of the center of the disk.

$$\bar{a} = 5 \text{ fps}^2 \qquad \textbf{\textit{Ans.}}$$

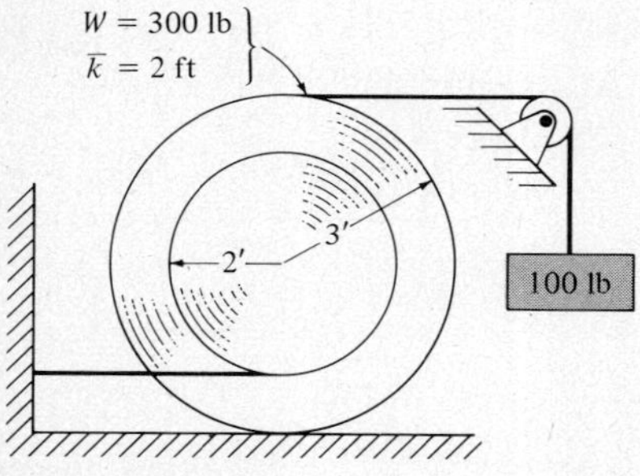

Figure P-13-4.14

13-4.15. The gear and concentric drum shown in Fig. P-13-4.15 rests on a rack *B*. The rack, which weighs 200 lb, is free to slide on smooth horizontal guides. A force $P = 280$ lb applied to the rack causes gear *A* to roll along the rack. Determine the tension in the cable and how much cable is wound on or off the drum in 2 sec starting from rest.

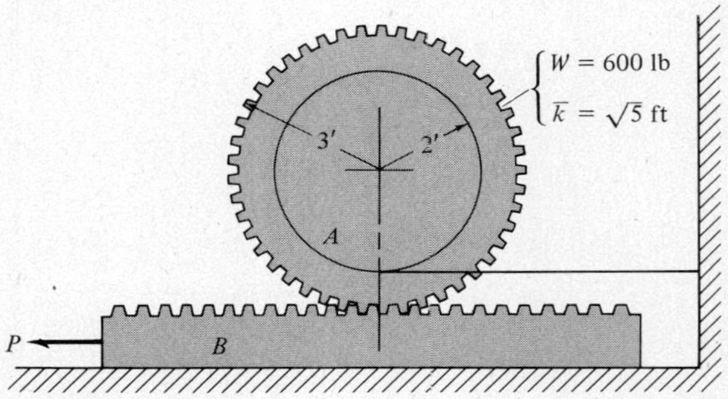

Figure P-13-4.15

13-4.16. Determine the value of the static friction force acting on disk *D* in Fig. P-13-4.16 if the disk rolls without slipping upon the incline.

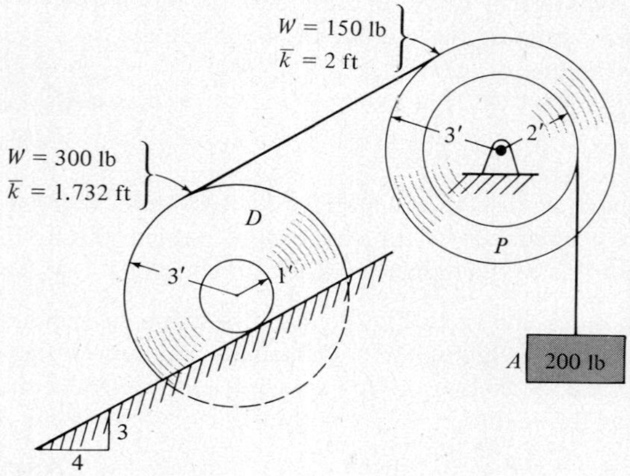

Figure P-13-4.16

13-4.17. Find the tension in the cord supporting body *A* in Fig. P-13-

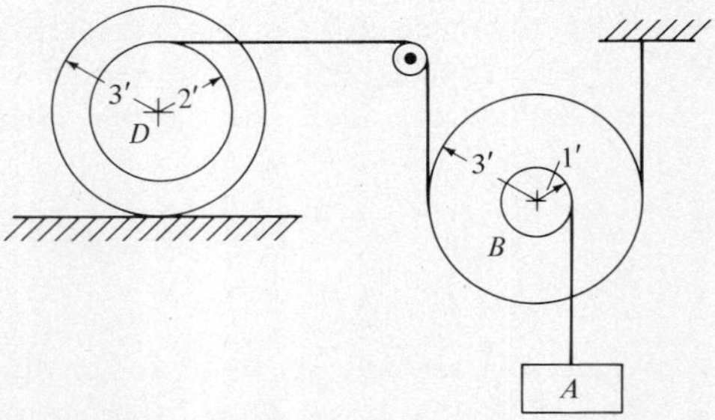

Figure P-13-4.17

4.17. Assume that $W_A = 100$ lb; for disk D, $W = 300$ lb and $\bar{k} = 2$ ft; for floating disk B, $W = 200$ lb and $\bar{k} = \sqrt{3}$ ft.

$$T_A = 81 \text{ lb} \qquad \textit{Ans.}$$

13-4.18. Determine the weight of body B that will make the compound wheel shown in Fig. P-13-4.18 roll freely up the 30° incline at 4 rad/sec². Neglect the backward inclination of B caused by inertia.

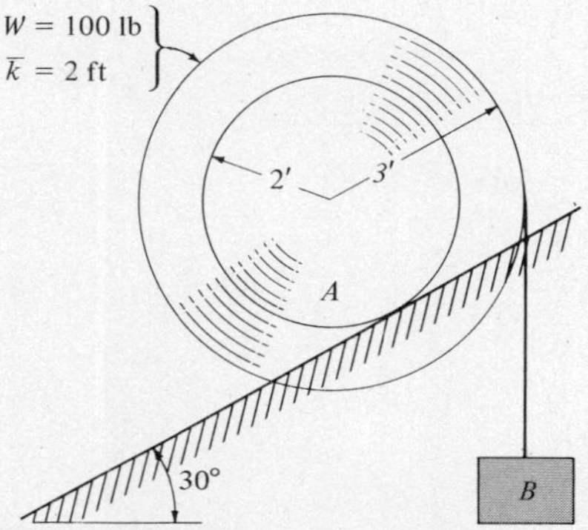

Figure P-13-4.18

13-4.19. In Fig. P-13-4.19 is shown the pitch circles of two gears A and B connected by the arm OC. If gear A is fixed and the system is released from rest at the given position, find the angular acceleration of gear B. Assume gear B is equivalent to a solid cylinder weighing 30 lb and that arm OC weighs 20 lb.

$$\alpha_B = 24.9 \text{ rad/sec}^2 \qquad \textit{Ans.}$$

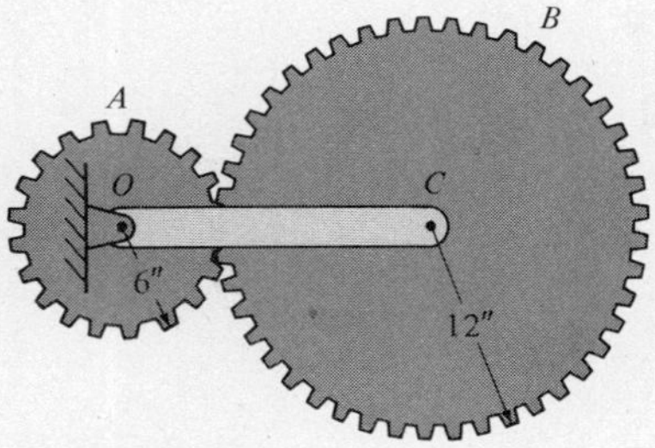

Figure P-13-4.19

13-4.20. As shown in Fig. P-13-4.20, a cord passes from the rim of a solid cylinder of weight W and radius r over a frictionless pulley of negligible weight and back to the center of the cylinder. Determine the tension in the cord.

$$T = W/3 \qquad \textit{Ans.}$$

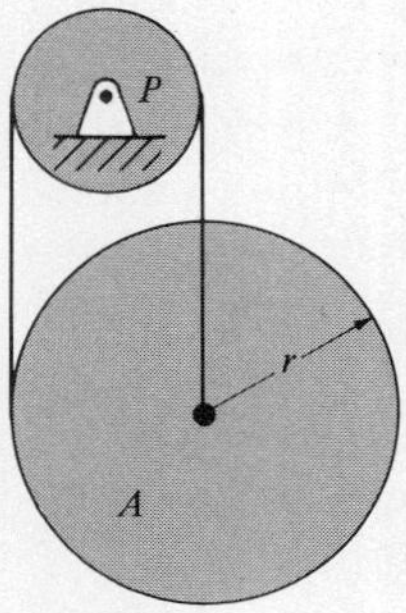

Figure P-13-4.20

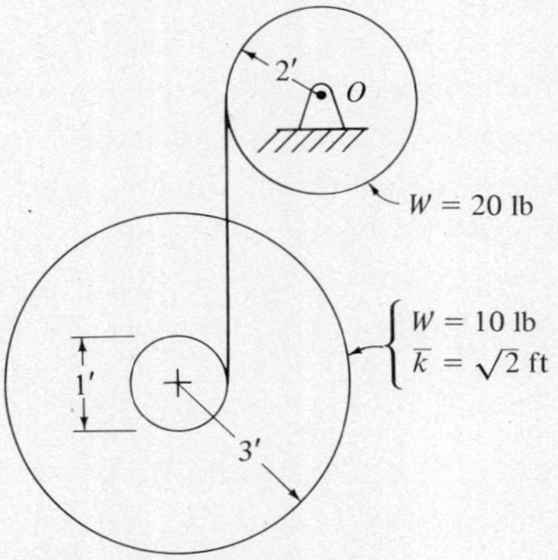

Figure P-13-4.21

13-4.21. Determine the tension in the cord connecting the solid cylinder in Fig. P-13-4.21 to the hub of the floating disk.

$$T = 4.43 \text{ lb} \qquad \textit{Ans.}$$

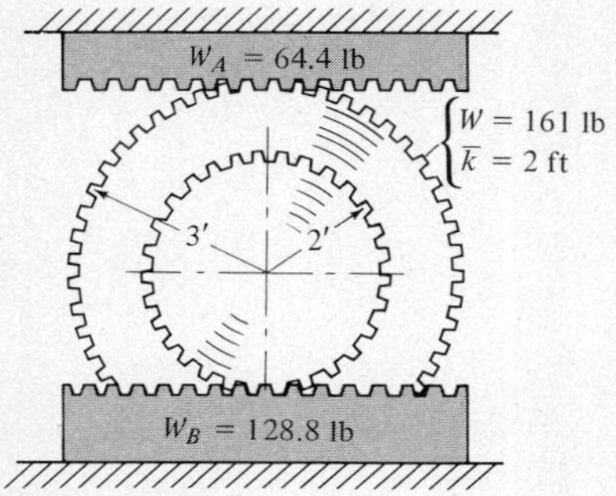

Figure P-13-4.23

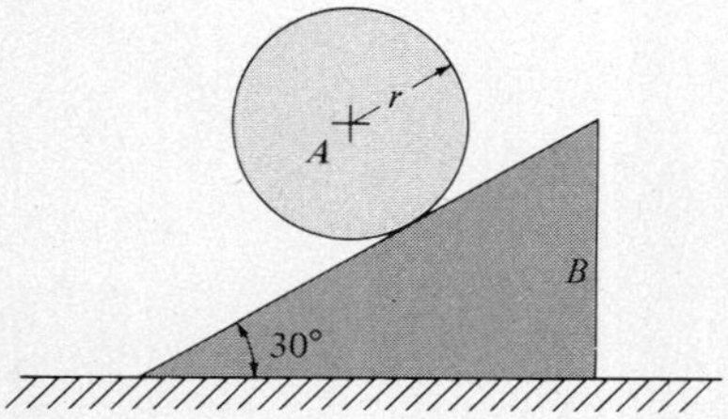

Figure P-13-4.24

13-4.22. If the cable in Prob. 13-4.15 is removed, show that the relative acceleration of the mass center of gear A with respect to the rack B is given by $a_r = r^2 a_B / r^2 + \bar{k}^2$. Also show that the acceleration of the rack will be

$$a_B = \frac{Pg}{W_B + \dfrac{W_A \bar{k}^2}{r^2 + \bar{k}^2}} \qquad \textit{Ans.}$$

13-4.23. In the compound gear and rack assembly shown schematically in Fig. P-13-4.23, a clockwise torque of 1180 ft-lb is applied to the gear. If the racks slide in frictionless guides, compute the acceleration $\bar{a}$ of the center of the compound gear.

$$\bar{a} = 4 \text{ fps}^2 \qquad \textit{Ans.}$$

13-4.24. As shown in Fig. P-13-4.24, a solid cylinder A rolls without slipping along wedge B which is on a smooth horizontal floor. If the cylinder and wedge each weigh W lb, determine the acceleration of wedge B.

$$a_3 = g/3\sqrt{3} \qquad \textit{Ans.}$$

13-5 GENERAL PLANE MOTION

For general plane motion, we use the same approach discussed previously for fixed-axis rotation and for rolling bodies. Conceptually, there is nothing new; we still equate a free-body diagram of the impressed forces to its dynamic equivalent of the effective force $\left(\frac{W}{g}\right)\bar{a}$ at the mass center and an effective couple $\bar{I}\alpha$, or we may apply these dynamic effects reversed on the free-body diagram and use the method of dynamic equilibrium.

There is an additional complication, however, in that generally no simple relation exists between $\bar{a}$ and α. Usually, a kinematic analysis of the type discussed in Section 12-4 is necessary in which the acceleration $\bar{a}$ of the mass center is found by combining the known acceleration a_A of any reference point A with the rotational components of the acceleration of the mass center about that reference point. Thus, temporarily denoting $\bar{a}$ by a_G, we apply

$$a_G = a_A \nrightarrow (a_{G/A} = r\omega^2 \nrightarrow r\alpha)$$

whence, on replacing a_G by $\bar{a}$ and r by the distance $\bar{r}$ between the mass center G and the reference point A, we obtain

$$\bar{a} = a_A \nrightarrow \bar{r}\omega^2 \nrightarrow \bar{r}\alpha \qquad (a)$$

On multiplying each term of Eq. (a) by W/g, we obtain

$$\frac{W}{g}\bar{a} = \frac{W}{g}a_A \nrightarrow \frac{W}{g}\bar{r}\omega^2 \nrightarrow \frac{W}{g}\bar{r}\alpha \qquad (b)$$

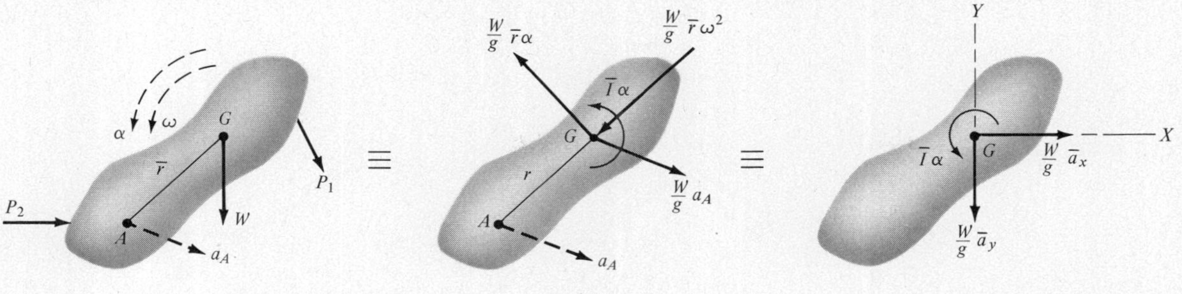

Figure 13-5.1 Equivalent sets of dynamic effects.

in which the right-side terms represent the components of the resultant effective force $\left(\frac{W}{g}\right)\bar{a}$. Although these components are not mutually perpendicular, they may be applied directly on the equivalent force diagram shown in part (b) of Fig. 13-5.1, or we may take the additional step of expressing $\bar{a}$ in terms of $\bar{a}_x$ and $\bar{a}_y$ and use $\left(\frac{W}{g}\right)\bar{a}_x$ and $\left(\frac{W}{g}\right)\bar{a}_y$ in their stead as shown in part (c).

As a study of Fig. 13-3.2 on p. 504 will show, part (b) of Fig. 13-5.1 actually combines the effective forces of a noncentroidal rotation about A with an effective force caused by the acceleration of the reference point A. Sometimes this set of effective forces may be more convenient than the set in part (c), particularly if A is the instant center of acceleration because then the moment relation takes the simple form $\Sigma M_A = I_A\alpha$. Unfortunately, it is difficult to locate the instant center of acceleration except in the special case when a body starts from rest. Then, as shown in Section 12-5, the instant center of acceleration coincides with the instant center of velocity which is usually easy to locate. Notice also that $\Sigma M_A = I_A\alpha$ also applies if the acceleration a_A is directed through G. Of course, this is really another example of applying $\Sigma \mathbf{M} = \dot{\mathbf{H}}$ when the acceleration of the reference point acts through the mass center.

ILLUSTRATIVE PROBLEMS

13-5.1. A uniform slender bar, 6 ft long and weighing 100 lb, moves with its ends in contact with the smooth surfaces shown in Fig. 13-5.2. If $\tan\theta = \frac{3}{4}$, determine its angular acceleration and the normal reactions N_A and N_B the instant after it starts from rest.

Solution

Since the bar starts from rest, the instant centers of acceleration and

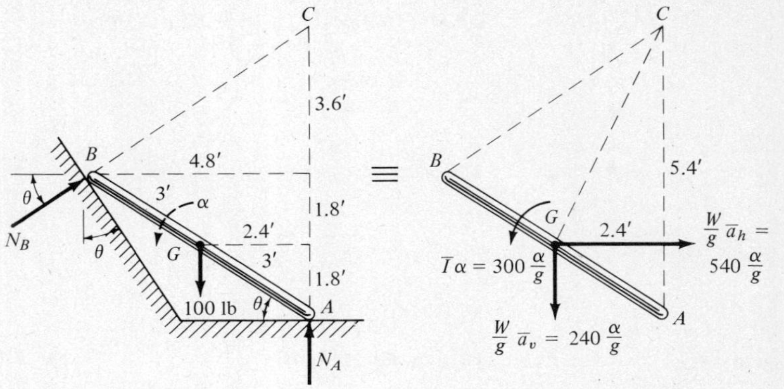

(a) Free-body diagram (b) Equivalent dynamic effects

Figure 13-5.2

of velocity coincide at C. Hence, in terms of α assumed counterclockwise, the components of $\bar{a} = a_G$, with G considered rotating about C, are $\bar{a}_h = 5.4\alpha \rightarrow$ and $\bar{a}_v = 2.4\alpha \downarrow$. The corresponding effective forces, shown in part (b) are $\frac{W}{g}\bar{a}_h = \frac{100}{g}(5.4\alpha) = 540\frac{\alpha}{g}$ lb, and $\frac{W}{g}\bar{a}_v = \frac{100}{g}(2.4\alpha) = 240\frac{\alpha}{g}$ lb. The effective couple $\bar{I}\alpha = \frac{1}{12}\frac{W}{g}L^2\alpha = \frac{1}{12}\left(\frac{100}{g}\right)(6)\alpha = 300\frac{\alpha}{g}$ ft-lb.

Taking moments about C, which eliminates the normal reactions N_A and N_B, we obtain

$[+\circlearrowleft(\Sigma M_C)_{\text{FBD}} = \Sigma(M_C)_{\text{dyn}}]$

$$100(2.4) = 300\frac{\alpha}{g} + 240\frac{\alpha}{g}(2.4) + 540\frac{\alpha}{g}(5.4)$$

whence

$$\frac{\alpha}{g} = 0.0633 \quad \text{and} \quad \alpha = 2.04 \text{ rad/sec}^2 \circlearrowleft \qquad \textit{Ans.}$$

Force summations now give

$[\overset{+}{\rightarrow}(\Sigma F_h)_{\text{FBD}} = \Sigma(F_h)_{\text{dyn}}]$ $\quad 0.8N_B = 540(0.0633)$

$$N_B = 42.7 \text{ lb} \qquad \textit{Ans.}$$

$[+\uparrow(\Sigma F_v)_{\text{FBD}} = \Sigma(F_v)_{\text{dyn}}]$

$$N_A + 0.6(N_B = 427) - 100 = -240(0.0633)$$

$$N_A = 59.1 \text{ lb} \qquad \textit{Ans.}$$

Observe that if the surfaces were rough with an angle of friction ϕ, the effective force-couple system of part (b) would not change, but the reactions at A and B would be inclined at the angle ϕ with their normals. Their intersection would locate a new moment center about which to eliminate them in order to find α directly.

13-5.2. One cylinder of an internal combustion engine is shown in Fig. 13-5.3. At the given position of the power stoke, a force $P = 1000$ lb acts on the 1.61-lb piston. Assuming the connecting rod AB to be a uniform bar 6 in. long and weighing 2 lb, determine the forces on the piston pin B and the crankpin A for a constant clockwise crank speed of 200 rad/sec (equivalent to 1910 rpm).

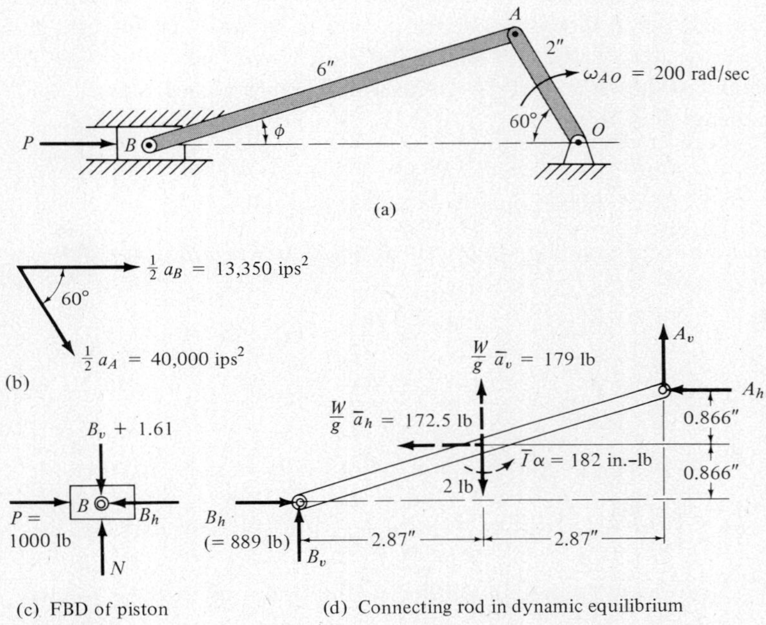

Figure 13-5.3

Preliminary

The kinematic properties of the connecting rod must first be determined. Using the methods discussed in Section 12-4, you will find them to be $\omega = 34.8$ rad/sec counterclockwise, $\alpha = 11{,}700$ rad/sec² clockwise, and $a_B = 26{,}700$ ips². Since $AB \sin \phi = AO \sin 60°$, we also have

$$\sin \phi = \frac{1.732}{6} = 0.289 \quad \text{and} \quad \cos \phi = \frac{\sqrt{33}}{6} = 0.957$$

which are used to compute the dimensions shown in part (d).

Solution

Here we illustrate the method of dynamic equilibrium in which the inertia components are inserted directly on the free-body diagram of the connecting rod. These inertia components act through the mass center G opposite to the acceleration components of G. Ordinarily, these acceleration components are found by assuming the plane motion of the connecting rod to be equivalent to a rotation about B as a reference point plus the translation of B. However, in this example where the mass center of the connecting rod is midway between its ends, it is simpler to compute the components of the mass center acceleration by using the result given in Prob. 12-4.15. There it was indicated that the midpoint acceleration of a rod is equal to one-half the vector sum of its end point accelerations. Here the end point accelerations are known to be $a_B = 26{,}700 \text{ ips}^2$ and $a_A = r\omega_{AO}^2 = 2(200)^2 = 80{,}000 \text{ ips}^2$, and their half-values are directed as shown in part (b). Hence, the horizontal and vertical components of $\bar{a}$ are

$$\bar{a}_h = 13{,}350 + 40{,}000 \cos 60° = 33{,}350 \text{ ips}^2 \text{ right}$$
$$\bar{a}_v = 40{,}000 \sin 60° = 34{,}600 \text{ ips}^2 \text{ down}$$

and the corresponding oppositely directed inertia components shown in part (d) are

$$\frac{W}{g}\bar{a}_h = \frac{2}{32.2}\left(\frac{33{,}350}{12}\right) = 172.5 \text{ lb left}$$
$$\frac{W}{g}\bar{a}_v = \frac{2}{32.2}\left(\frac{34{,}600}{12}\right) = 179 \text{ lb up}$$

The inertia couple $\bar{I}\alpha$, directed opposite in sense to α, is

$$\bar{I}\alpha = \left(\frac{1}{12}\frac{W}{g}L^2\right)\alpha = \frac{1}{12}\left(\frac{2}{32.2 \times 12}\right)(6)^2(11{,}700) = 182 \text{ in.-lb}$$

The kinetic equation applied to the piston gives B_h as follows:

$$\left[\overset{+}{\rightarrow}\Sigma X = \frac{W}{g}a\right] \qquad 1000 - B_h = \frac{1.61}{32.2}\left(\frac{26{,}700}{12}\right) = 111$$

$$B_h = 889 \text{ lb} \qquad \textbf{Ans.}$$

When we apply the conditions of dynamic equilibrium to the connecting rod, a moment summation about A yields B_v.

$$[\circlearrowleft + \Sigma M_A = 0] \qquad 5.74B_v - 889(1.732) - 2(2.87) + 172.5(0.866) + 179(2.87) - 182 = 0$$

from which

$$B_v = 185 \text{ lb} \qquad \textbf{Ans.}$$

Using this value of B_v in a vertical summation of forces we have

$$[+\uparrow \Sigma V = 0] \qquad 185 + 179 - 2 + A_v = 0$$
$$A_v = -362 \text{ lb} = 362 \text{ lb} \downarrow \qquad \textbf{Ans.}$$

while from a horizontal force summation, we obtain

$$[\overset{+}{\rightarrow}\Sigma H = 0] \qquad 889 - 172.5 - A_h = 0$$
$$A_h = 716.5 \text{ lb} \leftarrow \qquad \textbf{Ans.}$$

By combining their components, the total forces on the piston pin B and crankpin A are found to be

$$B = 909 \text{ lb} \quad \text{and} \quad A = 801 \text{ lb}$$

Supplementary

An independent check of A_v may be made by taking a moment summation about the intersection of A_h and B_v. Also, the crank effort (i.e., torque) exerted on the crankshaft is equal to the moment sum about O of the components of the crankpin force and equals $362(1) + 716.5(1.732) = 1600$ in.-lb.

13-5.3. In Fig. 13-5.4, a rod AB 6 ft long and weighing 60 lb hangs from the center of a solid cylinder of 2-ft radius and weighing 100 lb. If the cylinder can roll without slipping, determine the acceleration of its mass center A directly after a horizontal force $P = 66$ lb is applied at B.

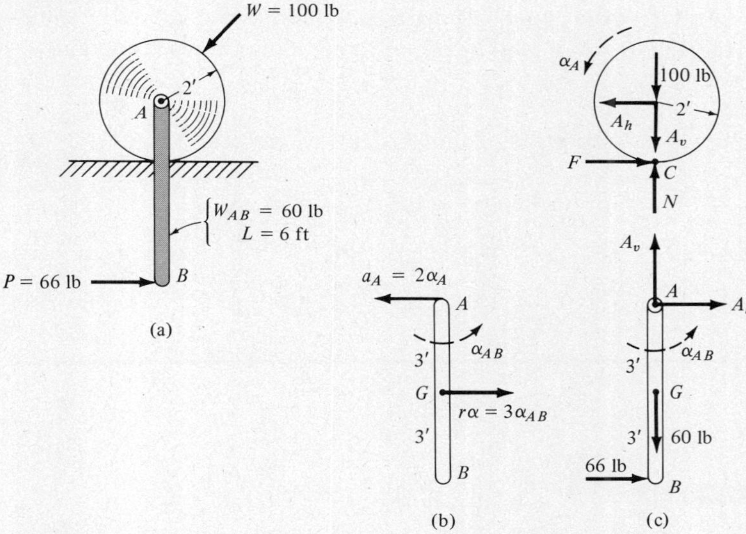

Figure 13-5.4

Solution

The rod AB simultaneously translates with and rotates about hinge A which, as part of the free-rolling cylinder, has a leftward acceleration $a_A = 2\alpha_A$ as shown in part (b). The acceleration of the mass center G of the rod is related to the acceleration of A by

$$\bar{a} = a_G = a_A \mathrel{+\!\!\!\!\rightarrow} (a_{G/A} = \overset{=0}{\bar{r}\omega^2} \mathrel{+\!\!\!\!\rightarrow} \bar{r}\alpha)$$

Note that the assumed counterclockwise rotation of G about A causes $\bar{r}\alpha$ to be $3\alpha_{AB}$, directed rightward as shown in part (b). Then the components of $\bar{a}$ are

$$\overset{+}{\rightarrow}\bar{a}_h = 3\alpha_{AB} - 2\alpha_A \quad \text{and} \quad \bar{a}_v = 0$$

Applying the kinetic equations to the free-body diagrams of the cylinder and rod shown in part (c), we have for the cylinder

$$\left[\circlearrowleft + \Sigma M_C = I_C\alpha = \frac{3}{2}mr^2\alpha\right] \quad 2A_h = \frac{3}{2}\left(\frac{100}{g}\right)(2)^2\alpha_A = \frac{600}{g}\alpha_A \quad (a)$$

and for the rod AB

$$\left[\circlearrowleft + \Sigma\bar{M} = \bar{I}\alpha = \frac{1}{12}mL^2\alpha\right]$$

$$3(66) - 3A_h = \frac{60}{12g}(6)^2\alpha_{AB} = \frac{180}{g}\alpha_{AB} \quad (b)$$

$$\left[\overset{+}{\rightarrow}\Sigma H = \frac{W}{g}\bar{a}_h\right] \quad 66 + A_h = \frac{60}{g}(3\alpha_{AB} - 2\alpha_A) \quad (c)$$

Solve for the common unknown A_h in these equations by multiplying Eq. (*a*) by $\frac{1}{5}$, Eq. (*b*) by -1, and adding them to obtain

$$A_h(0.4 + 3 + 1) + 66(1 - 3) = 0, \qquad A_h = 30 \text{ lb}$$

after which, substituting A_h in Eq. (*a*), we get

$$2(30) = \frac{600}{g}\alpha_A, \qquad \frac{\alpha_A}{g} = \frac{1}{10}$$

whence

$$a_A = 2\alpha_A = \frac{2g}{10} = 6.44 \text{ fps}^2 \qquad \textbf{Ans.}$$

PROBLEMS

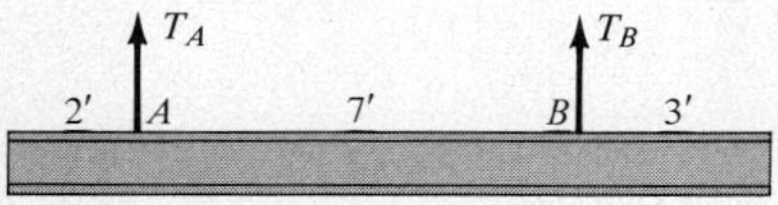

Figure P-13-5.4

13-5.4. A 12-ft girder weighing 980 lb is being lifted from rest by two separate cables as shown in Fig. P-13-5.4. If cable A is accelerated at 10 fps^2 and cable B at 3 fps^2, find the tension in each cable. Assume the girder is equivalent to a long slender bar.

$$T_A = 551 \text{ lb}; \; T_B = 612 \text{ lb} \qquad \textbf{Ans.}$$

13-5.5. In the previous problem, suppose both cables are being accelerated upward at the same rate of a fps^2 when cable A suddenly breaks. What will then be the tension in cable B?

$$T_B = 4W(a + g)/7g \qquad \textbf{Ans.}$$

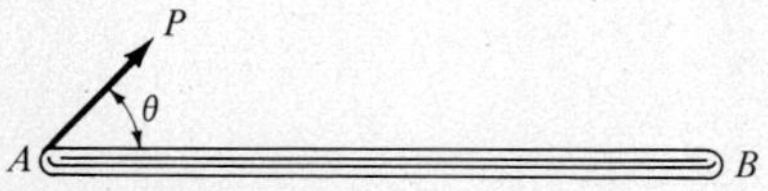

Figure P-13-5.6

13-5.6. A force P is applied as shown in Fig. P-13-5.6 to a bar of weight

W and length L resting on a smooth horizontal surface. Determine the acceleration of end B.

$$a_B = \frac{Pg}{W}\sqrt{\cos^2\theta + 4\sin^2\theta} \qquad \textbf{Ans.}$$

13-5.7. A uniform slender bar, 6 ft long and weighing 100 lb, slides along the smooth wall and floor shown in Fig. P-13-5.7. Find the force P that will give end A a starting acceleration of 9.6 fps² rightward when $\tan\theta = \frac{4}{3}$. Also find the reactions at A and B.

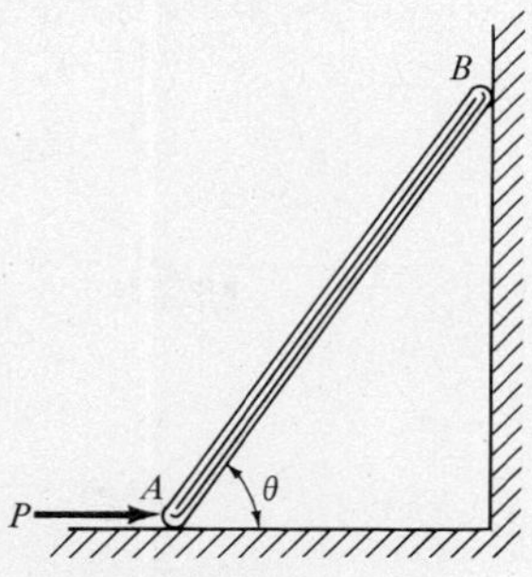

Figure P-13-5.7

13-5.8. Solve the preceding problem for a starting acceleration of $a_A = 10.8$ fps² rightward when $\tan\theta = \frac{3}{4}$.

$$P = 31.05 \text{ lb } N_A = 122.3 \text{ lb};\ N_B = 14.28 \text{ lb} \qquad \textbf{Ans.}$$

13-5.9. The system shown in Fig. P-13-5.9 starts from rest on the smooth surfaces at the given position. The bar AB is freely pinned to the mass centers of bodies A and B. Determine the angular acceleration of AB if AB weighs 100 lb and bodies A and B are of negligible weight.

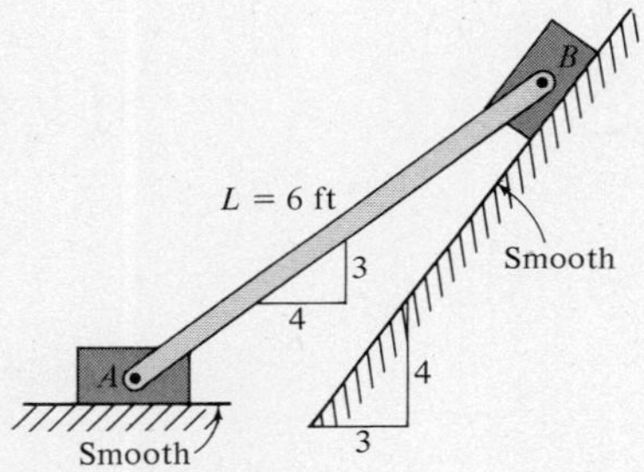

Figure P-13-5.9

13-5.10. Determine the initial acceleration of bodies A and B in the preceding problem if $W_{AB} = 100$ lb, $W_A = 50$ lb, and $W_B = 150$ lb.

$$a_A = 18.86 \text{ fps}^2;\ a_B = 15.73 \text{ fps}^2 \qquad \textbf{Ans.}$$

13-5.11. If the system shown in Fig. P-13-5.11 is released from rest, determine the initial acceleration of body C. Neglect friction at all surfaces. The uniform bar AB is 6 ft long and weighs 100 lb while body C weighs 150 lb.

$$a_C = 13.9 \text{ fps}^2 \qquad \textbf{Ans.}$$

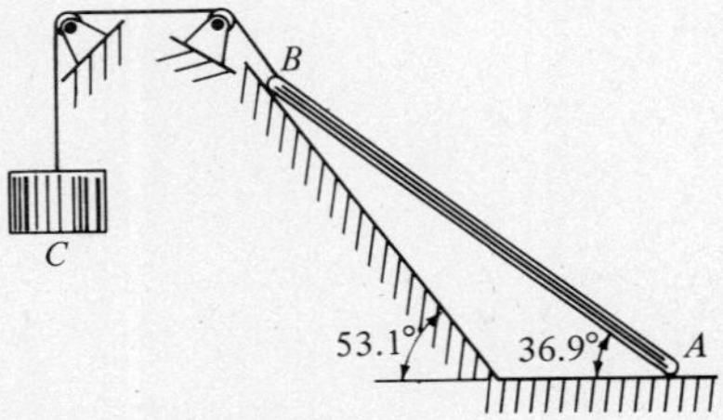

Figure P-13-5.11

13-5.12. The 6-ft bar AB, weighing 100 lb, is released from rest on the smooth surfaces shown in Fig. P-13-5.12. Compute the normal reaction at A.

13-5.13. The connecting rod of the steam engine shown schematically in Fig. P-13-5.13 is assumed to be a slender uniform rod 4 ft long weighing 322 lb. The crank AO is 1 ft long and rotates at a constant rate of 10 rad/sec. The force on the 64.4-lb cross head at the given instant is 2142 lb. Neglecting friction, determine the normal force on the cross head and the horizontal and vertical components of the crankpin force at A. (A kinematic analysis gives the following values: $\omega_{AB} = 1.8$ rad/sec; $\alpha_{AB} = 17.4$ rad/sec² clockwise; $a_B = 71.1$ fps² rightward.)

$$N = 402 \text{ lb};\ A_h = 1290 \text{ lb};\ A_v = 368 \text{ lb} \qquad \textbf{Ans.}$$

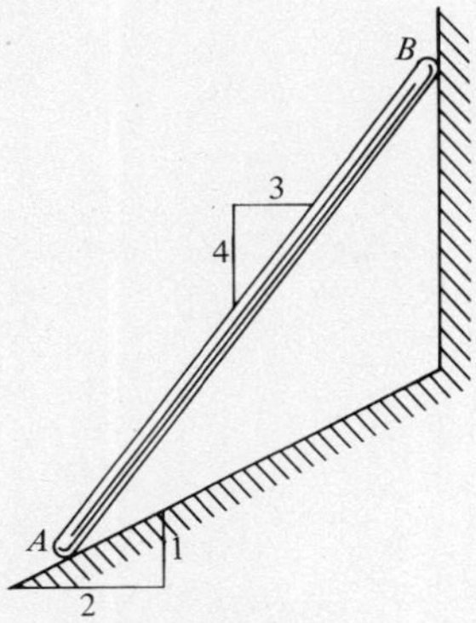

Figure P-13-5.12

13-5.14. Repeat Prob. 13-5.13 if the 4-ft connecting rod varies in cross section so that its mass center G is located 1.5 ft from A and its mass radius

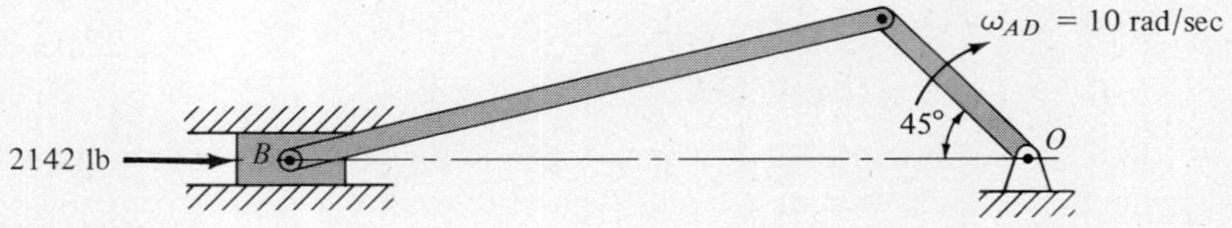

Figure P-13-5.13

of gyration is 1.0 ft about an axis through G perpendicular to the plane of motion. Assume all other data are unchanged.

13-5.15. When the piston of the engine described in Prob. 13-5.13 is at its extreme left position, the cross head force is zero. What is then the normal force on the cross head and the horizontal and vertical components of the crankpin force at A?

13-5.16. The circular disk shown in Fig. P-13-5.16 weighs 161 lb after the semicircular hole is cut out of it. Its radius of gyration about its mass center G is 1.0 ft. At the given instant, $\omega = 2$ rad/sec and $\alpha = 2$ rad/sec^2, both clockwise. If the disk does not slip, determine the value of the horizontal force P.

$$P = 65.3 \text{ lb} \qquad \textbf{\textit{Ans.}}$$

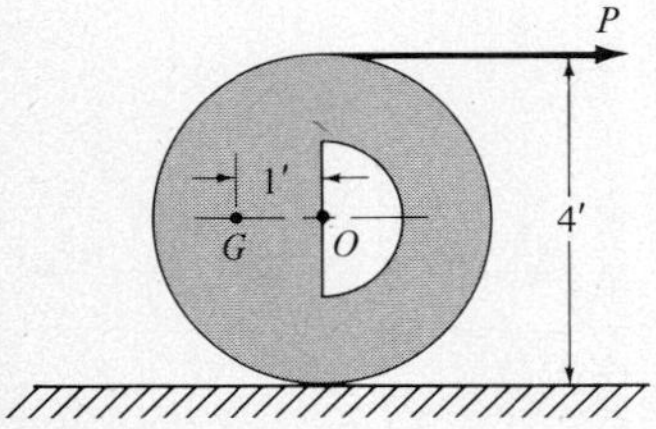

Figure P-13-5.16

13-5.17. A thin hoop of negligible weight has a 60-lb bar welded to its rim as shown in Fig. P-13-5.17. At the instant the bar is horizontal, the angular velocity and acceleration of the hoop are $\omega = \sqrt{0.4g}$ rad/sec and $\alpha = 0.1g$ rad/sec^2, both clockwise. Find the value of P at this instant. P is parallel to the incline and the hoop rolls without slipping.

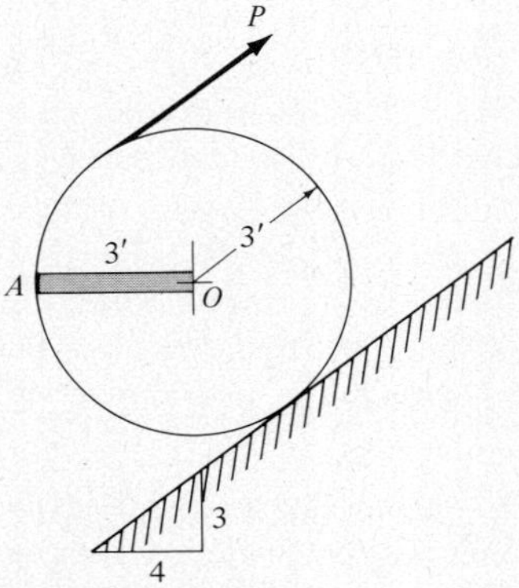

Figure P-13-5.17

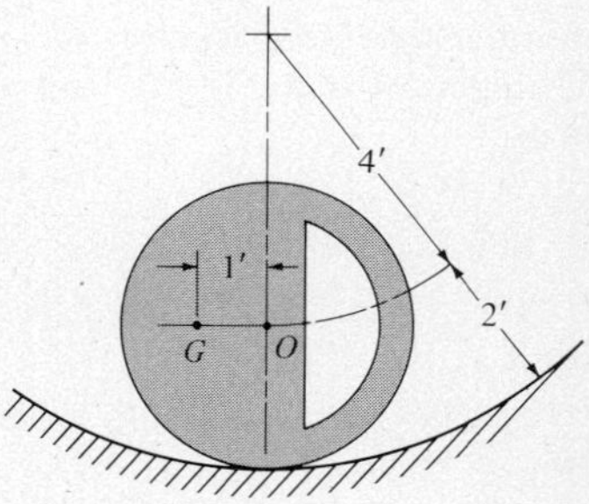

Figure P-13-5.19

13-5.18. In the preceding problem, determine the angular acceleration of the hoop and the moment exerted by the weld at A upon the bar at the given position when $P = 60$ lb and $\omega = \sqrt{0.4g}$ rad/sec.

$$M = 108.3 \text{ ft-lb} \qquad \textbf{\textit{Ans.}}$$

13-5.19. An unbalanced wheel rolls without slipping on the cylindrical surface shown in Fig. P-13-5.19. It weighs 200 lb and has a centroidal radius of gyration $\bar{k} = \sqrt{2}$ ft. Find the normal and friction forces on the wheel at the given position when its angular velocity is $\omega = \sqrt{0.4g}$ rad/sec.

Hint: Some useful results will be found in Prob. 12-5.11.

$$F = 45.6 \text{ lb};\ N = 217.2 \text{ lb} \qquad \textbf{\textit{Ans.}}$$

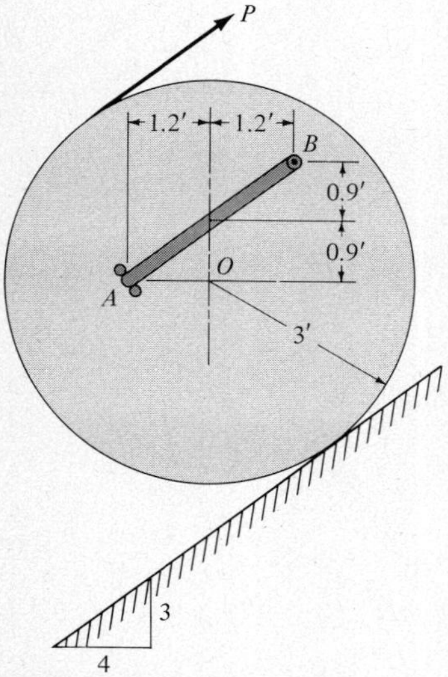

Figure P-13-5.20

13-5.20. The homogeneous wheel shown in Fig. P-13-5.20 rolls without slipping under the action of the pull P parallel to the incline. At the given position, $\omega = 3$ rad/sec and $\alpha = 4$ rad/sec^2 both clockwise. The wheel carries a uniform slender bar weighing 32.2 lb which is hinged to the wheel at *B* and rests between two smooth pins at *A*. At the given position, compute the horizontal and vertical components of the hinge force at *B* and the force exerted by the active pin at *A*.

13-5.21. Set up dynamic equilibrium for the combined system of wheel and bar described in the preceding problem and determine the value of the pull *P* at the given position. The wheel weighs 64.4 lb.

$P = 54.5$ lb **Ans.**

13-5.22. In Fig. P-13-5.22, a 6-ft rod weighing 100 lb hangs vertically from a smooth hinge at *A* on a 200-lb block at rest on a smooth horizontal surface. Find the acceleration of the block directly after a horizontal force $P = 60$ lb is applied to it. What will be the acceleration of end *B* of the rod?

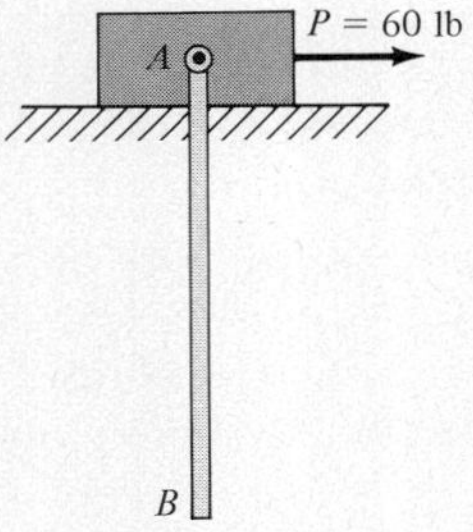

Figure P-13-5.22

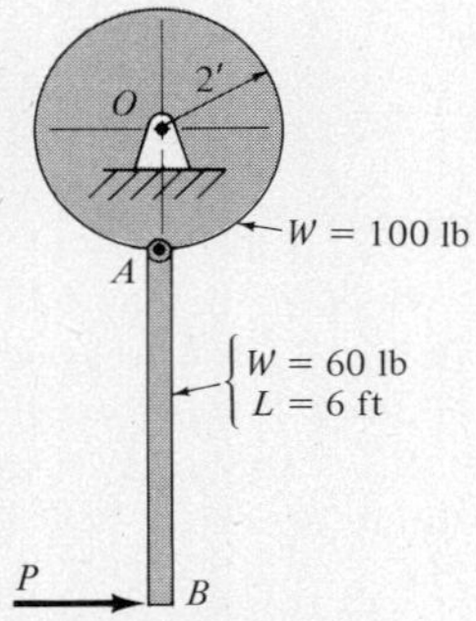

Figure P-13-5.23

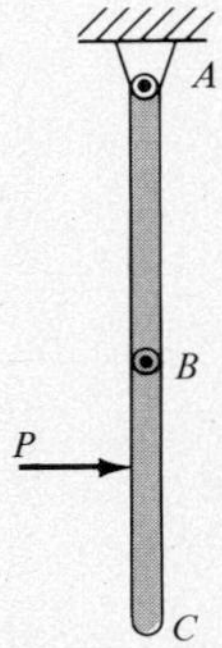

Figure P-13-5.24

13-5.23. In the system shown in Fig. P-13-5.23, a uniform rod *AB* is suspended from a frictionless hinge at the rim of a solid cylinder. Determine the acceleration of point *A* directly after a horizontal force $P = 52$ lb is applied at *B*. If *P* were applied at the mass center of the rod, what change occurs in the acceleration of *A*?

$a_A = 0.4g$ fps^2 **Ans.**

13-5.24. In Fig. P-13-5.24 is shown a double pendulum consisting of two equal rods, each of weight *W* and length *L*. If a rightward horizontal force *P* is applied at the midpoint of *BC*, determine the angular acceleration of each rod.

13-5.25. Solve the preceding problem if the rightward horizontal force *P* is applied at the midpoint of *AB*.

$$\alpha_{AB} = \frac{6Pg}{7WL} \circlearrowright; \quad \alpha_{BC} = \frac{9Pg}{7WL} \circlearrowleft$$ **Ans.**

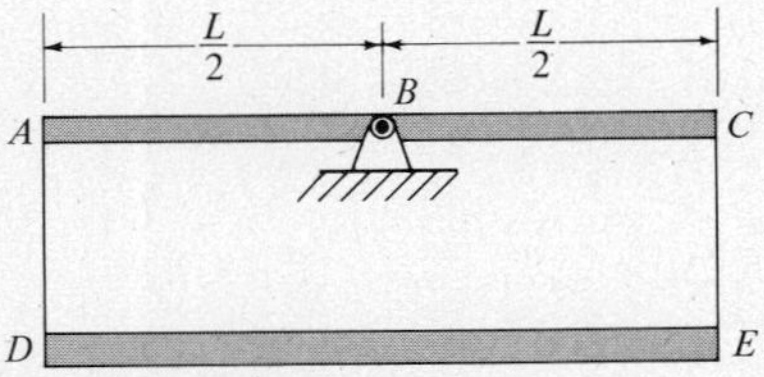

Figure P-13-5.26

13-5.26. Two identical slender bars, each of weight *W* and length *L*, are connected by two cords to form the assembly supported by a frictionless hinge at *B* as shown in Fig. P-13-5.26. Find the tension *T* in cord *CE* directly after cord *AD* is cut.

$T = W/7$ **Ans.**

13-5.27. The system shown in Fig. P-13-5.27 starts from rest. Body A weighs 50 lb; bar AB is 5 ft long and weighs 60 lb; and the solid cylinder B, which rolls freely, has a radius of 1 ft and weighs 100 lb. Determine the initial acceleration of body A.

$a_A = 9.87 \text{ fps}^2$ ***Ans.***

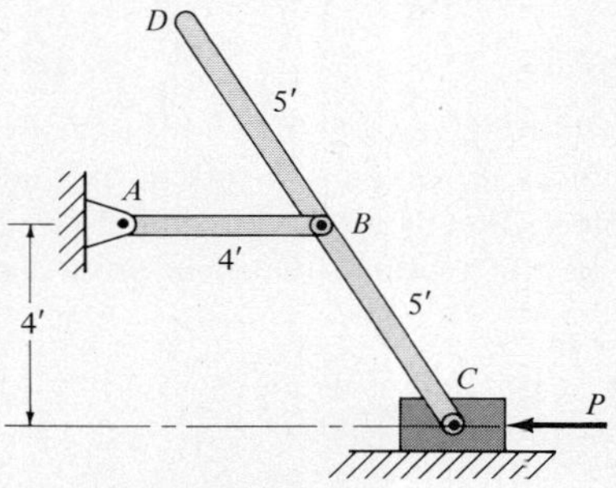

Figure P-13-5.28

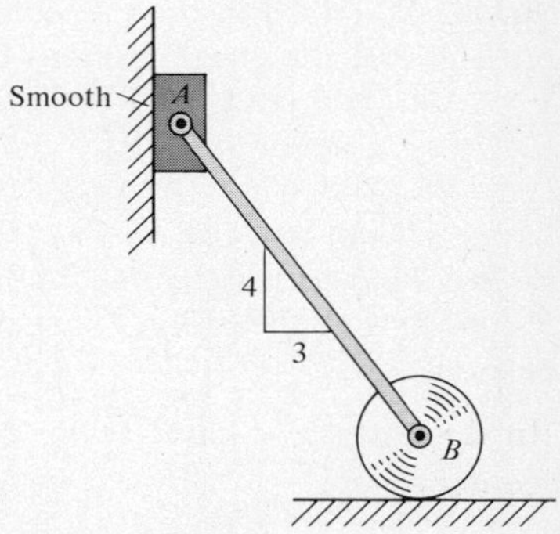

Figure P-13-5.27

13-5.28. In Fig. P-13-5.28, determine the force P that will give body C an initial leftward acceleration of 12 fps². The various parts of the system have the following weights: $W_C = 100$ lb; $W_{AB} = 60$ lb; $W_{CD} = 120$ lb.

Hint: Apply the virtual work method to diagrams of active forces and dynamic effects.

$P = 202.5$ lb ***Ans.***

SUMMARY

A common approach to all problems involving bodies symmetrical about their plane of motion is to apply the equations $\mathbf{R} = (W/g)\bar{\mathbf{a}}$ and $\Sigma \mathbf{M}_G = \dot{\mathbf{H}}_G$ in the scalar forms

$$\Sigma X = \frac{W}{g}\bar{a}_x; \qquad \Sigma Y = \frac{W}{g}\bar{a}_y; \qquad \Sigma \bar{M} = \bar{I}\alpha \tag{13-2.1}$$

where the bar superscripts denote reference to centroidal axes.

Two methods of applying these equations may be used. Either equate a free-body diagram of the impressed forces to its dynamic equivalent of the effective force $\left(\frac{W}{g}\right)\bar{a}$ at the mass center plus an effective couple $\bar{I}\alpha$, or apply these dynamic effects reversed on the free-body diagram and use dynamic equilibrium. Usually a kinematic analysis to relate $\bar{a}$ and α must also be used except in the case of fixed-axis rotation and homogeneous bodies rolling without slipping. Since a direct relation between $\bar{a}$ and α is known for these motions, convenient supplemental equations can be deduced and used as discussed in Sections 13-3 and 13-4.

Thus, for noncentroidal rotation, it is convenient to select X and Y axes through the mass center which are positive in the directions of $\bar{a}_n$ and $\bar{a}_t$ respectively. Also it is useful to supplement $\Sigma\bar{M} = \bar{I}\alpha$ with $\Sigma M_A = I_A\alpha$ which eliminates the generally unknown reaction at the axle A. The positive sense of $\Sigma\bar{M}$ and ΣM_A is that of the initial direction of rotation. Then Eqs. (13-2.1) are transformed into

$$\Sigma X = \frac{W}{g}\bar{a}_n = \frac{W}{g}\bar{r}\omega^2$$

$$\Sigma Y = \frac{W}{g}\bar{a}_t = \frac{W}{g}\bar{r}\alpha \qquad (13\text{-}3.2)$$

$$\Sigma\bar{M} = \bar{I}\alpha \quad \text{or} \quad \Sigma M_A = I_A\alpha$$

In the case of centroidal rotation where A coincides with the mass center, $\bar{r} = 0$ and the above equations reduce to

$$\Sigma X = 0; \qquad \Sigma Y = 0; \qquad \Sigma\bar{M} = \bar{I}_a \qquad (13\text{-}3.1)$$

For homogeneous rolling bodies (Section 3-4) in which the mass center has a rectilinear motion parallel to the surface on which it rolls, it is convenient to select reference axes at the mass center with the X axis parallel to the surface and positive in the initial direction of motion. Equations (13-2.1) then become

$$\Sigma X = \frac{W}{g}\bar{a}; \qquad \Sigma Y = 0; \qquad \Sigma\bar{M} = \bar{I}\alpha \qquad (13\text{-}4.1)$$

If the body is known to roll without slipping, an additional equation that can be used is

$$\Sigma M_C = I_C\alpha \qquad (13\text{-}4.2)$$

which eliminates both the unknown static frictional resistance and the normal force acting at the instant center C.

14

WORK-ENERGY METHOD

14-1 INTRODUCTION

In this chapter we present an alternate approach to dynamics which bypasses accelerations and directly relates forces, displacements, and velocities. This approach, known as the work-energy method, will usually be faster and easier to apply than the force-inertia method discussed in preceding chapters. As we shall see, the work-energy method involves only scalar quantities which are combined arithmetically and thereby eliminates the directional aspects of the vector quantities inherent in the force-inertia method. Moreover, the work-energy method can be applied directly to a complete system of connected bodies thereby eliminating forces which are internal to the system, and dispensing with drawing separate free-body diagrams of each part of the system. Finally, we deal with displacements and velocities which are much easier to visualize than are accelerations.

The work-energy method is based on integrating the fundamental equations $\mathbf{R} = m\bar{\mathbf{a}}$ and $\Sigma\mathbf{M} = \dot{\mathbf{H}}$ directly in terms of the displacement. Another approach, to be considered in the next chapter, is to integrate these equations with respect to time, thereby leading to the impulse-momentum method. Frequently, a combination of work-energy and impulse-momentum methods will solve problems that otherwise would be very difficult. After deriving the basic work-energy equations, we shall first apply them to rigid-body translation (both rectilinear and curvilinear) and then extend them to all other cases of rigid-body plane motion. As in the preceding chapter, we discuss only the plane motion of rigid bodies which are symmetrical with respect to the plane of motion through the mass center of the body. Other cases, including those of spatial motions, are presented in Chapter 16. In a sense, the work-energy method is an adjunct and introduction to Lagrangian and Hamiltonian mechanics.

14-2 WORK-ENERGY EQUATION FOR TRANSLATION

The general problem in dynamics is to determine the relation between the motion of a body (or a system of bodies) and the force system acting upon it. For a particle or a translating rigid body, this may be accomplished by solving $\mathbf{R} = m\mathbf{a}$ for the acceleration $\mathbf{a}$ and then integrating one of the differential kinematic equations ($a_t = dv/dt$, $v = ds/dt$, or $a_t\, ds = v\, dv$) to find the appropriate s-t, v-t, or v-s relation desired.

An alternate procedure is to integrate $\mathbf{R} = m\mathbf{a}$ directly either in terms of the time or in terms of the displacement. The time integral leads to the impulse-momentum method which we shall discuss in Chapter 15. The basis of the work-energy method is the displacement integral which is obtained by multiplying both sides of $\mathbf{R} = m\mathbf{a}$ by $d\mathbf{r}$. Since vector multiplication can either be the dot or the cross product, our choice is determined by expressing $\mathbf{a}$ as $\mathbf{a}_t + \mathbf{a}_n$ and noting that $d\mathbf{r}$ is tangent to the path. Since $\mathbf{a}_n$ is normal to the path, its dot product with $d\mathbf{r}$ will be zero. Therefore, using the dot product, we obtain

$$\mathbf{R} \cdot d\mathbf{r} = m\mathbf{a} \cdot d\mathbf{r} = ma_t\, ds = mv\, dv \qquad (a)$$

in which we have for convenience replaced the magnitude of $d\mathbf{r}$ by ds measured along the path and then substituted $v\, dv$ for $a_t\, ds$. The dot product $\mathbf{R} \cdot d\mathbf{r}$ defines the differential work done by the resultant force $\mathbf{R}$ during the displacement $d\mathbf{r}$. Similarly $\mathbf{F} \cdot d\mathbf{r}$ defines the differential work done by a component of $\mathbf{R}$.

Integrating both sides of Eq. (a), we obtain

$$\int_{\mathbf{r}_o}^{\mathbf{r}} \mathbf{R} \cdot d\mathbf{r} = m \int_{v_o}^{v} v\, dv = \tfrac{1}{2}m(v^2 - v_o^{\,2}) \qquad (b)$$

Of particular importance is the fact that these terms are scalar quantities which means that they can be combined arithmetically with similar equations without considering directional effects. The left-hand integral is known as resultant work (abbreviated as RW) done upon the particle while the term $\frac{1}{2}mv^2$ is called kinetic energy (abbreviated as KE). On replacing m in Eq. (b) by its equivalent W/g, we obtain

$$RW = \frac{W}{2g}(v^2 - v_o^2) = \Delta(\text{KE}) \tag{14-2.1}$$

which is the work-energy equation for a particle. Expressed in words, it states that the resultant work done upon a particle is equal to its change in kinetic energy.

An alternate derivation of the work-energy equation which is somewhat more direct is the following: Assume an X axis to be tangent to the path of the particle during each distance ds measured along the path. Then, by eliminating a_t between $\Sigma X = \frac{W}{g} a_t$ and $a_t\, ds = v\, dv$, we obtain

$$\Sigma X(ds) = \frac{W}{g} v\, dv$$

which is integrated between an initial velocity v_o at the initial position to the final velocity v at a distance s measured along the path to give

$$\int_o^s \Sigma X(ds) = \frac{W}{2g}(v^2 - v_o^2) \tag{14-2.2}$$

The left-hand integral is an alternate expression for resultant work in which ΣX represents the component of the resultant force along the path at each instant.

The unit of work is the dimensional product of force and distance; i.e., ft-lb, in.-lb, dyne-cm, etc. However, since we usually use the ft-lb-sec system of units in dynamics, work should be correspondingly expressed in ft-lb units to make it dimensionally consistent with kinetic energy whose unit is seen to be dimensionally

$$\text{KE} = \frac{1}{2}\frac{W}{g}v^2 = \frac{\text{lb}}{\text{ft/sec}^2}\left(\frac{\text{ft}^2}{\text{sec}^2}\right) = \text{ft-lb}$$

Then the work-energy equation will be dimensionally homogeneous which is required for the validity of any equation. Before considering the application of the work-energy equation, we examine next the meaning of work and how it is computed.

14-3 INTERPRETATION AND COMPUTATION OF WORK

Recalling that the dot product of two vectors is a scalar quantity equal to the product of the magnitudes of the two vectors by the cosine of their included angle, resultant work can be written as

$$\int \mathbf{R} \cdot d\mathbf{r} = \int (R \cos \theta) \, ds = \int R(ds \cos \theta)$$

in which we have replaced for convenience the magnitude of $d\mathbf{r}$ by ds measured along the path of motion.

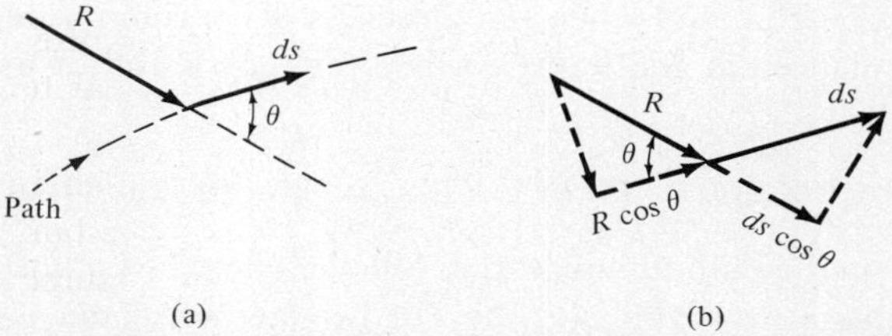

Figure 14-3.1 Differential work $\mathbf{R} \cdot d\mathbf{r}$ equals $(R \cos \theta) \, ds$ or $R(ds \cos \theta)$.

By referring to Fig. 14-3.1, we see that differential work may be interpreted as the product of the component $R \cos \theta$ of the resultant force in the direction of the displacement multiplied by the displacement ds. Also if differential work is rewritten as the product $R(ds \cos \theta)$, we see that resultant work may also be interpreted as the product of the resultant force multiplied by the component of displacement in the direction of the resultant force. Work done by a force may be positive or negative depending on whether the force component is directed respectively along or opposite to the direction of displacement.

Let us apply these concepts to the case of a body of weight W moving up an incline under the action of a constant force P as in Fig. 14-3.2. Interpreting work as the component of force in the direction of displacement multiplied by the displacement, the work done by P is $(P \cos \theta)s$. Note that in accordance with this concept, forces perpendicular to the displacement, such as N, do no work because they have no component in the direction of displacement.

(a) Work $= (P \cos \theta) s$
Work $= (-W \sin \theta) s$

(b) Work $= P(s \cos \theta)$
Work $= -W(s \sin \theta) = -Wh$

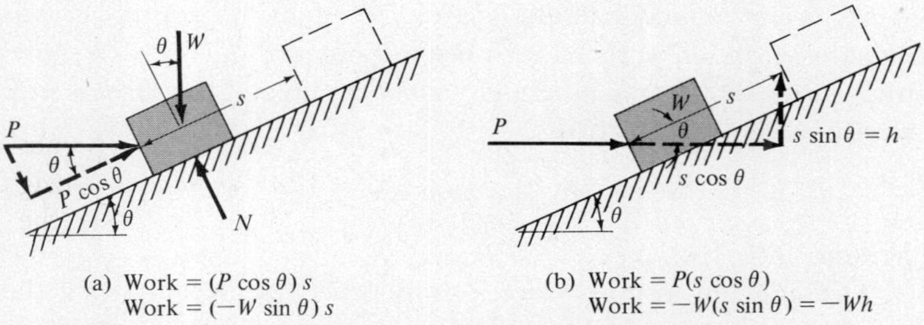

Figure 14-3.2 Two interpretations of work done by a force.

The scalar product $(P \cos \theta)s$ can also be written as $P(s \cos \theta)$ which expresses work as the product of a force multiplied by the component of displacement in the direction of the force. This concept is especially useful in defining the work done by gravity forces as the product of the weight multiplied by the change in vertical displacement, here being $-W(s \sin \theta) = -Wh$ where h is the vertical rise of the body. It is important to recognize that work done by gravity forces depends only on a change in elevation of a body, being negative (as in this case) when the direction of W is opposite to the vertical displacement and positive when they are both downward.

The general expression $\int_A^B \mathbf{R} \cdot d\mathbf{r}$ is a line integral meaning that its value depends on the path of integration followed between points A and B. To evaluate it, we expand the dot product in rectangular coordinates to obtain

$$\text{RW} = \int_A^B \mathbf{R} \cdot d\mathbf{r} = \int R_x\, dx + \int R_y\, dy + \int R_z\, dz \qquad (14\text{-}3.1)$$

A suitable selection of the reference axes will simplify the application of this result. Thus, for rectilinear motion, the X axis is chosen as positive in the initial direction of motion so that the components of the resultant force are $R_x = \Sigma X$ and $R_y = R_z = 0$. Also dx may be replaced by ds (equal to the magnitude of $d\mathbf{r}$). Then the resultant work in rectilinear motion reduces to

$$\text{RW} = \int_0^s \Sigma X(ds) \qquad (14\text{-}3.2)$$

For constant forces, this becomes $\Sigma X(s)$ while for variable forces, ΣX must be expressed in terms of s before resultant work can be evaluated.[1]

For cases in which the resultant force is not constant, and where a mathematical relation between ΣX and s is difficult to obtain (as for the variation of pressure with the position of a piston in an engine), experimental methods may be available to express graphically the variation of force with displacement. Figure 14-3.3 shows such a *force-displacement diagram*. Since the area of the shaded strip represents the term $\Sigma X(ds)$, it follows that the resultant work $\int_0^s \Sigma X(ds)$ is represented by the area under the force-displacement diagram.

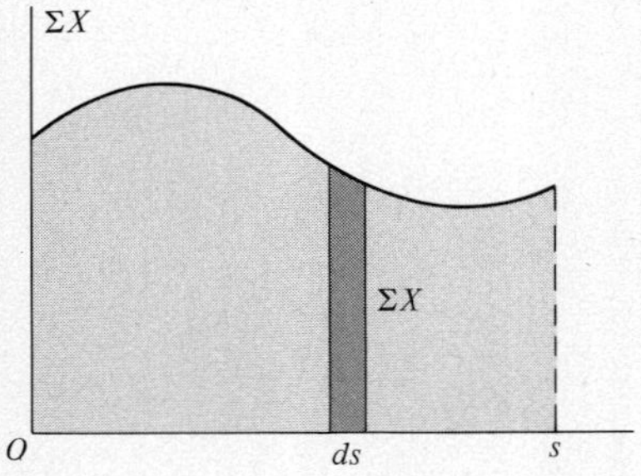

Figure 14-3.3 Force-displacement diagram.

This concept also provides a simple method of computing the work done by forces which are proportional to displacement, as in

[1] Refer also to the alternate derivation of the work-energy equation on p. 538.

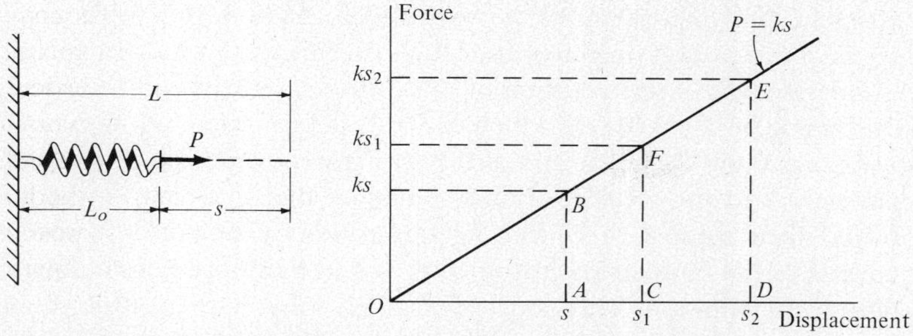

Figure 14-3.4 Force-displacement diagram for a spring.

springs. The force-deformation diagram for a linear spring[2] (Fig. 14-3.4) is a straight line as determined from Hooke's law and expressed by the equation

$$P = ks$$

in which k is known as the spring modulus. This modulus or spring constant represents the force required to deform a given spring through a unit distance. The spring deformation s is the difference between the original or free length of the spring and its deformed length.

Since resultant work is represented by the area under a force-displacement diagram, we see from Fig. 14-3.4 that the work done in deforming a spring from its free or unloaded length to an extension (or compression) of s units is the area of triangle OAB or

$$\int_0^s P\,ds = (\text{Area})_{F-s} = \tfrac{1}{2}(ks)(s) = \tfrac{1}{2}ks^2 \qquad (a)$$

or, alternatively, by direct integration, we again obtain

$$\int_0^s P\,ds = \int_0^s ks\,ds = \tfrac{1}{2}ks^2 \qquad (b)$$

The force-displacement diagram is especially useful in determining the work required to stretch a spring from an initial deformation s_1 to a larger deformation s_2. In this case the work done is the area of the trapezoid $CDEF$, which is equivalent to the average force multiplied by the change in deformation, or

$$\text{RW} = \int_{s_1}^{s_2} ks\,ds = \frac{ks_1 + ks_2}{2}(s_2 - s_1) \qquad (c)$$

Since the force exerted by a spring depends only on the spring

[2] There also are nonlinear springs classified as "hard" or "soft," depending on whether their stiffness increases or decreases with deformation.

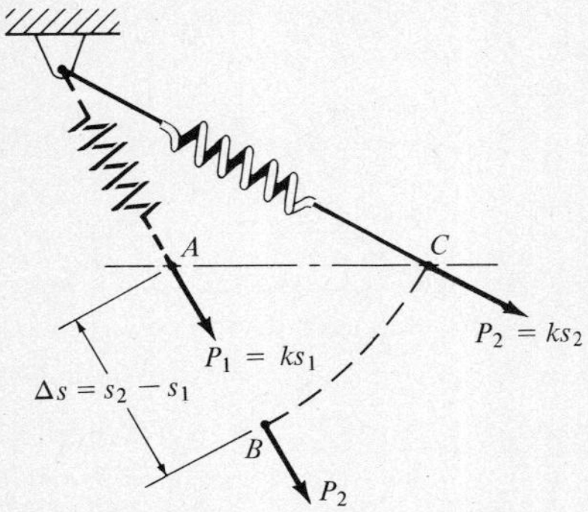

Figure 14-3.5 Work $= P_{ave}(\Delta S) = \frac{1}{2}(P_1 + P_2)(\Delta S)$.

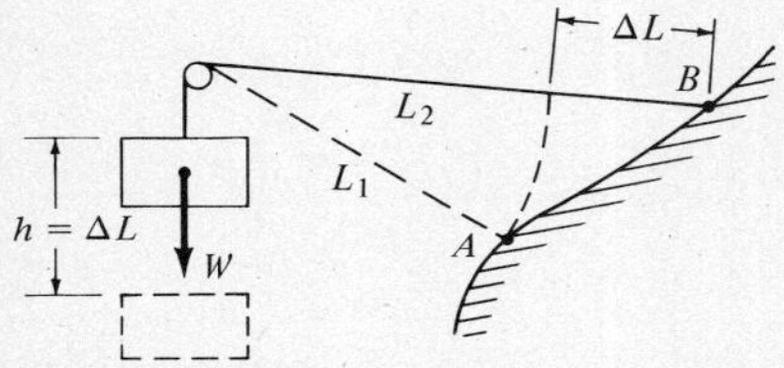

Figure 14-3.6 Work done on weight is $Wh = W(L_2 - L_1)$.

constant and the magnitude of spring deformation, the work done on a spring is due only to the change in length of the spring and is independent of any rotation of the spring.[3] For example, consider the spring in Fig. 14-3.5 which has the deformations s_1 and s_2 at positions A and C. As one end of this spring is moved from the dashed position A to the solid position C, imagine that it is first moved to B and then rotated to C. During this rotation, the spring force is normal to the circular path from B to C and therefore does no work during the rotation. This concept is much simpler than attempting an evaluation of the work integral in terms of the variable component of the spring force over the distance AC, and agrees with Eq. (c).

A similar situation is shown in Fig. 14-3.6. There a weight W at one end of an inextensible cord is raised as the other end moves along a curved path from A to B. Here the change in elevation of W is equal to the difference ΔL between the lengths L_1 and L_2.

Figures 14-3.5 and 14-3.6 illustrate cases in which the line integral $\int_A^B \mathbf{R} \cdot d\mathbf{r}$ is independent of the path of integration and depends only on the limits of integration. For this to be generally true, $\mathbf{R} \cdot d\mathbf{r}$ must be an exact differential of some function ϕ such that $d\phi = \mathbf{R} \cdot d\mathbf{r}$, and hence

$$\int_A^B \mathbf{R} \cdot d\mathbf{r} = \int_A^B d\phi = \phi \Big]_A^B = \phi(B) - \phi(A) \qquad (14\text{-}3.3)$$

If the function ϕ exists, the force may be derived from it as the two following expressions show:

$$\mathbf{R} \cdot d\mathbf{r} = R_x\, dx + R_y\, dy + R_z\, dz$$

$$d\phi = \frac{\partial \phi}{\partial x} dx + \frac{\partial \phi}{\partial y} dy + \frac{\partial \phi}{\partial z} dz$$

and hence

$$R_x = \frac{\partial \phi}{\partial x}; \qquad R_y = \frac{\partial \phi}{\partial y}; \qquad R_z = \frac{\partial \phi}{\partial z} \qquad (14\text{-}3.4)$$

The potential function ϕ used here is known as a force function and is assumed to be a function of space coordinates only. A spring is one example where the force is a function of position. A gravity field is another which we will discuss in detail later in Section 15-7 dealing with the motion of satellites. However, if the force is a function of velocity, it is not derivable from a potential and the line integral will not be independent of the path of integration.

[3] Any work done by the weight of the spring is neglected since its weight and change in elevation are usually negligible compared with other data.

14-4 WORK-ENERGY APPLIED TO PARTICLE MOTION

Particle motion includes rigid-body translation in which the body may be replaced by an equivalent particle having the mass and motion of the body. Basically, the work-energy method equates the total work done upon a body to the change in kinetic energy of the body. If a body is subjected to different sets of forces during different phases of its motion, the resultant work done during all these phases is equated directly to total change in kinetic energy expressed in terms of the initial and final velocity. It is *not* necessary to compute the velocity at the end of any phase so that it may be used as the initial velocity for the next phase.

The preceding section showed how to compute the work done on a body under various conditions. Specific details are explained in the illustrative problems below. Note especially that the work done by gravity[4] depends only on the change in vertical position. Also note that a spring does negative work upon a body whose motion stretches the spring. Sometimes the work absorbed by a spring is called potential energy since the work done in deforming the spring can be recovered if the spring is permitted to return to its undeformed position.

ILLUSTRATIVE PROBLEMS

14-4.1. As shown in Fig. 14-4.1, a 100-lb body moves along the two inclines for which the coefficient of friction is 0.20. If the body

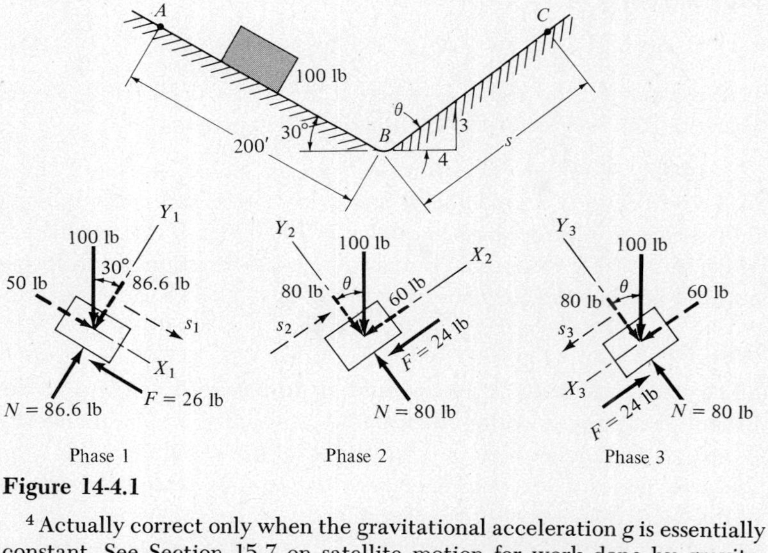

Figure 14-4.1

[4] Actually correct only when the gravitational acceleration g is essentially constant. See Section 15-7 on satellite motion for work done by gravitational attraction when g varies.

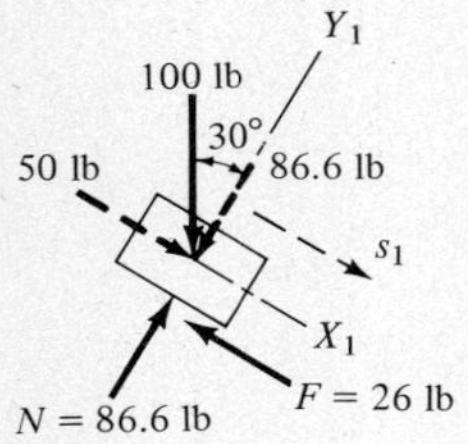

Phase 1

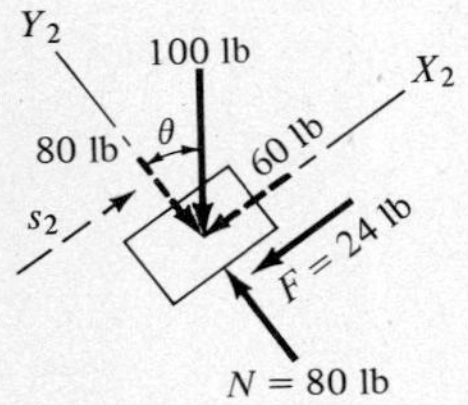

Phase 2

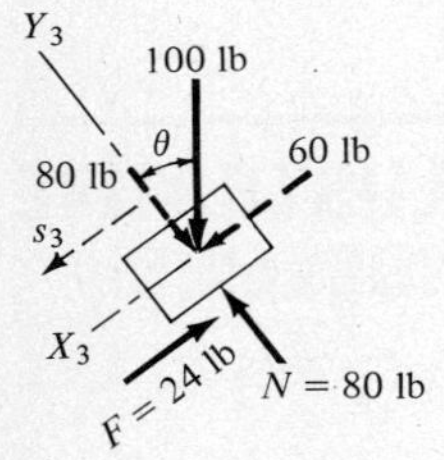

Phase 3

Figure 14-4.1

starts from rest at A and slides 200 ft down the 30° incline, how far will it then move along the other incline? What will be its velocity when it returns to B?

Solution

The FBD of the body during each phase of its motion is as shown. Computing the normal and frictional forces in the usual manner, we apply the work-energy equation to phases 1 and 2. Between A and C, the change in kinetic energy will be zero since the initial and final velocities are zero. We obtain

$$\left[\Sigma X_1(s_1) + \Sigma X_2(s_2) = \frac{W}{2g}(v^2 - v_o^2)\right]$$

$$(50 - 26)(200) - (60 + 24)s = 0$$

from which

$$s = 58.2 \text{ ft}$$

Note how simply the W-E method obtains this result directly even though two different accelerations are involved during the displacements s_1 and s_2.

On applying the work-energy equation to the motion from C back to B during phase 3, we have

$$\left[\Sigma X_3(s_3) = \frac{W}{2g}(v^2 - v_o^2)\right] \qquad (60 - 24)(58.2) = \frac{100}{64.4} v_B^2$$

from which

$$v_B = 36.8 \text{ fps} \qquad \textit{Ans.}$$

Is this velocity the same as when it first passed B? How would changing the weight of the body affect the answers?

14-4.2. A 600-lb block slides down an incline having a slope of 4 vertical to 3 horizontal. It starts from rest and, after moving 6 ft, strikes a spring whose modulus is 100 lb/in. If the coefficient of friction is 0.20, determine the maximum deformation of the spring and the maximum velocity of the block.

Solution

Since the block starts from rest and again has zero velocity at the instant of maximum spring deformation, its change in kinetic energy is zero. Referring to Fig. 14-4.2, we obtain the resultant work from the area under a force-displacement diagram in which the effects of the X components of the constant forces and of the variable spring force have been plotted separately. The negative force exerted by the spring on the body reaches its maximum value of $-1200s$ lb at

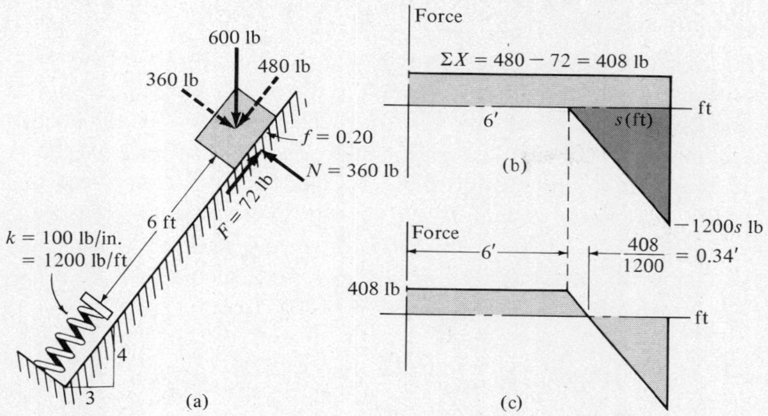

Figure 14-4.2

the instant the spring has reached its maximum deformation of s ft. We now obtain

$$[\text{RW} = \Delta\text{KE}] \qquad 408(6 + s) - \tfrac{1}{2}(1200s)(s) = 0$$

from which

$$s = 2.39 \text{ ft} \qquad \textit{Ans.}$$

The block will reach its maximum velocity when the net force on it is zero; not when it first contacts the spring which may have been your first thought. At the instant of maximum velocity, the acceleration of the block must be zero according to maxima-minima theory, or $a = dv/dt = 0$. You may understand this more clearly, however, by examining the net force-displacement diagram in part (c). This shows that positive work is being done on the block up to the position of zero net force, thereby increasing the KE and the velocity of the block; thereafter, negative work slows the body down. Since the spring force increases at 1200 lb/ft, the critical position occurs after a spring deformation of $408/1200 = 0.34$ ft. Now applying the W-E equation, we have

$$\left[\text{RW} = \frac{W}{2g}(v^2 - v_o{}^2)\right] \quad 408(6 + 0.34) - \frac{1}{2}(408)(0.34) = \frac{600}{64.4}v^2$$

from which

$$v^2 = 270.5 \quad \text{and} \quad v = 16.45 \text{ fps} \qquad \textit{Ans.}$$

14-4.3. A 100-lb weight is swung in a vertical circle at the end of a 4-ft cord. The topmost velocity of the weight is 12 fps. (a) Find the tension in the cord at 120° past top position. (b) What minimum velocity at the bottom will keep the weight in the circular path at the top?

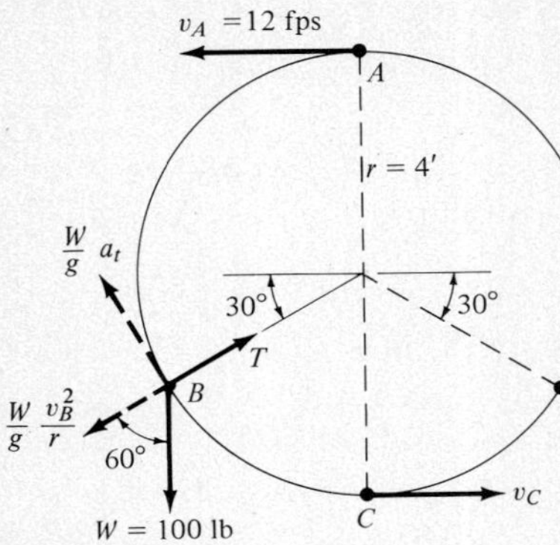

Figure 14-4.3

Solution: Part (a)

The tension is found by creating dynamic equilibrium at the 120° position at B as shown in Fig. 14-4.3. But first the centrifugal inertia force involving the velocity at B must be found. We could do this by computing the acceleration at any general position between A and B and then integrating to find v_B, but the work-energy method has the advantage of finding v_B directly. Note that the tension T is everywhere normal to the path and hence does no work. Only the weight W does work by dropping a vertical distance $h = 4 + 4 \sin 30°$. We obtain

$$\left[Wh = \frac{W}{2g}(v_B^2 - v_A^2)\right] \qquad 100(4 + 4 \sin 30°) = \frac{100}{64.4}(v_B^2 - 12^2)$$

$$v_B^2 = 530$$

It may be observed that if W is cancelled out of this equation, the terms may be rearranged to give $v_B^2 = v_A^2 + 2gh$ which is equivalent to the equation for free-falling bodies. This result will be obtained for *any frictionless translation* in which only the weight does work. Hence the velocity at D, which is at the same level as B, will be the same as at B. However, the time elapsed will, in general, differ from that required for free fall.

Now taking a summation of forces along the normal to the path at B gives

$$[\Sigma N = 0] \qquad T - \frac{W}{g}\frac{v_B^2}{r} - 100 \cos 60° = 0$$

$$T = \frac{100}{32.2}\left(\frac{530}{4}\right) + 50 = 461 \text{ lb} \qquad \textit{Ans.}$$

Part (b)

To find the minimum velocity at the bottom C to just keep the weight in the circular path at the top A, we note that the tension in the cord will be zero at A since the centrifugal force $\frac{W}{g}\frac{v_A^2}{r}$ will then just balance the weight W. Hence $v_A^2 = gr$. Applying the concept of the motion being equivalent to a free fall, we obtain

$$[v_C^2 = v_A^2 + 2gh] \qquad v_C^2 = gr + 2g(2r) = 5gr$$

from which

$$v_C = \sqrt{5(32.2)(4)} = 25.4 \text{ fps} \qquad \textit{Ans.}$$

PROBLEMS

14-4.4. In Eq. (*a*) on p. 537, the expression $\int m\mathbf{a} \cdot d\mathbf{r}$ may be rewritten as $\int m \frac{d\mathbf{v}}{dt} \cdot d\mathbf{r}$ to obtain $\int m\mathbf{v} \cdot d\mathbf{v}$. Show that this integral is always equivalent to $\int mv\, dv$ even though, as in curvilinear motion, $d\mathbf{v}$ is not necessarily collinear with $\mathbf{v}$.

14-4.5. After the block in Fig. P-14-4.5 has moved 10 ft from rest, the constant force P is removed. Find the velocity of the block when it returns to its initial position.

$v = 21.2$ fps **Ans.**

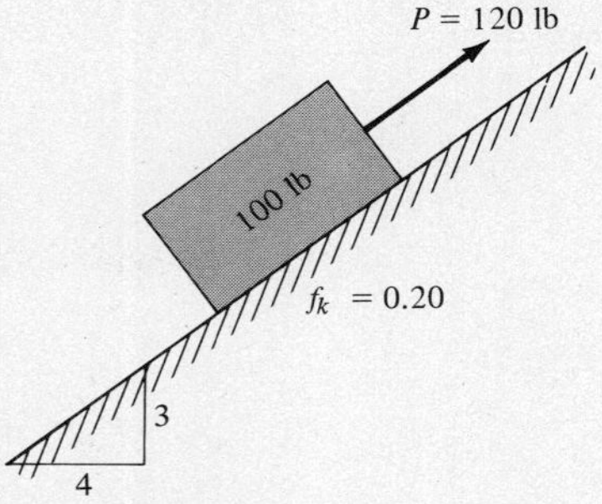

Figure P-14-4.5

14-4.6. A weight of W lb is suspended from a vertical spring (Fig. P-14-4.6) whose modulus is k lb/ft. The weight is pulled down s ft from its equilibrium position and then released. Determine its velocity when it returns to its equilibrium position.

$v = s\sqrt{kg/W}$ **Ans.**

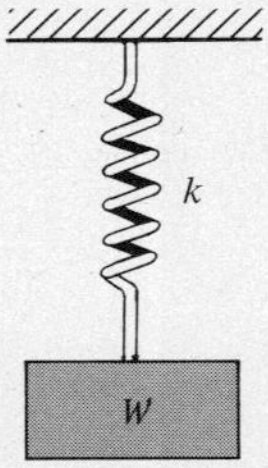

Figure P-14-4.6

14-4.7. Through what distance must the 600-lb block of Illus. Prob. 14-4.2 slide from rest before touching the spring if its velocity is 15 fps at the instant the spring is deformed 6 in.? Assume the spring constant is changed to 200 lb/in.

14-4.8. A weight is dropped from a position just above, but not touching, a spring. Show that the maximum deformation produced will be twice that if the same weight were gradually lowered upon the spring.

14-4.9. A weight W is dropped from a height h upon a spring-supported bar as shown in Fig. P-14-4.9. Determine h so that the maximum spring force is three times its final static equilibrium value. Neglect the weight of the bar.

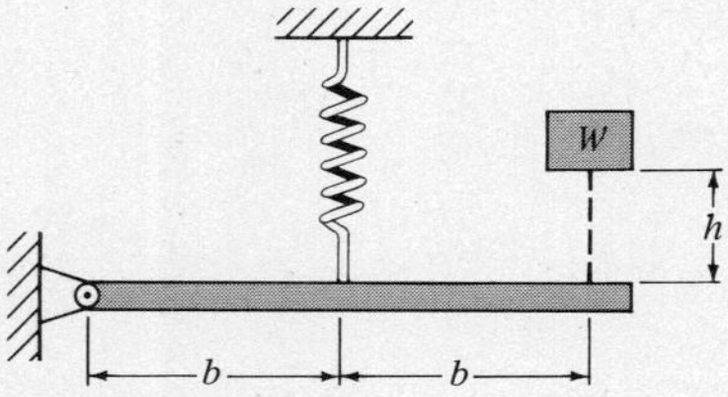

Figure P-14-4.9

14-4.10. An elevator weighing 4000 lb is being lowered at a constant rate of 10 fps when the hoisting drum is suddenly stopped. If the elastic properties of the supporting cable are such that it is equivalent to a spring with a modulus of 2000 lb/in., determine by how many times the sudden stop of the hoisting drum momentarily increases the cable tension. (This problem illustrates another insight to situations like that in Prob. 11-4.4.)

5.32 times **Ans.**

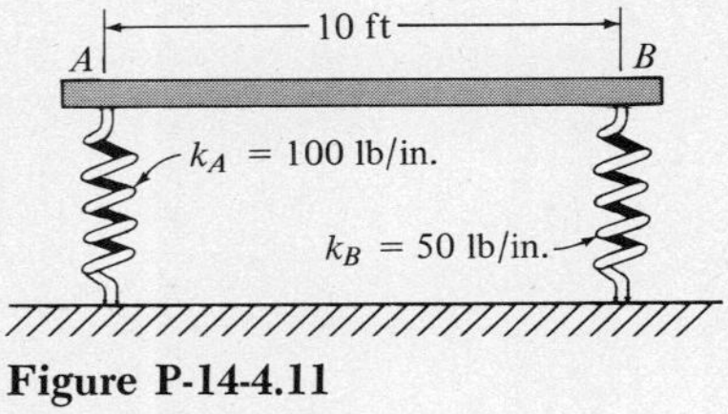

Figure P-14-4.11

14-4.11. The rigid horizontal bar in Fig. P-14-4.11 is supported by two springs. A 300-lb weight is then placed upon the bar at A. A sudden blow projects the weight toward B with an initial velocity of 6 fps. What is its velocity when it reaches B? Neglect friction and the weight of the bar.

$v = 7.22$ fps **Ans.**

14-4.12. A flexible rope, 80 ft long and weighing 0.5 lb/ft, passes over two smooth pegs as shown in Fig. P-14-4.12. The rope starts from rest when $d = 10$ ft. Determine the velocity of the rope at the instant when $d = 40$ ft.

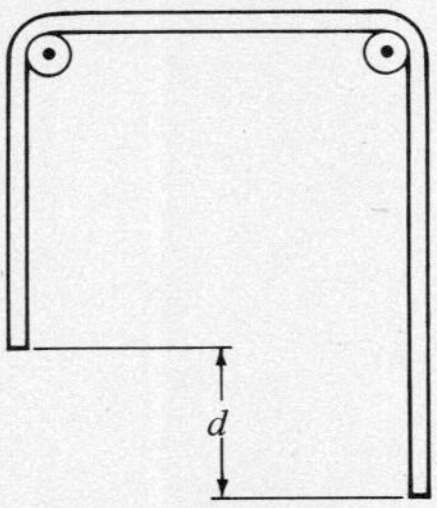

Figure P-14-4.12

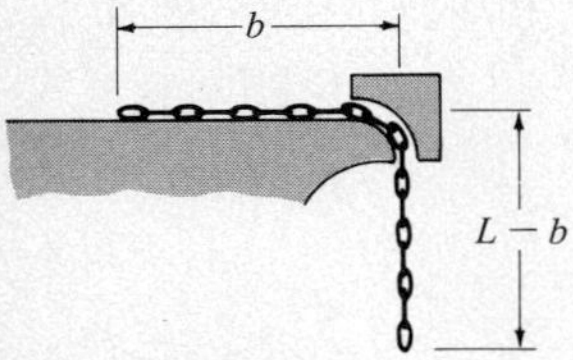

Figure P-14-4.13

14-4.13. A chain of length L and weighing w lb/unit length is released from rest on a smooth table when in the position shown in Fig. P-14-4.13. Determine the velocity of the chain as the last link leaves the table.

$$v = \sqrt{gb\left(2 - \frac{b}{L}\right)} \qquad \textbf{Ans.}$$

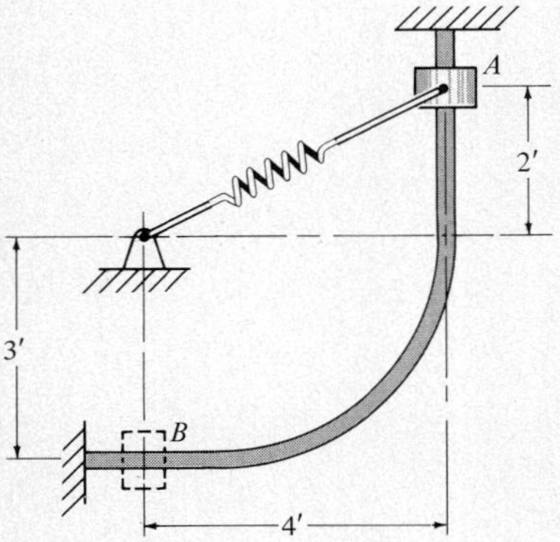

Figure P-14-4.14

14-4.14. A 100-lb weight moves along the smooth rigid guide from A to B as shown in Fig. P-14-4.14. Find its velocity at B if it starts from rest at A. The free length of the spring is 2 ft and its modulus is 5 lb/in.

14-4.15. A weight of 20 lb is swung in a vertical circle at the end of a 3-ft cord. At the lowest position of the weight, the tension in the cord is 80 lb. (a) How high above the lowest position will the weight rise on the circular path? (b) What would be this result if the cord is replaced by a stiff rod of negligible weight?

$$\text{(a) } h = 4 \text{ ft; (b) } h = 4.5 \text{ ft} \qquad \textbf{Ans.}$$

14-4.16. A weight W is attached to one end of a stiff rod of length L and negligible weight that is hinged to a horizontal axis at the other end. The rod is released from rest in a horizontal position and allowed to swing freely in a vertical arc. Through what angle must it swing to cause a tension in it of $1.5W$?

$$\theta = 30° \qquad \textbf{Ans.}$$

14-4.17. The rod in the previous problem is displaced an angle θ from its lowest position and released from rest. Find θ so that the tension in the rod at the lowest position is four times that just after release.

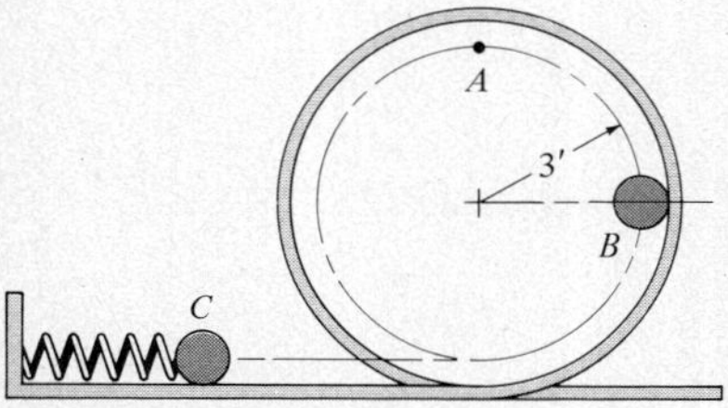

Figure P-14-4.18

14-4.18. In Fig. P-14-4.18, by how much should the spring be compressed so that it will cause the 1-lb pellet to travel completely around the frictionless vertical loop? What force is exerted by the track upon the pellet when it is at position B?

$$s = 0.5 \text{ ft; } N = 3 \text{ lb} \qquad \textbf{Ans.}$$

14-4.19. A particle of weight W, moving at a velocity of $\frac{1}{2}\sqrt{gr}$ fps at the top, slides vertically along the surface of a smooth cylinder of radius r. The axis of the cylinder is horizontal. Find the vertical distance the particle falls before it leaves the cylinder.

14-4.20. A particle starts from rest at the top and slides in a vertical plane along the smooth surface of a cylinder of 6-ft radius. The axis of the cylinder is horizontal. Measuring from the center of the cylinder, how far does the particle travel horizontally before it strikes the ground on which the cylinder rests?

$$x = 8.78 \text{ ft} \qquad \textbf{Ans.}$$

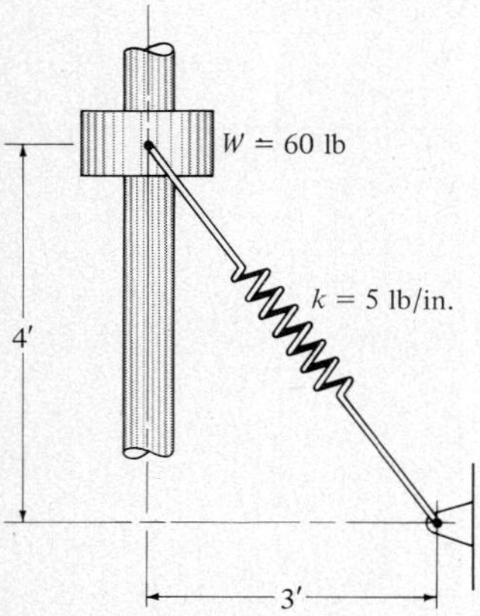

Figure P-14-4.21

14-4.21. Neglect friction of the 60-lb collar against its vertical guide and compute the velocity of the collar after it has fallen 7 ft, starting from rest in the position shown in Fig. P-14-4.21. The unstretched length of the spring is 2 ft.

$$v = 24.1 \text{ fps} \qquad \textbf{Ans.}$$

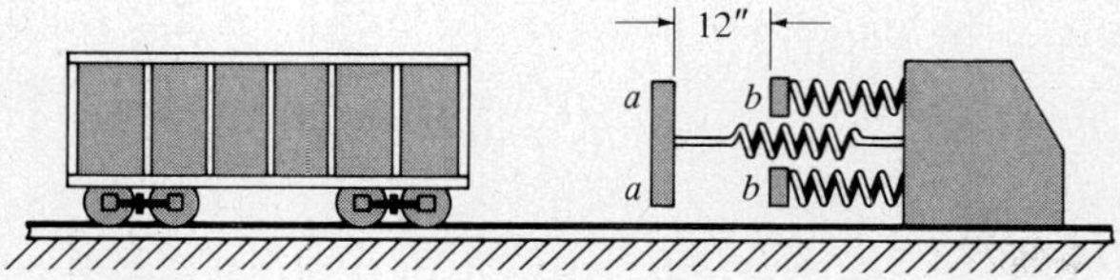

Figure P-14-4.22

14-4.22. The car in Fig. P-14-4.22 is moving toward the bumper spring and has a kinetic energy of 100,000 in.-lb. The main bumper shield (*a-a*) is connected to the main spring which has a modulus of 1000 lb/in. The two auxiliary bumper shields (b) are 12 in. behind *a-a* and are attached to secondary springs, each of which has a modulus of 500 lb/in. Determine the greatest movement of *a-a*. What percentage of the energy will then be absorbed by the main spring?

$s = 14$ in. **Ans.**

14-4.23. In the previous problem, what distance should separate the main and auxiliary bumper shields so that the main spring absorbs 80% of the kinetic energy of the car?

14-4.24. A 100-lb weight rotates in a vertical plane at the end of a 6-ft rod of negligible weight as shown in Fig. P-14-4.24. The spring, whose modulus is 10 lb/in., does not act on the rod until it exceeds its free length of 3 ft. Determine the velocity of the weight after it has swung from rest at the dashed position to the given position. Can the weight reach a position vertically below the pivot? If so, what will be its velocity then?

$v = 11.2$ fps **Ans.**

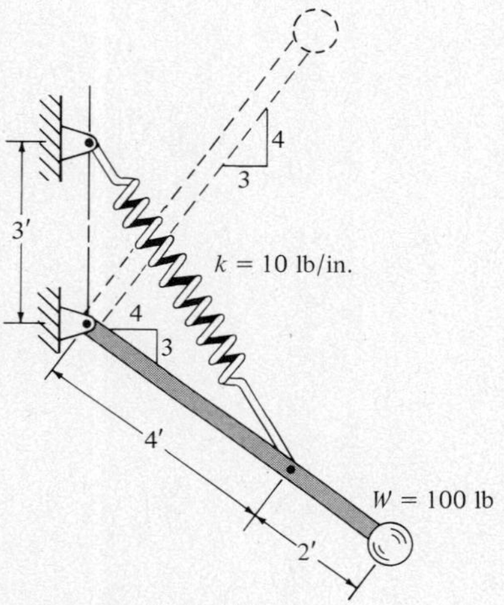

Figure P-14-4.24

14-4.25. Solve Prob. 14-4.24 if the spring is a "hard" spring in which the spring force in pounds is given by $P = 60s^2$ in which s is the spring deformation in feet.

14-5 POWER. EFFICIENCY

Power is the time rate at which work is done on a body. For example, if a train is being pulled by a locomotive, the work done on the train may be measured by the work done by the drawbar pull. Expressed mathematically,

$$\text{power} = \frac{\text{work}}{\text{time}}$$

This gives the average power. If F is the net force doing the work, then the work during a differential time is given by $F\,ds$, and the power exerted at any instant is

$$\text{power} = \frac{\text{work}}{\text{time}} = \frac{F\,ds}{dt} = Fv \tag{14-5.1}$$

i.e., the power exerted at any instant is the product of the net force multiplied by the instantaneous velocity.

The unit of power depends on the units of work and time. Common units are ft-lb per sec and kg-m per sec in the gravitational system, dyne-cm per sec (erg) or the joule per sec in the absolute system. These units are usually too small for use in engineering. The units commonly used here are the horsepower (hp) and the watt or kilowatt (kw). The horsepower is a traditional unit equivalent to 550 ft-lb of work per sec, or 33,000 ft-lb of work per min. The watt equals 10^7 ergs per sec, and the kilowatt equals 1000 watts.

The watt and kilowatt are used in electrical engineering. Relations between horsepower and kilowatts are as follows:

$$1\text{ hp} = 0.746\text{ kw}$$
$$1\text{ kw} = 1.34\text{ hp}$$

For large quantities of work, the units are the horsepower-hour (hp-hr) and the kilowatt-hour (kw-hr). These units indicate the amount of *work* done in one hour at the constant rate of 1 hp or 1 kw.

Because of losses resulting from friction and other causes, the power delivered from a machine or other device is never equal to the power put into it. Efficiency is the ratio of power output to power input. This ratio is usually multiplied by 100 so that efficiency may be given as a percentage. Since power output and power input are measured during the same time, efficiency may also be defined as the ratio of energy output to energy input or of work output to work input. The *overall efficiency* of a number of machines placed in series is equal to the product of their individual efficiencies.

ILLUSTRATIVE PROBLEM

14-5.1. A 1000-ton train is accelerated at a constant rate up a 2% grade. The train resistance is constant at 10 lb per ton. The velocity increases from 20 mph to 40 mph in a distance of 2000 ft. Determine the maximum horsepower developed by the locomotive.

Preliminary

The inclination of a grade is expressed as the rise in feet measured in a horizontal distance of 100 ft. Thus, a 2% grade means a rise of 2 ft in a horizontal distance of 100 ft. For grades up to 10%, the length of the incline is practically equal to its horizontal projection so that the cosine of the angle of the grade is practically equal to unity. Therefore the normal pressure may be considered equal to the weight of the train; hence, the train resistance remains constant. Furthermore, for small angles the sine may be considered equal to the tangent, so that in this case of a 2% grade, the sine will be equal to $\frac{2}{100}$.

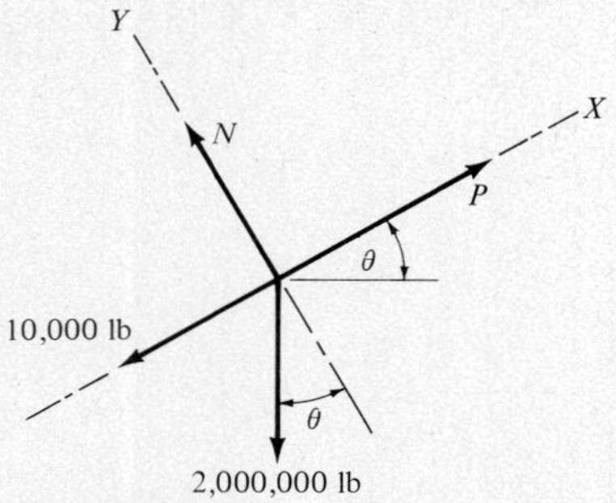

Figure 14-5.1

Solution

The constant drawbar pull exerted by the locomotive is found by applying the work-energy equation. From the point diagram of the forces acting on the train (Fig. 14-5.1) we have

$$\left[\Sigma X(s) = \frac{W}{2g}(v^2 - v_o^2)\right]$$

$$\left(P - 10{,}000 - 2{,}000{,}000 \times \frac{2}{100}\right) \times 2000$$

$$= \frac{2{,}000{,}000}{64.4} \times (58.6^2 - 29.3^2)$$

$$P = 90{,}000 \text{ lb}$$

The maximum power is developed when the train is moving at maximum speed. Therefore

$$\left[\text{hp} = \frac{Fv}{550}\right] \qquad \text{hp} = \frac{90{,}000 \times 58.6}{550} \qquad \text{hp} = 9600 \quad \textit{Ans.}$$

PROBLEMS

14-5.2. A train weighs 1200 tons. The train resistance is constant at 12 lb/ton. If 6000 hp are available to pull this train up a 3% grade, find its maximum speed in mph.

$v = 26.5$ mph **Ans.**

14-5.3. A train weighing 1200 tons is pulled up a 2% grade at 10 mph. If train resistance is constant at 10 lb/ton and the power exerted by the locomotive remains constant, determine the speed of the train on a level track. What horsepower does the locomotive develop?

14-5.4. A train weighing 1000 tons is being pulled up a 2% grade. The train resistance is constant at 10 lb/ton. The speed of the train is increased at a constant rate from 20 fps to 40 fps in a distance of 1000 ft. Find the maximum horsepower developed by the locomotive.

6350 hp **Ans.**

14-5.5. A conveyor belt moving at 5 fps transports 100 tons of material per hour through a vertical height of 150 ft. Assuming the mechanical efficiency of the system to be 85% and the drive motor to be 94% efficient, determine the power required in kilowatts.

14-5.6. Water flows through a nozzle of 1-in. diameter under a head of 400 ft to drive a turbine. The turbine is 90% efficient and drives a generator which is 94% efficient. Compute the power output in kilowatts.

25.1 kw **Ans.**

14-5.7. Water enters a hydraulic reaction turbine with a velocity of 12 fps and leaves it 4 ft lower with a velocity of 3 fps. If 100,000 lb of water flow through the turbine each second, compute the horsepower output. Assume the turbine is 90% efficient and water weighs 62.5 lb/ft^3.

994 hp **Ans.**

14-5.8. Determine the constant horsepower required to change the speed of a 4000-lb auto from 20 mph to 40 mph in a distance of 200 ft along a level road. What is the elapsed time? Assume road and air resistance are negligible.

Hint: With constant power, the acceleration is not constant.

66.6 hp; $t = 4.38$ sec **Ans.**

14-6 WORK-ENERGY APPLIED TO CONNECTED SYSTEMS

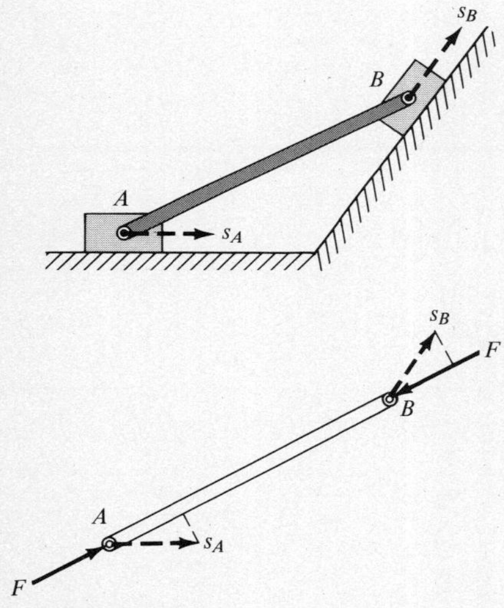

Figure 14-6.1

An outstanding advantage of the work-energy method is that it may be applied directly to connected systems of bodies without the necessity for considering forces which are internal to the system, such as those in inextensible connecting bars or cords. This is particularly useful in cases where the internal forces may vary (as in Illus. Prob. 14-6.2 below). To understand why internal forces are automatically eliminated, consider the system of two bodies in Fig. 14-6.1 which are connected by a rigid link of negligible weight. From Newton's second and third laws, the reactions of the bodies on the link are equal and oppositely directed. If the length of the link does not change, each end has the same component of displacement along the link and hence the positive work done by the connecting link on one of the bodies is equal to the negative work done on the other body. Consequently, the total change in the kinetic energy of a connected system of bodies is caused only by the resultant work of the external forces, provided that the length of the connecting members does not change and they are of negligible weight.

As an illustration of this principle, consider the system in Fig. 14-6.2. The work done by T on A is Ts_A, being positive because T is in the direction of s_A. On B, the work done by $2T$ is $-2Ts_B$, being negative because $2T$ is opposite to the direction of s_B. Since the work done by the internal tensions must be zero, we have

$$Ts_A - 2Ts_B = 0 \quad \text{or} \quad s_A = 2s_B$$

whence successive differentiation gives $v_A = 2v_B$ and $a_A = 2a_B$. We have already "borrowed" this principle for use in Illus. Prob. 11-3.2, but now we can also confirm these relations from our knowledge of kinematics. Thus, using the instant center of the floating pulley, it is evident that $s_A = 2s_B$, etc.

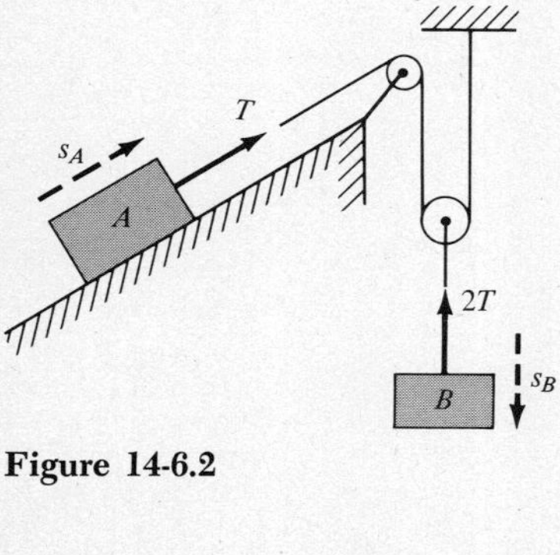

Figure 14-6.2

A general plan for applying the work-energy method consists of the following steps:

1. Determine the direction of motion. Alternatively, assume a direction of motion, confirming this assumption by noting that the resultant work must be positive to speed up a system, and vice versa.

2. Draw an *active-force diagram* for the system, showing only those forces that do work upon the system.

3. Determine the kinematic relations among the bodies composing the system.

4. Apply the work-energy equation by equating the resultant work on the system to the sum of the changes in kinetic energy of the bodies in the system.

5. If the internal force in a connecting member is desired, apply the work-energy equation to a free-body diagram of that part of the system on which this force is then exposed as an external force. If the internal force is not constant, however, this step will determine only its average value. The instantaneous value of a variable force must be found by the force-inertia method.

ILLUSTRATIVE PROBLEMS

14-6.1. If the system in Fig. 14-6.3 starts from rest in the given position, how much further will A move up the incline after B hits the ground? The coefficient of kinetic friction is 0.20 and the pulleys are frictionless and of negligible weight.

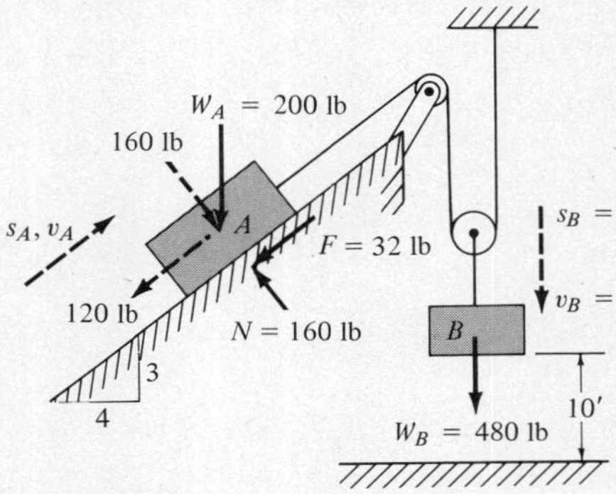

Figure 14-6.3

Solution

Since the initial and final velocities are zero, there is a strong temptation to likewise equate the total resultant work to zero. However, this approach would not account for the kinetic energy that is lost when B strikes the ground. Instead, we must consider two phases of the motion: (1) when A and B move as a connected system and (2) when A continues to move alone.

Phase (1). Only the external forces acting on the system which do work are shown. As noted on p. 552, the internal forces (tensions in the cords) do no work on the system and may be used to show that the displacement and velocity of B are one-half the corresponding values of A.

For convenience, we represent the normal and downplane components of the 200-lb weight of A by the dashed components, whence the friction force on A is $F = 0.20(160) = 32$ lb. If we now apply the work-energy equation to the system and note that the resultant work RW $= (\Sigma X\, s)_B + (\Sigma X\, s)_A$ where the X summations are positive in the indicated directions of motion, we obtain

$$\left[\text{RW} = \sum \frac{W}{2g}(v^2 - v_o^2)\right]$$

$$480(10) - (120 + 32)(20) = \frac{480}{2g}\left(\frac{1}{4}v_A^2\right) + \frac{200}{2g}v_A^2$$

$$4800 - 3040 = 1760 = \frac{v_A^2}{2g}(120 + 200)$$

whence

$$\frac{v_A^2}{2g} = \frac{1760}{320}$$

This result could be used to determine the initial velocity of A in phase (2); however, it is simpler to leave it in the form shown.

Phase (2). Note that after B hits the ground, the tension in the cord becomes zero; i.e., the cord goes slack. Applying now the work-energy equation to A alone, its further motion up the incline is given by

$$\left[\text{RW} = \frac{W}{2g}(v^2 - v_o^2)\right] \quad -152s = \frac{200}{2g}(0 - v_A^2) = -\frac{200(1760)}{320}$$

whence

$$s = \frac{200(1760)}{152(320)} = 7.24 \text{ ft} \qquad \textit{Ans.}$$

Some additional comments may be worthy of note. The tension in the cord supporting B during its descent can be found by applying the work-energy equation to a FBD of B as follows:

$$\left[\text{RW} = \frac{W}{2g}(v^2 - v_o^2)\right]$$

$$(480 - T_B)(10) = \frac{480}{2g}\left(\frac{1}{4}v_A^2\right) = 120\left(\frac{1760}{320}\right)$$

$$T_B = 480 - 66 = 414 \text{ lb} \qquad \textit{Ans.}$$

Also, since the forces and accelerations are constant, we may substitute the known value of v_A^2 in Eq. (9-3.6) to compute a_A during phases (1) and (2). Doing this gives

$$[v^2 = v_o^2 + 2as] \qquad \frac{1760}{320}(2g) = 2(a_A)_1(20) \qquad (a_A)_1 = 8.86 \text{ fps}^2$$

and

$$0 = \frac{1760}{320}(2g) + 2(a_A)_2(7.24) \qquad (a_A)_2 = -24.5 \text{ fps}^2$$

14-6.2. Body A is pulled along the frictionless level surface in Fig. 14-6.4 by a flexible, inextensible cord passing over smooth fixed pegs at C and D to the weight B. If A starts from rest at the given position, determine its velocity when it is in the dashed position.

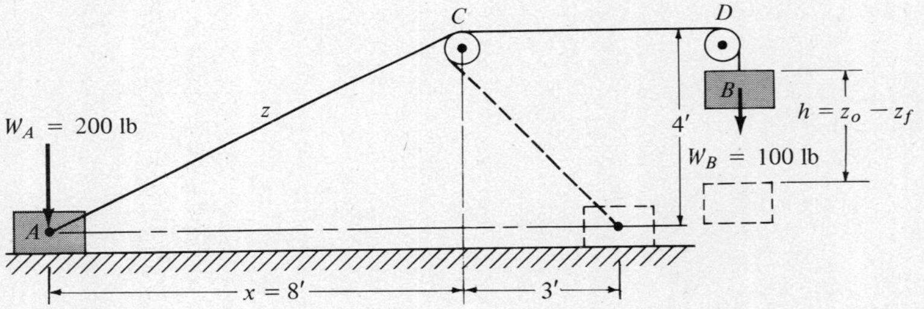

Figure 14-6.4

Solution

The only active force doing work on the system is W_B. The vertical movement of B is equal to the difference in length of AC as it moves between the two specified positions. Designating this variable length by z, and noting that the velocity of B is equal to the rate of change of this length $\left(\text{i.e., } v_B = \dfrac{dz}{dt}\right)$, we have

$$z^2 = x^2 + 4^2 \quad \text{whence} \quad z_o = \sqrt{8^2 + 4^2} = 8.95 \text{ ft}$$

$$\text{and} \quad z_f = \sqrt{3^2 + 4^2} = 5.00 \text{ ft}$$

To relate v_A to v_B, we differentiate the preceding expression with respect to the time. This gives

$$2z\frac{dz}{dt} = 2x\frac{dx}{dt} \quad \text{or} \quad zv_B = xv_A$$

whence for the dashed position

$$5v_B = 3v_A$$

Applying the work-energy equation, we obtain

$$\left[RW = \sum \frac{W}{2g}(v^2 - v_o^2) \right]$$

$$100(8.95 - 5) = \frac{200}{64.4}v_A^2 + \frac{100}{64.4}\left(v_B^2 = \frac{9}{25}v_A^2\right)$$

from which

$$v_A^2 = 108 \quad \text{and} \quad v_A = 10.4 \text{ ft per sec}$$ ***Ans.***

PROBLEMS

14-6.3. Determine the constant force P that will give the system of bodies shown in Fig. P-14-6.3 a velocity of 10 fps after moving 15 ft from rest.

$P = 250$ lb ***Ans.***

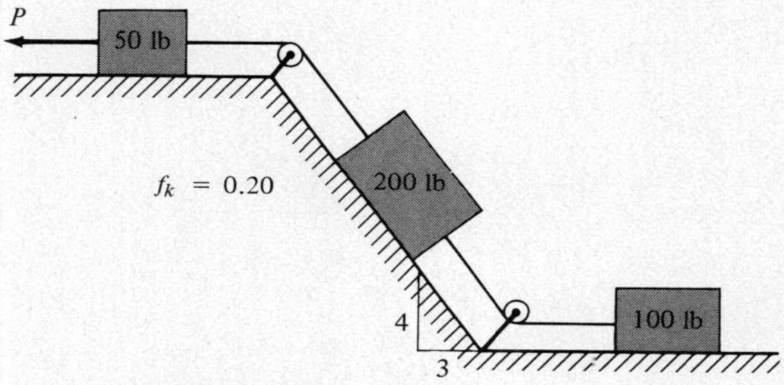

Figure P-14-6.3

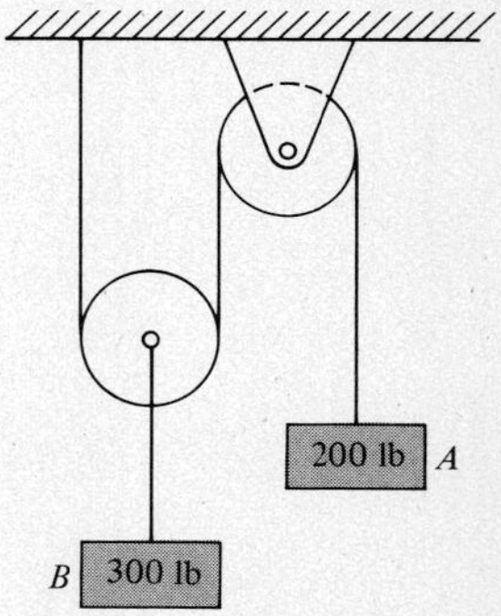

Figure P-14-6.4

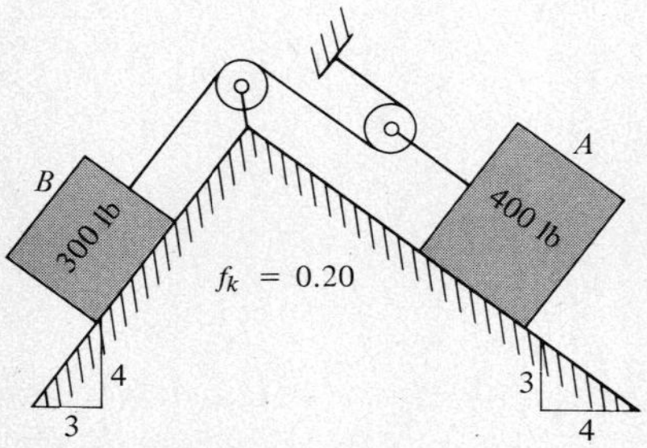

Figure P-14-6.5

14-6.4. Through what distance will body A in Fig. P-14-6.4 move in changing its velocity from 6 fps to 12 fps? Assume the pulleys to be frictionless and of negligible weight.

14-6.5. In what distance will body A of Fig. P-14-6.5 attain a velocity of 10 fps, starting from rest?

$s_A = 23.9$ ft ***Ans.***

14-6.6. Determine the velocity of body A in Fig. P-14-6.6 after it has moved 12 ft, starting from rest. Assume the pulleys to be frictionless and of negligible weight.

14-6.7. Assume the pulleys in Fig. P-14-6.7 to be frictionless and of negligible weight. Find the velocity of body B after it has moved 10 ft from rest. Then use this result to determine the acceleration of body B.

$v_B = 5.3$ fps; $a_B = 1.4$ fps^2 ***Ans.***

14-6.8. The system shown in Fig. P-14-6.8 is connected by flexible, inextensible cords. If the system starts from rest, find the distance d between

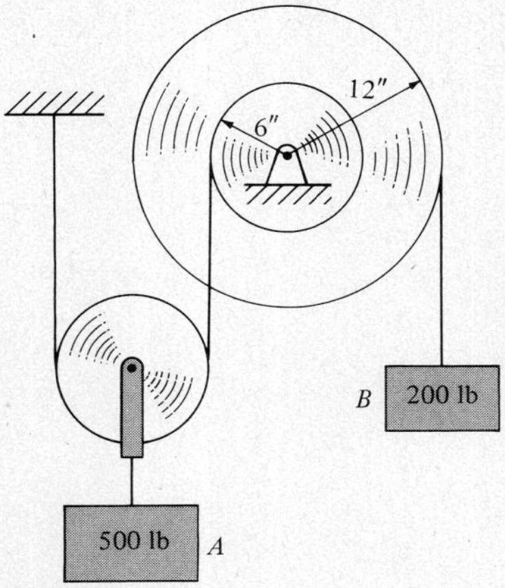

Figure P-14-6.6

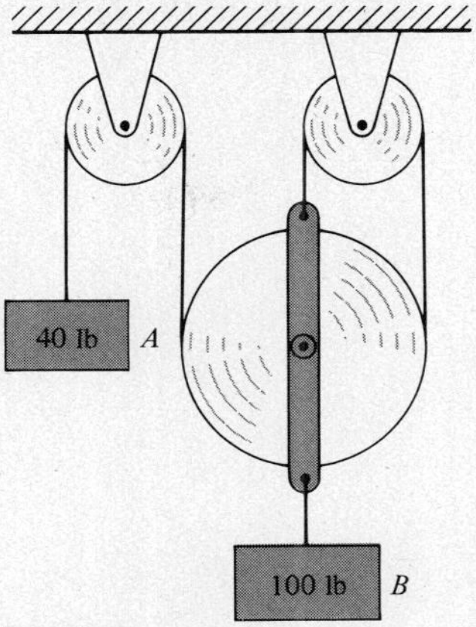

Figure P-14-6.7

A and the ground so that the system comes to rest with body B just touching A.

$$d = 2.67 \text{ ft} \qquad \textbf{Ans.}$$

14-6.9. If $d = 5$ ft in the preceding problem, determine the total distance moved by body C.

14-6.10. In Prob. 14-6.8, what should be the weight of body C so that for $d = 5$ ft the system stops with B just touching A?

$$W_C = 179 \text{ lb} \qquad \textbf{Ans.}$$

14-6.11. Body A starts from rest in the position shown in Fig. P-14-6.11. Determine its velocity after it has moved 15 ft along the frictionless surface.

14-6.12. Find the velocity of body A in Prob. 14-6.11 after it has moved, starting from rest at the given position, for 9 ft along the frictionless surface.

$$v_A = 15.57 \text{ fps} \qquad \textbf{Ans.}$$

14-6.13. Determine the tension in the cord connecting bodies A and B in Prob. 14-6.11 at the instant A has moved 9 ft starting from rest in the given position.

Hint: Use the result of Prob. 14-6.12.

$$T = 86.1 \text{ lb} \qquad \textbf{Ans.}$$

14-6.14. Two sliders, connected by a light rigid link 10 ft long, move in the frictionless guides shown in Fig. P-14-6.14. If B starts from rest when it is vertically below A, determine the velocity of B when $x = 6$ ft. Assume $W_A = W_B = 200$ lb and $W_C = 100$ lb.

$$v_B = 12.5 \text{ fps} \qquad \textbf{Ans.}$$

14-6.15. Repeat Prob. 14-6.14 when $x = 8$ ft.

14-6.16. A homogeneous triangular plate weighing 100 lb is supported by two parallel links of negligible weight as shown in Fig. P-14-6.16. The length of the links is $2b = 8$ ft. The system starts from rest at $\theta = 90°$.

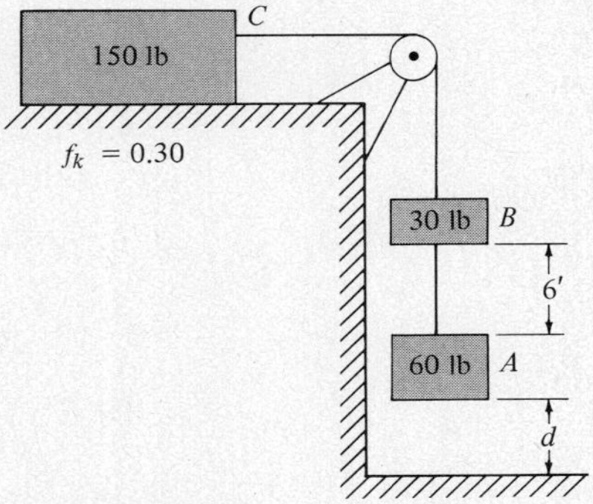

Figure P-14-6.8

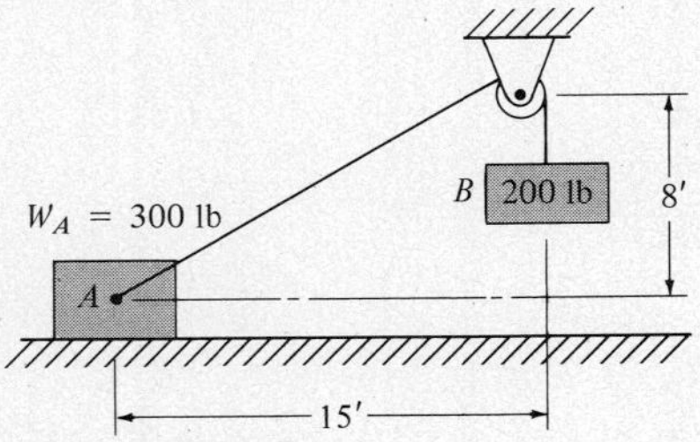

Figure P-14-6.11

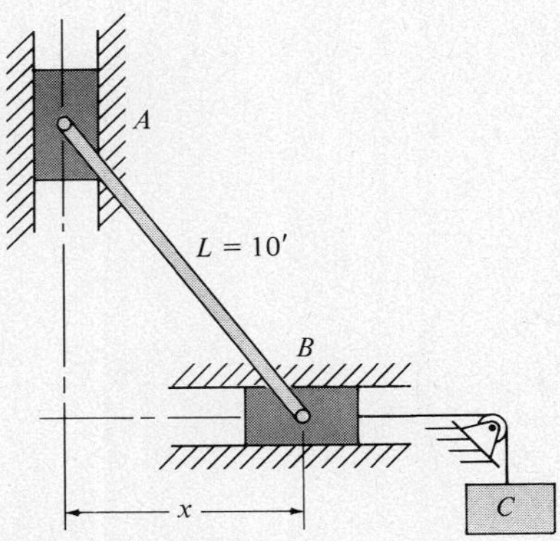

Figure P-14-6.14

Find the velocity of the 200-lb weight when $\theta = 60°$.

$v = 7.12$ fps **Ans.**

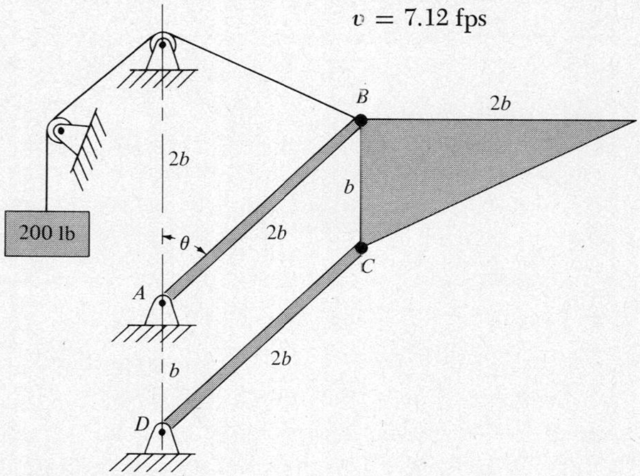

Figure P-14-6.16

14-6.17 Repeat Prob. 14-6.16 when $\theta = 30°$.

14-7 WORK-ENERGY METHOD APPLIED TO FIXED-AXIS ROTATION

The simplest and most direct way to obtain the work-energy equation for fixed-axis rotation is to eliminate α between the equations $\Sigma M = I\alpha$ and $\alpha\, d\theta = \omega\, d\omega$ to obtain

$$\Sigma M\, d\theta = I\omega\, d\omega \tag{a}$$

This is integrated between the limits or conditions that the angular displacement varies from an initial value θ_o to a final value of θ while the angular velocity changes from an initial value of ω_o to a final value of ω. We thus obtain

$$\int_{\theta_o}^{\theta} \Sigma M\, d\theta = I \int_{\omega_o}^{\omega} \omega\, d\omega$$

or

$$\int_{\theta_o}^{\theta} \Sigma M\, d\theta = \tfrac{1}{2}I(\omega^2 - \omega_o{}^2) = \tfrac{1}{2}I\omega^2 - \tfrac{1}{2}I\omega_o{}^2 \tag{14-7.1}$$

This is the fundamental work-energy equation for fixed-axis rotation. In it, both ΣM and I are computed about the axis of rotation. Hence, for centroidal rotation, these terms become $\Sigma \bar{M}$ and $\bar{I}$, respectively.

As in translation, this is a scalar equation.[5] The left-hand term

[5]Thus, by taking the dot product of $d\boldsymbol{\theta}$ with each side of $\Sigma \mathbf{M} = \dot{\mathbf{H}}$, we obtain

$$\Sigma \mathbf{M} \cdot d\boldsymbol{\theta} = \dot{\mathbf{H}} \cdot d\boldsymbol{\theta} = \frac{d\mathbf{H}}{dt} \cdot d\boldsymbol{\theta} = \boldsymbol{\omega} \cdot d\mathbf{H}$$

(continued on p. 559)

represents the resultant work of rotation during an angular displacement θ, expressed in foot-pounds. The right-hand terms represent the change in kinetic energy of the rotating body, also expressed in foot-pounds, during this angular displacement.

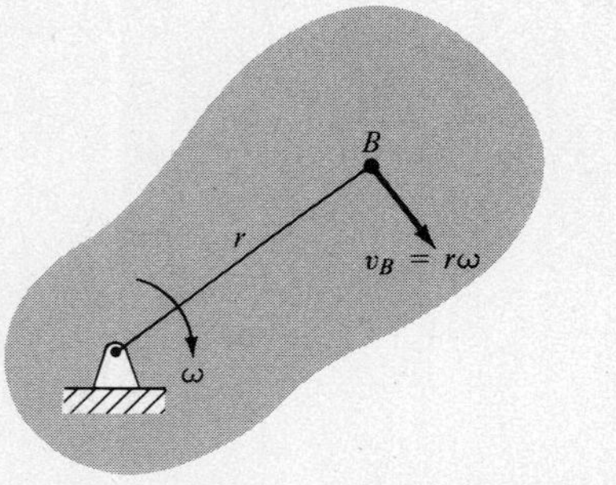

Figure 14-7.1 Velocity of a typical particle.

The expression for kinetic energy of a rotating body can also be developed from the fact that the total kinetic energy is the scalar summation of the kinetic energies of all particles of the body. Thus, in Fig. 14-7.1, if B is a typical particle, its kinetic energy is

$$\text{KE} = \tfrac{1}{2}\, dm\, v_B^2 = \tfrac{1}{2}\, dm\, r^2\omega^2$$

For the entire body, therefore,

$$\text{Total KE} = \int \tfrac{1}{2}\, dm\, r^2\omega^2 = \tfrac{1}{2}\omega^2 \int r^2\, dm$$

or, as before

$$\text{KE} = \tfrac{1}{2}I\omega^2$$

If ΣM is variable, the resultant work in rotation can be found only if ΣM can be expressed in terms of θ, or from the area under a ΣM-θ diagram. If ΣM is constant, the resultant work may be expressed as $\Sigma M(\theta)$, in which case the work-energy equation becomes

$$\Sigma M(\theta) = \tfrac{1}{2}I(\omega^2 - \omega_o^2) \tag{14-7.2}$$

Notice that, as in translation, moments in the initial direction of rotation do positive work on the body, and vice versa.

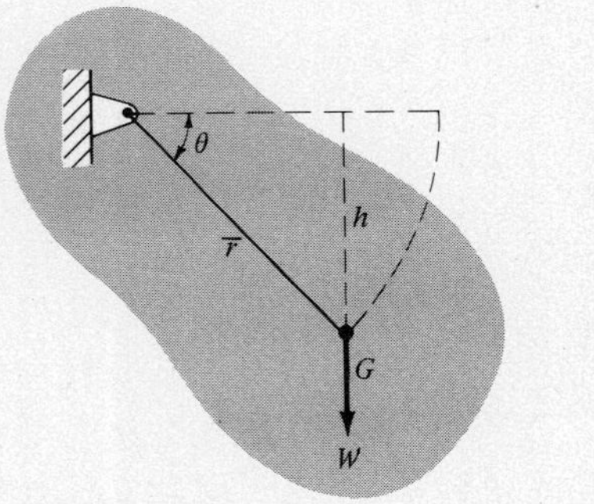

Figure 14-7.2 $RW = \int M\, d\theta = Wh$.

The work done by gravity forces on a rotating body will be the product of the weight W by its change in elevation h as discussed in Section 14-3. It is worthwhile to check this statement by considering Fig. 14-7.2 in which the body rotates through θ radians from the horizontal. The work done by W can be found by applying $\int_o^\theta \Sigma M\, d\theta$ which gives

$$\int_o^\theta \Sigma M\, d\theta = \int_o^\theta W\bar{r}\cos\theta\, d\theta = W\bar{r}[\sin\theta]_o^\theta = W\bar{r}\sin\theta$$

and, as $\bar{r}\sin\theta = h$, we obtain Wh as stated above.

As in the case of translation, the work-energy equation may be applied directly to a connected system of translating and rotating bodies without considering any of the internal forces in the connect-

---

Since all these vectors are perpendicular to the plane of motion, and (from p. 501) $\mathbf{H} = I\boldsymbol{\omega}$ so that $d\mathbf{H} = I\, d\boldsymbol{\omega}$, this reduces as before to Eq. (a).

ing elements. The resultant work (however computed) is always equal to the scalar changes in kinetic energy of the system, or

$$(\mathrm{RW})_{\mathrm{sys}} = \sum \frac{W}{2g}(v^2 - v_o^{\,2}) + \sum \frac{1}{2} I(\omega^2 - \omega_o^{\,2}) \tag{14-7.3}$$

Then after the relation between forces, displacements, and velocities for the original system has been determined, we may isolate any part of the system and reapply the work-energy equation to that part to determine constant internal forces.

The illustrative problems demonstrate several advantages of the work-energy method. In the first one, a single work-energy equation replaces a simultaneous solution of four equations resulting from applying a kinetic equation to each of four bodies, plus an additional kinematic equation. In the second problem, the work-energy equation leads directly to a velocity which otherwise would require integrating a kinematic equation involving a variable acceleration. In the third problem, we illustrate a solution combining both work-energy and force-inertia methods.

ILLUSTRATIVE PROBLEMS

14-7.1. Determine the distance that body D in Fig. 14-7.3 must move in order to reach a velocity of 12 fps starting from rest. What tension is acting in the cord joining the step pulleys B and C during this movement?

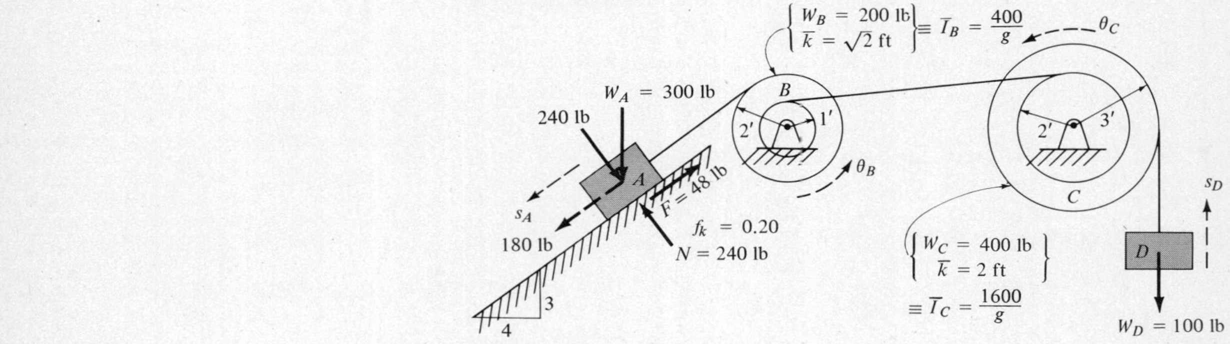

Figure 14-7.3

Solution

A preliminary calculation of the tensions in the various cords, assuming the system to be at rest, discloses an unbalanced upward force to act on D which therefore rises while the rest of the system moves as shown by the dashed displacement arrows. Using $s = r\theta$, the kinematic relations among the bodies are

$$s_A = 2\theta_B; \qquad \theta_B = 2\theta_C; \qquad 3\theta_C = s_D$$

Combining these relations gives

$$s_A = 2\theta_B = 4\theta_C = \tfrac{4}{3}s_D \qquad (a)$$

and since $v = r\omega$ is of the same mathematical form as $s = r\theta$, the velocities are similarly related as follows:

$$v_A = 2\omega_B = 4\omega_C = \tfrac{4}{3}v_D \qquad (b)$$

We are now ready to equate the resultant work of external forces on the system to the total change in kinetic energy of the system. We apply $\Sigma X(s)$ to the external forces on each translating body and $\Sigma M(\theta)$ to the moments of external forces on the rotating bodies (of which there are none in this case). Remember that ΣX and ΣM are positive in the initial direction of motion. The kinetic energy change in the system is $\sum \frac{W}{2g}(v^2 - v_o^2)$ for the translating bodies and is $\sum \frac{1}{2}\bar{I}(\omega^2 - \omega_o^2)$ for the rotating bodies. We then have, using Eqs. (a) and (b) to express values in terms of s_D and v_D,

$$(180 - 48)\left(\frac{4}{3}s_D\right) - 100s_D$$
$$= \frac{300}{2g}\left(\frac{16}{9}\right)v_D^2 + \frac{400}{2g}\left(\frac{4}{9}\right)v_D^2 + \frac{1600}{2g}\left(\frac{1}{9}\right)v_D^2 + \frac{100}{2g}v_D^2$$

which reduces to

$$76s_D = \frac{v_D^2}{18g}(4800 + 1600 + 1600 + 900) = \frac{v_D^2}{18g}(8900) \qquad (c)$$

whence on substituting the given value of $v_D = 12$ fps, Eq. (c) yields

$$s_D = 29.1 \text{ ft}$$ ***Ans.***

The advantage of using general symbols for s_D and v_D will be evident if we now differentiate Eq. (c) with respect to the time. Doing this gives

$$76v_D = \frac{8900}{18g}(2v_D)\frac{dv_D}{dt}$$

whence, cancelling the common term v_D, the acceleration of D is[6]

$$a_D = \frac{dv_D}{dt} = \frac{76(18)(32.2)}{8900(2)} = 2.48 \text{ fps}^2$$

Now, if desired, relations similar to Eqs. (a) or (b) can be used to find the accelerations of the other bodies. Similarly, the computed

[6] This method of finding instantaneous acceleration is a general procedure applicable to constant or variable accelerations. When the acceleration is constant, as is the case here when forces and moments are constant, the procedure outlined in Illus. Prob. 14-6.1 is more direct.

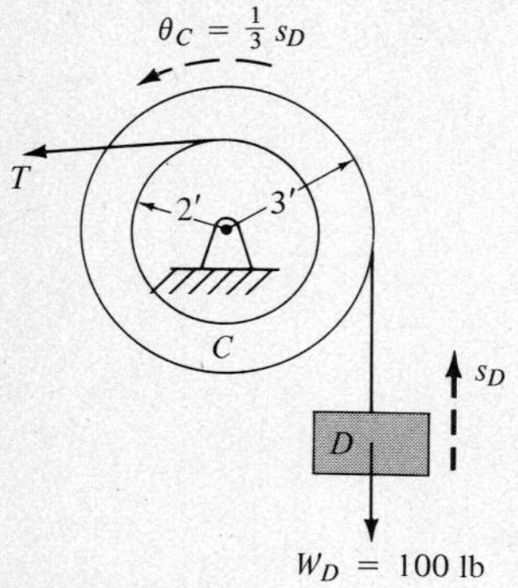

Figure 14-7.4

value of s_D can be used to determine the displacements of the other bodies.

To determine the tension in the cord between B and C, we cut this cord and isolate the system of bodies C and D as shown in Fig. 14-7.4. As this system moves through the distance s_D, its change in KE is the sum of the last two terms enclosed in the parentheses of Eq. (*c*). The tension T is an external force on this system and its work is computed from $\Sigma M(\theta_C)$ where the moment of T is $2T$ and $\theta_C = \frac{1}{3}s_D$. An equivalent way of finding the work done by T is to multiply T by the length of cord unwrapped from C which is $s = r\theta_C = 2(\frac{1}{3}s_D)$. Applying the work-energy equation to this system and using the value of $s_D = 29.1$ ft corresponding to a velocity of $v_D = 12$ fps, we obtain

$$[\text{RW} = \Delta\text{KE}] \qquad 2T\left(\frac{1}{3}\right)(29.1) - 100(29.1) = \frac{(12)^2}{18g}(2500)$$

$$T = 181 \text{ lb} \qquad \textit{Ans.}$$

14-7.2. At the position shown in Fig. 14-7.5, the uniform bar, 4 ft long and weighing 60 lb, has an angular velocity ω_o. The spring has a modulus of 5 lb/in. and a free length of 2 ft. Determine the value of ω_o so that the bar just reaches the horizontal dashed position. Is it possible for the bar to rebound from this horizontal position past its topmost position? If so, what will then be its angular velocity?

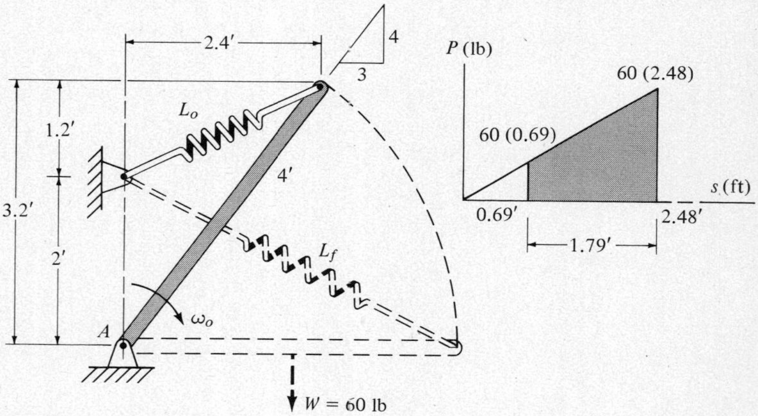

Figure 14-7.5

Solution

We first compute the initial and final lengths of the stretched spring to be respectively $L_o = 2.69$ ft and $L_f = 4.48$ ft. The corresponding deformation of the spring, found by subtracting the free length of 2 ft, is shown in the spring force versus deformation diagram. The corresponding forces, after converting the spring modulus to 60 lb/ft, are also shown. The shaded area represents the negative work done by the spring (stored in it as elastic potential energy).

We now apply the work-energy equation for rotation. Noting that

$$I_A = \frac{1}{3}\frac{W}{g}L^2 = \frac{1}{3}\left(\frac{60}{8}\right)(4)^2 = \frac{320}{g} \text{ ft-lb-sec}^2$$

and that the work done by gravity is Wh, we obtain

$[\text{RW} = \frac{1}{2}I_A(\omega^2 - \omega_o{}^2)]$

$$60\left(\frac{3.2}{2}\right) - \frac{60(0.69 + 2.48)}{2}(1.79) = \frac{1}{2}\left(\frac{320}{g}\right)(0 - \omega_o{}^2)$$

from which

$$\omega_o{}^2 = 14.95 \quad \text{and} \quad \omega_o = 3.87 \text{ rad/sec} \qquad \textit{Ans.}$$

To rebound past the topmost position, the potential energy stored in the spring must be larger than the negative work done by gravity. To check this and also determine the possible ω_2 at the topmost position, the work-energy equation gives

$[\text{RW} = \frac{1}{2}I_A(\omega_2{}^2 - \omega_1{}^2)]$

$$\frac{1}{2}(60 \times 2.48)(2.48) - 60(2) = \frac{1}{2}\left(\frac{320}{g}\right)(\omega_2{}^2 - 0)$$

$$184.6 - 120 = \frac{320}{64.4}\omega_2{}^2$$

Hence, there is a sufficient excess of potential energy over the work of gravity to give the bar an angular velocity at its topmost position of

$$\omega_2 = 3.61 \text{ rad/sec} \qquad \textit{Ans.}$$

If you are surprised at the small difference between ω_2 and ω_o, compute the resultant work done between the given position and the topmost position. You should obtain a value of -9.7 ft-lb.

14-7.3. A 150-lb uniform rod 6 ft long which carries a 30-lb weight at its end is bolted to a 200-lb solid cylinder as shown in Fig. 14-7.6. A cable wrapped around the cylinder passes horizontally over a small pulley to support a 400-lb body. The system starts from rest when the rod is in its lowest vertical position. At the instant the rod has rotated through 60° to the given position, determine the values of ω, α, and the tension T in the cable. Then find the instantaneous bearing reaction.

Solution

The angular velocity is found by equating the resultant work of external forces to the total change in kinetic energy of the system.

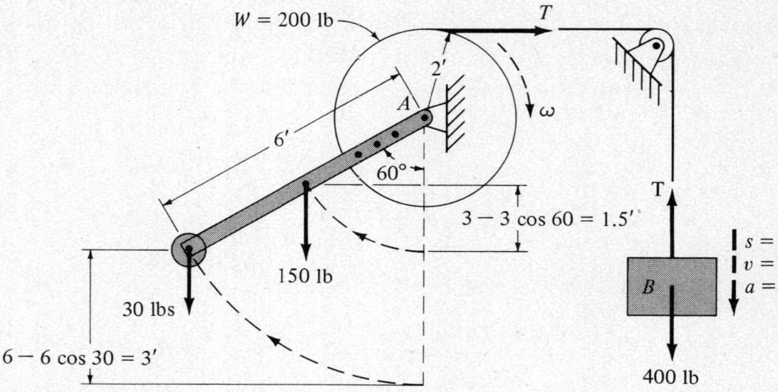

Figure 14-7.6

The cylinder rotates through $60(\pi/180) = \pi/3$ rad; hence, the vertical distance that body B falls is given by $s = r\theta = 2(\pi/3)$ ft. The vertical components of displacement of the gravity centers of the rod and attached weight are respectively 1.5 ft and 3 ft, computed as indicated in the figure.

The mass moment of inertia of the rotating elements about the axis A is the sum of the following values:

$$I_{\text{cyl}} = \frac{1}{2}mr^2 = \frac{200}{2g}(2)^2 = \frac{400}{g} \text{ ft-lb-sec}^2$$

$$I_{\text{rod}} = \frac{1}{3}mL^2 = \frac{150}{3g}(6)^2 = \frac{1800}{g} \text{ ft-lb-sec}^2$$

$$I_w = mk^2 = \frac{30}{g}(6)^2 = \frac{1080}{g} \text{ ft-lb-sec}^2$$

whence

$$I_A = I_{\text{cyl}} + I_{\text{rod}} + I_w = \frac{3280}{g} \text{ ft-lb-sec}^2$$

Using $v_B = r\omega = 2\omega$ to express the velocity of B in terms of ω, we now apply the work-energy equation to obtain

$[\text{RW} = \Delta\text{KE}]$

$$400(2)\left(\frac{\pi}{3}\right) - 150(1.5) - 30(3) = \frac{1}{2}\left(\frac{3280}{g}\right)\omega^2 + \frac{300}{2g}(2\omega)^2$$

which reduces to

$$\frac{\omega^2}{g} = 0.233 \quad \text{and} \quad \omega = 2.74 \text{ rad/sec} \qquad \textit{Ans.}$$

Note that by solving first for ω^2/g from which we found ω, we have the common term in the following expression for centrifugal inertia force on the rotating elements:

$$\text{CIF} = \sum \frac{W}{g}\bar{r}\omega^2 = \frac{\omega^2}{g}\Sigma W\bar{r} = 0.233(150 \times 3 + 30 \times 6) = 147 \text{ lb}$$

which acts radially outward from A.

If only the instantaneous angular acceleration were required, we could have used the method discussed in Illus. Prob. 14-7.1. However, since the instantaneous values of both α and T are asked for, it is simpler and more direct to apply the kinetic equations of translation and rotation respectively to the body B and the rotating elements. This gives

$$\left[+\downarrow \Sigma X = \frac{W}{g}a\right] \qquad 400 - T = \frac{400}{g}(2\alpha)$$

$$[\circlearrowleft + \Sigma M_A = I_A\alpha] \quad 2T - 150(3 \sin 60°) - 30(6 \sin 60°) = \frac{3280}{g}\alpha$$

which are solved to yield

$$\frac{\alpha}{g} = 0.052; \qquad \alpha = 1.675 \text{ rad/sec}^2 \circlearrowright; \qquad T = 358 \text{ lb} \qquad \textit{Ans.}$$

Note that α/g is the common term in the following expression for the total tangential inertia force on the rotating elements:

$$\text{TIF} = \sum \frac{W}{g}\bar{r}\alpha = \frac{\alpha}{g}\Sigma W\bar{r} = 0.052(150 \times 3 + 30 \times 6) = 32.8 \text{ lb}$$

Its direction is opposite to the tangential accelerations of the gravity centers of the rod and attached weight.

Finally, the components of the bearing reaction at A can be found by using these inertia effects to place the FBD of the rotating elements in dynamic equilibrium. However, the point diagram of dynamic equilibrium (shown in Fig. 14-7.7) is simpler to draw and just as effective since only force summations are needed to find the components of the bearing reaction. Note that the value of the resultant couple $\Sigma\bar{I}\alpha$ need not be computed because it has no effect on the following vertical and horizontal force summations of dynamic equilibrium:

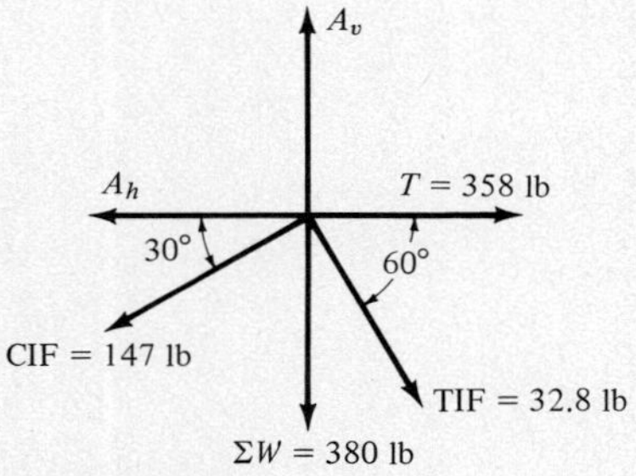

Figure 14-7.7

$$[+\uparrow \Sigma V = 0] \qquad A_v - 380 - 32.8 \sin 60° - 147 \sin 30° = 0$$
$$A_v = 482 \text{ lb} \uparrow$$

$$[\overset{+}{\leftarrow} \Sigma H = 0] \qquad A_h + 147 \cos 30° - 32.8 \cos 60° - 358 = 0$$
$$A_h = 247 \text{ lb} \leftarrow$$

When we combine these components, the instantaneous bearing reaction is

$$[A = \sqrt{A_h^2 + A_v^2}] \qquad A = 541 \text{ lb} \measuredangle 62.9° \qquad \textit{Ans.}$$

PROBLEMS

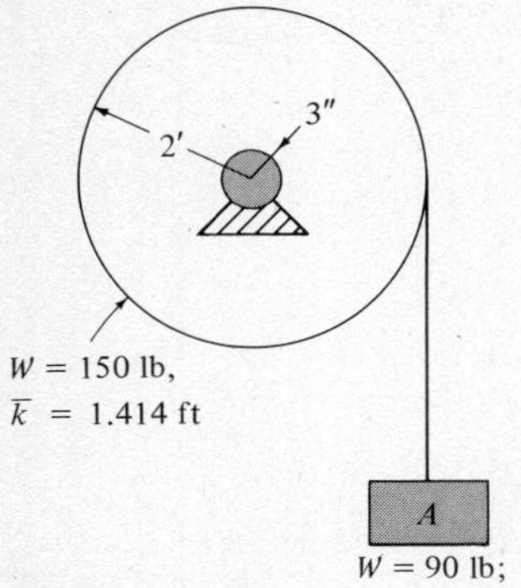

Figure P-14-7.4

14-7.4. Body A in Fig. P-14-7.4 drops 12 ft from rest before striking the ground. Pulley B is mounted on an axle 6 in. in diameter, and the axle friction is constant at 40 lb. How many turns will B make after A stops? Also compute the tension in the cord before A hits the ground.

$\theta_B = 6.41$ rev; $T = 49.7$ lb **Ans.**

14-7.5. Determine the distance moved by body A in Fig. P-14-7.5 in changing its velocity from 12 to 24 fps.

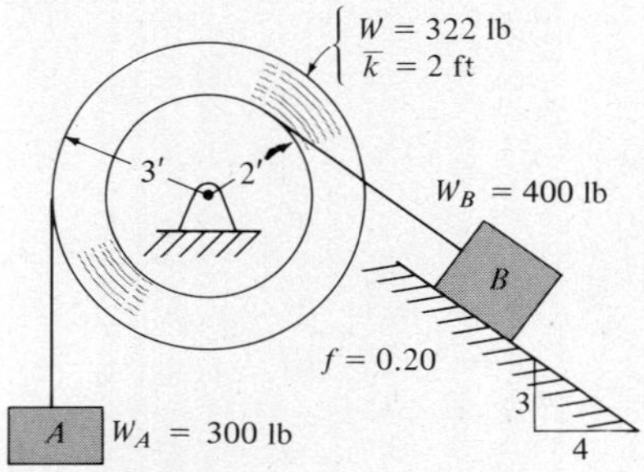

Figure P-14-7.5

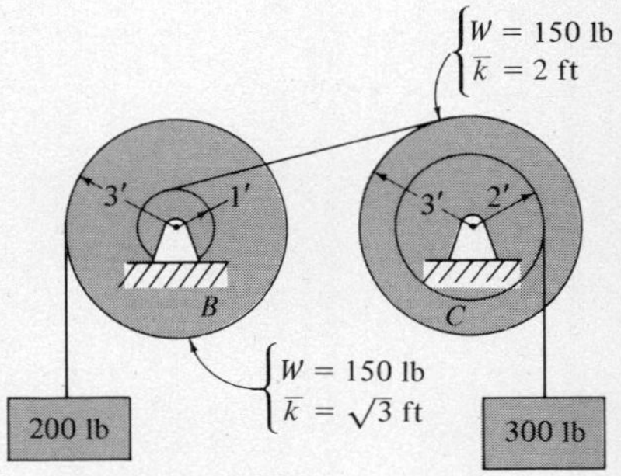

Figure P-14-7.6

14-7.6. For the system shown in Fig. P-14-7.6, what velocity will body D have after it has moved 10 ft, starting from rest? What is the tension in the cord joining pulleys B and C?

$v_D = 8.37$ fps; $T = 233$ lb **Ans.**

14-7.7. The compound pulleys in Fig. P-14-7.7 are connected by a crossed belt. Determine the velocity attained by body A after it has moved 12 ft from rest, and the tension in the cord supporting it.

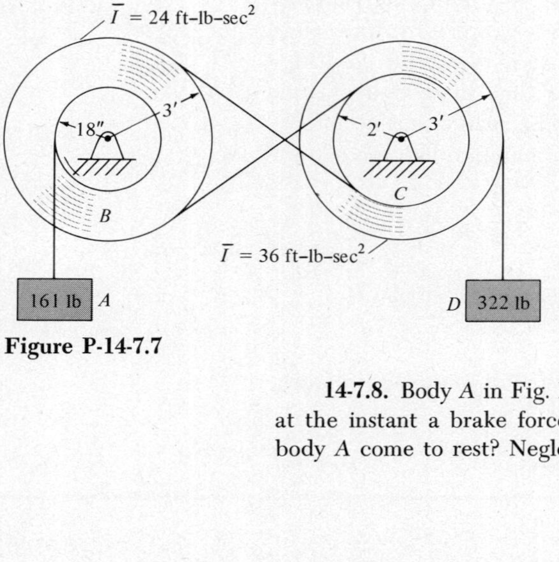

Figure P-14-7.7

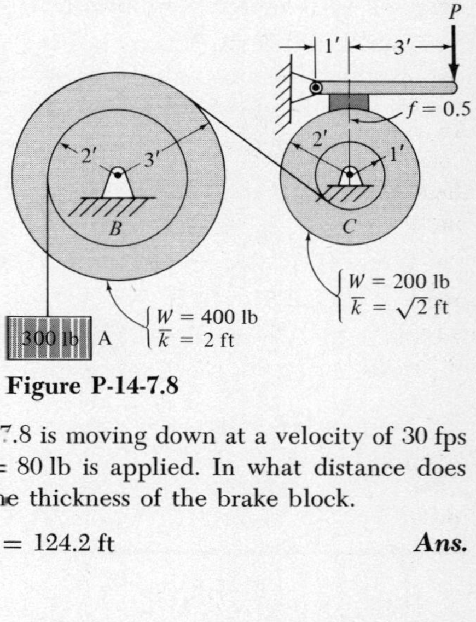

Figure P-14-7.8

14-7.8. Body A in Fig. P-14-7.8 is moving down at a velocity of 30 fps at the instant a brake force $P = 80$ lb is applied. In what distance does body A come to rest? Neglect the thickness of the brake block.

$s = 124.2$ ft **Ans.**

14-7.9. A heavy belt 80 ft long and weighing 10 lb/ft passes over the cylinder shown in Fig. P-14-7.9. At the start, $d = 10$ ft. Neglecting the thickness of the belt, determine its velocity at the instant when $d = 50$ ft.

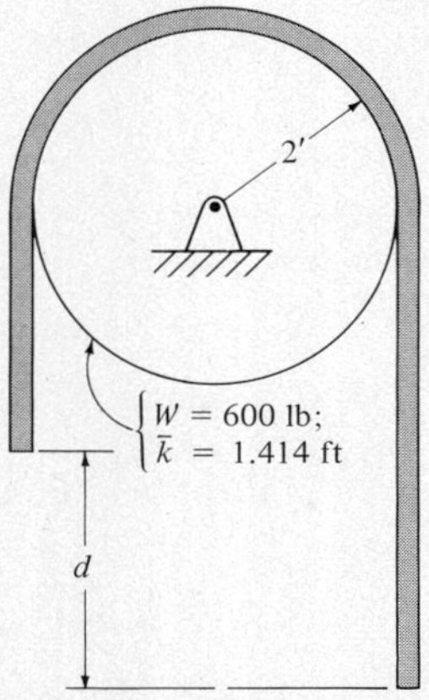

Figure P-14-7.9

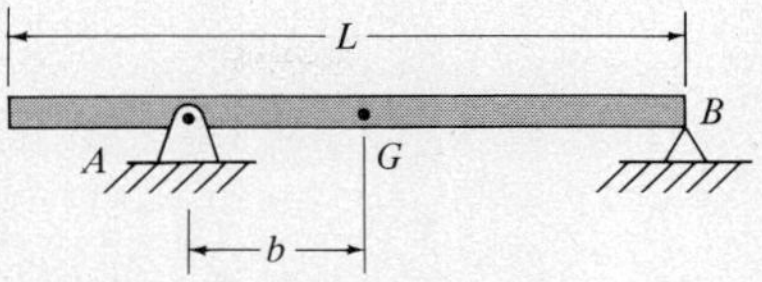

Figure P-14-7.10

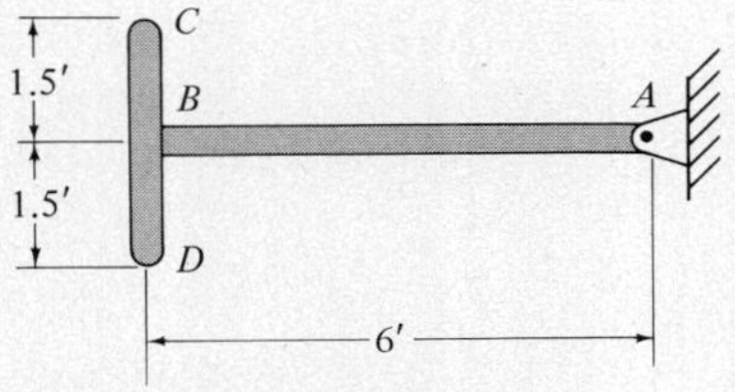

Figure P-14-7.11

14-7.10. As shown in Fig. P-14-7.10, a slender bar of weight W and length L is supported at one end and at a distance b from its mass center G. If the support at B is suddenly removed, determine b so that the bar attains a maximum angular velocity after a 90° rotation.

$$b = L/\sqrt{12} \qquad \textit{Ans.}$$

14-7.11. Two bars are welded together as shown in Fig. P-14-7.11. Segment AB weighs 60 lb and segment CD weighs 30 lb. The assembly starts from rest at the given position and rotates in a vertical plane about a horizontal axis at A. Compute the value of the bearing reaction at A after a 90° counterclockwise rotation.

$$A = 232 \text{ lb} \qquad \textit{Ans.}$$

14-7.12. A system, consisting of a 40-lb disk attached to a 30-lb rod AB as shown in Fig. P-14-7.12, is released from rest when AB is horizontal. Compute the angular velocity of AB after rotating 90° clockwise if (a) the disk is welded to the rod, (b) the disk is fastened to the rod by a smooth pin at B.

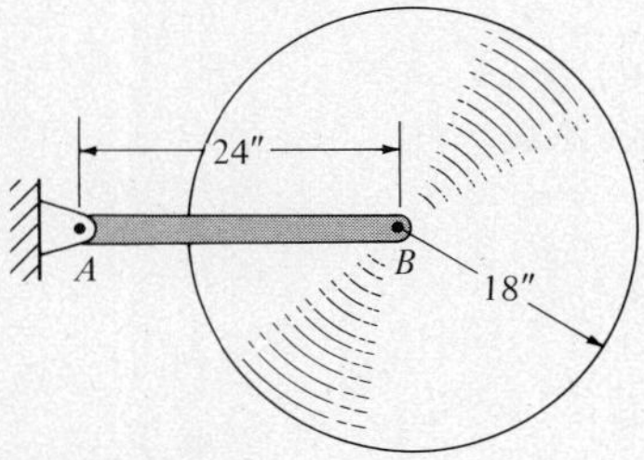

Figure P-14-7.12

14-7.13. The bent bar shown in Fig. P-14-7.13 weighs 10 lb/ft and is free to rotate in a vertical plane about a horizontal axis at A. If the bar is released from rest at the given position, determine its angular velocity at the instant it has rotated through 90°.

$$\omega = 2.18 \text{ rad/sec} \qquad \textit{Ans.}$$

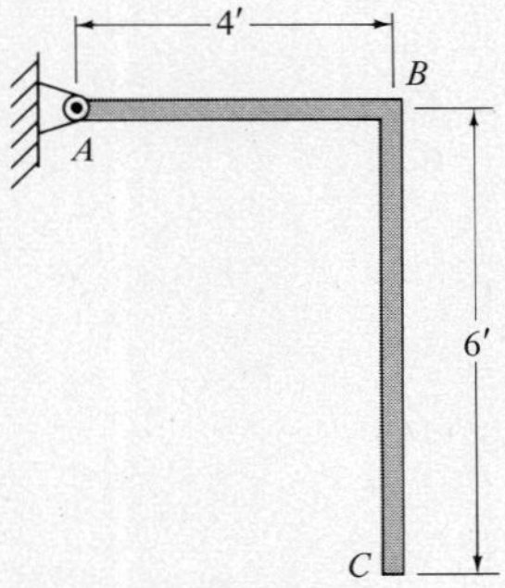

Figure P-14-7.13

14-7.14. A pole 6 ft long and weighing 96.6 lb hangs vertically from a horizontal axis through its upper end. A sudden blow gives the pole an initial angular velocity of 6 rad/sec. Determine the bearing reaction at the instant the pole has rotated through 60°.

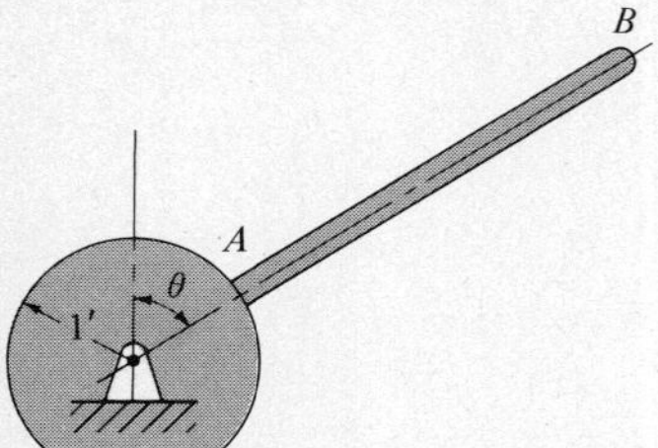

Figure P-14-7.16

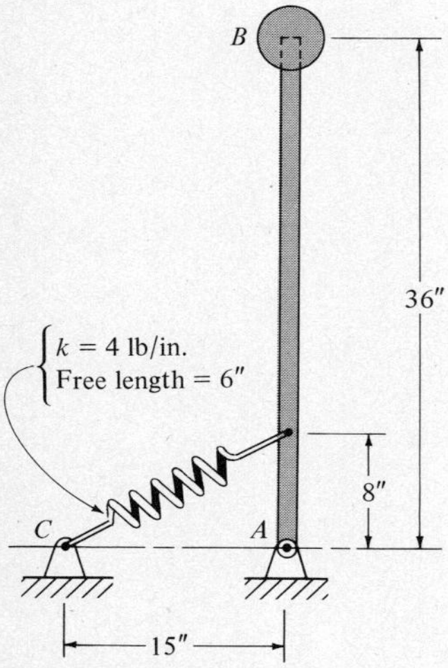

Figure P-14-7.17

14-7.15. A uniform rod with rounded ends is placed upright on a rough horizontal surface. It is permitted to fall from rest to the ground. Compute the coefficient of friction at the ground if the lower end slips when the rod has a slope of 4 vertical to 3 horizontal.

$$f = 0.367 \qquad \textit{Ans.}$$

14-7.16. A rod AB, 6 ft long and weighing 100 lb, is welded to the surface of a 200-lb cylinder as shown in Fig. P-14-7.16. If the system starts from rest when AB is vertically above the axle C, find the axial force (directed along AB) acting on the weld at A when $\tan\theta = \frac{4}{3}$.

$$F = 4.0 \text{ lb} \qquad \textit{Ans.}$$

14-7.17. A rod AB weighs 20 lb and carries a 10-lb weight at B as shown in Fig. P-14-7.17. It rotates in a vertical plane about a horizontal axis at A. If the rod is released from rest in the given position, find the velocity of end B when the rod is horizontal.

$$v_B = 17.6 \text{ fps} \qquad \textit{Ans.}$$

14-7.18. A 100-lb weight rotates in a vertical plane at one end of a 6-ft rod which weighs 90 lb. As shown in Fig. P-14-7.18, a spring, having a constant of 100 lb/ft, only acts on the rod after the spring exceeds its free length of 3.5 ft. If the rod starts from rest when it is vertically above the pivot A, find the angular velocity and angular acceleration of the rod when it has a downward slope of 3 vertical to 4 horizontal as shown.

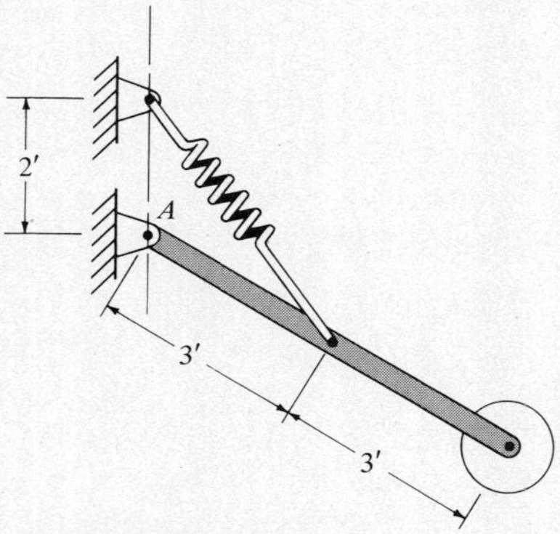

Figure P-14-7.18

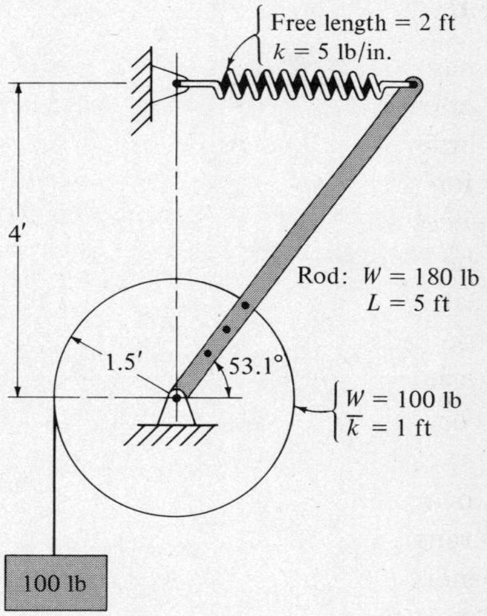

Figure P-14-7.19

14-7.19. At the instant shown in Fig. P-14-7.19, the spring is horizontal. Determine the clockwise angular velocity of the rod at the given position so that it will just reach a horizontal position.

$$\omega_o = 3.41 \text{ rad/sec} \qquad \textit{Ans.}$$

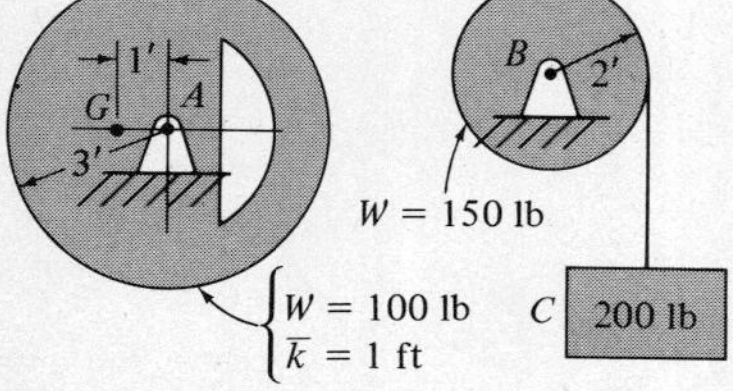

Figure P-14-7.20

14-7.20. In the system shown in Fig. P-14-7.20, A is an unbalanced wheel whose gravity center is at G, and B is a solid cylinder. If the system starts from rest at the given position, find the angular velocity of A after it has rotated 270° clockwise.

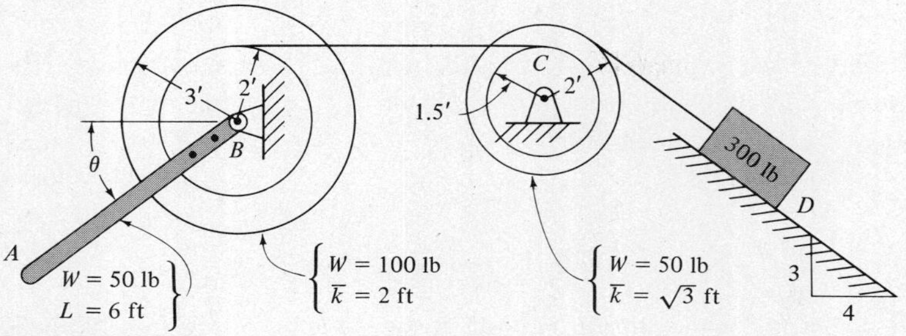

Figure P-14-7.21

14-7.21. In the system shown in Fig. P-14-7.21, AB is a 50-lb bar 6 ft long which is firmly attached to drum B. The system starts from rest when AB is in its lowest position. Determine the angular velocity of B at the instant $\theta = 30°$.

$\omega = 2.16$ rad/sec **Ans.**

14-8 WORK-ENERGY APPLIED TO PLANE MOTION

In the force analysis of plane motion problems, the dynamic effects are expressed in terms of a resultant force $\frac{W}{g}\bar{a}$ acting through the mass center and a resultant couple $\bar{I}\alpha$. In many problems the instantaneous angular velocity ω must be known as a preliminary step to finding the acceleration $\bar{a}$ of the mass center. Thus, recall from Section 13-5 that $\bar{a} = a_G = a_A \mathrel{+\!\!\!\rightarrow} (a_{G/A} = \bar{r}\omega^2 \mathrel{+\!\!\!\rightarrow} \bar{r}\alpha)$ where a_A is the acceleration of any reference point and $a_{G/A}$ is the acceleration of the mass center rotating about A. When α varies, as in noncentroidal rotation and general plane motion, it can be very laborious to use the differential equations of kinematics to determine an instantaneous value of ω. This is where work-energy methods become very useful since ω can be found directly without involving α.

In Section 13-2 we showed that a plane motion is equivalent to a combination of pure rotation about its centroidal axis and a translation of its mass center. We shall now show that the work-energy equation for plane motion is the sum of the work-energy expressions previously developed for the motions of translation and of centroidal rotation.

For the translational component of plane motion, taking the X axis tangent to the path of the gravity center during each displacement $d\bar{s}$ along the path, we have

$$\text{RW} = \int_0^s \Sigma X\, d\bar{s} = \frac{W}{2g}(\bar{v}^2 - \bar{v}_o^{\,2}) \qquad (a)$$

and for the rotational component,

$$\text{RW} = \int_0^\theta \Sigma \overline{M}\, d\theta = \tfrac{1}{2}\overline{I}(\omega^2 - \theta_o{}^2) \tag{b}$$

Equating the sum of the left-hand terms of these equations to the sum of the right-hand terms, we obtain the following work-energy equation for plane motion:

$$\text{RW} = \frac{W}{2g}(\overline{v}^2 - \overline{v}_o{}^2) + \frac{1}{2}\overline{I}(\omega^2 - \omega_o{}^2) \tag{14-8.1}$$

This arithmetical addition is possible because both work and kinetic energy are scalar quantities expressed in the same units. The resultant work (RW) of constant or variable forces and moments is computed as explained in previous sections. The kinetic energy is expressed in terms of the velocity $\overline{v}$ of the gravity center and $\overline{I}$ which is the centroidal mass moment of inertia.

If the instant center of plane motion can be located easily, it is usually more convenient to express the kinetic energy in terms of an instantaneous rotation about the instant center. Thus expressing the kinetic energy at any instant as

$$\text{KE} = \frac{1}{2}\frac{W}{g}\overline{v}^2 + \frac{1}{2}\overline{I}\omega^2 \tag{c}$$

we may replace $\overline{v}$ by $\overline{r}\omega$ where $\overline{r}$ is the distance from the instant center to the center of gravity. Doing this gives

$$\text{KE} = \frac{1}{2}\frac{W}{g}\overline{r}^2\omega^2 + \frac{1}{2}\overline{I}\omega^2 = \frac{1}{2}\omega^2\left(\overline{I} + \frac{W}{g}\overline{r}^2\right)$$

which reduces to

$$\text{KE} = \tfrac{1}{2}\, I_C\omega^2 \tag{d}$$

since, by the transfer formula $\overline{I} + \left(\frac{W}{g}\right)\overline{r}^2 = I_C$ which is the mass moment of inertia about the instant center C.

Except for homogeneous free-rolling wheels, the position of the instant center relative to the gravity center of the body usually changes, so that the change in kinetic energy must be expressed as

$$\Delta\text{KE} = \tfrac{1}{2}I_C\omega^2 - \tfrac{1}{2}I_{C_o}\omega_o{}^2 \tag{14-8.2}$$

in which I_C and I_{C_o} represent respectively the moments of inertia about the final and initial positions of the instant center C.

This last expression suggests a special but convenient form of the work-energy equation for *homogeneous* free-rolling wheels in which I_C is constant. Considering the plane motion as analogous to rotation about the instant center, we may write

$$\Sigma M_C(\theta) = \tfrac{1}{2} I_C(\omega^2 - \omega_o^{\ 2}) \qquad (14\text{-}8.3)$$

in which ΣM_C is the moment about C of the constant external forces acting during an angular displacement θ.[7] This moment summation about C automatically eliminates the static frictional force holding the instant center momentarily at rest and indicates that such friction forces (as the term "static" indicates) do no work on free-rolling wheels. Notice that the static friction force F at the instant center C may be replaced by an equal force at the mass center G plus a couple. See Fig. 14-8.1. If we do this, the sum of the translational work done by F at the mass center will cancel the rotational work of its couple, i.e., $F\bar{s} + (-F\bar{r})\theta = 0$, thereby verifying that the static friction does no work on free-rolling wheels. *If the wheel slips, Eq. (14-8.3) will not be valid;* we shall then have to use Eq. (14-8.1) and also consider the work done by kinetic friction forces.

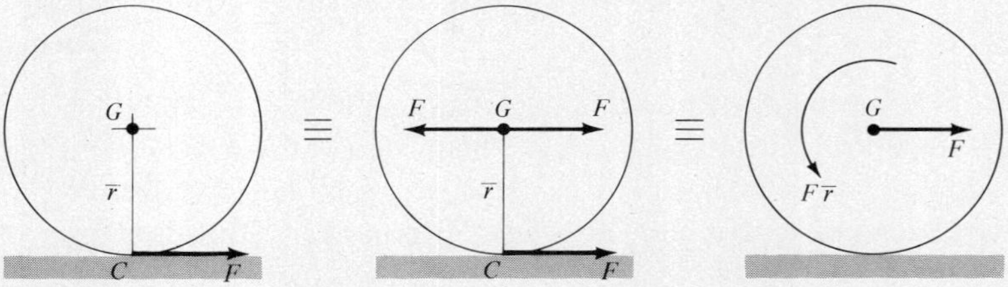

Figure 14-8.1

ILLUSTRATIVE PROBLEMS

14-8.1. The system in Fig. 14-8.2 has the properties shown. Assuming disk D to roll without slipping, determine the linear velocity of its center after body B has moved 12 ft starting from rest. Also determine the static friction force acting on disk D.

Solution

We first determine the mass moments of inertia.

$$[\bar{I} = m\bar{k}^2] \qquad \text{For } P\text{: } \bar{I}_P = \frac{161}{32.2}(1.414)^2 = 10 \text{ ft-lb-sec}^2$$

$$\text{For } D\text{: } \bar{I}_D = \frac{322}{32.2}(2)^2 = 40 \text{ ft-lb-sec}^2$$

[7] Equation (14-8.3) can also be obtained by eliminating α from between $\Sigma M_C = I_C\alpha$ and $\alpha\, d\theta = \omega\, d\omega$ and then integrating according to the procedure outlined in Section 14-7. Naturally, the ensuing result is valid only for homogeneous free-rolling wheels to which $\Sigma M_C = I_C\,\alpha$ may be applied.

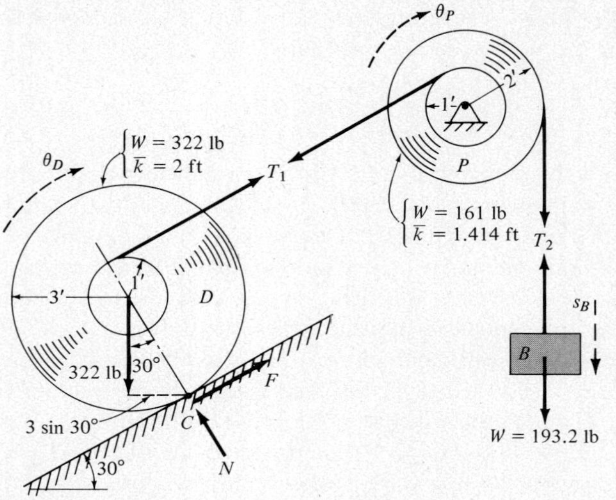

Figure 14-8.2

$$[I_C = \bar{I} + md^2] \quad \text{For } D\text{: } I_C = 40 + \left(\frac{322}{32.2}\right)(3)^2 = 130 \text{ ft-lb-sec}^2$$

The directions of motion may be found by the method described in Illus. Prob. 13-4.2 on p. 518. These directions are shown by the dashed arrows, adjacent to each body, which specify the positive directions of forces and moments. Alternatively, the directions of motion may be assumed and then confirmed by observing that, for a system starting from rest, the resultant work based on these assumptions must be positive to speed up the system.

The kinematic relations among the bodies are

$$[s = r\theta] \qquad s_{T_1} = 4\theta_D = 1(\theta_P) \quad \text{also} \quad s_{T_2} = s_B = 2\theta_P$$

which are combined into the following relation:

$$4\theta_D = \theta_P = \tfrac{1}{2}s_B \tag{a}$$

whence, by differentiation, the velocities are similarly related by

$$4\omega_D = \omega_P = \tfrac{1}{2}v_B \tag{b}$$

We now apply the work-energy equation to the system, considering only the work of active external forces. There are several alternate methods of computing the work done by the weight W of disk D. Since the center of D moves the distance $\bar{s}_D = 3\theta_D$, the work done by W is either $-W \sin 30°(\bar{s}_D)$ or $-Wh = -W(\bar{s}_D \sin 30°)$, both reducing to $-W(3 \sin 30°)\theta_D$. But probably the best method is to apply $\Sigma M_C(\theta_D)$ to obtain $-W(3 \sin 30°)\theta_D$, giving the same result directly. The value of θ_D, using Eq. (*a*) and the given value of $s_B = 12$ ft, is $\theta_D = \frac{12}{8} = 1.5$ rad. The resultant work done on the system is the sum of the left-hand terms in Eqs. (14-2.2), (14-7.2), and (14-8.3) used respectively for the translational, rotational, and plane motion parts of the system. The total change in kinetic energy is the sum of the right-hand terms of these equations. Note that although no forces external to the system do work upon the pulley, it nevertheless gains kinetic energy. Thus we obtain

$$[\Sigma X(s)]_B + [\Sigma \bar{M}(\theta)]_P + [\Sigma M_C(\theta)]_D$$
$$= \left[\frac{W}{2g}(v^2 - v_o^2)\right]_B + \left[\frac{1}{2}\bar{I}(\omega^2 - \omega_o^2)\right]_P + \left[\frac{1}{2}I_C(\omega^2 - \omega_o^2)\right]_D$$

$$193.2(12) - (322)(3 \sin 30°)(1.5)$$
$$= \frac{193.2}{64.4} v_B^2 + \frac{1}{2}(10)\omega_P^2 + \frac{1}{2}(130)\omega_D^2 \qquad (c)$$

Substituting $v_B = 8\omega_D$ and $\omega_P = 4\omega_D$ from Eq. (*b*) in Eq. (*c*) gives

$$2320 - 724 = 3(64\omega_D^2) + 5(16\omega_D^2) + 65\omega_D^2 = 337\omega_D^2$$

whence

$$\omega_D^2 = 4.73 \quad \text{and} \quad \omega_D = 2.18 \text{ rad/sec}$$

The velocity of the center of disk D therefore is

$[\bar{v} = \bar{r}\omega]$ $\qquad \bar{v}_D = 3(2.18) = 6.54$ fps **Ans.**

The friction force acting on D (and also the tension T_1) is found by applying the following work-energy relations to the FBD of the disk:

$$[\circlearrowleft + \Sigma M_C(\theta) = \tfrac{1}{2}I_C(\omega^2 - \omega_o^2)]$$
$$[4T_1 - 322(3 \sin 30°)](1.5) = \tfrac{1}{2}(130)(4.73) \qquad (d)$$

$$[\circlearrowleft + \Sigma \bar{M}(\theta) = \tfrac{1}{2}\bar{I}(\omega^2 - \omega_o^2)] \qquad [1(T_1) - 3F](1.5) = \tfrac{1}{2}(40)(4.73) \qquad (e)$$

which are solved to yield

$T_1 = 172$ lb and $F = 36.3$ lb **Ans.**

In place of Eq. (*e*) representing the rotational component of the work-energy equation for the disk, we could have used the translational component in terms of the motion of its mass center, i.e., $\Sigma X(\bar{s}) = \frac{W}{2g}(\bar{v}^2 - \bar{v}_o^2)$. Try this to see which is easier.

14-8.2. The circular disk shown in Fig. 14-8.3 weighs 200 lb after the semicircular hole is cut out of it. Its radius of gyration about the gravity center G is 1.2 ft. Determine the angular velocity of the

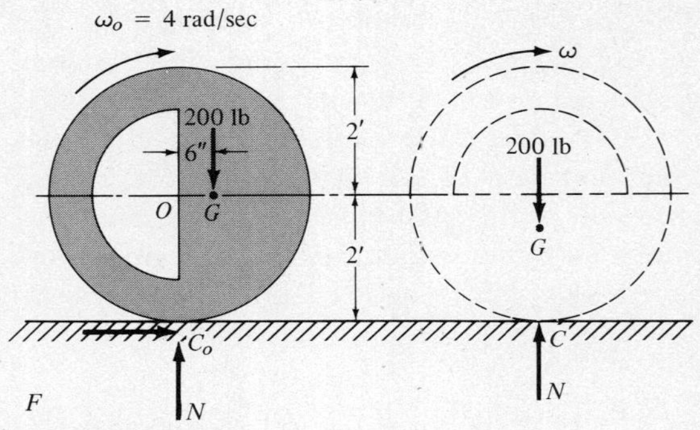

Figure 14-8.3

disk after it has rolled without slipping for 450° or $1\frac{1}{4}$ turns clockwise, starting with an initial clockwise angular velocity of 4 rad/sec. What is then the normal reaction N?

Solution

During the movement from the initial to the dashed final position, the work done is caused only by the center of gravity dropping a net vertical distance of 6 in. = 0.5 ft. Since free-rolling is assumed, the static friction F does no work. Before equating the resultant work to the change in kinetic energy given by Eq. (14-8.2), let us compute the values of I_{C_o} and I_C. Using the transfer formula, we obtain

$[I = \bar{I} + md^2]$

$$I_{C_o} = \frac{200}{g}(1.2)^2 + \frac{200}{g}(\overline{0.5}^{\,2} + 2^2) = \frac{1113}{g}\text{ ft-lb-sec}^2$$

$$I_C = \frac{200}{g}(1.2)^2 + \frac{200}{g}(1.5)^2 = \frac{738}{g}\text{ ft-lb-sec}^2$$

The work-energy equation now gives

$$[RW = \tfrac{1}{2}I_C\omega^2 - \tfrac{1}{2}I_{C_o}\omega_o{}^2] \quad 200(0.5) = \frac{1}{2}\left(\frac{738}{g}\right)\omega^2 - \frac{1}{2}\left(\frac{1113}{g}\right)(4)^2$$

from which we get

$$\omega^2 = 32.9 \quad \text{and} \quad \omega = 5.74 \text{ rad/sec} \qquad \textbf{\textit{Ans.}}$$

The value of N at the final position will be found by using $R = (W/g)\bar{a}$. At this instant $\Sigma\bar{M} = \bar{I}\alpha$ gives $\alpha = 0$ and hence the acceleration of the geometric center O is $a_o = r\alpha = 0$. Then, if we use O as a reference point, the acceleration of the gravity center is $\bar{a} = a_G = a_o +\!\!\!\!\rightarrow (a_{G/o} = r\omega^2 +\!\!\!\!\rightarrow r\alpha)$ which reduces to $\bar{a} = r\omega^2 = 0.5(32.9)$ fps²↑. Hence, we obtain

$$\left[+\uparrow R = \frac{W}{g}\bar{a}\right] \qquad N - 200 = \frac{200}{g}(0.5)(32.9) \quad N = 302 \text{ lb} \qquad \textbf{\textit{Ans.}}$$

14-8.3. In Fig. 14-8.4, the slider A, weighing 200 lb, moves in the vertical frictionless guide. It is connected to the center of a solid cylinder B, 4 ft in diameter and weighing 300 lb, by a uniform slender rod AB which is 10 ft long and weighs 180 lb. If the system starts with negligible velocity when B is vertically below A, and B rolls freely, compute the velocity of A in the given position.

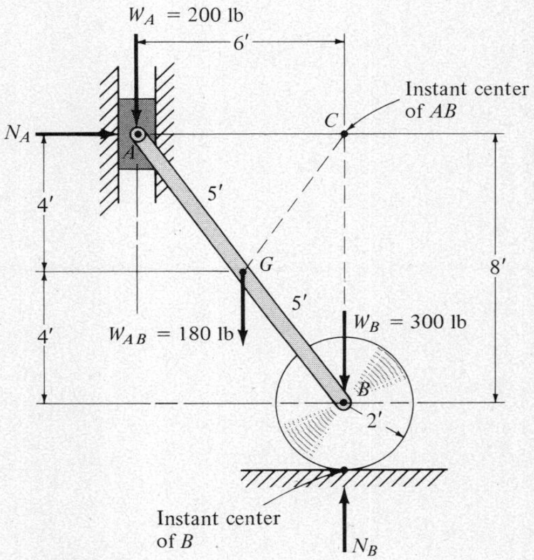

Figure 14-8.4

Solution

The reactive forces N_A and N_B, which do no work, may be omitted from the active-force diagram. The resultant work done on the system is due only to W_A and W_{AB}, dropping respectively through $10 - 8 = 2$ ft and $5 - 4 = 1$ ft.

If we use the instant center C of AB, the kinematic relations between the bodies at the instant shown are

$$[v = r\omega] \qquad v_A = 6\omega_{AB} \quad \text{and} \quad \bar{v}_B = 8\omega_{AB} = 2\omega_B \tag{a}$$

Since the system starts with negligible velocity, the change in kinetic energy is the sum of $\dfrac{W}{2g}v^2$ for the translating slider A and $\Sigma\frac{1}{2}I_C\omega^2$ for the plane motions of bar AB and cylinder B. About their respective instant centers, their moments of inertia are

$[I = \bar{I} + md^2]$

$$\text{For } AB\text{: } I_{C_{AB}} = \frac{1}{12}\left(\frac{180}{g}\right)(10)^2 + \frac{180}{g}(5)^2 = \frac{6000}{g} \text{ ft-lb-sec}^2$$

$$\text{For } B\text{: } I_{C_B} = \frac{1}{2}\left(\frac{300}{g}\right)(2)^2 + \frac{300}{g}(2)^2 = \frac{1800}{g} \text{ ft-lb-sec}^2$$

Applying the work-energy equation, and using the kinematic relations (a) to express v_A and ω_B in terms of ω_{AB}, we have

$[RW = \Delta KE]$

$$200(2) + 180(1) = \frac{200}{2g}(6\omega_{AB})^2 + \frac{1}{2}\left(\frac{6000}{g}\right)\omega_{AB}^2 + \frac{1}{2}\left(\frac{1800}{g}\right)(4\omega_{AB})^2$$

whence

$$580 = \frac{\omega_{AB}^2}{2g}[7200 + 6000 + 28{,}800] = \frac{\omega_{AB}^2}{2g}(42{,}000)$$

$$\omega_{AB}^2 = 0.889 \quad \text{and} \quad \omega_{AB} = 0.943 \text{ rad/sec}$$

Hence, the required velocity of slider A is

$[v_A = 6\omega_{AB}]$ $\qquad v_A = 6(0.943) = 5.67$ fps **Ans.**

PROBLEMS

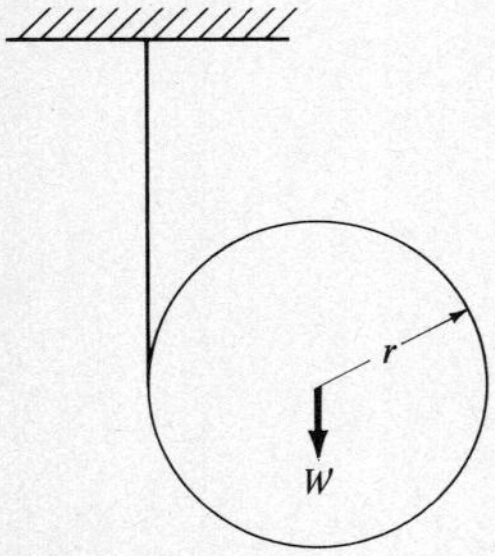

Figure P-14-8.4

14-8.4. As shown in Fig. P-14-8.4, a cord from a fixed support is wrapped around a solid homogeneous cylinder of radius r and weight W. Find the velocity of the mass center of the cylinder after it has dropped h ft, starting from rest.

$$\bar{v} = 2\sqrt{gh/3}$$ **Ans.**

14-8.5. The disk shown in Fig. P-14-8.5 rolls freely on a horizontal track. Compute the angular velocity of the disk after its center has moved 12 ft from rest. Neglect the backward inclination of the block caused by its inertia.

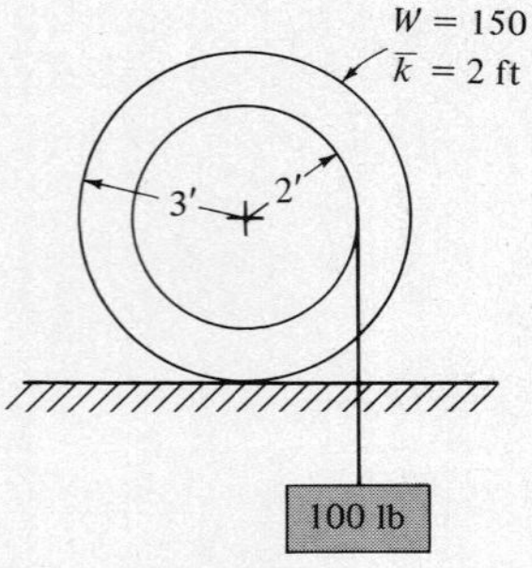

Figure P-14-8.5

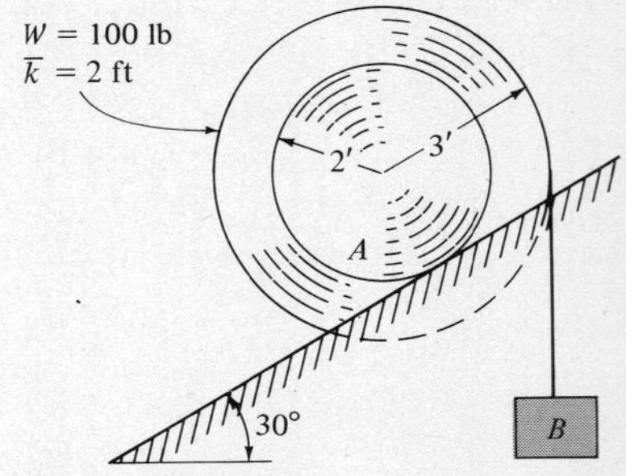

Figure P-14-8.6

14-8.6. Disk A is composed of a 3-ft radius cylinder sandwiched between two cylinders each of 2-ft radius. It rolls without slipping between two parallel rails as shown in Fig. P-14-8.6. Determine the angular velocity of disk A after its center has moved 12 ft, starting from rest. Assume body B weighs 150 lb and neglect the effect of inertia on its inclination with the vertical.

$$\omega = 6.46 \text{ rad/sec}$$ **Ans.**

14-8.7. Assume the body A in Fig. P-14-8.7 moves 12 ft, starting from rest. After determining the velocity of A corresponding to this displacement, compute the tension in the cord joining disk D and pulley P by applying the work-energy method to a FBD of disk D.

$v_A = 15.6$ fps; $T = 58.2$ lb **Ans.**

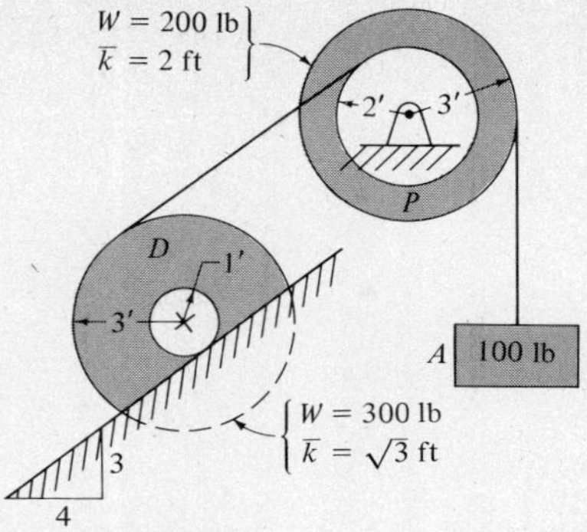

Figure P-14-8.7

14-8.8. The compound drum A and the floating pulley B shown in Fig. P-14-8.8 roll without slipping. Compute the velocity of the center of A after it moves 12 ft from rest.

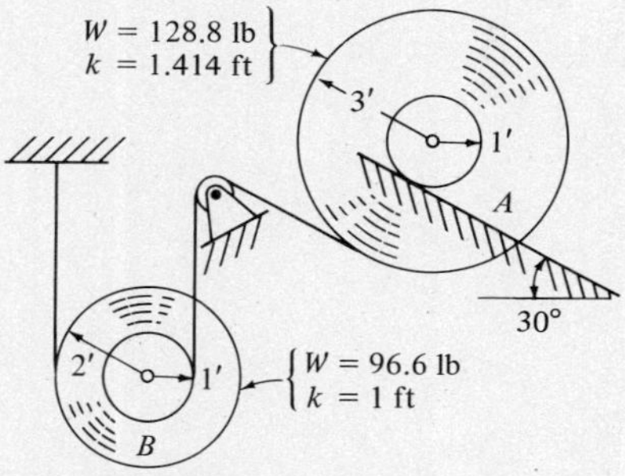

Figure P-14-8.8

14-8.9. The floating disk D in Fig. P-14-8.9 rolls without slipping on the cord passing from the fixed support and around it to pulley P. Compute the velocity of body A after A has moved 12 ft from rest. What is the tension in the cord connecting pulley P with disk D?

$v_A = 15.6$ fps; $T = 144$ lb **Ans.**

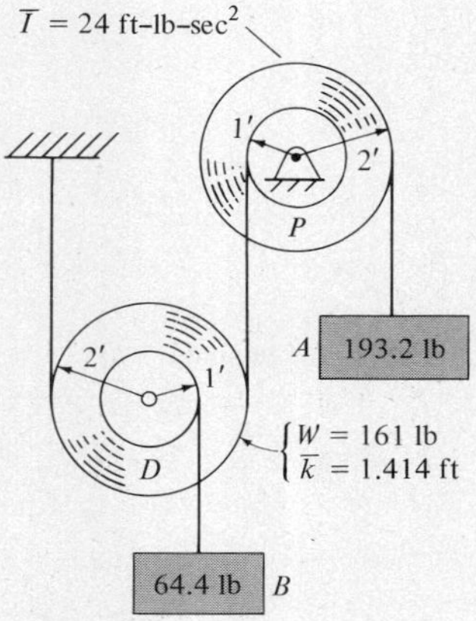

Figure P-14-8.9

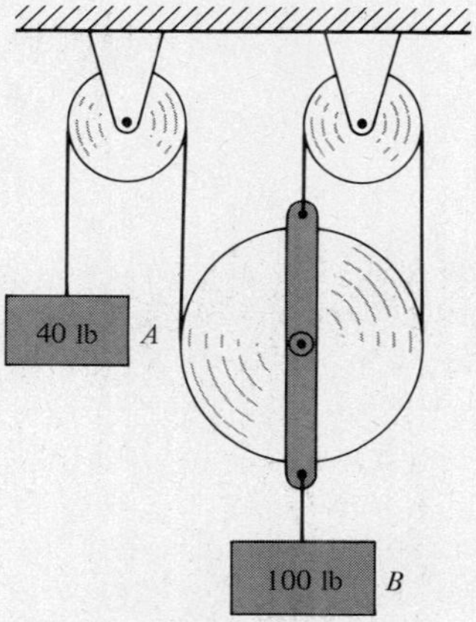

Figure P-14-8.10

14-8.10. The floating pulley in Fig. P-14-8.10 is equivalent to a 1-ft diameter solid cylinder weighing 80 lb. The other pulleys have negligible weight. Determine the velocity of body B after it has moved 10 ft from rest.

$v_B = 7.43$ fps **Ans.**

14-8.11. Disk B rolls without slipping on the incline shown in Fig. P-14-8.11. Compute the velocity of its center after the rod attached to cylinder A has rotated from rest at its lowest position to the horizontal position shown.

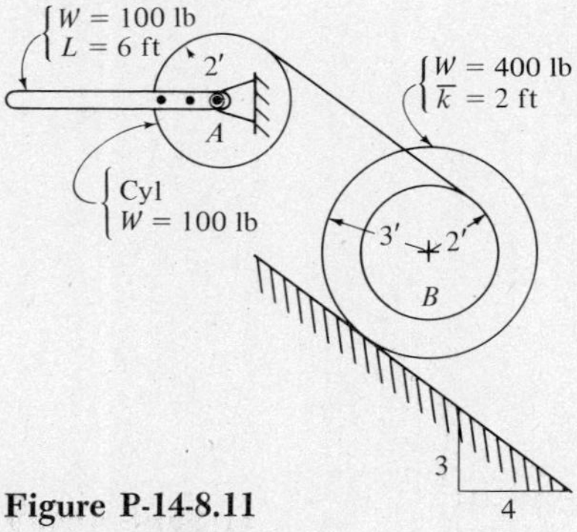

Figure P-14-8.11

14-8.12. At the position shown in Fig. P-14-8.12, the 100-lb unbalanced wheel has an angular velocity of $\sqrt{g/2}$ rad/sec. About its gravity center G, the radius of gyration is 1 ft. Determine the angular velocity of the wheel after it has rolled freely down the incline for $\frac{1}{2}$ turn.

$\omega = 14.87$ rad/sec **Ans.**

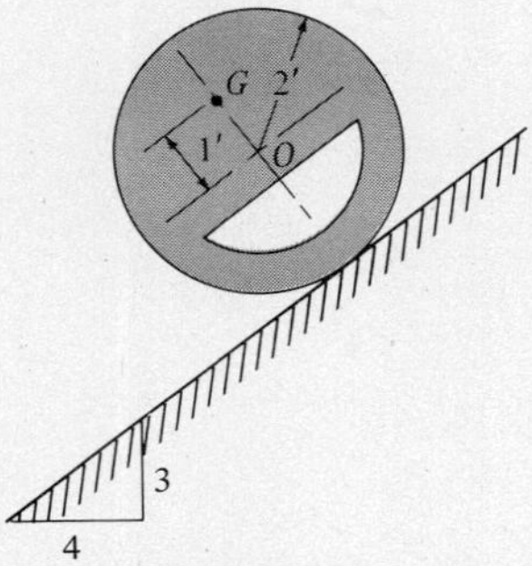

Figure P-14-8.12

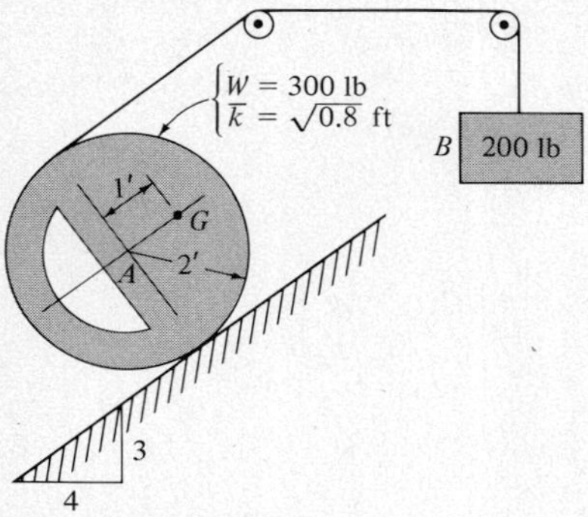

Figure P-14-8.13

14-8.13. For the system shown in Fig. P-14-8.13, find the angular velocity of the unbalanced wheel after it has rolled freely for $\frac{1}{4}$ turn, starting from rest at the given position.

$$\omega = 3.97 \text{ rad/sec} \qquad \textbf{\textit{Ans.}}$$

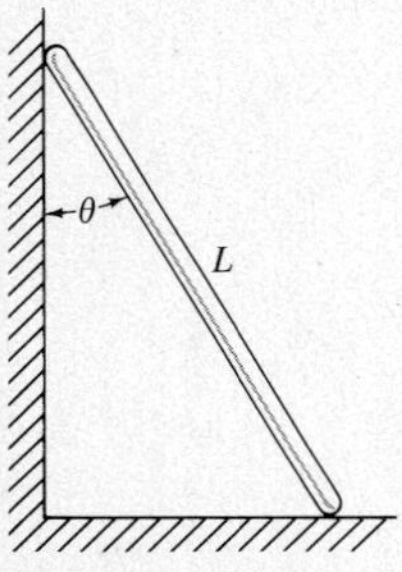

Figure P-14-8.14

14-8.14. A homogeneous slender rod L ft long and weighing W lb slides down a smooth vertical wall and along a smooth horizontal floor as shown in Fig. P-14-8.14. Assume that the ends of the rod remain in contact with the wall and the floor. If the rod is at rest when $\theta = 0°$, determine the value of the instantaneous angular velocity ω in terms of the angular displacement θ. Then by differentiating ω, obtain the value of α in terms of θ. What is the effect upon α if the lower end of the rod is not at rest when $\theta = 0°$?

$$\omega = \sqrt{\frac{3g}{L}(1 - \cos\theta)}; \; \alpha = \frac{3g}{2L}\sin\theta \qquad \textbf{\textit{Ans.}}$$

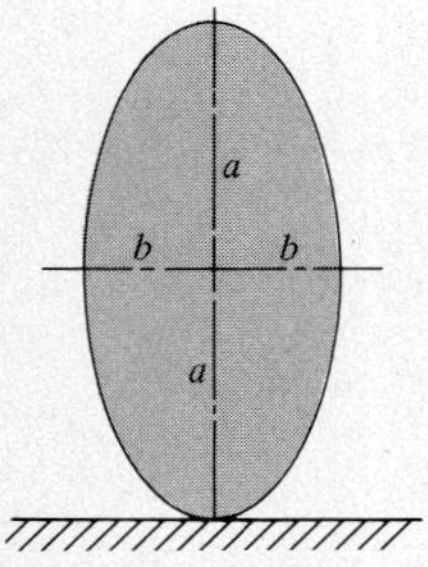

Figure P-14-8.15

14-8.15. An elliptical plate has semiaxes a and b. The plate is released from rest with the major axis vertical as in Fig. P-14-8.15, and rolls freely on the horizontal surface shown. Determine the angular velocity at the instant the major axis is horizontal.

$$\omega = \sqrt{\frac{8g(a - b)}{a^2 + 5b^2}} \qquad \textbf{\textit{Ans.}}$$

14-8.16. The system shown in Fig. P-14-8.16 consists of a 60-lb bar 60 in. long smoothly pinned to two identical disks, each weighing 40 lb. If the system starts with negligible velocity at the given position, compute the normal reaction under each disk when the bar is at its lowest position.

14-8.17. A solid cylinder weighing 100 lb starts from rest at the position shown in Fig. P-14-8.17 and rolls without slipping down the incline and on to the circular path. Determine the vertical reaction upon the cylinder at its lowest position.

$$N = 184.5 \text{ lb} \qquad \textbf{\textit{Ans.}}$$

14-8.18. Bar AB, 10 ft long and weighing 200 lb, is fastened at A to a roller of negligible weight moving in a vertical slot as shown in Fig. P-14-8.18. The other end B is smoothly pinned to a cylinder weighing 300 lb and of 2-ft radius. The spring has a free length of 2 ft and a modulus of

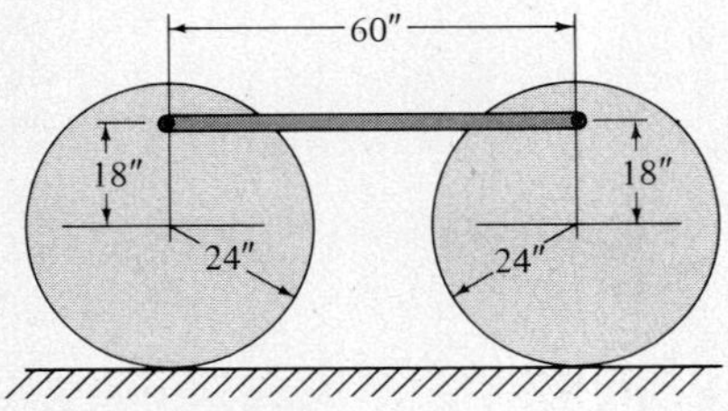

Figure P-14-8.16

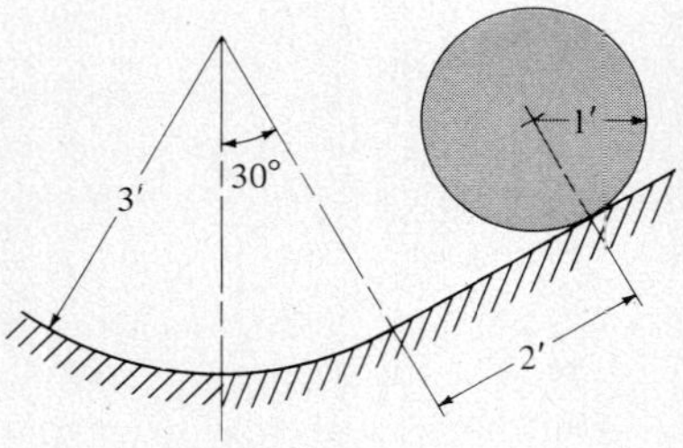

Figure P-14-8.17

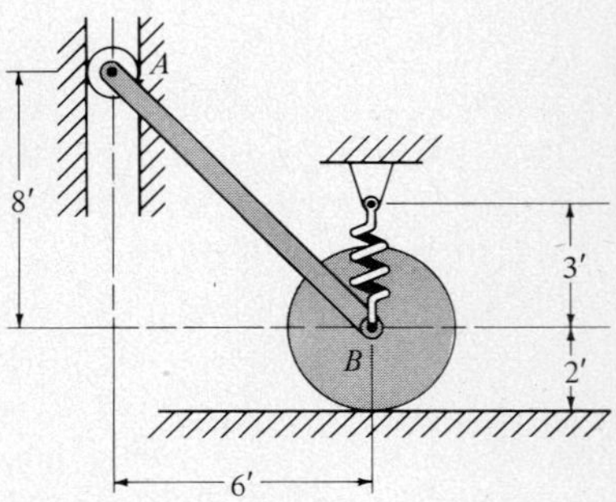

Figure P-14-8.18

60 lb/ft. If the system starts from rest when B is vertically below A, compute the linear velocity of point B when in the given position.

$$v_B = 9.86 \text{ fps} \qquad \textbf{Ans.}$$

14-8.19. At the position shown in Fig. P-14-8.19, the given system is released from rest on the smooth surfaces. The uniform bar AB, weighing 100 lb, is smoothly pinned to the mass centers of bodies A and B which weigh respectively 50 lb and 150 lb. Determine the velocity of body B at the instant when bar AB is horizontal.

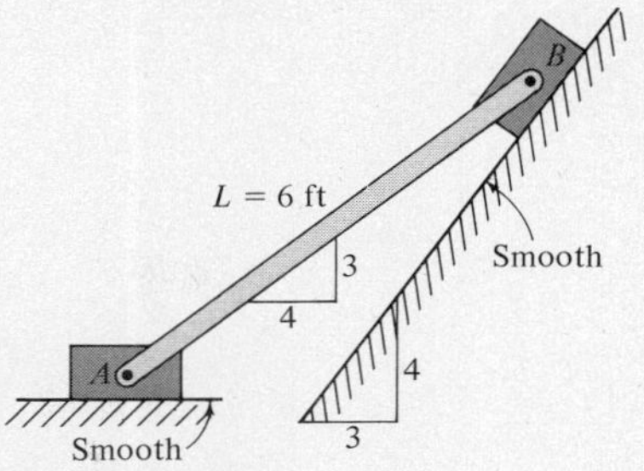

Figure P-14-8.19

14-8.20. The pitch circles of two gears, A and B, connected by the link OC are shown in Fig. P-14-8.20. If gear A is held stationary and the system is released from rest at the given position, find the angular velocity of gear B when it is at its lowest position. Assume gear B is equivalent to a solid cylinder weighing 30 lb and that link OC weighs 20 lb.

Figure P-14-8.20

14-8.21. A uniform rod 3 ft long and weighing 15 lb is welded to the rim of a 6-ft diameter hoop weighing 10 lb. At the position shown in Fig. P-14-8.21, the hoop has a clockwise angular velocity of 4 rad/sec. Compute the angular velocity of the hoop after it has rolled without slipping through a half-turn.

$$\omega = 6.94 \text{ rad/sec} \qquad \textbf{Ans.}$$

14-8.22. Determine the angular velocity of the hoop in the preceding problem after the hoop has rolled freely through three-quarters of a turn.

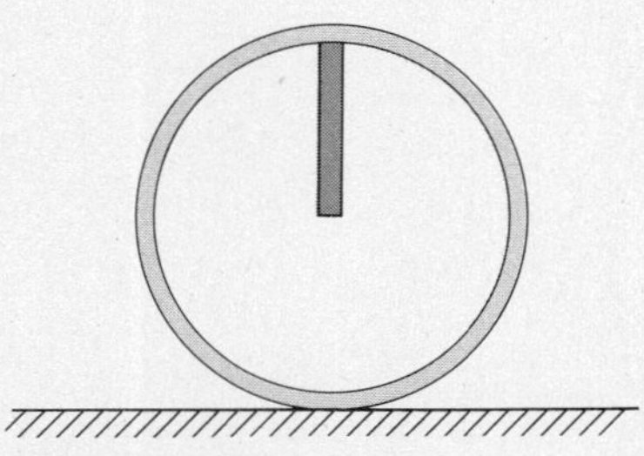

Figure P-14-8.21

14-8.23. The system shown in Fig. P-14-8.23 consists of a crank OA joined to a solid cylinder B by a connecting rod AB. The crank and connecting rod may be considered to be uniform slender rods, and the cylinder rolls without slipping. When crank OA is vertical, its angular velocity is 2 rad/sec clockwise. Determine the angular velocity of the crank when it is horizontal.

$$\omega_{OA} = 7.69 \text{ rad/sec} \qquad \textbf{Ans.}$$

14-8.24. The solid cylinder A in Fig. P-14-8.24 weighs 322 lb and is 3 ft in diameter. Assume the connecting cord to be attached to the cylinder at its center. Determine the linear velocity of the center of the cylinder after it has rolled without slipping for 9 ft, starting from rest at the given position.

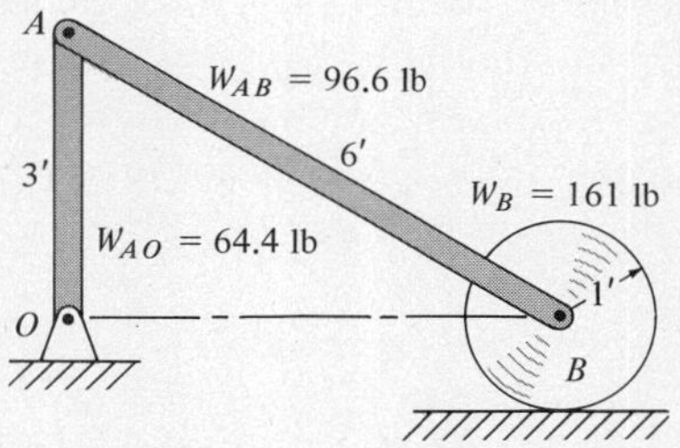

Figure P-14-8.23

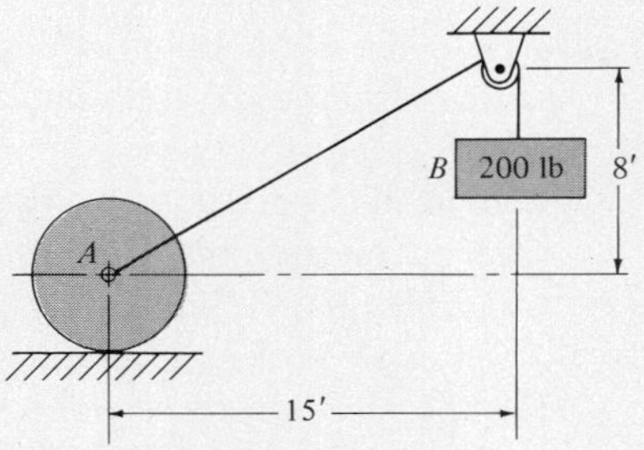

Figure P-14-8.24

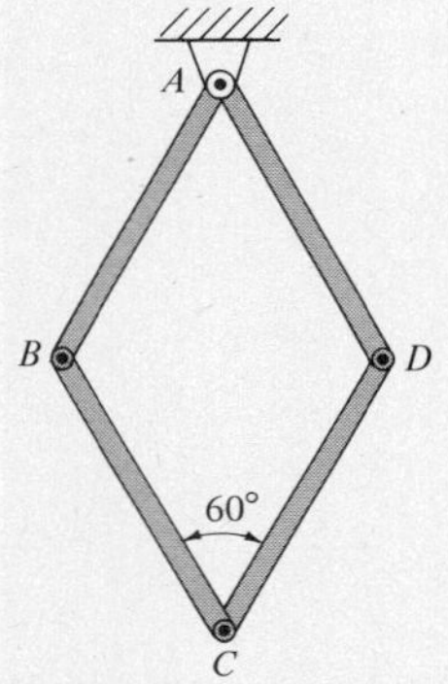

Figure P-14-8.25

14-8.25. The linkage shown in Fig. P-14-8.25 consists of four identical bars, each of weight W and length L, connected by smooth pins. The linkage is released from rest when D coincides with A. Compute the velocity of D at the given position.

$$v_D = 2\sqrt{\frac{6gL\cos 30^\circ}{7}}$$ ***Ans.***

SUMMARY

By eliminating the acceleration from the equations $\Sigma X = \frac{W}{g} a_t$ and $a_t\, ds = v\, dv$, we obtain the relation

$$\int_0^s \Sigma X(ds) = \frac{W}{2g}(v^2 - v_0^2) \qquad (14\text{-}2.2)$$

which is the fundamental work-energy equation for a particle or a translating rigid body. The X axis is assumed to be tangent to the path of the particle during each distance ds measured along the path. Hence ΣX represents the component of the resultant force along the path at each instant.

The expression $\int_0^s \Sigma X(ds)$ is defined as resultant work [or as $\Sigma X(s)$ if the forces are constant]. Work done by a force may be positive or negative according to whether the force component is directed respectively along or opposite to the direction of displacement. The expression $\frac{W}{2g} v^2$ is defined as kinetic energy. The usual unit for work and energy is the foot-pound.

In Section 14-3 it is shown that the geometric significance of work is the product of the component of force in the direction of displacement multiplied by the displacement, or the product of displacement in the direction of the force multiplied by the force. Forces perpendicular to the displacement do no work.

In the application of the work-energy method, the work done during all phases of a motion is equated to the total change in kinetic energy. Only the work done by *active* external forces need be considered; the work done by internal forces (whether constant or variable) in inextensible links or cords connecting a system of bodies is automatically eliminated.

Power is defined mathematically by the expression Fv where F is the resultant force doing work at any instant and v is the corresponding value of the instantaneous velocity. The unit of power (foot-pounds per second) is too small for use in engineering; the more common unit is the horsepower (hp) defined as equivalent to 550 ft-lb per sec or 33,000 ft-lb per min.

The work-energy equation for rotation and the methods employed are analogous to those applied to translation. The fundamental equation is obtained by eliminating the angular acceleration from $\Sigma M = I\alpha$ and $\alpha\, d\theta = \omega\, d\omega$; this gives

$$\int_o^\theta \Sigma M(d\theta) = \frac{1}{2} I(\omega^2 - \omega_o^2) \tag{14-7.2}$$

The expression $\int_o^\theta \Sigma M(d\theta)$ [or $\Sigma M(\theta)$ if the moments are constant] is defined as resultant work in rotation. Moments in the sense of initial angular motion do positive work, and vice versa. The expression $\frac{1}{2}I\omega^2$ is defined as kinetic energy in rotation. The unit for both expressions is usually the foot-pound.

As in the case of translation, the work-energy method may be applied directly to a system of connected bodies, some of which translate and others rotate, without considering any of the internal forces in the connecting elements. The resultant work done by *active* external forces on the system is equal to the sum of changes in kinetic energy of each body of the system, or

$$(\mathrm{RW})_{\mathrm{sys}} = \sum \frac{W}{2g}(v^2 - v_o^2) + \sum \frac{1}{2} I(\omega^2 - \omega_o^2) \tag{14-7.3}$$

The work-energy equation applied to plane motion is the sum of the work-energy expressions for translation of the mass center plus that of centroidal rotation, namely,

$$\mathrm{RW} = \frac{W}{2g}(\bar{v}^2 - \bar{v}_o^2) + \frac{1}{2}\bar{I}(\omega^2 - \omega_o^2) \tag{14-8.1}$$

If the instant center of zero velocity can easily be located, it is usually more convenient to express the change in kinetic energy as

$$\Delta \mathrm{KE} = \tfrac{1}{2} I_C \omega^2 - \tfrac{1}{2} I_{C_o} \omega_o^2 \tag{14-8.2}$$

in which I_C and I_{C_o} are the moments of inertia about the final and initial positions of the instant center C.

For homogeneous free-rolling bodies, the most convenient form of the work-energy equation is

$$\Sigma M_C(\theta) = \tfrac{1}{2} I_C(\omega^2 - \omega_o^2) \tag{14-8.3}$$

in which ΣM_C is the moment about the instant center C of the constant external forces acting during an angular displacement θ.

15

IMPULSE AND MOMENTUM

15-1 INTRODUCTION

We have seen that the work-energy method was based on integrating the kinetic equations of motion with respect to the displacement. The resulting displacement integrals were scalar quantities which were particularly useful in problems relating force, velocity, and displacement. In this chapter we present another approach to dynamics, known as the impulse-momentum method, which directly relates force, velocity, and time. It is based on integrating the kinetic equations of motion with respect to the time.

The impulse-momentum method is particularly convenient in situations when forces act for very small time intervals during which the forces may vary, as in an impact or sudden blow. The method is also useful for solving problems in which a system gains or loses mass. Still other cases will involve a combination of work-energy and impulse-momentum methods, as in satellite motion.

We start with the application of the impulse-momentum method to particle motion or the equivalent rigid-body motion of translation, and then extend its application in subsequent sections to fixed-axis rotation and general plane motion.

15-2 LINEAR IMPULSE-MOMENTUM

As was shown in Section 10-5, the motion of the mass center of any body, rigid or nonrigid, is governed by $\mathbf{R} = m\bar{\mathbf{a}}$. On replacing $\bar{\mathbf{a}}$ by its equivalent $d\bar{\mathbf{v}}/dt$, we obtain

$$\mathbf{R}\,dt = m\,d\bar{\mathbf{v}} \qquad (a)$$

which is the differential form of the linear impulse-momentum equation for all systems which neither gain nor lose mass. If we assume that the velocity has the value $\bar{\mathbf{v}}_o$ when t is zero and the value $\bar{\mathbf{v}}$ at any other value of t, then integration of Eq. (a) between these limits gives the following general form of the linear impulse-momentum equation:

$$\int_o^t \mathbf{R}\,dt = m\bar{\mathbf{v}} - m\bar{\mathbf{v}}_o \qquad (15\text{-}2.1)$$

Note that this equation refers to the motion of the mass center of any system of constant mass. When it is applied to the motion of a single particle or a translating rigid body, there is no need to retain the bar sign over the velocity terms.

RESULTANT LINEAR IMPULSE

The expression $\int_o^t \mathbf{R}\,dt$ is known as resultant linear impulse. Examining it dimensionally shows the unit of linear impulse to be lb-sec. Obviously, linear impulse is a vector quantity since it represents a vector force multiplied by scalar time. In analytical solutions, linear impulse is usually evaluated in terms of its rectangular components, namely,

$$\int_o^t \mathbf{R}\,dt = \hat{\mathbf{i}}\int_o^t \Sigma X\,dt + \hat{\mathbf{j}}\int_o^t \Sigma Y\,dt + \hat{\mathbf{k}}\int_o^t \Sigma Z\,dt \qquad (15\text{-}2.2)$$

Clearly, these integrals can only be evaluated if the components of $\mathbf{R}$ are constant or known functions of time, or are available experimentally from the areas under curves representing the variation of these components with time. Review now the discussion on p. 405 concerning the use of force-time curves to determine velocity and displacement.

LINEAR MOMENTUM

The quantity $m\bar{\mathbf{v}} = (W/g)\bar{\mathbf{v}}$ is known as the linear momentum of a body at any instant and may be symbolized by $\bar{\mathbf{p}}$, and the quantity $m\,d\bar{\mathbf{v}}$ may be represented by $d\bar{\mathbf{p}}$. Substituting dimensional equivalents in the expression for linear momentum gives

$$\frac{W}{g}\bar{v} = \frac{\text{lb}}{\text{ft/sec}^2}\left(\frac{\text{ft}}{\text{sec}}\right) = \frac{\text{lb(sec}^2)}{\text{ft}}\left(\frac{\text{ft}}{\text{sec}}\right) = \text{lb-sec}$$

This shows the dimensional unit of linear momentum to be identical with that of linear impulse.

As in the case of impulse, it is important to note that linear momentum consists of a vector $\bar{v}$ multiplied by a scalar coefficient; this results in a vector having the same sense and inclination as $\bar{v}$ but a different magnitude. In analytical solutions, the change in linear momentum is best expressed as the difference between the components of final and initial momentum.

Equation (15-2.1) may now be summarized in the following principle: *The component of the resultant linear impulse along any axis is equal to the change in the components of the linear impulse along that axis.* The explicit form of this statement is

$$\int_o^t \Sigma X\, dt = \frac{W}{g}(v_x - v_{o_x}), \text{ etc.} \qquad (15\text{-}2.3)$$

In applying the component form of the impulse-momentum principle, it is convenient to assume that the positive sense of each term is directed along the components of the initial direction of motion. Alternatively, we may rearrange Eq. (15-2.1) to $m\bar{v}_o + \int \mathbf{R}\, dt = m\bar{v}$ and denote it diagrammatically as in Fig. 15-2.1. Then choosing any convenient direction along which to equate the components of each term in this diagrammed equation, we revert to Eq. (15-2.3). The advantage of this approach is to emphasize the vector nature of the impulse-momentum equation and to ensure that all terms are properly included.

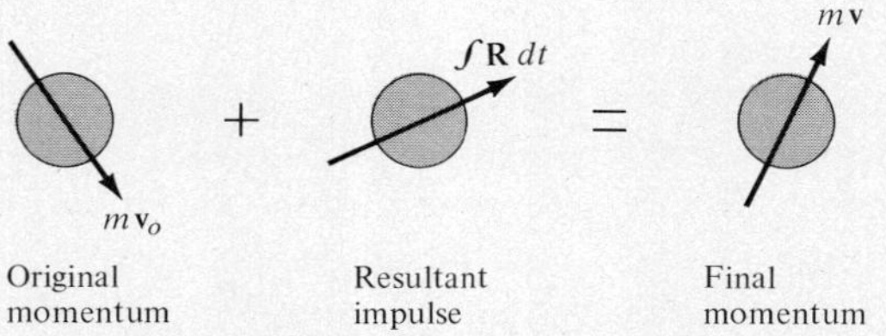

Figure 15-2.1 Diagram of impulse-momentum equation.

But there is a class of problems for which impulse momentum is uniquely suitable. These problems involve what is called impulsive motion; that is, motion resulting from the application of large forces acting for short periods of time due to a sudden impact. Such forces and their moments are known as impulsive forces and impulsive moments and are so much larger than ordinary forces such as weight that impulsive effects of the latter can be considered negligibly small in comparison. In situations involving impact, significant changes in velocity occur even though the time interval involved is too short to cause significant changes in position. Observe that to produce an impulse, a force need only exist for a time interval.

ILLUSTRATIVE PROBLEMS

15-2.1. Determine the velocity of body B in Fig. 15-2.2 after moving for 5 sec (a) starting from rest and (b) starting with a downward velocity of 6 fps.

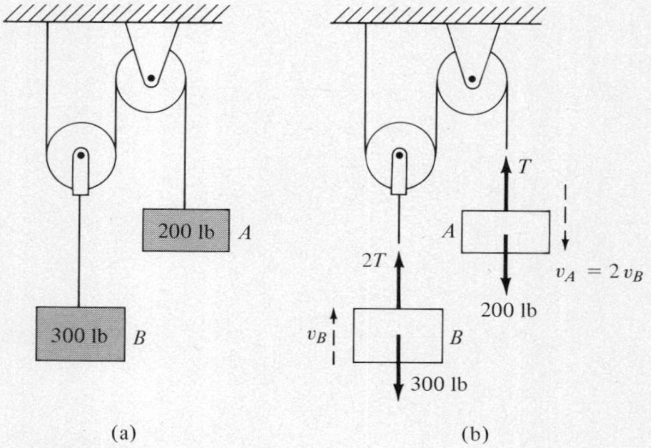

Figure 15-2.2

Solution

The FBD of each body is shown in part (b) of the figure. Noting that $v_A = 2v_B$ and applying the impulse-momentum equation (15-2.3) (with ΣX constant) to each body gives

$$\left[\Sigma X(t) = \frac{W}{g}(v - v_o)\right] \qquad (200 - T)(5) = \frac{200}{g}(2)(v_B - v_{B_o})$$

$$(2T - 300)(5) = \frac{300}{g}(v_B - v_{B_o})$$

Multiplying the first of these equations by 2 and adding them eliminates the tension T and gives

$$500 = \frac{1100}{g}(v_B - v_{B_o})$$

whence, for part (a) with $v_{B_o} = 0$, we obtain $v_B = 14.64$ fps up and for part (b) with $v_{B_o} = -6$ fps (minus because down), we obtain $v_B = 8.64$ fps up.

Note that T is not automatically eliminated by adding the I-M equations for each body as it would be in a W-E solution. It would have been just as simple in this problem to apply $\Sigma X = \left(\frac{W}{g}\right)a$ to each body to find the acceleration of B, and then substitute this in the kinematic relation $v = v_o + at$ to find the desired velocity. The advantage of the I-M method to eliminate acceleration will be more apparent in the next example where the acceleration varies during the time interval considered.

15-2.2. The system shown in Fig. 15-2.3 has a rightward velocity of 10 fps. Determine the constant value of P that will give it a leftward velocity of 20 fps in a time interval of 20 sec.

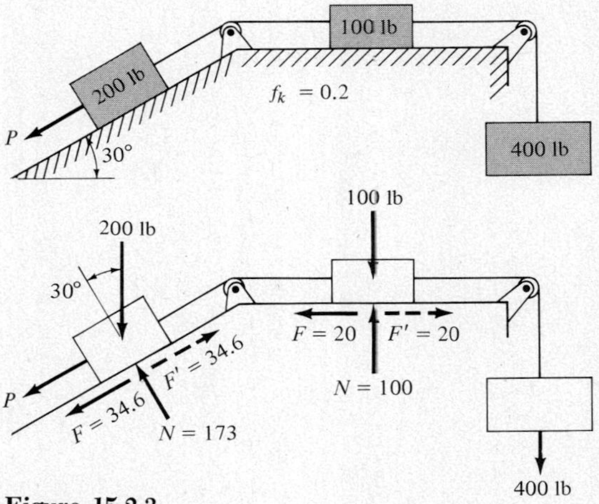

Figure 15-2.3

Solution

Since all bodies have the same displacement in the same time, the impulses of internal tensions in the connecting cords will be eliminated. Hence, using a free-body diagram of the system showing only the external forces, we apply

$$(\Sigma X)_{ext}(t) = \sum \frac{W}{g}(v - v_o)$$

Two sets of equal friction forces are shown: F (drawn solid) which acts during rightward motion and equal but reversed values F' (shown dashed) which act during leftward motion. Using the friction values F which resist the rightward motion during the time t_1 to come to rest, we have

$$(400 - 20 - 34.6 - 200 \sin 30° - P)t_1 = \frac{700}{32.2}(0 - 10)$$

which reduces to

$$(P - 245.4)t_1 = 217 \qquad (a)$$

Using the dashed friction values F' during the time t_2 to move leftward from rest to a velocity of 20 fps, we have

$$(P + 200 \sin 30^\circ - 34.6 - 20 - 400)t_2 = \frac{700}{32.2}(20 - 0)$$

which reduces to

$$(P - 354.6)t_2 = 434 \qquad (b)$$

Also, from the given data, we have

$$t_1 + t_2 = 20 \qquad (c)$$

These equations may be solved simultaneously, but it is easier and faster to use the following method of successive approximations. From Eq. (*b*), P must be larger than 354.6 lb. Let us try $P = 400$ lb as a start. Then from Eq. (*a*), we obtain

$$t_1 = \frac{217}{400 - 245} = 1.4 \text{ sec}$$

which is substituted into Eq. (*c*) to give $t_2 = 20 - 1.4 = 18.6$ sec. Using this value of t_2 in Eq. (*b*) yields

$$P = 354.6 + \frac{434}{18.6} = 377.9 \text{ lb}$$

Repeat the cycle of operations. From Eq. (*a*),

$$t_1 = \frac{217}{377.9 - 245.4} = 1.64 \text{ sec}, \qquad \therefore t_2 = 18.36 \text{ sec}$$

From Eq. (*b*),

$$P = 354.6 + \frac{434}{18.36} = 378.25 \text{ lb} \approx 378 \text{ lb} \qquad \textit{Ans.}$$

This result may be accepted as the answer since there has been no appreciable change from the previous value.

PROBLEMS

15-2.3. A 4-oz baseball is thrown with a velocity of 60 fps toward a batter. After being struck by the bat B, the ball has a velocity of 160 fps directed as shown in Fig. P-15-2.3. Find the average force exerted on the ball if the impact lasts for 0.02 sec.

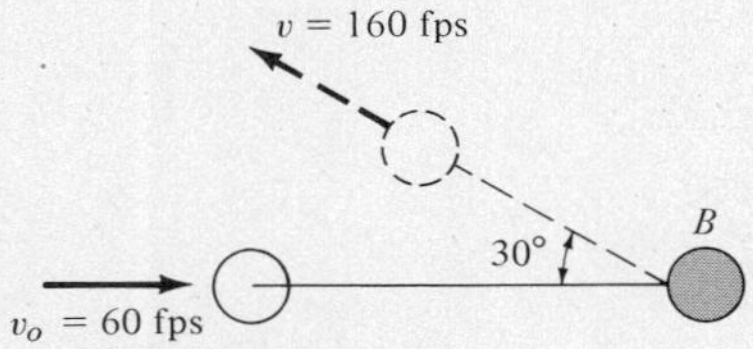

Figure P-15-2.3

15-2.4. A 40-ton railroad car moving at 2 mph along a level track strikes and is coupled to a stationary 60-ton car in 0.50 sec. Determine the average force acting on each car during the coupling.

$$F = 8750 \text{ lb} \qquad \textit{Ans.}$$

15-2.5. During an Apollo mission to the moon, the lunar excursion module (LEM) is separated from the command module by an average thrust P lasting for 1 sec. As a result, the LEM has a velocity 20 fps slower than the command module. Determine P assuming the mass of the LEM and the command module to be respectively 120 lb-sec²/ft and 300 lb-sec²/ft.

$P = 1720$ lb **Ans.**

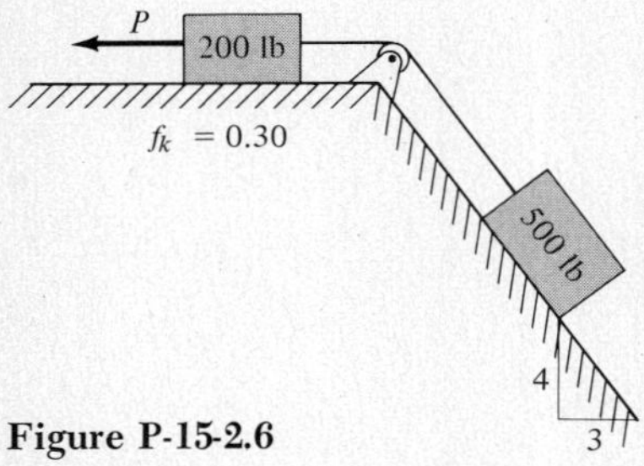

Figure P-15-2.6

15-2.6. The system shown in Fig. P-15-2.6 is moving rightward at a velocity of 15 fps when a constant horizontal force P is applied as shown. Determine the value of P that will give the system a leftward velocity of 30 fps in a time interval of 10 sec.

15-2.7. Repeat the previous problem if the velocity of the system is to be changed from a rightward velocity of 30 fps to a leftward velocity of 15 fps in a time interval of 10 sec.

$P = 590$ lb **Ans.**

15-2.8. Oil with a specific gravity of 0.9 is flowing at 10 fps in a straight pipe 2 ft in diameter and 2000 ft long. How long will it take to close a valve at the discharge end without raising the average pressure more than 40 psi?

$t = 6.07$ sec **Ans.**

15-3 DYNAMIC ACTION OF JET STREAMS

The force produced by a jet of water or other system of particles impinging upon a stationary or moving object is easily obtained by applying the concept that the component of resultant impulse is equal to the change in the corresponding components of momentum. Details of application are discussed in the illustrative problems below.

Impulse-momentum methods are also useful to solve situations in which a moving body absorbs or ejects material. These cases are often called variable mass problems[1] since the mass of the moving body does not stay constant. Consider Fig. 15-3.1 showing a thruster ejecting an elemental mass Δm in the time Δt at a relative velocity $\mathbf{u}$ from a body of mass m moving with an absolute velocity $\mathbf{v}$. The velocity of Δm changes in the time Δt from an initial velocity $\mathbf{v}$ to the absolute final value $\mathbf{v}_A$, equal to the sum of $\mathbf{v}$ and $\mathbf{u}$ as shown. The force $\mathbf{P}$ exerted on the moving body is equal and opposite to the force acting on the ejected mass. This force is easily found by equating the impulse of P to the change in momentum of the ejected mass, thereby giving

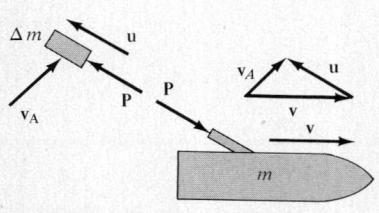

Figure 15-3.1 Jet thrust **P** on a moving body.

$$\mathbf{P}\,\Delta t = \Delta m(\mathbf{v}_A - \mathbf{v}) = \Delta m(\mathbf{u})$$

[1] We exclude problems of relativistic mechanics in which the mass is a function of velocity.

or

$$\mathbf{P} = \frac{\Delta m}{\Delta t}\mathbf{u} \tag{a}$$

If mass is ejected continuously, although not necessarily at the same rate, the instantaneous value of **P** is found by letting Δt approach zero as a limit. We thus obtain

$$\mathbf{P} = \mathbf{u}\frac{dm}{dt} \tag{15-3.1}$$

Observe that **P** and **u** are collinear. Usually, both **u** and dm/dt are constant so that **P** remains at a constant value with respect to the moving body. A similar analysis gives the same result for the retarding force exerted on a moving body absorbing material at a relative velocity **u.**

The subsequent motion of the moving body under the influence of **P** and all other external forces that may be acting is determined by applying the kinetic equations as we have done previously. However, there is one important difference. The mass of the moving body is now time-dependent and may change appreciably as in the case of a rocket in which the amount of ejected fuel is a significant portion of its original total weight.

ILLUSTRATIVE PROBLEMS

15-3.1. A jet of water 2 in. in diameter issues from a nozzle with a velocity of 100 fps and impinges tangentially upon a perfectly smooth stationary vane which deflects it through an angle of 30° without loss of velocity as shown in Fig. 15-3.2. Determine the total force exerted by the jet upon the vane.

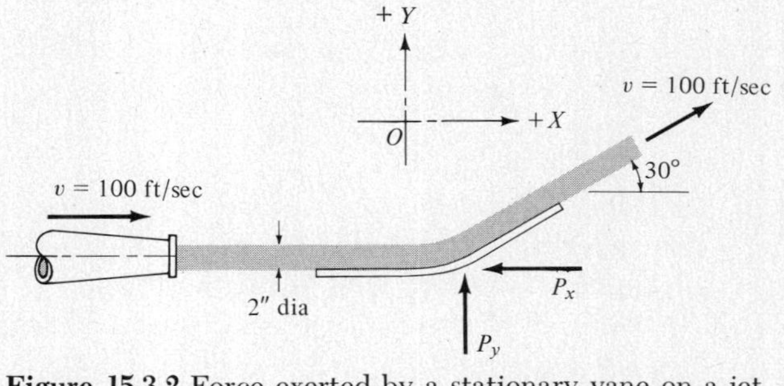

Figure 15-3.2 Force exerted by a stationary vane on a jet.

Solution

Since it is more convenient to treat the jet of water as the free body, we represent the effect of the stationary vane upon the jet by means of its components P_x and P_y. These forces are equal but opposite to the effect of the jet upon the vane and are constant for a steady flow of water. The rate of flow in pounds per second is $Q = wAv$ where w is the specific weight of water (taken as 62.5 lb/ft^3 for fresh water), A is the cross-sectional area of the jet in square feet, and v is the velocity of the jet in feet per second. In an interval of t sec, the weight of water whose momentum is changed thus becomes

$$[W = Qt] \qquad W = 62.5\left(\frac{\pi \times 2^2}{4 \times 144}\right)(100)t = 136.3t \text{ lb}$$

The weight of water in contact with the vane at any instant is relatively small compared with the impulsive forces P_x and P_y and may be neglected in applying the impulse-momentum relation.

Taking X and Y axes as shown (with the X axis positive in the initial direction of motion), and applying the condition that the X component of resultant impulse equals the change in X components of momentum, we have

$$\left[\int \Sigma X\, dt = \frac{W}{g}(v_x - v_{o_x})\right] \qquad -P_x(t) = \frac{136.3t}{32.2}(100 \cos 30° - 100)$$

$$P_x = 56.7 \text{ lb}$$

Similarly, considering the Y components of the impulse-momentum equation, we have

$$\left[\int \Sigma Y\, dt = \frac{W}{g}(v_y - v_{o_y})\right] \qquad P_y(t) = \frac{136.3t}{32.2}(100 \sin 30° - 0)$$

$$P_y = 212 \text{ lb}$$

Combining these components, the resultant force on the vane is

$$[P = \sqrt{P_x^2 + P_y^2}] \quad P = \sqrt{(56.7)^2 + (212)^2} \quad P = 219 \text{ lb} \qquad \textit{Ans.}$$

15-3.2. Assume that the vane in the previous problem is moving to the right at a velocity $u = 40$ fps; all other data remain unchanged.

Solution

In this problem, the total weight of water whose momentum is changed during a time interval t is expressed by $W' = wA(v - u)t$ where $(v - u)$ is the relative velocity with which the water approaches the vane. In terms of W, the total weight of water issuing from the nozzle in this time, we also have $W' = W\left(\frac{v - u}{v}\right)$ so that the weight of water whose momentum is being changed may be found from either

$$W' = 62.5\left(\frac{\pi \times 2^2}{4 \times 144}\right)(100 - 40)t = 81.8t \text{ lb}$$

or, using the result of the preceding problem,

$$W' = W\left(\frac{v-u}{v}\right) = 136.3t\left(\frac{100-40}{100}\right) = 81.8t \text{ lb} \qquad \textbf{\textit{Check}}$$

In the case of a vane moving with constant velocity, it is convenient to use reference axes which are fixed in and moving with the vane. With respect to these axes, we proceed as in the previous problem to equate the components of impulse to the corresponding changes in the components of momentum. However, the velocities to be used now are the relative velocities of approach to and departure from the vane, each having the magnitude $v - u = 60$ fps and directed as shown in Fig. 15-3.3.

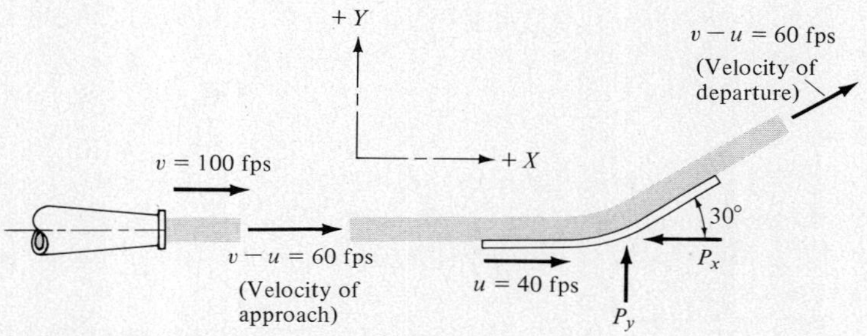

Figure 15-3.3 Jet impinging on a moving vane.

From the X component of the impulse-momentum equation we obtain

$$\left[\int_0^t \Sigma X\,dt = \frac{W'}{g}(v_x - v_{o_x})\right] \qquad -P_x(t) = \frac{81.8t}{32.2}(60\cos 30° - 60)$$

$$P_x = 20.3 \text{ lb}$$

Similarly, the Y component of the impulse-momentum equation gives

$$\left[\int_0^t \Sigma Y\,dt = \frac{W'}{g}(v_y - v_{o_y})\right] \qquad P_y(t) = \frac{81.8t}{32.2}(60\sin 30° - 0)$$

$$P_y = 76.3 \text{ lb}$$

As before, the resultant force on the vane is found from

$$\left[P = \sqrt{P_x^2 + P_y^2}\right] \qquad P = \sqrt{(20.3)^2 + (76.3)^2} \qquad P = 79 \text{ lb} \qquad \textbf{\textit{Ans.}}$$

Power is developed by the X component of P which acts in the direction of motion of the vane. The power developed is

$$\left[\text{hp} = \frac{Fv}{550}\right] \qquad \text{hp} = \frac{20.3(40)}{550} \qquad \text{hp} = 1.48 \qquad \textbf{\textit{Ans.}}$$

This result is the power developed by one vane. If a series of moving vanes are used in which the next vane (as in a rotating Pelton turbine) intercepts the jet just after the first vane has moved out of the way, all the water from the jet is utilized, and we should use $wAvt$ instead of $wA(v - u)t$ in the preceding computations. Doing this changes the power developed to

$$\text{hp} = \frac{wAvt}{wA(v - u)t} \times 1.48 = \frac{100}{60} \times 1.48 = 2.47 \qquad \textbf{\textit{Ans.}}$$

15-3.3. The total weight of a rocket including its fuel is W lb. The fuel is ejected at a constant rate of w lb/sec with a constant velocity u fps relative to the longitudinal axis of the rocket. Determine the velocity and height of the rocket after t sec, assuming vertical flight and that t is less than t_b at burnout of fuel. Neglect air resistance and the variation in g with altitude.

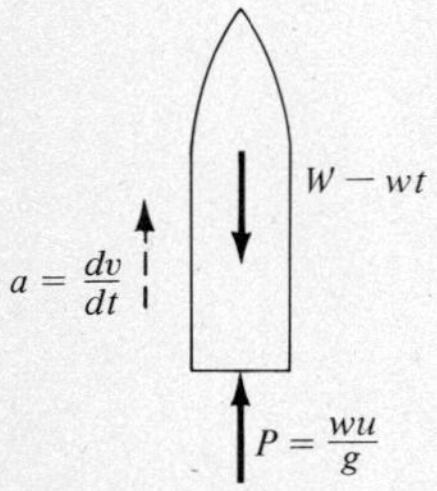

Figure 15-3.4

Solution

In Fig. 15-3.4 is shown the free-body diagram of the rocket. At any instant before burnout, the net weight of the rocket is $W - wt$ where wt is the weight of fuel expelled at time t. From Eq. (15-3.1), the jet thrust $P = wu/g$. The acceleration of the rocket at this instant is

$$\left[+\uparrow \Sigma Y = \frac{W}{g} a\right] \qquad \frac{wu}{g} - (W - wt) = \frac{W - wt}{g} a$$

whence replacing a by dv/dt and multiplying by $g/(W - wt)$, we obtain

$$a = \frac{dv}{dt} = \frac{wu}{W - wt} - g$$

Now separate the variables v and t, and integrate between appropriate limits. This procedure determines v as follows:

$$\int_o^v dv = \int_o^t \frac{wu}{W - wt}\, dt - g \int_o^t dt$$

$$v = -u \ln(W - wt)\Big]_o^t - gt$$

$$v = u \ln \frac{W}{W - wt} - gt \qquad \textbf{\textit{Ans.}}$$

To find the height h at time t, now replace v by dy/dt and integrate between appropriate limits to obtain

$$v = \frac{dy}{dt} = u \ln \frac{W}{W - wt} - gt$$

$$\int_0^h dy = u\int_0^t \ln\frac{W}{W-wt}\,dt - g\int_0^t t\,dt$$

$$h = -u\int_0^t \ln\frac{W-wt}{W}\,dt - \frac{1}{2}gt^2$$

The integral in this relation is in the form $\int \ln x\,dx = x\ln x - x$ where $x = \dfrac{W-wt}{W}$ and $dx = -\dfrac{w\,dt}{W}$. (Can you see now why the integrand was inverted in the last step?) Making these substitutions results in

$$h = u\frac{W}{w}\int_0^t \ln\frac{W-wt}{W}\left(-\frac{w}{W}\,dt\right) - \frac{1}{2}gt^2$$

which reduces to

$$h = \frac{uW}{w}\left[\frac{W-wt}{W}\ln\frac{W-wt}{W} - \frac{W-wt}{W} + 1\right] - \frac{1}{2}gt^2 \qquad \textit{Ans.}$$

Some supplementary remarks on rocket motion are pertinent. Observe that liftoff will occur only if $P = wu/g$ is greater than the "wet" weight W. For example, liftoff of a 10-ton rocket with an exhaust velocity of 6000 fps would require a rate of fuel consumption of at least 3.22 tons/min. Consequently, "exotic" fuels are required which have both a high combustion rate and a large exhaust velocity.

PROBLEMS

15-3.4. What force would a 2-in. diameter jet of water flowing at 2 ft³/sec exert on a stationary plate placed as in parts (a) and (b) of Fig. P-15-3.4?

(a) $P = 356$ lb; (b) $P = 178$ lb **Ans.**

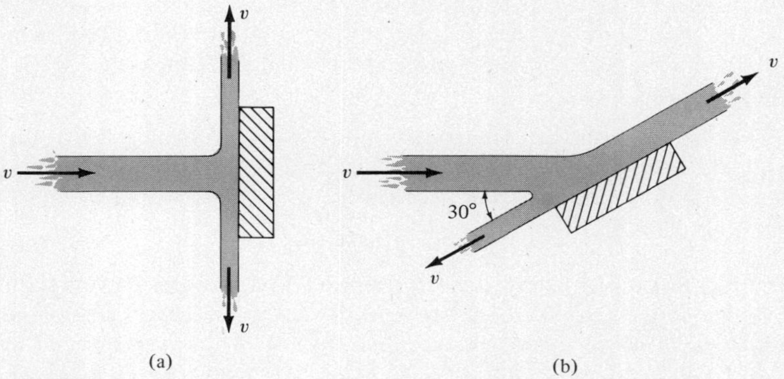

Figure P-15-3.4

15-3.5. A jet of water issuing from a nozzle at the rate of 20 lb/sec at a velocity of 100 fps strikes a series of vanes and is deflected through 120°. If the speed of the vanes is 60 fps, compute the power delivered to the vanes.

4.07 hp **Ans.**

15-3.6. A tank weighing 1000 lb rests on platform scales. It is being filled with water from a vertical jet having a velocity of 150 fps and a cross-sectional area of 0.15 sq ft. What will be the total scale reading after 5 sec?

15-3.7. A 12-in. pipe carries water with a velocity of 10 fps around a 60° elbow. The pressure intensity at both ends of the elbow is 40 psi. Compute the resultant force exerted by the water on the elbow. Neglect the weight of water in the elbow.

$P = 4680$ lb **Ans.**

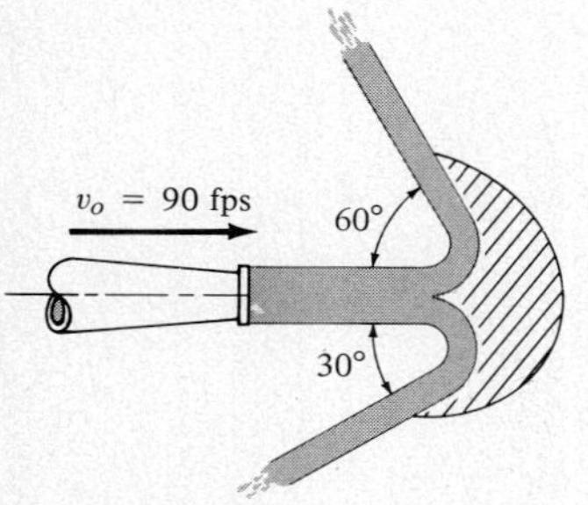

Figure P-15-3.8

15-3.8. A stream of water flowing at 100 lb/sec and moving rightward at 90 fps strikes the vane shown in Fig. P-15-3.8. If half of the water flows across each part of the vane, determine the force required to hold it in place.

$P = 474$ lb **Ans.**

15-3.9. If the vane of Prob. 15-3.8 is moving rightward at 40 fps and one-third of the water flows over the upper portion while two-thirds flow across the lower portion, determine the power developed.

10.95 hp **Ans.**

15-3.10. Solve Prob. 15-3.9 if two-thirds of the water flow across the upper portion while one-third flows across the lower portion of the vane.

15-3.11. A jet of water flowing at w lb/sec and a velocity of v fps is deflected through θ° by a series of vanes on the periphery of an impulse turbine. Determine the optimum peripheral speed of the vanes to generate maximum power and the corresponding power output.

$\text{hp} = wv^2(1 - \cos\theta)/2200\,g$ **Ans.**

15-3.12. A rocket shell weighs 1000 lb and contains 3000 lb of fuel. The fuel is consumed at 200 lb/sec and ejected at a relative speed of 1610 fps. Find the acceleration of the rocket (a) at liftoff; (b) 10 sec after liftoff; (c) just before burnout. Assume vertical flight and neglect any variation in the value of g.

15-3.13. A chain 80 ft long that weighs 10 lb/ft is suspended vertically with the lower end just touching a platform. If the chain is dropped, determine the reaction of the platform at the instant the upper end of the chain has dropped 60 ft.

$R = 1800$ lb **Ans.**

15-3.14. A 1-oz ball is placed in a vertical jet of water flowing from a nozzle at the rate of 0.30 lb/sec with a velocity of 15 fps. Determine the height h above the nozzle at which the ball will stay suspended in the jet.

$h = 2.80$ ft **Ans.**

15-3.15. After a jet plane has touched down, reverse thrust to decelerate it is obtained by means of exhaust vanes which partially reverse the exhaust stream as shown in Fig. P-15-3.15. For a two-engine plane weighing 10 tons and moving at a ground speed of 150 mph, determine the maximum deceleration if each engine scoops in air at a rate of 100 lb/sec and discharges it at a relative velocity of 2000 fps while consuming 2 lb of fuel per sec.

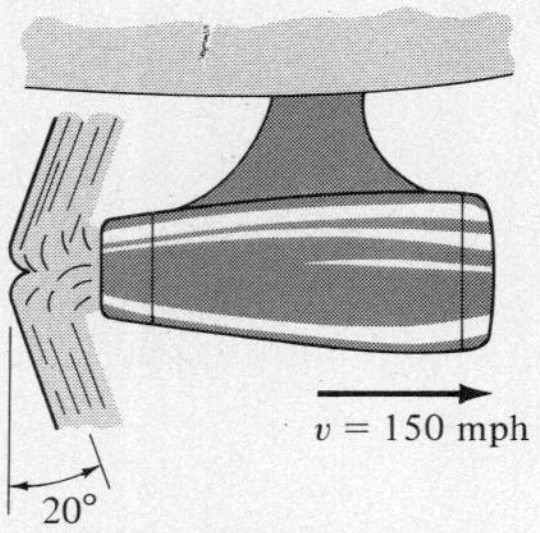

Figure P-15-3.15

15-3.16. A snowplow, moving at a constant speed of 20 mph, clears a 10-ft path from an airport runway. The average depth of snow is 1.5 ft and its density is 7 lb/ft^3. The snow is diverted sideways at a 120° angle from the forward motion of the plow and, because the snow compacts somewhat, at a relative speed of 15 mph. Determine the power required beyond that needed to move the snowplow itself.

hp = 93.3 ***Ans.***

15-3.17. The total weight of a rocket including its fuel is 10 tons. The fuel is consumed at the rate of 120 lb/sec and ejected with a constant rearward relative velocity of 6000 fps. If burnout occurs at 150 sec after launch, what will then be the velocity and height of the rocket? Assume vertical flight and neglect both air resistance and any variation in g.

v = 8970 fps; h = 58.2 miles ***Ans.***

15-3.18. The rocket of the previous problem is redesigned as a two-stage rocket. Each stage weighs 10,000 lb, including 9000 lb of fuel. When the first stage reaches burnout, it is detached and the second stage fires. If the fuel is again consumed at the rate of 120 lb/sec and ejected at a constant relative velocity of 6000 fps, determine the velocity when the first stage is released and the maximum velocity reached by the second stage.

15-3.19. A 6-ton airplane is climbing at an angle of 10° at a constant air speed of 300 mph. Its jet engine scoops in air at the rate of 100 lb/sec and discharges it at a relative velocity of 1500 fps while burning 2 lb of fuel per sec. (a) Determine the lift and drag forces perpendicular and parallel to the flight path. (b) What will be the acceleration of the plane as it levels off?

(b) a = 5.59 fps^2 ***Ans.***

15-4 CONSERVATION OF LINEAR MOMENTUM

The integral form of the linear impulse-momentum equation for a body or a system of particles is

$$\int \mathbf{R}\, dt = \int m\, d\bar{\mathbf{v}} = \int d\mathbf{p}$$

in which $d\mathbf{p}$ represents differential momentum. Observe that if the resultant external force $\mathbf{R}$ is zero during any time interval, this equation reduces to $d\mathbf{p} = 0$ and therefore the total momentum $\mathbf{p}$ must be constant during this time interval.

The condition that **p** (the momentum of a system of particles) be constant applies either to static equilibrium of a system moving with constant velocity, or, more importantly, when **R** is zero because of mutual action and reaction between the particles composing the system. Although the total resultant external force may be zero, this does not mean that the resultant force on any single particle is zero. For example, when a shell is discharged from a gun, the force acting on the shell is always equal and opposite to the force acting on the gun. The resultant force acting on the system composed of shell and gun will be zero although a propulsive force is acting on both shell and gun if either is considered as a free body. From another viewpoint, for a system subjected to a zero resultant force, the motion of the mass center of the system remains unchanged although there may be changes in the motion of its individual parts. *The condition that the momentum be constant applies only to the system as a whole, never to its component parts.* This is known as *the principle of conservation of momentum.*

Therefore, if a system is composed of particles of weight W_1, W_2, etc., having velocities $\mathbf{v}_1$, $\mathbf{v}_2$, etc. and after a mutual reaction between the particles they possess new velocities $\mathbf{v}_1'$, $\mathbf{v}_2'$, etc., the condition that the momentum of the system be constant is expressed by means of the following equation:

$$\frac{W_1}{g}\mathbf{v}_1 + \frac{W_2}{g}\mathbf{v}_2 + \cdots = \frac{W_1}{g}\mathbf{v}_1' + \frac{W_2}{g}\mathbf{v}_2' + \cdots \tag{15-4.1}$$

Note that it is possible to have a zero component of impulse along one coordinate direction, thereby resulting in conserving the components of linear momentum along that direction even though the components of momentum along other directions are not conserved.

Another approach which may give more meaning to conservation of momentum is the example shown in Fig. 15-4.1. Here we have the FBDs of two bodies moving in the same direction which have collided because it is assumed that $v_1 > v_2$. Assume the bodies to be either perfectly or partially elastic and to continue in the same direction with the primed velocities after the impact. Obviously $v_2' > v_1'$ for separation to occur.

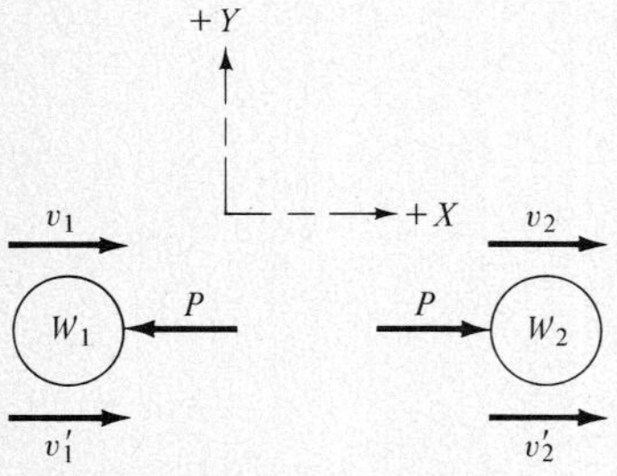

Figure 15-4.1 Free-body diagrams during impact.

The velocity changes are caused by the pair of equal and opposite reaction forces P which vary in an unknown manner during the impact. Significant information about the velocities can be obtained by writing the impulse-momentum equation for each body even though the variation of P with time is unknown. Using the same coordinate system shown for each body, and neglecting the impulse of the weights or of friction as negligibly small compared with that of P, we obtain

For W_1: $$-\int P\,dt = \frac{W_1}{g}(v_1' - v_1)$$

For W_2: $$\int P\,dt = \frac{W_2}{g}(v_2' - v_2)$$

Adding these equations eliminates the impulse $\int P\,dt$ and gives

$$0 = \frac{W_1}{g}(v_1' - v_1) + \frac{W_2}{g}(v_2' - v_2)$$

Rearranging the terms gives the following result which represents the conservation of the components of momentum along the X direction:

$$\frac{W_1}{g}v_1 + \frac{W_2}{g}v_2 = \frac{W_1}{g}v_1' + \frac{W_2}{g}v_2'$$

ILLUSTRATIVE PROBLEMS

15-4.1. A 160-lb man, moving horizontally with a velocity of 10 fps, jumps off the end of a pier into a 640-lb boat. Determine the horizontal velocity of the boat (a) if it had no initial velocity and (b) if it was approaching the pier with an initial velocity of 3 fps.

Solution: Part (a)

Since the boat and the man exert mutual forces on each other, the principle of conservation of momentum may be applied to the system of boat and man. The final horizontal velocity of the boat and man will be the same; hence, applying the horizontal components of Eq. (15-4.1) with the common factor g cancelled out, we have for part (a)

$$[W_1\mathbf{v}_1 + W_2\mathbf{v}_2 = (W_1 + W_2)\mathbf{v}]$$
$$160(10) + 640(0) = (160 + 640)v$$
$$v = 2 \text{ fps} \qquad \textit{Ans.}$$

The positive value of the final velocity indicates that the boat and man move in the direction of the man's initial velocity. If the vertical component of the man's velocity were known, a similar analysis would yield the common initial vertical velocity with which the boat and man would start to bob up and down in the water.

Part (b)

In this case the initial velocity of the boat is opposite to that of the man and hence is considered negative when we consider the horizontal components of momentum.

$$[W_1\mathbf{v}_1 + W_2\mathbf{v}_2 = (W_1 + W_2)\mathbf{v}]$$
$$160(10) + 640(-3) = (160 + 640)v$$
$$v = -0.4 \text{ fps} \qquad \textit{Ans.}$$

The minus sign indicates a final velocity directed toward the pier.

15-4.2. A ballistic pendulum consists of a sand box weighing 59 lb that is suspended from a cord 10 ft long. A 1-lb shell is fired horizontally into the box and remains embedded in it. Because of impact, the sand box swings through a maximum angle of 30°, as shown in Fig. 15-4.2. Determine the velocity with which the shell strikes the box and the loss in energy of the system.

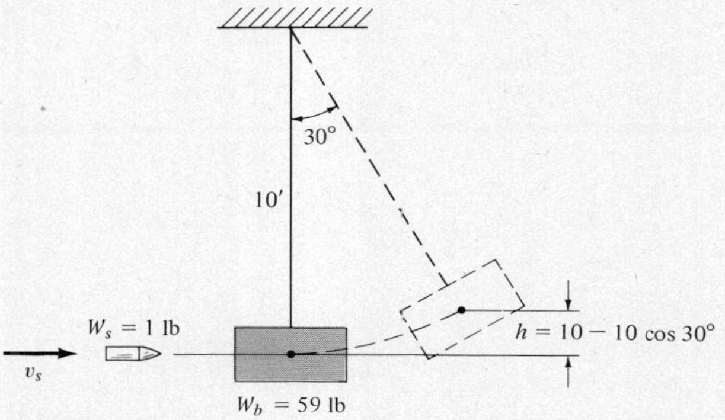

Figure 15-4.2

Solution

The initial velocity of the sand box with the embedded shell is found by the work-energy method. The work done is negative because the gravity force acts opposite to the upward rise.

$$\left[-Wh = \frac{W}{2g}(v^2 - v_0^2)\right] \quad -W(10 - 10\cos 30°) = \frac{W}{64.4}(0 - v_0^2)$$
$$v_0 = 9.3 \text{ fps} \qquad \textit{Ans.}$$

This value of velocity represents the common velocity of the shell and box directly after impact. The velocity of the shell before impact can be found now by applying the principle of conservation of momentum.

$$[W_1\mathbf{v}_1 + W_2\mathbf{v}_2 = W_1 + W_2)\mathbf{v}] \qquad 1(v_s) + 0 = (1 + 59)(9.3)$$
$$v_s = 558 \text{ fps} \qquad \textit{Ans.}$$

To determine the loss in energy of the system, we have

$$\text{initial energy of shell} = \frac{1}{2}mv^2 = \frac{1}{2}\left(\frac{1}{32.2}\right)(558)^2 = 4830 \text{ ft-lb}$$

$$\text{final energy of system} = \frac{1}{2}mv^2 = \frac{1}{2}\left(\frac{60}{32.2}\right)(9.3)^2 = 80.5 \text{ ft-lb}$$

Their difference represents an energy loss of 4749.5 ft-lb which is dissipated in heat.

15-4.3. In Fig. 15-4.3, bodies *A* and *B* are connected by a flexible, inextensible cord. Body *A* is lifted vertically and then dropped from rest. If it acquires a velocity v_o just before the cord becomes taut, find its velocity right after body *B* is jerked into motion.

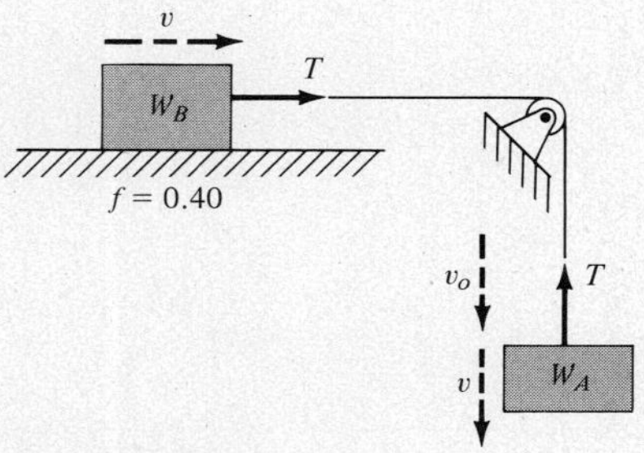

Figure 15-4.3

Solution

In this problem, neglect the impulse of the weights and friction as negligibly small compared to that of *T*. Since equal and opposite impulses act on each body due to the tension *T*, momentum of the system is conserved. We therefore obtain for the common final velocity *v* of both bodies,

$$[W_1\mathbf{v}_1 + W_2\mathbf{v}_2 = (W_1 + W_2)\mathbf{v}] \qquad W_A v_o + 0 = (W_A + W_B)v$$

$$v = \frac{W_A v_o}{W_A + W_B} \qquad \textit{Ans.}$$

Alternatively, the same result can be found by a direct application of the impulse-momentum equation to each body as follows:

$$\text{On A:} \quad -\int T\,dt = m_A v - m_A v_o$$
$$\text{On } B\text{:} \quad \int T\,dt = m_B v - 0$$

Adding these equations eliminates the impulsive forces and, after converting mass to weight by cancelling out the common factor *g*, we again obtain

$$v = \frac{W_A v_o}{W_A + W_B} \qquad \textit{Ans.}$$

PROBLEMS

15-4.4. A 1000-lb shell is fired from a 200,000-lb cannon with a velocity of 2000 fps. Find the modulus of a nest of springs that will limit the recoil of the cannon to 3 ft.

$$k = 5750 \text{ lb/in.} \qquad \textit{Ans.}$$

15-4.5. A cannon is mounted on one end of a railroad car and a rigid target is mounted on the other end. The car is free to roll without resistance on a level track. A shell is fired from the gun when the car is at rest. Discuss the motion of the car and what happens when the shell hits the target. Assume that the shell is a solid ball which remains embedded in the target.

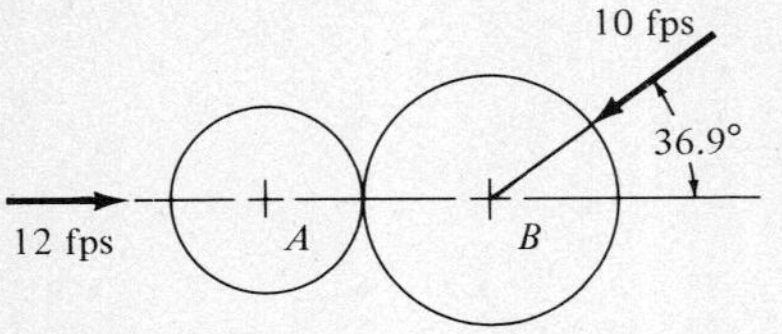

Figure P-15-4.6

15-4.6. Just before they collide, two disks on a horizontal surface have the velocities shown in Fig. P-15-4.6. Knowing that the 20-lb disk A rebounds to the left with a velocity of 6 fps, determine the rebound velocity of the 30-lb disk B. Assume the colliding surfaces are smooth.

$v = 7.21$ fps $\measuredangle$ 56.25° ***Ans.***

15-4.7. Solve the previous problem if the coefficient of friction at the colliding surfaces is 0.40 and the collision lasts for 0.01 sec.

15-4.8. Two moving balls of putty collide and cling together. Determine the magnitude of their common final velocity and the loss in kinetic energy. The first ball A weighs 1 lb and has a velocity $\mathbf{v}_A = 10\hat{\mathbf{i}} + 15\hat{\mathbf{j}}$ fps. The second ball B weighs 2 lb and has a velocity $\mathbf{v}_B = 4\hat{\mathbf{i}} - 3\hat{\mathbf{j}} + 6\hat{\mathbf{k}}$ fps.

$v = 7.81$ fps; KE loss $= 4.10$ ft-lb ***Ans.***

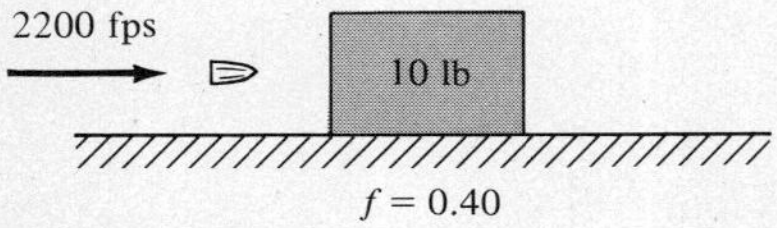

Figure P-15-4.9

15-4.9. A bullet weighing 1 oz and moving at 2200 fps penetrates the 10-lb body in Fig. P-15-4.9 and emerges with a velocity of 600 fps. How far and how long does the body then move?

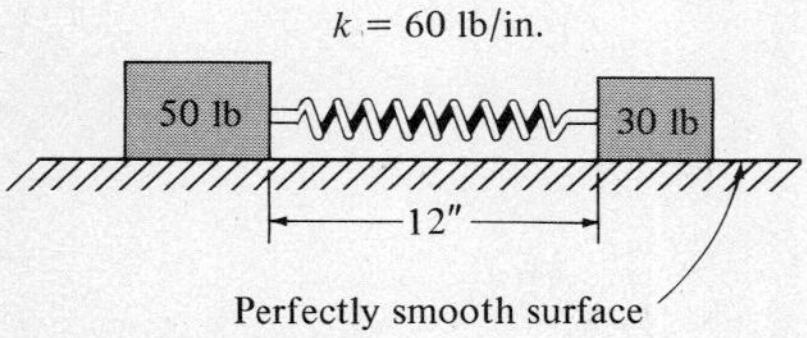

Figure P-15-4.10

15-4.10. The spring shown in Fig. P-15-4.10 has a free length of 12 in. It is compressed to half its length and the blocks are suddenly released from rest. Determine the velocity of each block when the spring is again 12 in. long.

$v_{30} = +11$ fps; $v_{50} = -6.6$ fps ***Ans.***

15-4.11. Determine the velocity of each block in the preceding problem at the instant the spring is 9 in. long.

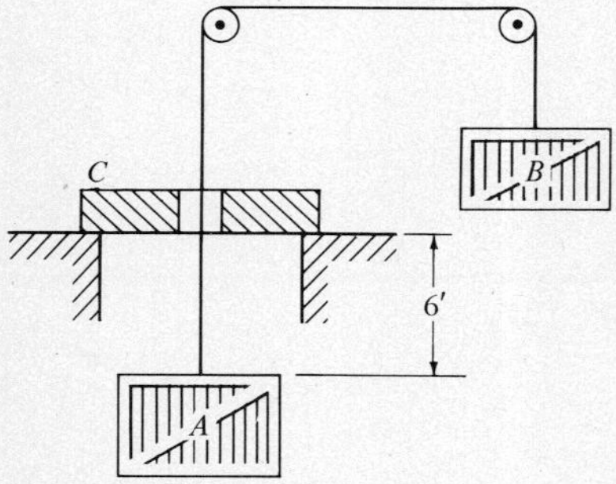

Figure P-15-4.12

15-4.12. After body A in Fig. P-15-4.12 moves 6 ft from rest, it picks up body C. How much further does A move before reversing its direction? Assume $W_A = 10$ lb, $W_B = 20$ lb, $W_C = 15$ lb.

8 ft ***Ans.***

15-4.13. A 200-lb man and a 100-lb boy jump to a pier from a 300-lb rowboat with a horizontal velocity of 10 fps relative to the rowboat. If the rowboat is initially motionless in the water, under which of the following conditions will the rowboat acquire the largest velocity: (a) man and boy jump simultaneously; (b) man jumps first, followed by boy; (c) boy jumps first, followed by man.

15-4.14. A wooden pile that weighs 500 lb is driven vertically into the ground by successive blows from a piledriver whose 1000-lb hammer falls freely through a height of 6 ft upon the head of the pile. How deep does a single blow of the hammer drive the pile when the ground has developed an average resistance to penetration of 6500 lb? Assume that the hammer and pile cling together after impact.

Hint: Resultant work done equals kinetic energy lost in impact.

$s = 0.80$ ft ***Ans.***

15-4.15. A 1200-lb hammer falling freely through 3 ft drives a 600-lb pile 6 in. vertically into the ground. Assuming the hammer and pile to cling together after impact, determine the average ground resistance to penetration by the pile.

15-5 ELASTIC IMPACT

When two elastic bodies collide, they are deformed at first; then they spring apart because of the action of restoring elastic forces. Throughout this elastic impact, there exist mutual action and reaction forces. (Refer back to Fig. 15-4.1.) The magnitude of these impact forces during the short time they exist is so large that nonimpulsive forces are negligible in comparison. Hence, the resultant force on the system is zero and conservation of momentum may be applied to express one relation between the unknown final velocities of the bodies. Another relation between these velocities is obtained from the definition of the coefficient of restitution. The coefficient of restitution (symbolized by e) is defined as the ratio of the relative velocity of colliding bodies after impact to their relative velocity before impact. The relative velocities are measured along the line of impact which is the common normal to the colliding surfaces.

If the line of impact passes through the mass centers of the bodies, the collision is known as central impact; otherwise the impact is eccentric. Several possibilities are shown in Fig. 15-5.1. In each case, the relative velocities involved in the coefficient of restitution are the components of velocity along the line of impact. Since the velocity components involved in both the definition of the coefficient of restitution and in the components of momentum along the line of impact are all along the same line, we may use their scalar values.

Now consider direct central impact of the bodies shown in Fig. 15-5.2a where the unprimed values are initial velocities and the

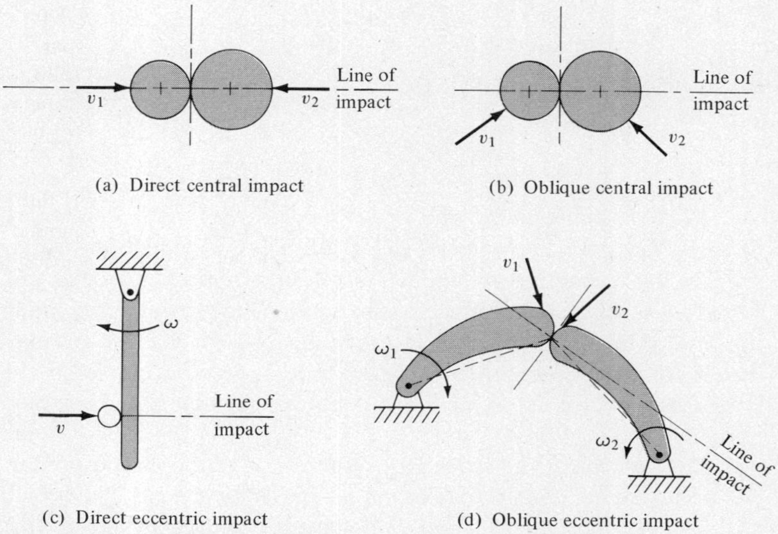

Figure 15-5.1 Various types of impact.

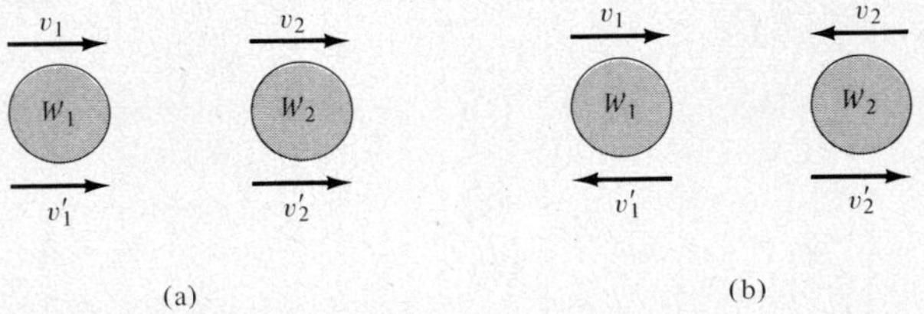

Figure 15-5.2 Two extreme cases of direct central impact.

primed values are their velocities after impact. Body W_1 is overtaking body W_2; hence, v_1 must be larger than v_2 and the relative velocity before impact is $v_1 - v_2$. After impact, assume that the bodies continue moving in their original directions. In order for them to separate, body W_2 must have a larger final velocity than body W_1; therefore, the relative velocity after impact is $v_2' - v_1'$. Figure 15-5.2b shows another extreme case of direct central impact in which the bodies approach each other before impact and have opposite velocities after impact. In this case, the relative velocity again is the algebraic difference $v_1 - v_2$ which corresponds to the arithmetic sum of their magnitudes. Thus, since the sign of v_2 is minus (i.e., opposite to v_1 assumed as plus)

$$v_1 - v_2 = v_1 - (-v_2) = v_1 + v_2$$

Similarly, the relative velocity after impact is also given by

$$v_2' - v_1' = v_2' - (-v_1') = v_2' + v_1'$$

A general equation defining e can therefore be written as

$$e = \frac{\text{Relative velocity after impact}}{\text{Relative velocity before impact}} = \frac{v_2' - v_1'}{v_1 - v_2}$$

the preferred form of which is

$$e(v_1 - v_2) = v_2' - v_1' \qquad (15\text{-}5.1)$$

In direct impact in which the velocities of both bodies are directed along the line of impact, Eq. (15-5.1) is the algebraic difference of the respective velocities. For oblique impact, however, the velocities in Eq. (15-5.1) must be the components of velocities along the line of impact. Note that the components of velocities perpendicular to the line of impact are unchanged by the impact if the bodies are smooth; that is, no forces act perpendicular to the common normal.

The coefficient of restitution is a measure of the elastic properties of a pair of bodies. For example, perfectly elastic bodies will have exactly the same relative velocity after impact as before; that is, the value of e will be unity. Perfectly inelastic bodies (those which cling together after impact) will have the same final velocity and the value of e is obviously zero. The coefficient of restitution will therefore always lie somewhere between the limits of zero and unity.

Impact problems usually imply point contact as in colliding spherical bodies. The value of e in any particular case depends not only on the two materials involved, but also on the impact velocity and the size and shape of the bodies. Handbook values of e for different pairs of materials usually refer to spherical bodies and should be used with prudence for other shapes.

Imperfect elastic action causes a loss in kinetic energy which can be found by taking the difference between the kinetic energies before and after impact. The kinetic energy lost is partially dissipated by heat generated at the contact surfaces. The rise in temperature of a steel hammer repeatedly struck against an anvil demonstrates this. Energy is also dissipated by vibrations within the bodies, hysteresis effects, and sound waves produced in the air.

ILLUSTRATIVE PROBLEMS

15-5.1. The 10-lb and 20-lb bodies in Fig. 15-5.3 are approaching each other with the velocities shown. If $e = 0.60$, determine the velocity of each body directly after impact.

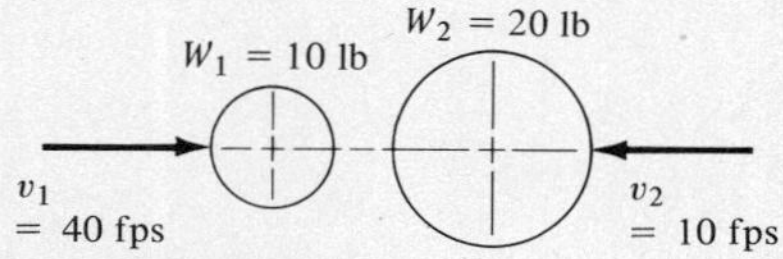

Figure 15-5.3

Solution

Applying the equation of conservation of momentum with the factor g canceled out gives

$$[W_1\mathbf{v}_1 + W_2\mathbf{v}_2 = W_1\mathbf{v}_1' + W_2\mathbf{v}_2']$$

$$10(40) + 20(-10) = 10v_1' + 20v_2'$$

or

$$20 = v_1' + 2v_2' \qquad (a)$$

Using Eq. (15-5.1) to obtain another relation between these velocities, we have

$$[e(v_1 - v_2) = v_2' - v_1'] \qquad 0.6[40 - (-10)] = v_2' - v_1'$$

or

$$30 = v_2' - v_1' \qquad (b)$$

Solving Eqs. (a) and (b) yields

$$\left.\begin{aligned} v_1' &= -13.33 \text{ fps} \\ v_2' &= +16.67 \text{ fps} \end{aligned}\right\} \qquad \textit{Ans.}$$

The minus sign before v_1' indicates that the 10-lb body rebounds to the left after impact.

15-5.2. A ball is dropped from a height of 8 ft upon a 15° incline and rebounds as shown in Fig. 15-5.4. If $e = 0.80$, find the distance x at which the ball again strikes the incline.

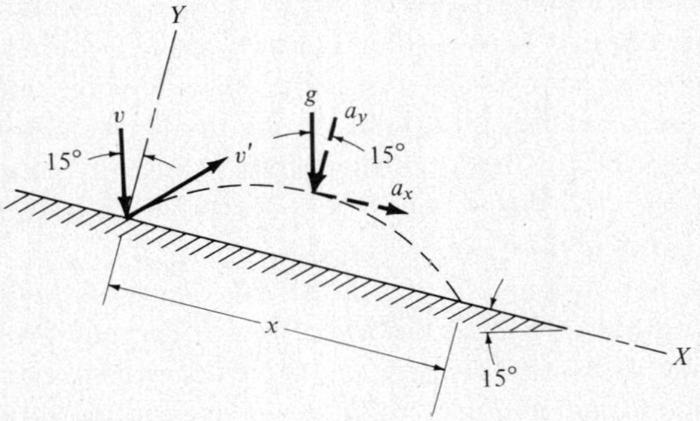

Figure 15-5.4

Solution

This problem is an example of oblique central impact. The initial velocity with which the ball strikes the incline is

$$[v^2 = v_0{}^2 + 2gh] \qquad v = \sqrt{64.4(8)} = 22.7 \text{ fps}$$

and its components parallel and perpendicular to the incline are

$$v_x = v \sin 15° = 5.87 \text{ fps} \qquad v_y = v \cos 15° = 21.9 \text{ fps}$$

Assuming smooth contact surfaces, the X component of the rebound velocity is $v'_x = v_x = 5.87$ fps. The value of the rebound component v'_y, as found from Eq. (15-5.1), is

$$[e(v_1 - v_2) = v'_2 - v'_1] \qquad 0.8(21.9 - 0) = 0 - (-v'_y)$$

$$v'_y = 17.52 \text{ fps}$$

The rest of the problem involves the trajectory of the ball under the influence of a constant gravitational acceleration g fps^2. Using the X-Y axes shown, the constant components of acceleration which affect the rebound components v'_x and v'_y are

$$a_x = g \sin 15° = 8.33 \text{ fps}^2 \quad \text{and} \quad a_y = -g \cos 15° = -31.1 \text{ fps}^2$$

Since these components of acceleration are constant, we may use $s = v_0 t + \frac{1}{2}at^2$ to determine the corresponding rectangular components of displacement. Since $y = 0$ when the ball again strikes the incline, the time of flight is found to be

$$[y = v_{o_y}t + \tfrac{1}{2}a_y t^2] \qquad 0 = 17.52t - \tfrac{1}{2}(31.1t^2) \qquad t = 1.127 \text{ sec}$$

When we use this value of t, the distance x is

$$[x = v_{o_x}t + \tfrac{1}{2}a_x t^2] \qquad x = 5.87(1.127) + \tfrac{1}{2}(8.33)(1.127)^2$$

$$x = 11.91 \text{ fps}^2 \qquad \textit{Ans.}$$

Notice the relative ease of this approach to the trajectory problem as compared with that of Illus. Prob. 9-6.3. To use horizontal and vertical components of acceleration as we did there would here require first transforming v'_x and v'_y into corresponding horizontal and vertical components of velocity, and then solving a pair of simultaneous equations.

PROBLEMS

15-5.3. A ball is thrown at an angle θ with the normal to a smooth floor as shown in Fig. P-15-5.3. It rebounds at an angle θ' with the normal. Show that the coefficient of restitution is expressed by $e = \tan\theta/\tan\theta'$

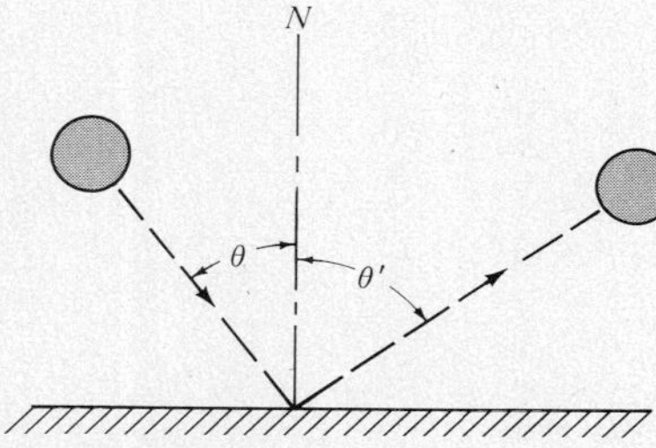

Figure P-15-5.3

15-5.4. Direct central impact occurs between a 60-lb body moving rightward at 10 fps and a 30-lb body moving leftward at 20 fps. If the coefficient of restitution is $e = 0.6$, determine the average impact force for a time of impact lasting 0.02 sec.

$F = 1490$ lb **Ans.**

15-5.5. Direct central impact occurs between a 30-lb body moving to the right at 6 fps and a body of weight W moving to the left at 4 fps. The coefficient of restitution is $e = 0.5$. After impact the 30-lb body has a leftward velocity of 3 fps. (a) Find the value of W. (b) If the impact lasts for 0.02 sec, find the average impact force.

15-5.6. A golf ball is dropped from a height of 10 ft upon a concrete floor. The coefficient of restitution is 0.984. Find the height to which the ball rebounds on the first, second, and third bounces.

15-5.7. Show that a general expression for the answer to Illus. Prob. 15-5.2 is $x = 4eh(1 + e)\sin\theta$.

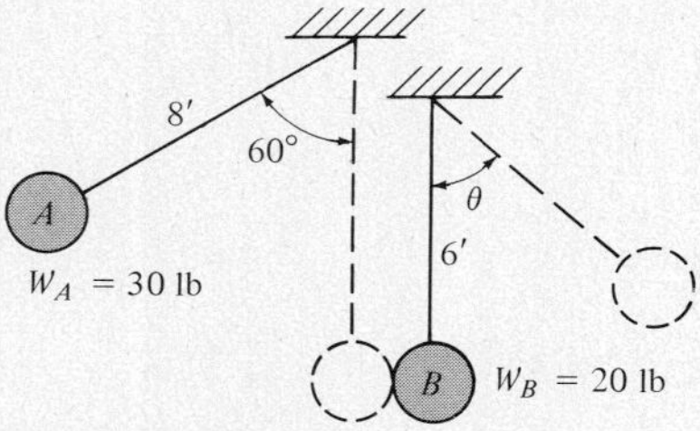

Figure P-15-5.8

15-5.8. The balls A and B in Fig. P-15-5.8 are attached to stiff rods of negligible weight. Ball A is released from rest and allowed to strike B. If $e = 0.6$, determine the maximum angle θ through which ball B will swing. What is the maximum and minimum tension in the rod attached to B? If the impact lasts for 0.01 sec, also find the average impact force.

$\theta = 67.3°$; $T_{max} = 44.6$ lb; $T_{min} = 7.72$ lb; $P = 957$ lb **Ans.**

15-5.9. The system shown in Fig. P-15-5.8 is used to determine experimentally the coefficient of restitution. If ball A is released from rest and ball B swings through $\theta = 53.1°$ after being struck, determine e.

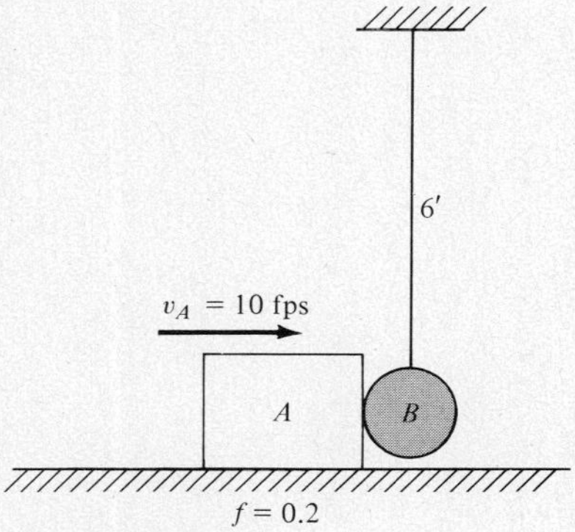

Figure P-15-5.10

15-5.10. The 10-lb block A in Fig. P-15-5.10 has a velocity of 10 fps when it strikes a 20-lb ball B suspended from a 6-ft cord. If $e = 0.80$, determine the final position of block A and the maximum and minimum tension in the cord supporting B.

15-5.11. As shown in Fig. P-15-5.11, a ball is thrown against a smooth vertical wall. The ball is released from a position 40 ft from the wall and

6 ft above the level ground and travels in a plane perpendicular to the wall. Its initial velocity is 50 fps upward at an angle of 60° with the horizontal. The coefficient of restitution between ball and wall is 0.6. How far from the wall does the ball hit the ground?

$$x = 18.35 \text{ ft} \qquad \textit{Ans.}$$

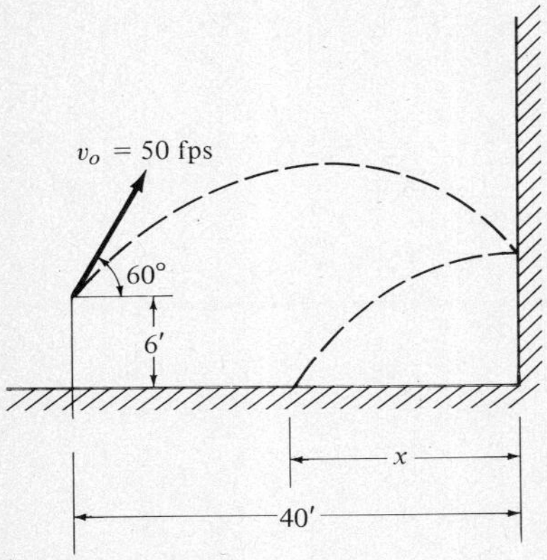

Figure P-15-5.11

15-5.12. Solve Prob. 15-5.11 if the ball travels in a vertical plane which makes an angle of 45° with the wall.

$$5.90 \text{ ft} \qquad \textit{Ans.}$$

15-5.13. A 6-lb ball moving at 10 fps up to the right at 60° to the horizontal collides with a 10-lb ball moving horizontally leftward at 10 fps. At the moment of impact, the line joining the centers of the balls is horizontal. If $e = 0.6$, determine the amount and direction of the velocity of each ball directly after impact.

$$v_{10} = 1 \text{ fps to left; } v_6 = 13.23 \text{ fps} \measuredangle 41° \qquad \textit{Ans.}$$

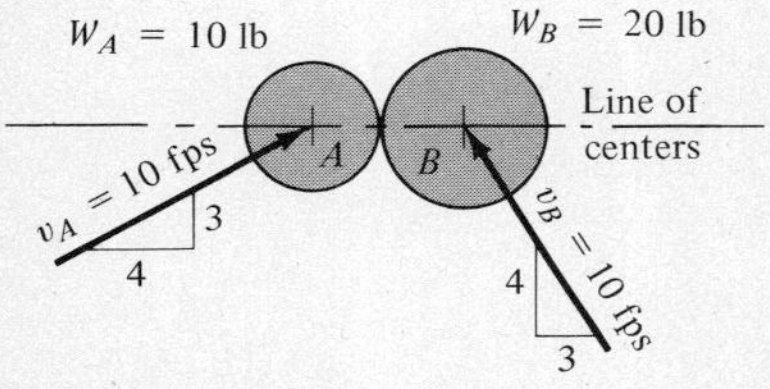

Figure P-15-5.14

15-5.14. In the oblique central impact shown in Fig. P-15-5.14, the coefficient of restitution is 0.6. The flat disks slide on a smooth horizontal surface. Compute the final velocity of each disk directly after impact.

$$v_A = 9.16 \text{ fps} \measuredangle 40.9°; \; v_B = 8.13 \text{ fps} \measuredangle 79.6° \qquad \textit{Ans.}$$

15-5.15. If the surfaces of contact between the colliding disks of Prob. 15-5.14 were rough with a coefficient of friction of 0.40, determine the final velocity of each disk.

$$v_A = 13.83 \text{ fps} \measuredangle 59.9°; \; v_B = 5.22 \text{ fps} \measuredangle 73.7° \qquad \textit{Ans.}$$

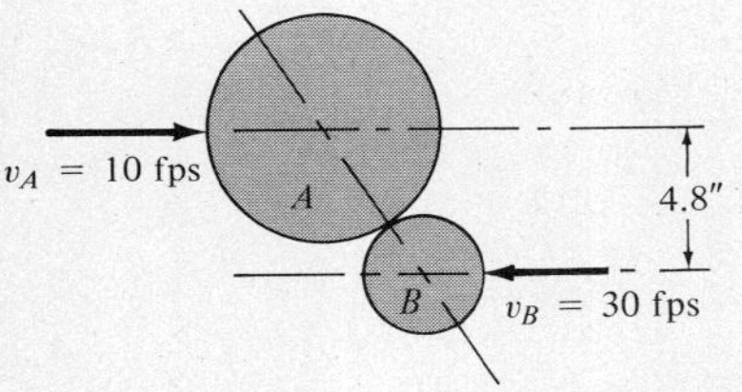

Figure P-15-5.16

15-5.16. Two spheres are moving along parallel paths with the velocities shown in Fig. P-15-5.16. Sphere A has a weight of 60 lb and a radius of 4 in., whereas sphere B has a weight of 20 lb and a radius of 2 in. If $e = 0.6$, determine the velocity of each sphere directly after impact.

15-5.17. Two bodies of masses m_1 and m_2 moving in opposite directions with velocities v_1 and v_2 collide. If e is the coefficient of restitution, show that the energy loss due to direct central impact is given by

$$\text{Loss in KE} = \frac{1}{2}\frac{m_1 m_2}{m_1 + m_2}(1 - e^2)(v_1 - v_2)^2 \qquad \textit{Ans.}$$

15-6 IMPULSE MOMENTUM IN PLANE MOTION

The impulse-momentum equations directly relating force, velocity, and time for any body are obtained by taking the time integrals of the fundamental equations $\mathbf{R} = m\bar{\mathbf{a}}$ and $\Sigma\mathbf{M} = \dot{\mathbf{H}} = d\mathbf{H}/dt$. Recall that the moment equation is valid with respect to either the mass center G or a center A which is fixed in space. Thus, by multiplying each side of the fundamental equations by dt and integrating, we obtain

$$\int_{t_1}^{t_2} \mathbf{R}\, dt = \int_{t_1}^{t_2} m\bar{\mathbf{a}}\, dt = \int_{\bar{\mathbf{v}}_1}^{\bar{\mathbf{v}}_2} m\, d\bar{\mathbf{v}} = m\bar{\mathbf{v}}_2 - m\bar{\mathbf{v}}_1 \qquad (a)$$

$$\int_{t_1}^{t_2} \Sigma \mathbf{M}\, dt = \int_{\mathbf{H}_1}^{\mathbf{H}_2} d\mathbf{H} = \mathbf{H}_2 - \mathbf{H}_1 \qquad (b)$$

For the plane motion of bodies which are symmetrical about the plane of motion, we showed in Section 13-2 that the value of $\mathbf{H}$ about the mass center is $\bar{\mathbf{H}} = \bar{I}\boldsymbol{\omega}$ while for rotation about a fixed axis A, the value of $\mathbf{H}$ is $\mathbf{H}_A = I_A\boldsymbol{\omega}$. Hence, the scalar form of the impulse-momentum equations for the plane motion of symmetrical bodies becomes

$$\left.\begin{aligned}
\int_{t_1}^{t_2} \Sigma X\, dt &= m\bar{v}_{2_x} - m\bar{v}_{1_x} = m(\bar{v}_{2_x} - \bar{v}_{1_x})\\
\int_{t_1}^{t_2} \Sigma Y\, dt &= m\bar{v}_{2_y} - m\bar{v}_{1_y} = m(\bar{v}_{2_y} - \bar{v}_{1_y})\\
\int_{t_1}^{t_2} \Sigma \bar{M}\, dt &= \bar{I}\omega_2 - \bar{I}\omega_1 = \bar{I}(\omega_2 - \omega_1)\\
\int_{t_1}^{t_2} \Sigma M_A\, dt &= I_A\omega_2 - I_A\omega_1 = I_A(\omega_2 - \omega_1)
\end{aligned}\right\} \qquad (15\text{-}6.1)$$

RESULTANT ANGULAR IMPULSE

The expression $\int_{t_1}^{t_2} \Sigma M\, dt$ is called resultant angular impulse. In terms of the ft-lb-sec system of units, the dimensional unit of this expression obviously is ft-lb-sec. If $\Sigma\mathbf{M}$ is variable, the resultant angular impulse can be found only if $\Sigma\mathbf{M}$ can be expressed mathematically in terms of time t, or if a graph can be obtained showing its variation with time. In the latter case, the resultant angular impulse is equivalent to the area under the proper limits of the $\Sigma\mathbf{M}$-t diagram.[2] In many cases, however, $\Sigma\mathbf{M}$ is constant and the resultant angular impulse is expressed simply as $\Sigma\mathbf{M}(t)$. It is important to note that resultant angular impulse is a vector quantity having the same sense and inclination as $\Sigma\mathbf{M}$.

ANGULAR MOMENTUM

In plane motion of symmetrical bodies, the angular momentum $\mathbf{H}$ about either the mass center or a fixed axis is $I\boldsymbol{\omega}$ and is also expressed in units of ft-lb-sec. The right-hand terms of the last two of Eqs. (15-6.1) represent the difference between the final and initial angular momenta and is called the change in angular momentum. The change in angular momentum depends only on the final and initial angular

[2] Except for a change in symbols, this concept may be used to determine angular velocity and angular displacement analogous to the corresponding use of force-time curves as discussed on p. 405 and again on p. 583.

velocities since I is constant with respect to axes fixed in the body. Likewise, as in angular impulse, it is important to note that angular momentum is a vector quantity which, in plane motion, has the same sense and inclination as $\boldsymbol{\omega}$.

The vector properties of angular impulse and angular momentum are usually of little importance in plane motion since they have a constant direction perpendicular to the plane of motion. Hence, in plane motion, they are usually written in scalar form as in Eqs. (15-6.1). Although scalar quantities are involved in Eqs. (15-6.1) the linear and angular terms have different units and directions. This fact discloses an important difference between work-energy and impulse-momentum methods. It was possible to apply the work-energy equations directly to a connected system of bodies having different motions because the terms involved were scalar quantities expressed in the same units which could be added arithmetically. In order to combine *linear* impulse-momentum terms with *angular* impulse-momentum terms, the linear terms must be transformed into angular terms by multiplying them by the appropriate moment arms. Usually this transformation is accomplished automatically if the impulse-momentum equations [Eqs. (15-6.1)] are applied to each separate body of the system. Notice also that all external forces on a system, including unknown constraint forces, have impulses which affect the momenta, whereas in work-energy the constraint forces may be neglected since they generally do no work on the system.

An alternate development which emphasizes the physical application of the impulse-momentum equations in plane motion is to start with the equivalence between applied forces and their dynamic effects as discussed in Section 13-2 and shown again here as Fig. 15-6.1. Multiplying both sides of these equivalents by dt gives Fig. 15-6.2 which relates the differential impulses to the differential momenta. This, after integrating over the time interval involved, then becomes Fig. 15-6.3 which is finally rearranged to give Fig. 15-6.4. Although the diagrams show only a single body, this body may also represent a system of bodies. Expressed in words, Fig. 15-6.4 states the impulse-momentum principle as follows:

Initial system momenta + Resultant ext. impulses
= Final system moments (15-6.2)

The advantage of this diagrammed approach is that angular momenta are seen to be couples, while the vector nature of the linear momenta is clearly shown by the different directions that $\bar{\mathbf{v}}_1$ and $\bar{\mathbf{v}}_2$ have after the elapsed time interval. Equations (15-6.1) may be obtained directly from Fig. 15-6.4 by summing and equating the X and Y components of the linear momenta and impulses, and by equating the

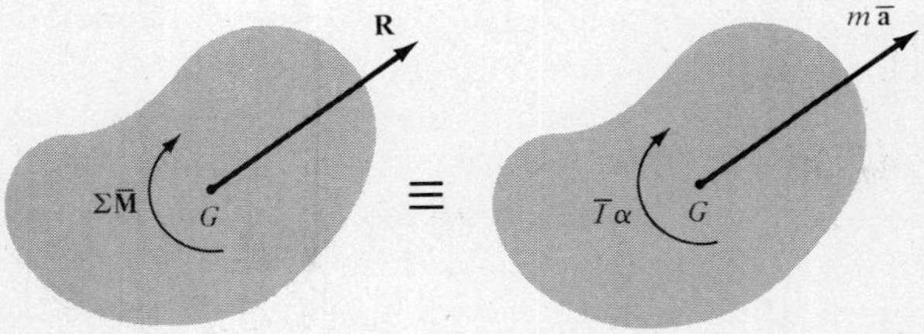

Figure 15-6.1 Equivalence of applied forces to their dynamic effects.

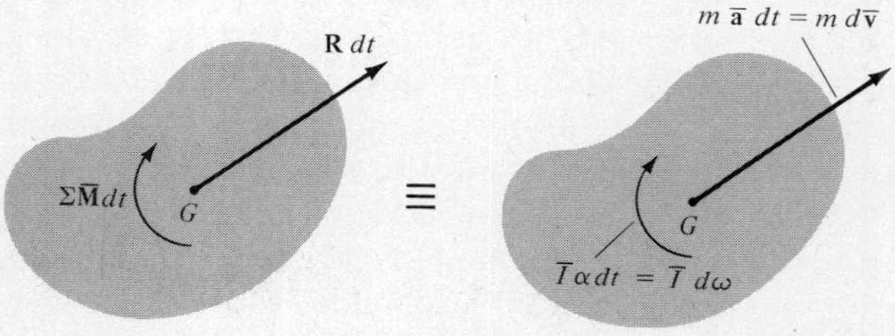

Figure 15-6.2 Equivalence of differential impulses to differential momenta.

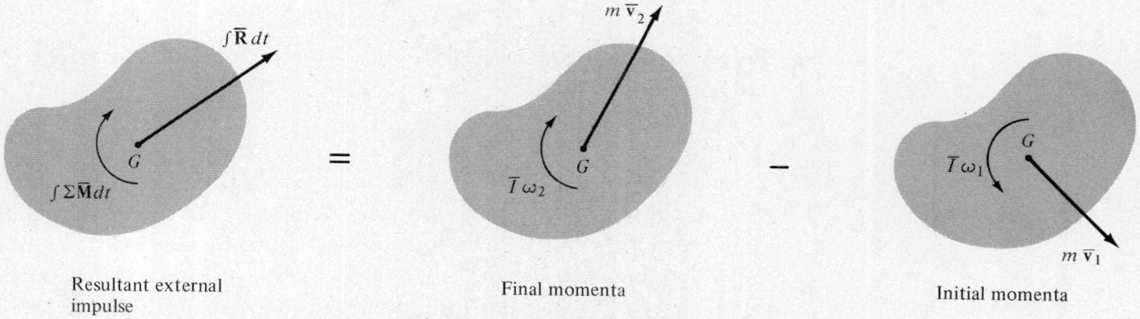

Figure 15-6.3 Effect of integrating differential impulse-momentum relations.

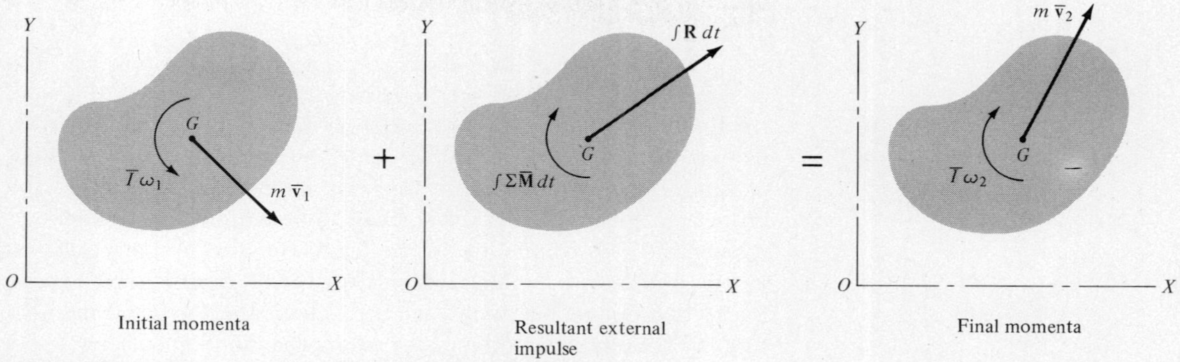

Figure 15-6.4 Rearranged impulse-momentum relations.

moments of these vectors and the angular momentum couples about the same point in each part of the diagrammed equivalents. The coordinate axes may be fixed in space or may move with the mass center while maintaining fixed directions.

The impulse-momentum approach may be used to relate force, time, and velocity for various types of plane motion. However, there is little advantage in doing so since a combination of force-inertia methods plus the equations of kinematics yield the same results almost as rapidly. The real advantages of impulse-momentum occur in systems where either or both the resultant linear impulse or resultant angular impulse is zero over a time interval, thereby resulting in conservation of linear momentum or angular momentum, or both. Important also are situations involving impact where impulsive forces or impulsive moments act over so short a time interval that there is no significant change in position although significant changes in velocity do take place. Usually, impulsive forces and impulsive moments due to impact are so much larger in comparison to those due to ordinary forces such as weight that the impulsive effects of the latter can be ignored.

CONSERVATION OF ANGULAR MOMENTUM

Referring to Eq. (*b*) on p. 607, the integral form of the impulse-momentum equation, we have

$$\int \Sigma \mathbf{M}\, dt = \int d\mathbf{H}$$

from which we see that during the time interval over which $\Sigma \mathbf{M} = 0$, the angular momentum $\mathbf{H}$ must be constant. The condition that angular momentum be constant applies either to rotation at a uniform angular velocity, or, more importantly, to a system in which the resultant moment of external forces is zero about the *same* fixed axis because of mutual action and reaction between the particles or bodies composing a rotating system. Conservation of angular momentum applies only to the system as a whole, never to its component parts.

The principle of conservation of angular momentum can also be applied to a translating body striking a rotating body. In this case, the linear momentum of the translating body must be multiplied by its moment arm about the axis of rotation to convert it into moment of linear momentum; i.e., its "angular momentum." To avoid the error of neglecting to convert linear momentum into moment of momentum, it is recommended the diagrammatic approach outlined in Fig. 15-6.4 and summarized by Eq. (15-6.2) be used in all cases. Recall that the momenta for each body consists of the linear momentum $m\bar{\mathbf{v}}$ acting through the mass center plus the moment couple $\bar{I}\boldsymbol{\omega}$. This diagrammatic approach is very useful to obtain moment summations about a center that eliminates resultant angular impulse, thereby leading automatically to conservation of angular momentum.

ILLUSTRATIVE PROBLEMS

15-6.1. As indicated in Fig. 15-6.5, a 1-oz bullet moving at 1500 fps strikes and pierces a 6-ft rod weighing 10 lb which is hanging motionless from a frictionless pivot at A. If the bullet emerges with a speed of 600 fps, determine the maximum angle through which the rod will swing.

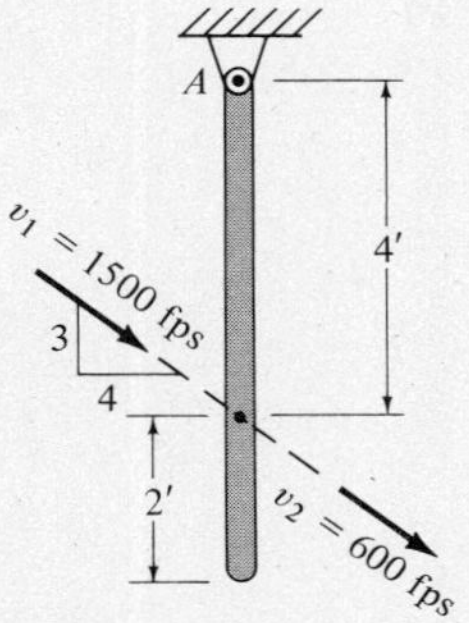

Figure 15-6.5

Solution

In Fig. 15-6.6, we equate the sum of the initial momenta and resultant external impulse to the final momenta. The initial linear momentum $m_b \mathbf{v}_1$ of the bullet is represented by its components. The only external impulse is that acting at the pivot A since those caused by mutual action and reaction of the bullet and rod cancel each other and the impulsive effect of the weights are negligibly small. The final momenta consist of the components of the final linear momentum $m_b \mathbf{v}_2$ of the bullet and the momenta $m\bar{v}$ and $\bar{I}\omega_2$ acquired by the rod.

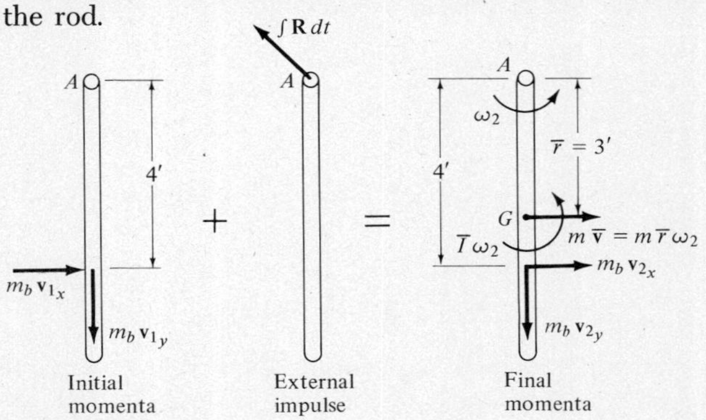

Figure 15-6.6

Equating moment summations about A eliminates the impulse at A and yields

$$m_b v_{1_x} r = m_b v_{2_x} r + \bar{I}\omega_2 + m\bar{v}\bar{r}$$

which represents conservation of angular momentum about the fixed pivot A. Notice how the moment summation of the diagrammed impulse-momentum terms automatically converted linear momentum of the bullet into moment of linear momentum about A. Now substituting numerical data and noting that $\bar{I} = \frac{1}{12}\frac{W}{g}L^2$, we have

$$\frac{1}{16g}(1500)(0.8)(4) = \frac{1}{16g}(600)(0.8)(4) + \frac{1}{12}\frac{10}{g}(6)^2\omega_2 + \frac{10}{g}(3\omega_2)(3)$$

from which

$$\omega_2 = 1.5 \text{ rad/sec} \circlearrowleft$$

Notice that the sum of $\bar{I}\omega$ and $m\bar{v}\bar{r}$ equals $(\bar{I} + m\bar{r}^2)\omega_2 = I_A\omega_2$.

We now apply the work-energy relation RW = ΔKE to determine the maximum swing angle of the rod. At the final dashed position in Fig. 15-6.7, the initial KE due to impact has been dissipated by the negative work done by gravity so that

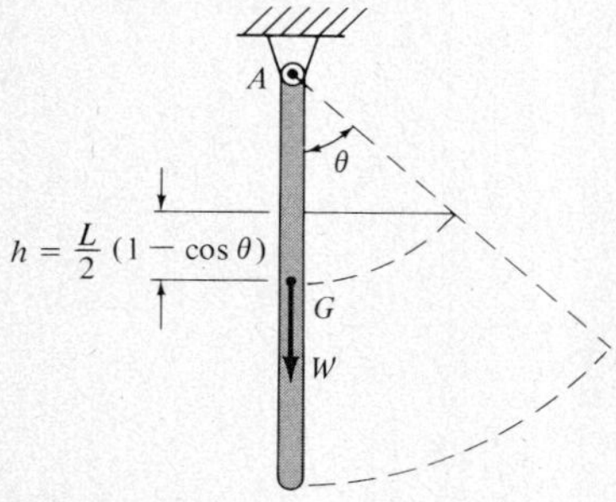

Figure 15-6.7

$$\left[-Wh = \frac{1}{2}I_A(\omega^2 - \omega_o^2)\right] \quad -\frac{WL}{2}(1 - \cos\theta) = \frac{1}{2}\left(\frac{W}{3g}L^2\right)(0 - \omega_2^2)$$

which reduces to

$$\cos\theta = 1 - \frac{L\omega_2^2}{3g} = 1 - \frac{6(1.5)^2}{3g} = 0.86 \qquad \theta = 30.7° \qquad \textbf{\textit{Ans.}}$$

If the passage of the bullet occurs in a time assumed to be 0.002 sec, the average reaction R at A can be found by equating the X and Y components of linear momenta in Fig. 15-6.6. Thus,

$$[m_b v_{1_x} - R_x\,\Delta t = m_b v_{2_x} + m\bar{r}\omega_2]$$

$$\frac{1}{16g}(1500)(0.8) - R_x(0.002) = \frac{1}{16g}(600)(0.8) + \frac{10}{g}(3)(1.5)$$

$$R_x = 0$$

$$[-m_b v_{1_y} + R_y\,\Delta t = -m_b v_{2_y}]$$

$$-\frac{1}{16g}(1500)(0.6) + R_y(0.002) = -\frac{1}{16g}(600)(0.6)$$

$$R_y = 524 \text{ lb}\uparrow$$

$$\therefore \quad R = \sqrt{R_x^2 + R_y^2} = 524 \text{ lb}\uparrow$$

Observe that while angular momentum about A is conserved, there is not conservation of linear momentum in the X or Y directions. If desired, the loss in KE is readily found by taking the difference between the initial and final KE as follows:

$$\text{Loss in KE} = \frac{1}{2}mv_1^2 - \frac{1}{2}mv_2^2 - \frac{1}{2}I_A\omega_2^2$$

$$= \frac{1}{2}\left(\frac{1}{16g}\right)(1500)^2 - \frac{1}{2}\left(\frac{1}{16g}\right)(600)^2 - \frac{1}{2}\left(\frac{10}{3g}\right)(6)^2(1.5)^2$$

$$= 2180 - 349 - 4.19 \approx 1827 \text{ ft-lb}$$

Notice how small is the KE absorbed by the rod; the rest is largely dissipated by heat.

15-6.2. A rod 5 ft long and weighing 9 lb is rotating in a vertical plane about one end. At the instant the rod is in the vertical position shown in Fig. 15-6.8, it has a clockwise angular velocity $\omega_1 = 2$ rad/sec when it is hit by a 0.5-lb ball moving horizontally at 200 fps. If the coefficient of restitution is $e = 0.6$, determine the angular velocity of the rod directly after impact. Assuming the impact lasts for 0.005 sec, find the average impact force P between the ball and the rod.

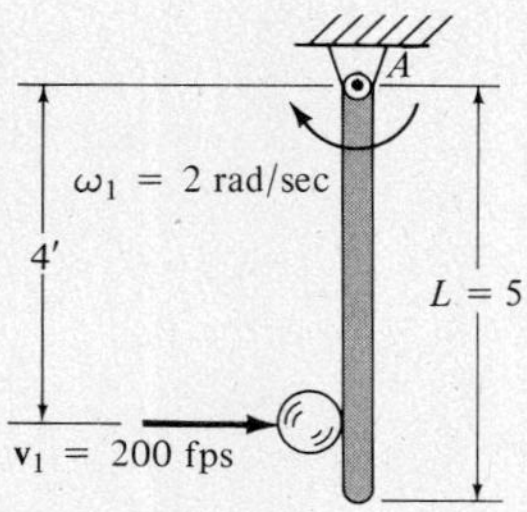

Figure 15-6.8

Solution

This problem illustrates direct eccentric impact in which we assume the ball and the rod rebound in opposite directions as shown in Fig. 15-6.9 by the diagrammed impulse-momentum equation. Noting that $\bar{I}\omega + m\bar{v}\bar{r} = I_A\omega$ as shown in the previous problem, we equate mo-

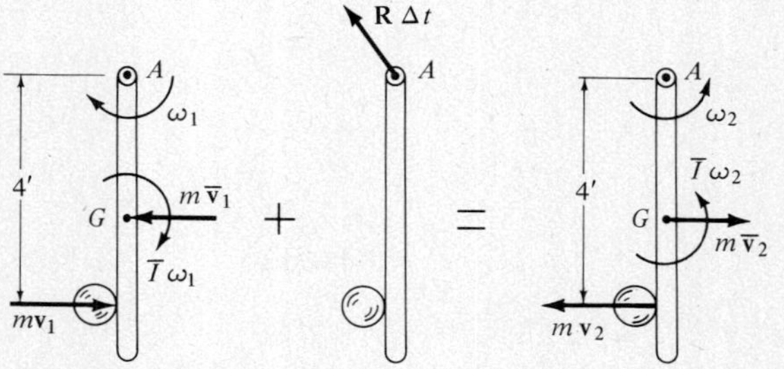

Figure 15-6.9

ment summations about A to eliminate the impulse at A, thereby giving the following equation of conservation of angular momentum:

$$mv_1r - I_A\omega_1 = I_A\omega_2 - mv_2r$$

Substituting numerical data and noting that

$$I_A = \frac{1}{3}\frac{W}{g}L^2 = \frac{9(5)^2}{3g} = \frac{75}{g} \text{ ft-lb-sec}^2$$

we have

$$\frac{0.5}{g}(200)(4) - \frac{75}{g}(2) = \frac{75}{g}\omega_2 - \frac{0.5}{g}(v_2)(4)$$

which reduces to

$$250 = 75\omega_2 - 2v_2 \qquad (a)$$

Another equation relating ω_2 and v_2 is available from the definition of the coefficient of restitution, Eq. (15-5.1). Along the line of impact, the original velocity of the impact point on the rod is $r\omega = 4(2) = 8$ fps leftward and its final velocity is $4\omega_2$ rightward. Hence, taking the original rightward velocity of the ball as the positive sense of velocity, we obtain

$$[e(v_1 - v_2) = v_2' - v_1'] \qquad 0.6[200 - (-8)] = 4\omega_2 - (-v_2)$$

which reduces to

$$125 = 4\omega_2 + v_2 \tag{b}$$

Solving Eqs. (*a*) and (*b*) simultaneously, we obtain

$$\omega_2 = 6.02 \text{ rad/sec} \circlearrowleft \quad \text{and} \quad v_2 = 100.7 \text{ fps} \leftarrow$$

The positive results confirm the assumed directions to be correct.

The contact force P can be found either by applying angular impulse momentum to the rod or applying linear impulse momentum to the ball. Using both methods will serve as a check. For the rod, we obtain

$$[+\circlearrowleft \Sigma M_A\, \Delta t = I_A(\omega - \omega_o)] \qquad 4P(0.005) = \frac{75}{g}[6.02 - (-2)]$$

$$P = 934 \text{ lb} \qquad \textbf{Ans.}$$

while for the ball, we have

$$[\overset{+}{\rightarrow}\Sigma X\, \Delta t = m(v_2 - v_1)] \qquad -P(0.005) = \frac{0.5}{g}(-100.7 - 200)$$

$$P = 934 \text{ lb} \qquad \textbf{Check}$$

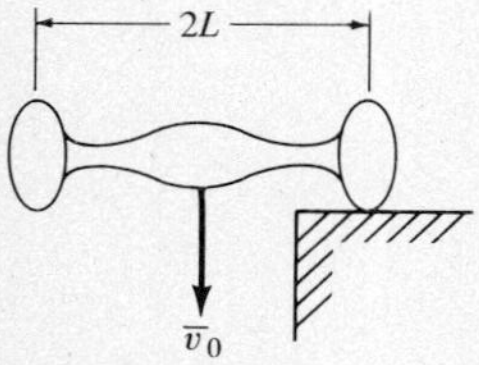

Figure 15-6.10

15-6.3. A dumbbell has a mass m and a centroidal radius of gyration k about an axis perpendicular to its axis of symmetry. At the instant before one end of the dumbbell strikes the corner of a table as in Fig. 15-6.10, its angular velocity is zero but its mass center has a downward velocity $\bar{v}_o$. Determine its angular velocity directly after (a) plastic impact; (b) partially elastic impact with a coefficient of restitution e.

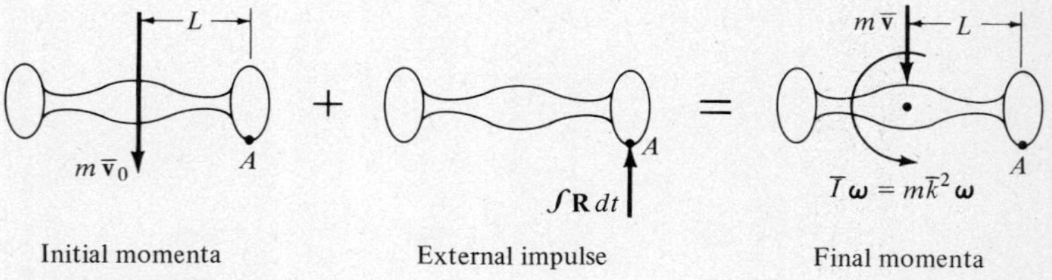

Figure 15-6.11

Solution: Part (a)

Applying the impulse-momentum equation shown in Fig. 15-6.11 and equating moments about A, we have

$$m\bar{v}_o L = m\bar{v}L + mk^2\omega \tag{a}$$

After plastic impact, the body is instantaneously rotating about A and the kinematic relation between $\bar{v}$ and ω is

$$[v = r\omega] \qquad \bar{v} = L\omega \tag{b}$$

Solving Eqs. (*a*) and (*b*) then results in

$$\omega = \frac{\bar{v}_o L}{L^2 + k^2} \circlearrowright \qquad \textbf{\textit{Ans.}}$$

Part (b)

With partially elastic impact, end A rebounds upward with a velocity $e\bar{v}_o$. Using kinematics, the velocity $\bar{v}$ of the mass center now is

$$[v_G = v_A \mathop{+}\!\!\!\rightarrow (v_{G/A} = r\omega)] \qquad \bar{v} = e\bar{v}_o\uparrow + L\omega\downarrow = (L\omega - e\bar{v}_o)\downarrow$$

Substituting this value of $\bar{v}$ in Eq. (*a*) above now results in

$$\omega = \frac{\bar{v}_o L(1 + e)}{L^2 + k^2} \circlearrowright \qquad \textbf{\textit{Ans.}}$$

Observe that for perfectly elastic impact ($e = 1$), the final value of ω is twice that for plastic impact.

PROBLEMS

15-6.4. A pulley of 2.5-ft radius is rotated about its centroidal axis by a tangentially applied force $P = 24t - 4t^2$ where P is in pounds and t is in seconds. If the pulley has a centroidal mass moment of inertia of 12 ft-lb-sec^2, find the number of revolutions through which the pulley rotates from rest before starting to reverse its direction. Also find its angular velocity when it returns to its original position.

Hint: See footnote on p. 607.

$$\theta = 24.1 \text{ rev};\ \omega = -120 \text{ rad/sec} \qquad \textbf{\textit{Ans.}}$$

15-6.5. Repeat the previous problem but change the radius of the pulley to 3 ft and the value of the tangentially applied force to $P = 6t^2 - 0.5t^3$ where P is in pounds and t is in seconds.

15-6.6. A wheel 6 ft in diameter is rotated about its centroidal axis by a tangentially applied force $P = 40 + 6t^2$ where P is in pounds and t is in seconds. The rotation is resisted by a brake ($f = 0.40$) which exerts a constant normal force of 100 lb on the rim. If P acts for 6 sec and is then removed, through how many revolutions will the wheel turn before stopping? Assume the mass moment of inertia is 6 ft-lb-sec^2.

$$\theta = 238 \text{ rev} \qquad \textbf{\textit{Ans.}}$$

15-6.7. As a golf ball of radius r hits a green as in Fig. P-15-6.7, it

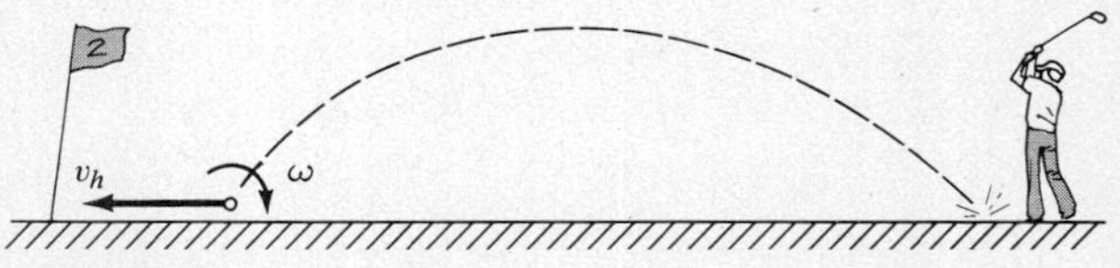

Figure P-15-6.7

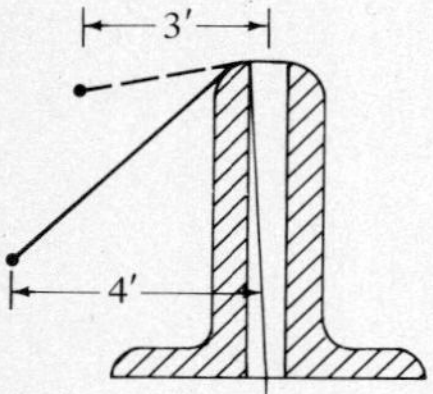

Figure P-15-6.8

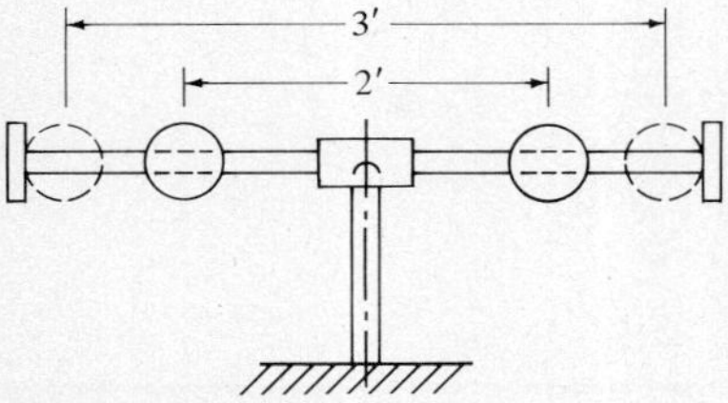

Figure P-15-6.9

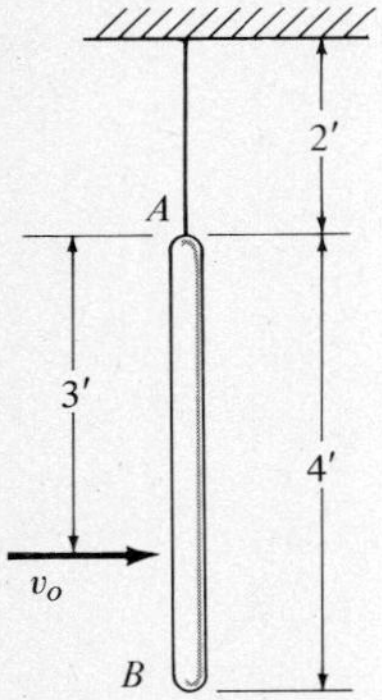

Figure P-15-6.13

has a horizontal component of velocity v_h and a backspin angular velocity ω. Determine the relation between v_h and ω that will cause the golf ball to stop dead upon hitting the green. Assume plastic impact.

15-6.8. A 2-lb weight is attached to one end of a cord as shown in Fig. P-15-6.8. At the given position, the weight is rotating about the vertical axis of the hollow support in a circular path of 4-ft radius at 9 rad/sec. If the cord is slowly drawn into the support so that the radius of rotation is 3 ft, determine the new angular velocity of the weight and the tension in the cord. Further, determine and explain the increase in kinetic energy.

$\omega = 16$ rad/sec; $T = 47.7$ lb; $\Delta\text{KE} = 67.1$ ft-lb **Ans.**

15-6.9. When the speed becomes critical, the balls in a certain type of governor are released by a catch, and they move to the positions shown by the dashed outlines in Fig. P-15-6.9. Each ball weighs 20 lb and $\bar{I}$ for the rod is 3 ft-lb-sec². If the speed of the assembly is 20 rad/sec when the governor operates, determine the final speed of the system. What would be the final speed if only one ball moved to the dashed position?

15-6.10. Two disks A and B are mounted loosely on a horizontal shaft. The mass moment of inertia for disk A is 6 ft-lb-sec², and for B it is 2 ft-lb-sec². A spring wound loosely around the shaft connects the disks. One disk is used to wind up 300 ft-lb of energy in the spring, after which both disks are released from rest. Neglecting the mass of the spring, compute the angular velocity of each disk at the instant when the spring has released all its stored energy to the disks.

$\omega_A = \pm 5$ rad/sec; $\omega_B = \mp 15$ rad/sec **Ans.**

15-6.11. A uniform bar, 3 ft long and weighing 20 lb, is hanging motionless from a horizontal axis at its upper end. It is struck at 2 ft below this axis by a 2-lb ball of putty moving horizontally at 60 fps. If all the putty clings to the bar, compute the maximum angle through which the bar will swing and the loss in kinetic energy.

$\theta = 52.2°$; KE loss $= 98.6$ ft-lb **Ans.**

15-6.12. A uniform rod, 6 ft long and weighing 10 lb, is suspended vertically from a horizontal axis at its upper end. A 1-oz bullet is fired at 1000 fps along a horizontal line 5 ft below and perpendicular to the axis from which the rod is suspended. The bullet pierces the rod which is observed to swing through 30°. Compute the velocity of the bullet directly after it emerges from the rod.

$v = 436$ fps **Ans.**

15-6.13. A uniform 4-ft bar weighing 30 lb is hanging motionless from a 2-ft cord as shown in Fig. P-15-6.13. A 1-oz bullet moving horizontally at $v_o = 2000$ fps pierces the bar and emerges with a velocity of 800 fps. Determine the velocity of end B of the bar directly after impact.

$v_B = 5.31$ fps **Ans.**

15-6.14. Determine the velocity of end A of the previous problem if the path of the bullet is downward at a slope of 3 vertical to 4 horizontal.

15-6.15. A friction drive consists of two disks keyed to parallel shafts. They can be separated or brought into contact by moving disk B in the horizontal slot as shown in Fig. P-15-6.15. Initially, they are separated and rotate in opposite directions at $\omega_A = 6$ rad/sec and $\omega_B = 30$ rad/sec. Their radii and moments of inertia are $r_A = 9$ in., $r_B = 6$ in., $I_A = 1.2$ ft-lb-sec^2, and $I_B = 0.4$ ft-lb-sec^2. Find the angular velocity of the disks after they are brought into contact and slipping stops.

$\omega_A = 12$ rad/sec ↻; $\omega_B = 18$ rad/sec ↺ **Ans.**

15-6.16. Solve the previous problem if both disks are initially rotating clockwise at $\omega_A = 8$ rad/sec and $\omega_B = 30$ rad/sec.

15-6.17. In Fig. P-15-6.17, block A weighs 100 lb and the homogeneous cylinder B weighs 200 lb. An inextensible cord passes through a hole in the block and is wrapped around the cylinder which is at rest. Block A is released from rest in the given position and drops onto the weightless stopper C. Determine the velocity of A after it has descended an additional 3 ft after striking C. Neglect any rebound and assume cord remains taut.

$v = 12.7$ fps **Ans.**

15-6.18. The system shown in Fig. P-15-6.18 starts from rest in the given position. After rising $h = 4$ ft, body B picks up body C. How much higher will B move?

2.54 ft **Ans.**

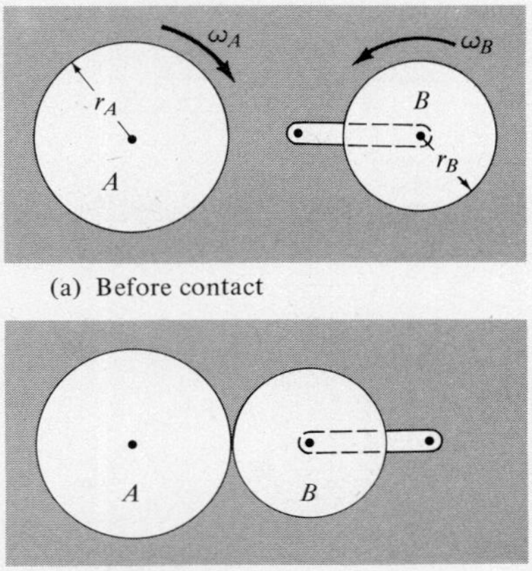

(a) Before contact

(b) After contact

Figure P-15-6.15

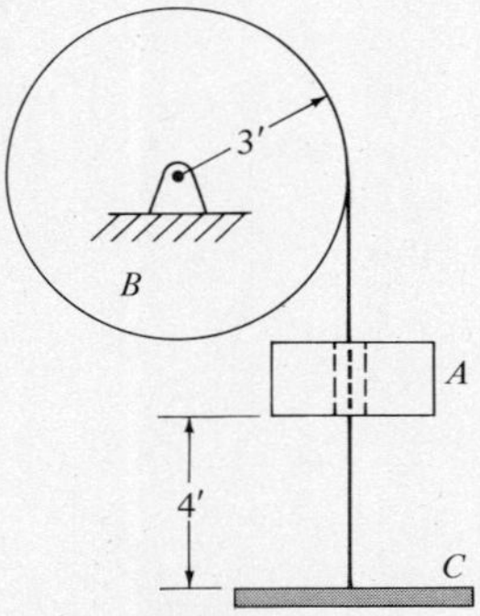

Figure P-15-6.17

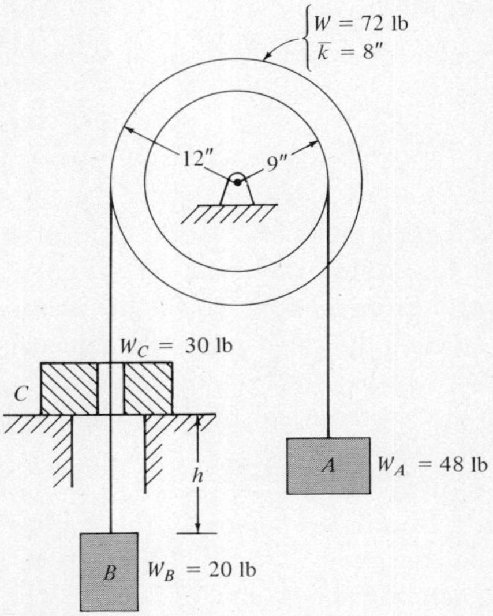

Figure P-15-6.18

15-6.19. The uniform sphere in Fig. P-15-6.19 rolls without slipping down the incline. Its center has a velocity $\mathbf{v}_o$ just before it touches the level floor. Determine the velocity of the center of the sphere as it starts rolling without slipping on the floor. Assume there is no rebound.

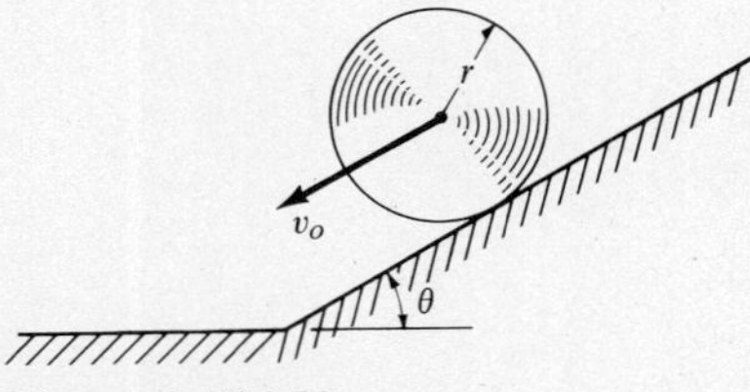

Figure P-15-6.19

15-6.20. A uniform sphere 30 in. in diameter and weighing 10 lb is rolling without slipping on a level floor when it hits a step 6 in. high as

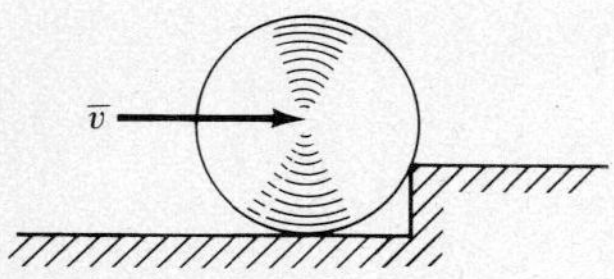

Figure P-15-6.20

shown in Fig. P-15-6.20. Determine the smallest velocity $\bar{v}$ of the sphere so that it will roll over the step, assuming plastic impact.

$\bar{v} = 8.23$ fps **Ans.**

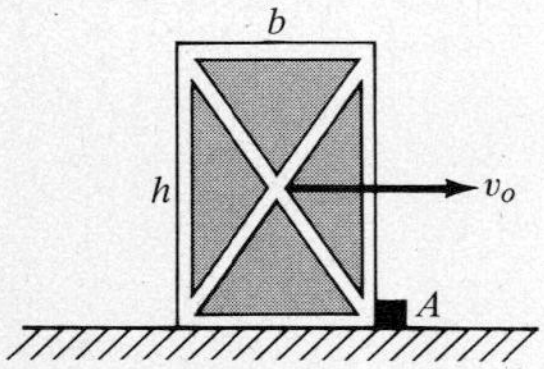

Figure P-15-6.21

15-6.21. As shown in Fig. P-15-6.21, a box weighing W lb is moving with a velocity v_o along a level floor when it hits a cleat at A. Determine the angular velocity of the box directly after impact if (a) $e = 0$ and (b) for any value of e. Ignore the height of the cleat.

15-6.22. In Fig. P-15-6.22, the uniform rods AB and CD are free to swing in a vertical plane about A and C. Each rod is 6 ft long and weighs 30 lb. Rod AB is released from rest at $\theta = 90°$ and a hard rubber knob at its end strikes CD. If $e = 0.60$, through what angle will CD swing, starting from rest?

$\theta = 62.9°$ **Ans.**

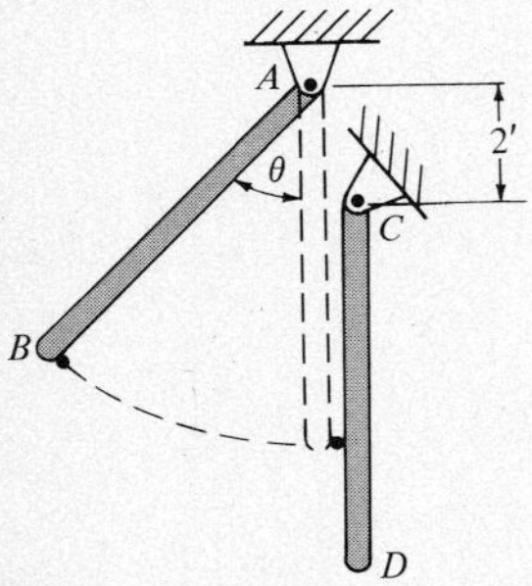

Figure P-15-6.22

15-6.23. Determine the coefficient of restitution for the system of Prob. 15-6.22 if CD swings through 45° after AB swings through $\theta = 60°$.

15-6.24. The uniform bars in Fig. P-15-6.24 rotate on a horizontal surface about vertical axes at A and C. At the given position, AB is rotating counterclockwise at 2 rad/sec when it strikes CD which is at rest. Each bar is 6 ft long and weighs 30 lb. Assuming plastic impact, determine the angular velocity of each bar directly after impact and the loss in KE.

$\omega_{AB} = 0.82$ rad/sec; KE loss = 19.8 ft-lb **Ans.**

15-6.25. Solve the preceding problem for partially elastic impact with $e = 0.60$.

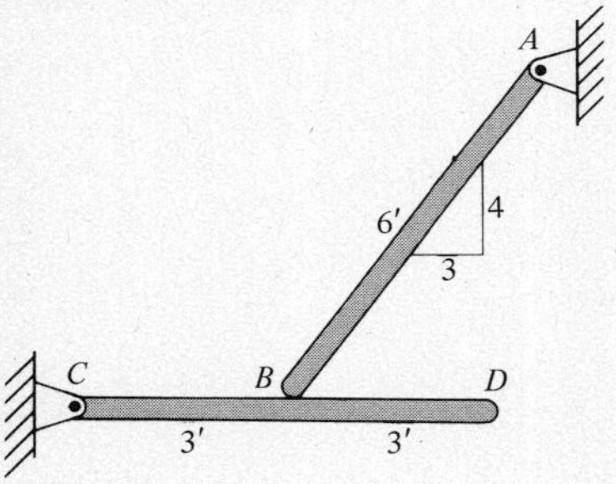

Figure P-15-6.24

*15-7 SATELLITE MOTION

When a body moves in a curved path under the action of a force directed through its mass center toward or away from a fixed point, we have what is known as central force motion. This is the condition which applies to satellites orbiting the earth under the influence of gravitational attraction.[3] With respect to the center O of the earth, therefore, the motion of a satellite is governed by a gravitational pull $F = \dfrac{GMm}{r^2}$ [see Eq. (10-2.1)] and a moment $\mathbf{M}_O = 0$. Hence, as discussed in Section 15-6, with zero moment about O, the angular momentum $\mathbf{H}_O$ about O must be constant.

Refer now to the orbiting satellite shown in Fig. 15-7.1. The angular momentum at any instant is the moment about O of the linear momentum $m\mathbf{v}$ and hence, in terms of the position vector $\mathbf{r}$ from O,

$$\mathbf{H}_O = \mathbf{r} \times m\mathbf{v}$$

[3] By using appropriate gravitational pulls, the discussion applies also to satellites orbiting other planets. For an extended discussion of satellite motion, see W. T. Thomson, *Introduction to Space Dynamics*, John Wiley & Sons (1961).

from which we conclude that, since $\mathbf{H}_O$ is constant in magnitude and direction, the satellite moves in a fixed plane, defined by $\mathbf{r} \times m\mathbf{v}$, which always is perpendicular to the constant direction of $\mathbf{H}_O$.

We now discuss how a combination of angular impulse-momentum and work-energy principles will solve various problems associated with satellite motion.

Assume that the path of an orbiting satellite is an ellipse as shown in Fig. 15-7.2. Other possible paths will be discussed later. The velocity of the satellite varies with changes in its position vector from O because of the work done on it by gravitational attraction.

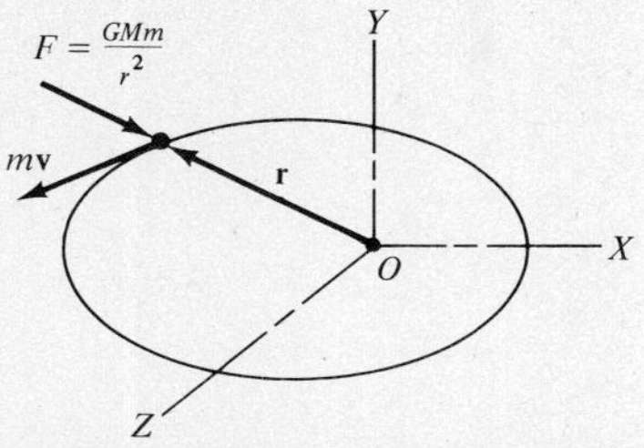

Figure 15-7.1 Satellite orbiting about the center O of the earth.

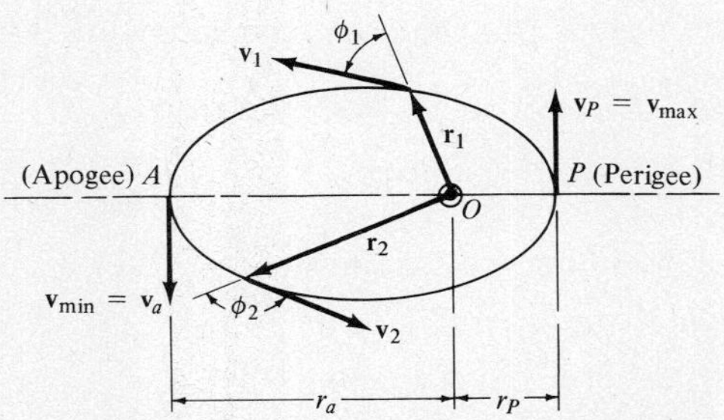

Figure 15-7.2 Velocity variations of a satellite in an elliptical orbit.

From Eq. (10-2.1) we recall that the magnitude of the gravitational force on the satellite is $F = GMm/r^2$. Noting that $\mathbf{F}$ is always directed toward O and hence opposite to $\mathbf{r}$, the resultant work done on the satellite as it moves from any position 1 to any other position 2 is

$$\text{RW} = \int \mathbf{F} \cdot d\mathbf{r} = -\int F\,dr = -\int_{r_1}^{r_2} \frac{GMm}{r^2}\,dr = GMm\left[\frac{1}{r}\right]_{r_1}^{r_2}$$

which may be rewritten in either of the following forms:

$$\text{RW} = GMm\left(\frac{1}{r_2} - \frac{1}{r_1}\right) = \frac{GMm}{r_2}\left(1 - \frac{r_2}{r_1}\right)$$

On equating the change in kinetic energy to this resultant work, we obtain

$$\frac{1}{2}m(v_2^2 - v_1^2) = GMm\left(\frac{1}{r_2} - \frac{1}{r_1}\right) = \frac{GMm}{r_2}\left(1 - \frac{r_2}{r_1}\right)$$

or

$$v_2^2 - v_1^2 = 2GM\left(\frac{1}{r_2} - \frac{1}{r_1}\right) = \frac{2GM}{r_2}\left(1 - \frac{r_2}{r_1}\right) \qquad (15\text{-}7.1)$$

Another relation between these velocities is obtained by recalling that the angular momentum is constant, thereby giving

$$mv_1r_1 \sin\phi_1 = mv_2r_2 \sin\phi_2$$

or

$$v_1 r_1 \sin \phi_1 = v_2 r_2 \sin \phi_2 \tag{15-7.2}$$

Observe also in Fig. 15-7.2 that $\sin \phi = 1$ at both the apogee A, the position furthest from O where the velocity is a minimum, and at the perigee P, the position closest to O where the velocity is a maximum.

With some additional information, Eqs. (15-7.1) and (15-7.2) are sufficient to solve most problems of satellite motion. The value of GM is readily found by noting that the gravitational pull on a body at the earth's surface is its weight W. Denoting the radius of the earth by R, we then have

$$W = mg = \frac{GMm}{R^2}$$

whence substituting $R = 3960$ miles, we obtain

$$\begin{aligned} GM &= gR^2 = 32.2(3960 \times 5280)^2 \\ &= 1.408 \times 10^{16}\ \text{ft}^3/\text{sec}^2 = 124 \times 10^{10}\ \text{mi}^3/\text{hr}^2 \end{aligned} \tag{15-7.3}$$

In order for a satellite to move in a circular orbit of radius r at the constant velocity v, the gravitational pull at every position will be balanced by the centrifugal inertia force, resulting in the following relation:

$$\frac{GMm}{r^2} = m\frac{v^2}{r} \quad \text{or} \quad v^2 = \frac{GM}{r} \tag{15-7.4}$$

For smaller or larger velocities, the various orbits possible are shown in Fig. 15-7.3. It is evident that multiplying a circular orbiting veloc-

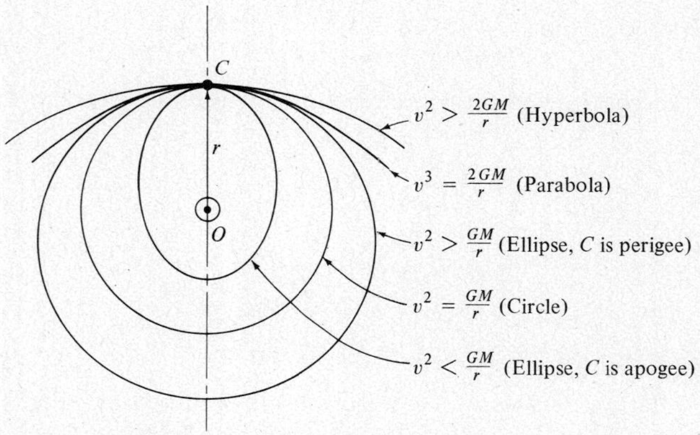

Figure 15-7.3 Various possible orbital paths.

ity by $\sqrt{2}$ will determine the velocity v_e to escape from the earth's gravitational field. At the surface of the earth, this value is

$$v_e^2 = \frac{2GM}{R} = \frac{2(124 \times 10^6)}{3960} \quad \text{giving} \quad v_e = 25{,}020 \text{ mph} = 36{,}700 \text{ fps}$$

while at larger radii where the gravitational pull is smaller, the escape velocity will decrease.

We conclude our discussion of satellite motion by showing how to determine the major and minor semiaxes of an elliptical orbit and how they are related to the period of the orbit; i.e., the time τ required to complete an orbit. If we refer to Fig. 15-7.4 in which

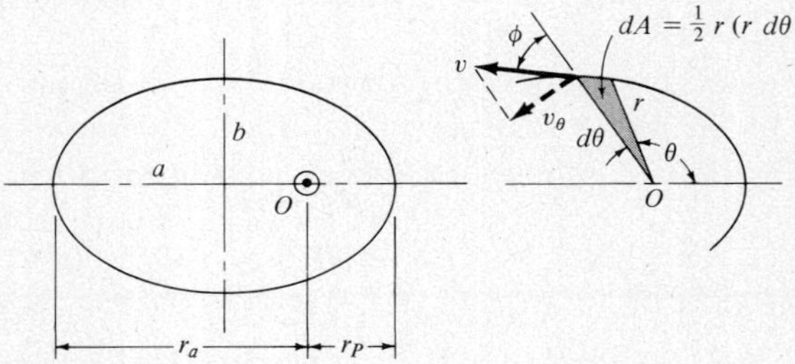

Figure 15-7.4

the earth is at one focus of the elliptical path, we find that analytic geometry expresses the major and minor semiaxes, a and b, in terms of the apogee and perigee distances, r_a and r_p, as follows:

$$a = \frac{1}{2}(r_a + r_p) \quad \text{and} \quad b = \sqrt{r_a r_p} \tag{15-7.5}$$

As a preliminary to determining the period of the orbit, we first define *areal velocity* as the time rate of the area swept out by the position vector. Thus, in Fig. 15-7.4, the shaded area dA generated by r in the time dt is $\frac{1}{2}r(r\,d\theta)$ from which

$$\text{areal velocity} = \frac{dA}{dt} = \frac{1}{2}r\left(r\frac{d\theta}{dt}\right) \tag{a}$$

Recalling from Eq. (15-7.2) that $v \sin \phi$ is a constant and that $v \sin \phi$ is the transverse component of velocity $v_\theta = r\theta$, we see that Eq. (*a*) can be rewritten as

$$\frac{dA}{dt} = \frac{1}{2} rv \sin \phi \tag{b}$$

whence separating the variables in Eq. (*b*) and integrating, the period is found to be

$$\tau = \frac{2A}{rv \sin \phi}$$

which, in terms of the area πab of the ellipse and the identities $rv \sin \phi = r_a v_a = r_p v_p$, becomes

$$\tau = \frac{2\pi ab}{r_a v_a} = \frac{2\pi ab}{r_p v_p} \tag{15-7.6}$$

It is remarkable that from only astronomical observations, Johann Kepler (1571–1630) was able to deduce his famous laws of planetary motion, two of which we have verified by Newtonian mechanics. Kepler's laws may be stated as:

1. Each planet describes an ellipse about the sun as a focus.

2. The position vector from the sun to a planet sweeps out equal areas in equal times.

3. For each planet, the square of its periodic time is proportional to the cube of the major semiaxis of its orbit. [The precise relation is $(\tau/2\pi)^2 = a^3/GM$ whose derivation constitutes Prob. 15-7.7.]

ILLUSTRATIVE PROBLEMS

15-7.1. The distances from the center of the earth to the apogee and perigee of an orbiting satellite are respectively r_a and r_p. Show that the corresponding velocities v_a and v_p at apogee and perigee are given by

$$v_a^2 = \frac{2GM}{r_a + r_p}\left(\frac{r_p}{r_a}\right) \quad \text{and} \quad v_p^2 = \frac{2GM}{r_a + r_p}\left(\frac{r_a}{r_p}\right)$$

Compute v_a and v_p in mph if the maximum and minimum altitudes of the orbit are respectively 2140 miles and 240 miles, and the radius of the earth is $R = 3960$ miles.

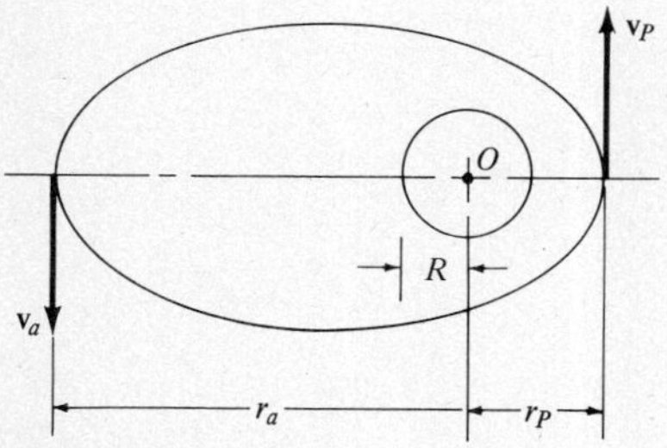

Figure 15-7.5

Solution

At apogee and perigee, the velocities are perpendicular to their position vectors as shown in Fig. 15-7.5. Applying conservation of angular momentum, we have

$$mv_a r_a = mv_p r_p \quad \text{or} \quad v_p = \frac{r_a}{r_p} v_a \tag{a}$$

From Eq. (15-7.1) we obtain

$$v_a^2 - v_p^2 = 2GM\left(\frac{1}{r_a} - \frac{1}{r_p}\right) = \frac{2GM}{r_a}\left(1 - \frac{r_a}{r_p}\right)$$

which, on replacing v_p by its value from Eq. (a), becomes

$$v_a^2\left(1 - \frac{r_a^2}{r_p^2}\right) = v_a\left(1 + \frac{r_a}{r_p}\right)\left(1 - \frac{r_a}{r_p}\right) = \frac{2GM}{r_a}\left(1 - \frac{r_a}{r_p}\right)$$

whence

$$\left.\begin{aligned} v_a^2 &= \frac{2GM}{r_a + r_p}\left(\frac{r_p}{r_a}\right) \\ \text{and similarly,} \qquad & \\ v_p^2 &= \frac{2GM}{r_a + r_p}\left(\frac{r_a}{r_p}\right) \end{aligned}\right\} \qquad (15\text{-}7.7)$$

Noting from Eq. (15-7.3) that $GM = 124 \times 10^{10}$ mi³/hr² and that the apogee and perigee distances are $r_a = 3960 + 2140 = 6100$ miles and $r_p = 3960 + 240 = 4200$ miles, we obtain

$$v_a^2 = \frac{2(124 \times 10^{10})}{6100 + 4200}\left(\frac{4200}{6100}\right) = 166.8 \times 10^6, \qquad v_a = 12{,}880 \text{ mph}$$

$$v_p^2 = \frac{2(124 \times 10^{10})}{6100 + 4200}\left(\frac{6100}{4200}\right) = 350 \times 10^6, \qquad v_p = 18{,}700 \text{ mph}$$

15-7.2. A satellite is launched into orbit at an altitude of 1740 miles when it has a velocity of 18,000 mph directed at 75° with the local vertical as shown in Fig. 15-7.6. Determine the maximum and minimum altitudes of the orbit.

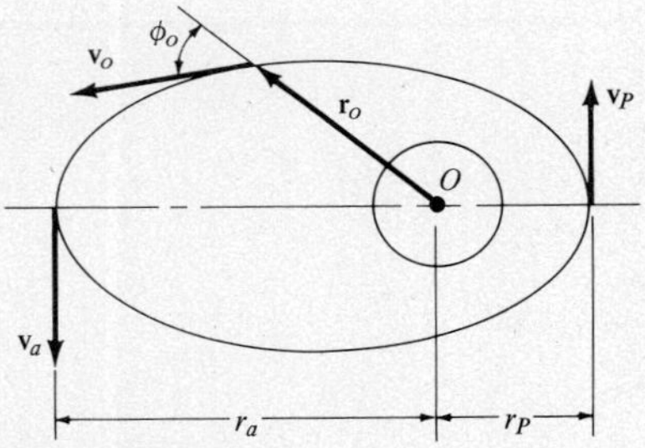

Figure 15-7.6

Solution

Using Eqs. (15-7.2) and (15-7.1), we have

$$r_p v_p = r_a v_a = r_o v_o \sin\phi_o \qquad (a)$$

and

$$v_a^2 - v_o^2 = 2GM\left(\frac{1}{r_a} - \frac{1}{r_o}\right) = \frac{2GM}{r_o}\left(\frac{r_o}{r_a} - 1\right) \qquad (b)$$

After replacing v_a by its value from Eq. (*a*), Eq. (*b*) becomes

$$\frac{r_o^2}{r_a^2}v_o^2\sin^2\phi_o - v_o^2 = v_o^2\left(\frac{r_o^2}{r_a^2}\sin^2\phi_o - 1\right) = \frac{2GM}{r_o}\left(\frac{r_o}{r_a} - 1\right)$$

which, when divided by v_o^2, may be rearranged into the following quadratic equation in terms of (r_o/r_a):

$$\sin^2\phi_o\left(\frac{r_o}{r_a}\right)^2 - \frac{2GM}{r_o v_o^2}\left(\frac{r_o}{r_a}\right) + \frac{2GM}{r_o v_o^2} - 1 = 0 \qquad (c)$$

Using the given data, we have

$$\frac{2GM}{r_o v_o^2} = \frac{2(124 \times 10^{10})}{(3960 + 1740)(18{,}000)^2} = 1.343 \quad \text{and} \quad \sin^2 75° = 0.933$$

which is substituted into Eq. (*c*) to yield

$$0.933\left(\frac{r_o}{r_a}\right)^2 - 1.343\left(\frac{r_o}{r_a}\right) + 0.343 = 0$$

of which the two roots are found to be

$$\frac{r_o}{r_a} = 0.332 \quad \text{and} \quad \frac{r_o}{r_a} = 1.107$$

The corresponding values of r_a are (with $r_o = 3960 + 1740 = 5700$ miles)

$$r_a = \frac{5700}{0.332} = 17{,}170 \text{ miles} \quad \text{and} \quad r_a = \frac{5700}{1.107} = 5150 \text{ miles}$$

of which, in view of Eq. (*a*), the larger value is the apogee distance while the smaller value is the perigee distance. The corresponding maximum and minimum altitudes of the orbit are

Max. altitude $= 17{,}170 - 3960 = 13{,}210$ miles **Ans.**
Min. altitude $= 5150 - 3960 = 1190$ miles **Ans.**

PROBLEMS

15-7.3. Determine the period of a satellite moving at 15,000 mph in a circular orbit about the earth. If air resistance is neglected, what is the minimum time in which a satellite can complete a circular orbit about the earth?

$t = 138.5$ min; $t_{\min} = 84.4$ min **Ans.**

15-7.4. Determine the velocity and altitude of a communications satellite that will occupy a fixed position relative to the earth's equator.

15-7.5. A satellite is in a circular polar orbit at an altitude of 340 miles. Find the change in longitude of its orbit during each pass as observed at the equator. What result would be observed at a latitude of 40° North?

15-7.6. Compute the time required for the command module of an Apollo mission to complete one orbit of the moon at a constant altitude of 80 miles. The radius of the moon is 1080 miles and its mass is 0.0123 that of the earth.

$t = 120.6$ min **Ans.**

15-7.7. Derive Kepler's third law: $\left(\frac{\tau}{2\pi}\right)^2 = \frac{a^3}{GM}$.

15-7.8. Compute the maximum and minimum velocities of a satellite whose period is 120 min if its altitude at perigee is known to be 240 miles.

15-7.9. A lunar excursion module (LEM) is launched at an altitude of 100 miles with a velocity v parallel to the moon's surface. Determine the minimum value of v so that the LEM will go into orbit.

$v = 3520$ mph **Ans.**

15-7.10. A satellite orbiting the earth has maximum and minimum altitudes of 640 miles and 240 miles. If it is launched at an altitude of

340 miles, determine its launch velocity v and its angle ϕ with the local vertical. What is the period of the satellite?

$v = 17{,}170$ mph; $\phi = 87.8°$; $t = 98.8$ min **Ans.**

15-7.11. Show that the kinetic energy required to change a circular orbit of radius r to a circular orbit of radius $3r$ is $GMm/3r$.

15-7.12. A satellite whose earth weight is 966 lb is circling the earth at a constant altitude of 440 miles. Its thruster engines can exert a force of 100 lb. In order to place the satellite into a new orbit at a constant altitude of 2040 miles, for what time intervals should its engines be fired? Neglect mass of fuel consumed and the change in altitude during thrust.

$t_1 = 548$ sec; $t_2 = 506$ sec **Ans.**

15-7.13. As it orbits the moon, the command module of an Apollo mission is moving at 2600 mph at its apogee altitude of 600 miles. Determine the velocity change at its perigee to put it into a circular orbit of 60 miles above the moon. The radius of the moon is 1080 miles and its mass is 0.0123 that of the earth.

$v = 170$ mph **Ans.**

15-7.14. As shown in Fig. P-15-7.14, a space shuttle at an altitude of 1040 miles has a velocity $v_A = 17{,}000$ mph directed at $\phi_A = 70°$. At B where its altitude is 1540 miles, a course correction θ is made so that the space shuttle at its apogee C will be tangent to the orbit of a skylab circling the earth at an altitude of 3540 miles. Determine the course correction angle θ.

$\theta = 7.8°$ **Ans.**

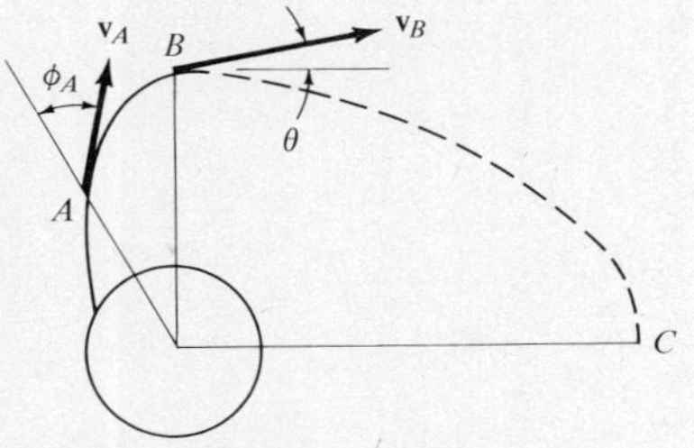

Figure P-15-7.14

* 15-8 INTRODUCTION TO GYROSCOPIC ACTION

We now apply angular impulse-momentum principles to some elementary but important aspects of rigid-body rotation about a fixed point as contrasted to fixed-axis rotation. A more general discussion of rotation about a fixed point appears in the next chapter on spatial dynamics. Here we concentrate on the gyroscopic effect that is present whenever a body that rotates about one axis also rotates about another axis that is perpendicular to the first one. This effect appears, for example, in the wheels of an automobile which is rounding a curve or in the rotor of a jet engine of an airplane making a turn.

The usual concept of a gyroscope is a solid of revolution that rotates rapidly about its axis of summetry. Let us consider the case in which the gyroscope is a disk on an axle supported at point O by a ball-and-socket joint in Fig. 15-8.1. The disk is spinning with a large angular velocity $\boldsymbol{\omega}$ about its axis of symmetry. This axis will henceforth be called the *spin axis*. The angular momentum of the disk about this axis is $\mathbf{H}_O = \bar{I}\omega\hat{\mathbf{i}}$.

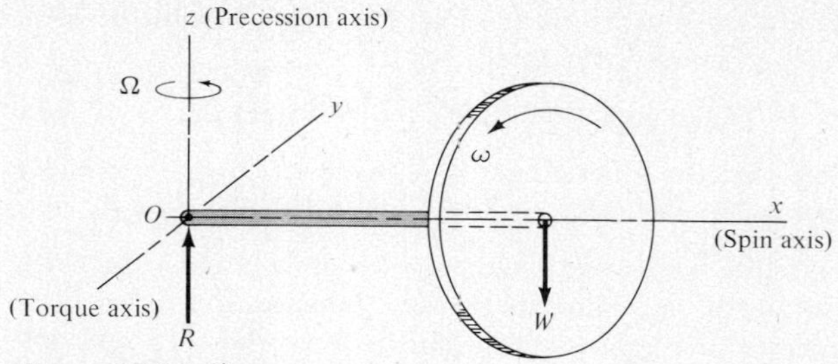

Figure 15-8.1 Gyroscope.

The external forces, consisting of the weight W of the disk and the vertical reaction R at O, create a torque $\mathbf{M} = M\hat{\mathbf{j}}$ about the y axis. This axis is called the *torque axis*. Since the axle is free to rotate in any direction about O, it seems as though the disk will fall. Strangely enough, as we shall soon show, the angular momentum of the spinning disk will not permit it to fall but will cause it instead to rotate about the vertical axis z at the rate $\boldsymbol{\Omega}$. This rotation is known as *precession* and the z axis is called the *precession axis*. Our discussion of the gyroscope is limited here to the case in which the spin axis, the torque axis, and the precession axis are mutually perpendicular.

Note that the spin, torque, and precession axes constitute a rotating reference frame xyz which rotates about a fixed Z axis through O at the rate $\boldsymbol{\Omega}$. Usually, the spin rate $\boldsymbol{\omega}$ is constant so that the angular momentum $\mathbf{H}_O$ may be considered to be a vector fixed in this rotating reference frame. Then since O is a fixed point about which the equation $\mathbf{M}_O = \dot{\mathbf{H}}_O$ is valid, and using the omega theorem (Section 12-6) to obtain the time derivative of $\mathbf{H}_O$, we have

$$[\mathbf{M}_O = \dot{\mathbf{H}}_O] \qquad M_O\hat{\mathbf{j}} = \boldsymbol{\Omega} \times \mathbf{H} = \Omega\hat{\mathbf{k}} \times \bar{I}\omega\hat{\mathbf{i}} = \bar{I}\omega\Omega\hat{\mathbf{j}}$$

which in scalar form is

$$M_O = \bar{I}\omega\Omega \tag{15-8.1}$$

Generally the rate of precession Ω is small compared with ω so that angular momentum associated with Ω may be neglected.

An alternate development which gives the physical significance of this result is obtained by expressing the angular impulse-momentum relation in the following form:

Final angular momentum
= Angular impulse + Initial angular momentum

These vector quantities, represented by the right-hand rule, are combined vectorially as shown in Fig. 15-8.2. In the differential time

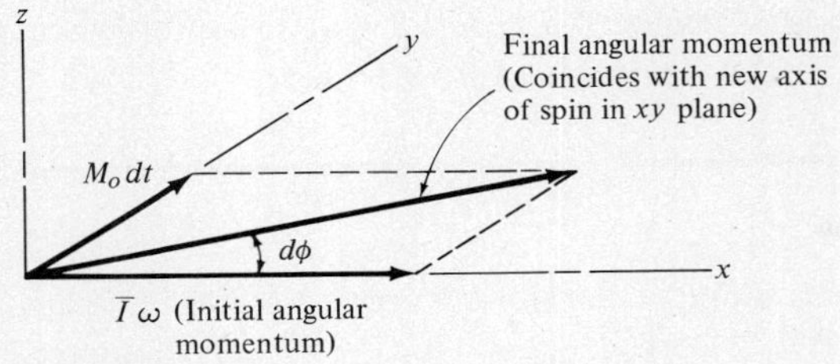

Figure 15-8.2 Precession.

dt, the angular impulse caused by the external torque is $\mathbf{M}_O\, dt$. Since the final angular momentum vector coincides with the axis about which the disk tries to spin, the spin axis precesses through the angle $d\phi$. Noting that $d\phi$ is a small angle whose tangent is equal to the angle in radians, we have

$$M_O\, dt = \bar{I}\omega\, d\phi$$

whence dividing by dt and noting that $\Omega = d\phi/dt$ is the rate of change of angular displacement of the spin axis, we obtain, as before,

$$M_O = \bar{I}\omega\Omega$$

If we attempt to hurry the rate of precession by applying a force in the xy plane, we shall merely make the rotating disk and axle rise. This rise is caused by the angular impulse of the hurrying torque represented by a vector directed along the z axis as shown in Fig. 15-8.3. Similarly, a force in the xy plane exerted to retard the precession will make the axle descend.

We have shown that a torque about an axis perpendicular to the spin axis causes a precession about a third axis perpendicular to the other two. Conversely, a forced precession about one axis will induce a torque about a third axis perpendicular to the spin and precession axes. For example, the wheels of an automobile going around a horizontal curve are forced to precess about a vertical axis through the center of curvature. This forced precession causes a torque about a third axis whose direction is along the tangent to

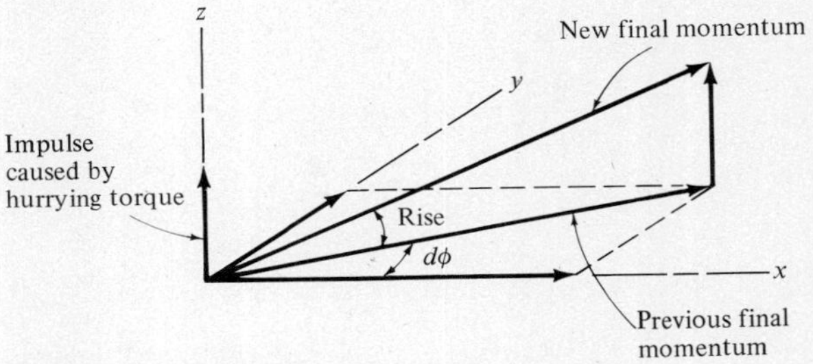

Figure 15-8.3 Effect of hurrying torque.

the curve. Because of this torque, the reaction on the outer wheel is increased and that on the inner wheel is decreased.

ILLUSTRATIVE PROBLEM

15-8.1. A car is rounding a horizontal curve of 200-ft radius at a speed of 30 fps. Each wheel weighs 60 lb, is 30 in. in diameter, and has a radius of gyration of 12 in. If the distance separating the wheels on the axles is 5 ft, compute the change in ground reaction on each pair of wheels caused by the gyroscopic effect.

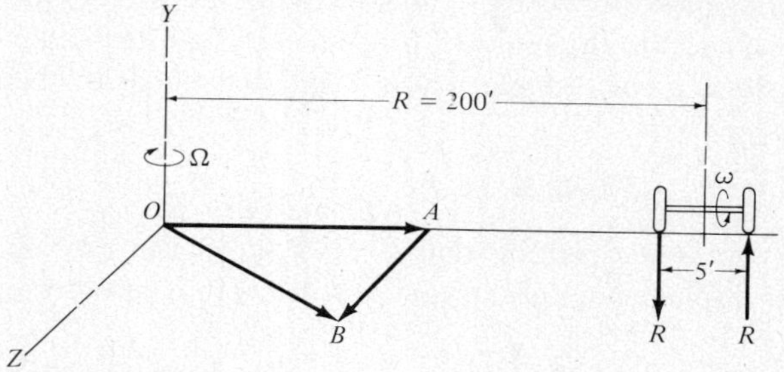

Figure 15-8.4

Solution

The vector *OA* in Fig. 15-8.4 represents the angular momentum of a pair of wheels as the car moves in the *OZ* direction. As the car rounds the curve, this vector rotates at the rate Ω about the vertical *Y* axis to position *OB*. The corresponding angular impulse *AB* is caused by an induced torque acting on the car wheels. If we apply the right-hand rule and extend our thumb in the direction *AB*, our fingers will curl about the vector *AB* in a counterclockwise sense, indicating that the gyroscopic reactions on the wheels must act as shown.

The moment of inertia of a pair of wheels is

$$\left[\bar{I} = \frac{W}{g}\bar{k}^2\right] \qquad \bar{I} = \frac{2(60)}{32.2}\left(\frac{12}{12}\right)^2 = 3.73 \text{ ft-lb-sec}^2$$

and the values of ω and Ω are

$$\left[\omega = \frac{v}{r}\right] \qquad \omega = \frac{30}{15/12} = 24 \text{ rad per sec}$$

$$\left[\Omega = \frac{v}{R}\right] \qquad \Omega = \frac{30}{200} = 0.15 \text{ rad per sec}$$

Applying Eq. (15-8.1), we now obtain

$$[M_O = \bar{I}\omega\Omega] \qquad 5R = 3.73(24)(0.15) \qquad R = 2.68 \text{ lb} \qquad \textit{Ans.}$$

PROBLEMS

15-8.2. A solid disk 12 in. in diameter weighing 322 lb rotates at 1800 rpm. It is keyed to a shaft 10 in. long which is supported by a frictionless pivot attached to the outer race of a ball bearing, as shown in Fig. P-15-8.2. If the disk rotates in a clockwise sense when viewed from the right, determine the angular velocity and sense of the precession.

$\Omega = 1.14$ rad per sec clockwise when viewed from above **Ans.**

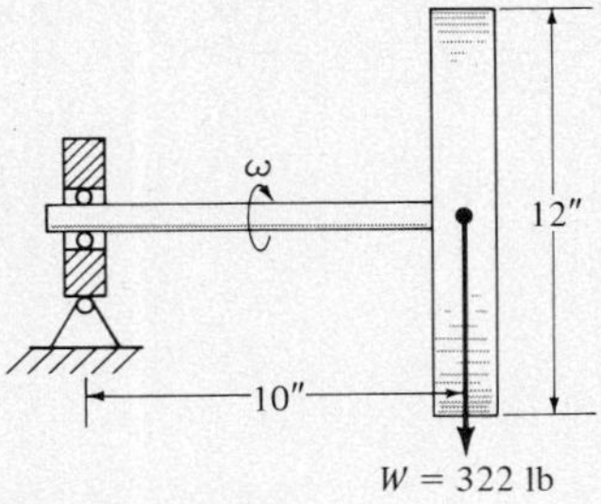

Figure P-15-8.2

15-8.3. A pair of locomotive driving wheels and their axle weigh 4830 lb. Their diameter is 6 ft and their radius of gyration is 2.5 ft. Determine the changes in wheel reactions caused by the gyroscopic effect when the locomotive is rounding a curve of 3000 ft radius at 60 mph. Assume the rails to be 5 ft apart.

$R = 161$ lb **Ans.**

15-8.4. The wheel in Fig. P-15-8.4 weighs 322 lb and has a radius of gyration of 1.732 ft. It rotates about its horizontal axis AB with a constant speed of 200 rad per sec. Axis AB is mounted in a yoke which when viewed from above rotates at 1 rad per sec clockwise about the Y axis passing through the center of gravity of the wheel. Compute the forces on the bearings at A and B.

$R_A = 1161$ lb up; $R_B = 839$ lb down **Ans.**

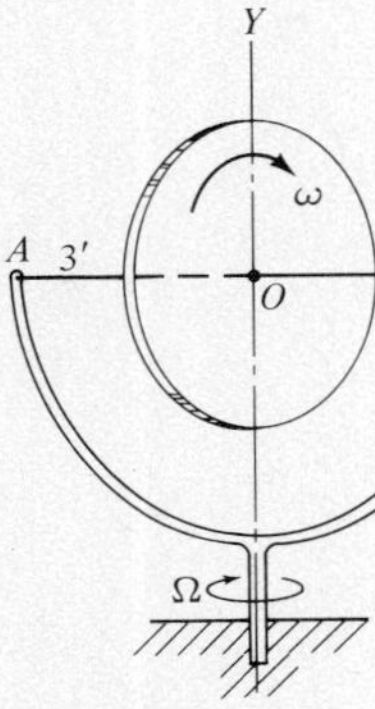

Figure P-15-8.4

15-8.5. Compute the magnitude of the force P acting 2 ft to the right of O in Fig. P-15-8.4 that is required to increase the rate of precession in Prob. 15-8.4 to 2 rad per sec. In what direction must it be applied?

$P = 3000$ lb acting vertically down **Ans.**

15-8.6. A steam turbine in a ship is mounted with its axis parallel to the propeller shaft. The turbine weighs 3220 lb, rotates at 1800 rpm, and is mounted in bearings 4 ft apart. If it has a radius of gyration of 3 ft, compute the changes in bearing reactions caused by the gyroscopic effect of the ship making a turn of 2000-ft radius at 20 knots.

Note: 1 knot = 1.69 ft per sec.

$R = 717$ lb **Ans.**

15-8.7. A grinding mill consists of two wheels free to spin about axles which are independently pivoted at A to the vertical shaft AB in Fig. P-15-8.7. If each wheel weighs W lb, determine the ground reaction on each wheel when AB rotates at a constant rate of Ω rad/sec.

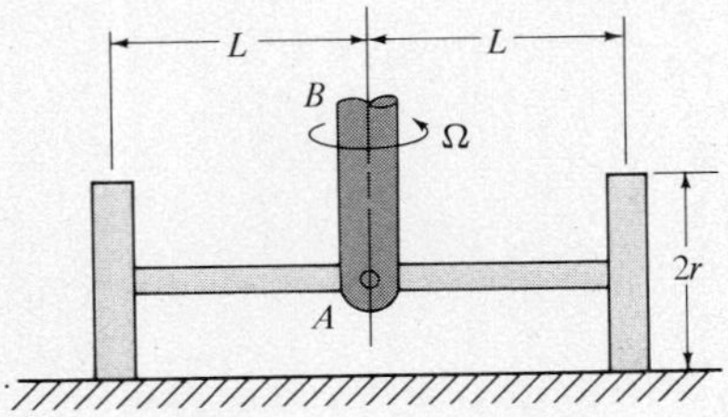

$$N = W\left(1 + \frac{r\Omega^2}{2g}\right) \qquad \textbf{Ans.}$$

SUMMARY

By taking the time integral of $\mathbf{R} = m\bar{\mathbf{a}}$, we obtain

$$\int_o^t \mathbf{R}\, dt = m\bar{\mathbf{v}} - m\bar{\mathbf{v}}_o \qquad (15\text{-}2.1)$$

The equation refers to the motion of the mass center of a body of constant mass, rigid or nonrigid. When applied to a particle or to a translating rigid body, the bar sign can be omitted from the velocity terms. The expression $\int_o^t \mathbf{R}\,dt$ is known as resultant linear impulse. Its unit is lb-sec. The quantity $m\bar{\mathbf{v}} = (W/g)\bar{\mathbf{v}}$ is defined as linear momentum and also has the dimensional unit of lb-sec.

In most applications such as those involving the dynamic action of jet streams, the impulse-momentum equation is applied in terms of its scalar components, namely,

$$\int_o^t \Sigma X\,dt = \frac{W}{g}(v_x - v_{ox}),\ \text{etc.} \qquad (15\text{-}2.3)$$

which state that the component of resultant impulse along any axis is equal to the change in the components of linear momentum along that axis.

For variable mass situations in which a moving body absorbs or ejects material, the force **P** exerted on the moving body is given by

$$\mathbf{P} = \mathbf{u}\frac{dm}{dt} \qquad (15\text{-}3.1)$$

where **u** is the velocity of the absorbed or ejected mass relative to the moving body and $\frac{dm}{dt}$ is the rate at which the moving body gains or loses mass. Note that the mass of the moving body is time-dependent and may change appreciably in the case of a rocket in which the amount of ejected fuel is a significant portion of its original total weight.

The principle of conservation of linear momentum may be applied to any system of particles acted upon only by mutual reactive forces. This principle is expressed by the equation

$$\frac{W_1}{g}\mathbf{v}_1 + \frac{W_2}{g}\mathbf{v}_2 + \cdots = \frac{W_1}{g}\mathbf{v}_1' + \frac{W_2}{g}\mathbf{v}_2' + \cdots \qquad (15\text{-}4.1)$$

Problems involving elastic impact between particles are solved by means of the principle of conservation of linear momentum in combination with the definition of the coefficient of restitution e. The coefficient of restitution is defined by the equation

$$e = \frac{\text{Relative velocity after impact}}{\text{Relative velocity before impact}}$$

and applied in the form

$$e(v_1 - v_2) = v_2' - v_1' \qquad (15\text{-}5.1)$$

in which the velocities are the components of velocities along the common normal to the impacting bodies.

For rigid bodies in fixed-axis rotation or general plane motion, impulse-momentum methods are the only practical way of treating bodies which react mutually upon each other, or are subjected to impact. Starting with $\mathbf{R} = m\bar{\mathbf{a}}$ and $\Sigma\mathbf{M}_G = \dot{\mathbf{H}}_G$ which relate the resultant force-couple system at the mass center G to their equivalent dynamic effects on a body, the time-integrated forms of these equations become

$$\int_{t_1}^{t_2} \mathbf{R}\, dt = m\bar{\mathbf{v}}_2 - m\bar{\mathbf{v}}_1$$

$$\int_{t_1}^{t_2} \Sigma\bar{M}\, dt = \bar{\mathbf{H}}_2 - \bar{\mathbf{H}}_1 = \bar{I}(\boldsymbol{\omega}_2 - \boldsymbol{\omega}_1)$$

Both of these equations are combined diagrammatically in Fig. 15-6.4 which expresses the impulse-momentum principle as

Initial system momenta + Resultant ext. impulses = Final system momenta (15-6.2)

The expression $\int_{t_1}^{t_2} \Sigma\mathbf{M}\, dt$ is called resultant angular impulse and has the dimensional unit of ft-lb-sec. The expression $I\boldsymbol{\omega}$ represents angular momentum of plane motion and is also dimensionally in units of ft-lb-sec. Notice that both angular impulse and angular momentum correspond to couples whose vectors are perpendicular to the plane of motion whereas linear impulse and linear momentum correspond to forces in the plane of motion.

The advantage of the diagrammed approach of Eq. (15-6.2) is that it clearly distinguishes between linear and angular impulses and momenta. Moreover, a moment center which eliminates impulsive forces is readily apparent. Moment summations about such a center leads automatically to conservation of angular momentum. From Fig. 15-6.4, there also may be obtained the following scalar forms of the impulse-momentum equations:

$$\int_{t_1}^{t_2} \Sigma X\, dt = m(\bar{v}_{2_x} - \bar{v}_{1_x})$$

$$\int_{t_1}^{t_2} \Sigma Y\, dt = m(\bar{v}_{2_y} - \bar{v}_{1_y}) \qquad (15\text{-}6.1)$$

$$\int_{t_1}^{t_2} \Sigma\bar{M}\, dt = \bar{I}(\omega_2 - \omega_1)$$

The following form of the angular impulse-momentum equation for fixed-axis rotation about an axle A is also useful:

$$\int_{t_1}^{t_2} \Sigma M_A \, dt = I_A(\omega_2 - \omega_1)$$

Observe that here the moment sum about A of $m\bar{v}$ and $\bar{I}\omega$ reduces to $I_A\omega$.

A combination of conservation of angular momentum and work-energy will solve many problems involving the motion of satellites. The specific equations used are

$$v_1 r_1 \sin \phi_1 = v_2 r_2 \sin \phi_2 \tag{15-7.2}$$

and

$$v_2{}^2 - v_1{}^2 = 2GM\left(\frac{1}{r_2} - \frac{1}{r_1}\right) \tag{15-7.1}$$

where

$$\begin{aligned} GM &= 1.408 \times 10^{16} \text{ ft}^3/\text{sec}^2 \\ &= 124 \times 10^{10} \text{ mi}^3/\text{hr}^2 \end{aligned} \tag{15-7.3}$$

The velocity at which a satellite will remain in a circular orbit is found from

$$v^2 = \frac{GM}{r} \tag{15-7.4}$$

In terms of the apogee and perigee distances r_a and r_p, the major and minor semiaxes of the elliptical path of an orbiting satellite are

$$a = \tfrac{1}{2}(r_a + r_p) \qquad \text{and} \qquad b = \sqrt{r_a r_p} \tag{15-7.5}$$

which are used to determine the period of the orbiting satellite from

$$\tau = \frac{2\pi ab}{r_a v_a} = \frac{2\pi ab}{r_p v_p} = \frac{2\pi a^{3/2}}{\sqrt{GM}} \tag{15-7.6}$$

The introduction to gyroscopic action is limited to the case in which the spin axis, the torque axis, and the precession axis are mutually perpendicular. The applied torque about a fixed point O is given by

$$M_O = \bar{I}\omega\Omega \tag{15-8.1}$$

where $\bar{I}$ is the centroidal moment of inertia about the spin axis of the rotor spinning at the rate ω and precessing at the rate Ω. A more extended discussion of gyroscopic action appears in Chapter 16.

16

INTRODUCTORY SPATIAL DYNAMICS OF RIGID BODIES*

* 16-1 INTRODUCTION

In Chapter 10 we developed the two force and moment equations which are fundamental to rigid-body dynamics. These equations are $\mathbf{R} = m\bar{\mathbf{a}}$ [Eq. (10-5.2)] which relates the resultant external force on a rigid body to the acceleration of its mass center, and the general moment equation $\Sigma\mathbf{M} = \dot{\mathbf{H}}$ [Eq. (10-6.2)] which relates the moment sum of the external forces to the angular motion of the body. Recall that the moment equation is valid only when the moment center is (a) fixed in space, (b) the mass center, or (c) a point whose acceleration is directed through the mass center.

So far, we have limited the application of these equations to plane motion of rigid bodies which are symmetrical about the plane of motion. As we showed in Section 13-2, this restriction resulted in an angular momentum vector $\mathbf{H} = I\boldsymbol{\omega}$ which is always perpendicular to the plane of motion. Consequently, the general moment equation reduced to the simplified form $\Sigma\mathbf{M} = I\boldsymbol{\alpha}$, valid about any center fulfilling the restrictions mentioned above. The principles of rigid-body kinematics were also necessary in order to relate the acceleration $\bar{\mathbf{a}}$ of the mass center and the angular acceleration $\boldsymbol{\alpha}$ of the body.

In this chapter, we shall extend the application of the force and moment equations to the general spatial motion of rigid bodies of *any* shape. As in plane motion, kinematic principles are used to correlate the linear and angular motions of the body. You should review and be thoroughly familiar with the use of the general omega theorem developed in Section 12-9 (namely, $\dfrac{d\mathbf{A}}{dt} = \dfrac{\delta\mathbf{A}}{\delta t} + \boldsymbol{\omega} \times \mathbf{A}$) and the use of relative motion relations involving moving reference frames.

In general spatial motion, the development of a general expression for angular momentum $\mathbf{H}$ and its time derivative $\dot{\mathbf{H}}$ is greatly simplified by using vector notation and the general omega theorem. As we shall see, this development proceeds in a straightforward manner, even though the results are quite complex. Fortunately, practical application of these results are considerably simplified by a judicious choice of reference axes as exemplified by Euler's equations (Section 16-4). The impulse-momentum and work-energy methods developed for plane motion are also valid for general spatial motion, but more general formulations for angular momentum and kinetic energy are required.

*16-2 GENERAL ANGULAR MOMENTUM

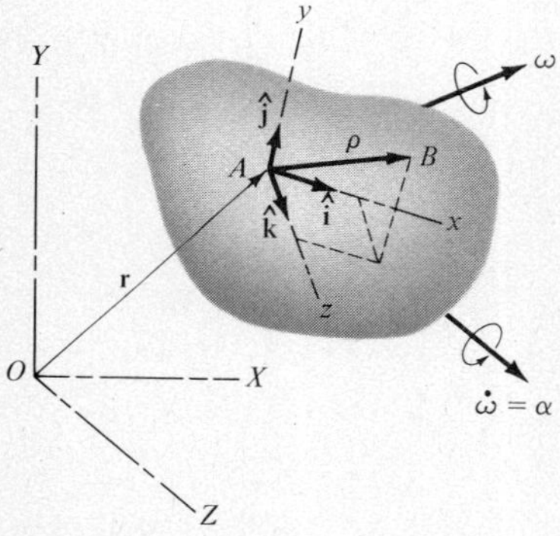

Figure 16-2.1

Consider now the rigid body shown in Fig. 16-2.1. Let a set of body axes xyz be fixed in it which rotate with the angular velocity $\boldsymbol{\omega}$ and the angular acceleration $\dot{\boldsymbol{\omega}} = \boldsymbol{\alpha}$ of the body as observed from the fixed reference axes XYZ. Further, let the origin of the body axes be at a point A of the body. The position vector $\boldsymbol{\rho}$ from A to any particle B of mass dm of the body has a constant length since the body is assumed to be rigid. Hence, the relative velocity and acceleration of B about A is due solely to the directional changes of $\boldsymbol{\rho}$ that are caused by $\boldsymbol{\omega}$ and $\dot{\boldsymbol{\omega}}$.

The total moment of momentum (or the angular momentum of a rigid body) is the sum of the moments of linear momenta about A. Noting that the absolute linear momentum of a typical element at B is $\mathbf{v}\,dm = (\dot{\mathbf{r}} + \dot{\boldsymbol{\rho}})\,dm$ and its moment of momentum about A is $\boldsymbol{\rho} \times (\dot{\mathbf{r}} + \dot{\boldsymbol{\rho}})\,dm$, we therefore obtain as the total absolute angular momentum about A,

$$\begin{aligned}\mathbf{H}_A &= \int \boldsymbol{\rho} \times \dot{\mathbf{r}}\,dm + \int \boldsymbol{\rho} \times \dot{\boldsymbol{\rho}}\,dm \\ &= \left(\int \boldsymbol{\rho}\,dm\right) \times \dot{\mathbf{r}} + \int \boldsymbol{\rho} \times \dot{\boldsymbol{\rho}}\,dm \\ &= \bar{\boldsymbol{\rho}} m \times \dot{\mathbf{r}} + \int \boldsymbol{\rho} \times \dot{\boldsymbol{\rho}}\,dm \qquad (a)\end{aligned}$$

in which the transformation of the first right-hand integral was accomplished by first rewriting it as $(\int \boldsymbol{\rho}\, dm) \times \dot{\mathbf{r}}$ since $\dot{\mathbf{r}}$ is independent of $\boldsymbol{\rho}$. This then becomes $\bar{\boldsymbol{\rho}} m \times \dot{\mathbf{r}}$ when the total moment of mass, $\int \boldsymbol{\rho}\, dm$, is replaced by the equivalent form $\bar{\boldsymbol{\rho}} m$.

Usually the point A selected as the origin of the body axes is a body point that is either fixed in space or else is the mass center because these are two of the conditions under which $\Sigma \mathbf{M}_A = \dot{\mathbf{H}}_A$. If A is fixed in space, we observe that $\mathbf{r}$ is constant and hence $\dot{\mathbf{r}} = 0$, or if A coincides with the mass center G, then $\bar{\boldsymbol{\rho}} = 0$. Under either of these restrictions, therefore, Eq. (a) reduces to

$$\mathbf{H} = \mathbf{H}_A = \mathbf{H}_G = \int \boldsymbol{\rho} \times \dot{\boldsymbol{\rho}}\, dm = \int \boldsymbol{\rho} \times (\boldsymbol{\omega} \times \boldsymbol{\rho})\, dm \qquad (b)$$

where we have used the omega theorem [Eq. (12-6.2)] to replace $\dot{\boldsymbol{\rho}}$ by $\boldsymbol{\omega} \times \boldsymbol{\rho}$.

Noting that $\boldsymbol{\rho} = x\hat{\mathbf{i}} + y\hat{\mathbf{j}} + z\hat{\mathbf{k}}$ and $\boldsymbol{\omega} = \omega_x\hat{\mathbf{i}} + \omega_y\hat{\mathbf{j}} + \omega_z\hat{\mathbf{k}}$ with respect to the body axes xyz of Fig. 16-2.1, we now use the vector identity $\mathbf{a} \times (\mathbf{b} \times \mathbf{c}) = (\mathbf{a} \cdot \mathbf{c})\mathbf{b} - (\mathbf{a} \cdot \mathbf{b})\mathbf{c}$ [Eq. (2-8.6)] to evaluate the integrand of Eq. (b) as follows:

$$\begin{aligned}
\boldsymbol{\rho} \times (\boldsymbol{\omega} \times \boldsymbol{\rho}) &= (\boldsymbol{\rho} \cdot \boldsymbol{\rho})\boldsymbol{\omega} - (\boldsymbol{\rho} \cdot \boldsymbol{\omega})\boldsymbol{\rho} \\
&= (x^2 + y^2 + z^2)(\omega_x\hat{\mathbf{i}} + \omega_y\hat{\mathbf{j}} + \omega_z\hat{\mathbf{k}}) \\
&\quad - (x\omega_x + y\omega_y + z\omega_z)(x\hat{\mathbf{i}} + y\hat{\mathbf{j}} + z\hat{\mathbf{k}})
\end{aligned}$$

Collecting terms,

$$\begin{aligned}
\boldsymbol{\rho} \times (\boldsymbol{\omega} \times \boldsymbol{\rho}) &= \hat{\mathbf{i}}[\omega_x(y^2 + z^2) - \omega_y xy - \omega_z xz] \\
&\quad + \hat{\mathbf{j}}[\omega_y(z^2 + x^2) - \omega_z yz - \omega_x yx] \\
&\quad + \hat{\mathbf{k}}[\omega_z(x^2 + y^2) - \omega_x zx - \omega_y zy]
\end{aligned}$$

On substituting this evaluation of the integrand into Eq. (b) and noting that the components of $\boldsymbol{\omega}$ are independent of the position of B (and therefore may be written outside of the integral sign), we find the components of $\mathbf{H}$ to be

$$\begin{aligned}
H_x &= \omega_x \int (y^2 + z^2)\, dm - \omega_y \int xy\, dm - \omega_z \int xz\, dm \\
H_y &= \omega_y \int (z^2 + x^2)\, dm - \omega_z \int yz\, dm - \omega_x \int yx\, dm \qquad (c) \\
H_z &= \omega_z \int (x^2 + y^2)\, dm - \omega_x \int zx\, dm - \omega_y \int zy\, dm
\end{aligned}$$

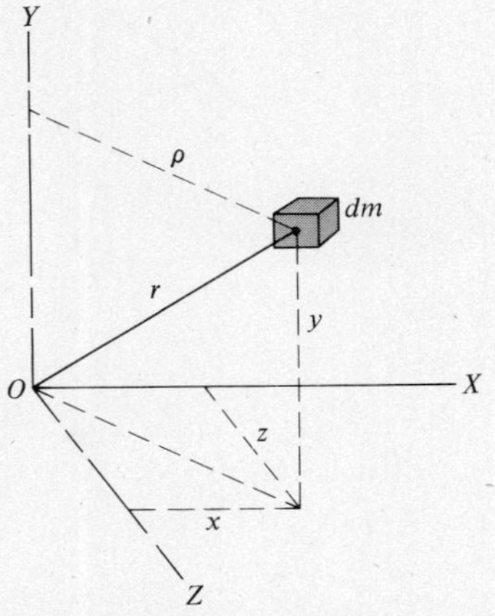

Figure 16-2.2 To compute mass moment of inertia.

Recall that the definition of mass moment of inertia is the integral of each differential mass multiplied by the square of its moment arm ρ about the reference axis; i.e., $I = \int \rho^2\, dm$. Thus, using Fig. 16-2.2 as an aid, we see that

$$\int (z^2 + x^2)\, dm = \int (r^2 - y^2)\, dm = \int \rho^2\, dm = I_y, \text{ etc.}$$

The products of inertia are defined as $P_{xy} = \int xy\, dm$, etc. Using this notation, the components of **H** given by Eq. (*c*) are rewritten finally as

$$\begin{aligned} H_x &= I_x\omega_x - P_{xy}\omega_y - P_{xz}\omega_z \\ H_y &= I_y\omega_y - P_{yz}\omega_z - P_{yx}\omega_x \\ H_z &= I_z\omega_z - P_{zx}\omega_x - P_{zy}\omega_y \end{aligned} \tag{16-2.1}$$

Note that the inertia integrals are defined with respect to axes fixed in and rotating with the body. Hence, the inertia integrals have constant values with respect to such axes. This fact will be very useful whenever we need to compute the time derivative of **H** in applying $\Sigma\mathbf{M} = \dot{\mathbf{H}}$.

By taking advantage of body symmetry or by the use of transformation equations analogous to those for area moments of inertia, it is always possible to select a set of body axes originating at the reference point with respect to which the products of inertia will be zero. Such axes are called *principal axes of inertia*. In this case, Eqs. (16-2.1) reduce to

$$H_x = I_x\omega_x \qquad H_y = I_y\omega_y \qquad H_z = I_z\omega_z \tag{16-2.2}$$

For the special case in which $\boldsymbol{\omega}$ is directed along a principal axis of inertia, **H** and $\boldsymbol{\omega}$ will be coincident vectors. This is the situation prevailing in the plane motion of symmetrical bodies which results in $\mathbf{H} = I\boldsymbol{\omega}$ where both **H** and $\boldsymbol{\omega}$ are perpendicular to the plane of motion.

Finally, it is noteworthy that the spatial equivalent of Fig. 15-6.2 is that the total momenta of a body consists of the linear momenta $m\bar{\mathbf{v}}$ acting at the mass center G and an angular momentum $\mathbf{H}_G$ about the mass center as shown here in Fig. 16-2.3. With respect to any other body point A, the angular momentum $\mathbf{H}_A$ about A will be the sum of the moment of the linear momentum vector $m\bar{\mathbf{v}}$ and the couple vector $\mathbf{H}_G$. Thus, denoting by $\boldsymbol{\rho}_{G/A}$ the position vector of G from A, the total angular momentum about A is given by

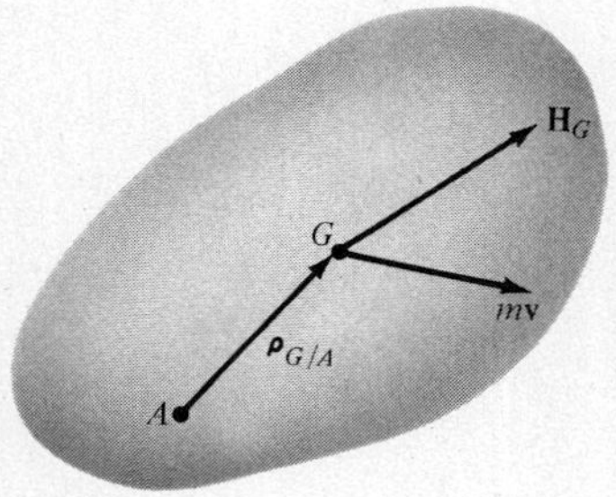

Figure 16-2.3

$$\mathbf{H}_A = \mathbf{H}_G + \boldsymbol{\rho}_{G/A} \times m\bar{\mathbf{v}} \tag{16-2.3}$$

which is often useful as a transfer formula for angular momentum.

In most applications of spatial dynamics, we shall deal with angular momentum $\mathbf{H}_G$ about the mass center G. Its components, expressed by Eq. (16-2.2) when the directions of the centroidal principal axes of inertia are obvious, are merely the respective centroidal moments of inertia multiplied by the corresponding components of $\boldsymbol{\omega}$. If the principal axes are not obvious, we use Eq. (16-2.1) to compute the components of $\mathbf{H}_G$ in terms of the moments and products of inertia with respect to arbitrary centroidal axes.

Occasionally we shall need angular momentum $\mathbf{H}_A$ about a noncentroidal fixed point A. Usually the directions of principal axes at A are not obvious so that both moments and products of inertia about A are required in order to compute $\mathbf{H}_A$ by means of Eq. (16-2.1). However, a convenient alternate approach to finding $\mathbf{H}_A$ is to start with centroidal angular momentum $\mathbf{H}_G$ which may be converted into $\mathbf{H}_A$ by adding the transfer term $\boldsymbol{\rho}_{G/A} \times m\bar{\mathbf{v}}$ of Eq. (16-2.3).

Both of these methods of finding noncentroidal angular momentum are applied independently to the same situation in the following first two illustrative problems so that you may compare them. Usually it will be preferable to compute $\mathbf{H}_A$ by the combination of $\mathbf{H}_G$ and the transfer term $\boldsymbol{\rho}_{G/A} \times m\bar{\mathbf{v}}$.

ILLUSTRATIVE PROBLEMS

16-2.1. Determine the components of the angular momentum about the given body axes at O of the homogeneous rectangular parallelepiped shown in Fig. 16-2.4. Consider the three situations in which the body rotates at ω rad/sec about a fixed axis coincident with (a) the edge OA, (b) the face diagonal OB, and (c) the body diagonal OC.

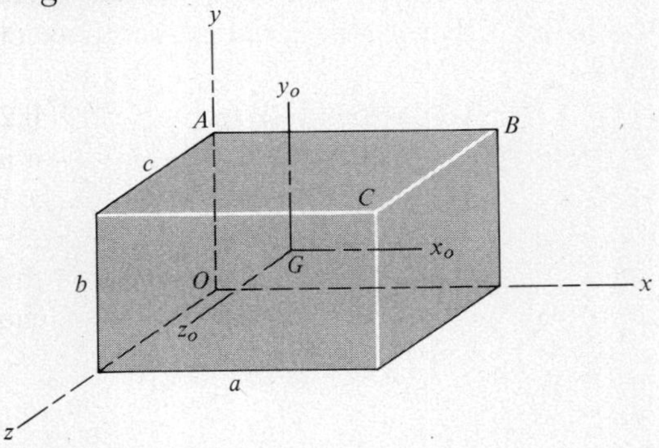

Figure 16-2.4

Solution

Here we shall use the approach of applying Eq. (16-2.1) directly. As a preliminary, we must compute the moments and products of inertia about the axes $Oxyz$. From Table 7-17.1, we note that the moments of inertia about the parallel centroidal axes $Gx_oy_oz_o$ are

$$\bar{I}_x = \frac{m}{12}(b^2 + c^2); \qquad \bar{I}_y = \frac{m}{12}(c^2 + a^2); \qquad \bar{I}_z = \frac{m}{12}(a^2 + b^2)$$

These values in conjunction with the transfer formula determine the moments of inertia about $Oxyz$ as follows:

$$[I = \bar{I} + md^2] \quad I_x = \frac{m}{12}(b^2 + c^2) + m\left(\frac{b^2 + c^2}{4}\right) = \frac{m}{3}(b^2 + c^2)$$

and similarly,

$$I_y = \frac{m}{3}(c^2 + a^2); \qquad I_z = \frac{m}{3}(a^2 + b^2)$$

Since the products of inertia are zero about the centroidal axes of symmetry, the transfer formula determines the products of inertia about $Oxyz$ as follows:

$$[P_{xy} = \bar{P}_{xy} + m\bar{x}\bar{y}] \qquad P_{xy} = 0 + m\left(\frac{a}{2}\right)\left(\frac{b}{2}\right) = \frac{mab}{4}$$

and similarly,

$$P_{yz} = \frac{mbc}{4}; \qquad P_{zx} = \frac{mca}{4}$$

Part (a)

For fixed-axis rotation about the edge OA, the components of $\boldsymbol{\omega}$ are

$$\omega_x = 0; \qquad \omega_y = \omega; \qquad \omega_z = 0$$

If we substitute these values and the preceding moments and products of inertia directly into Eq. (16-2.1), the required components of angular momentum are

$$H_x = 0 - \frac{mab}{4}(\omega) - 0 \qquad = -\frac{m\omega ab}{4}$$

$$H_y = \frac{m}{3}(c^2 + a^2)(\omega) - 0 - 0 = \frac{m\omega}{3}(c^2 + a^2)$$

$$H_z = 0 - 0 - \frac{mbc}{4}(\omega) \qquad = -\frac{m\omega bc}{4}$$

Part (b)

For fixed-axis rotation about the face diagonal OB, the angular velocity is $\boldsymbol{\omega} = \omega\hat{\mathbf{n}}_{OB}$ where $\hat{\mathbf{n}}_{OB} = \dfrac{a\hat{\mathbf{i}} + b\hat{\mathbf{j}}}{d}$ is a unit vector along the diagonal OB of length $d = (a^2 + b^2)^{1/2}$. Hence, the components of $\boldsymbol{\omega}$ along $Oxyz$ are

$$\omega_x = \frac{\omega a}{d}; \qquad \omega_y = \frac{\omega b}{d}; \qquad \omega_z = 0$$

Substituting these values and the moments and products of inertia into Eq. (16-2.1) gives the following components of angular momentum:

$$H_x = \frac{m}{3}(b^2 + c^2)\frac{\omega a}{d} - \frac{mab}{4}\left(\frac{\omega b}{d}\right) - 0 = \frac{m\omega a}{12d}(b^2 + 4c^2)$$

$$H_y = \frac{m}{3}(c^2 + a^2)\frac{\omega b}{d} - 0 - \frac{mab}{4}\left(\frac{\omega a}{d}\right) = \frac{m\omega b}{12d}(a^2 + 4c^2)$$

$$H_z = 0 - \frac{mca}{4}\left(\frac{\omega a}{d}\right) - \frac{mbc}{4}\left(\frac{\omega b}{d}\right) \quad = -\frac{m\omega c}{4d}(a^2 + b^2)$$

Part (c)

For fixed-axis rotation about the body diagonal OC, the angular velocity is $\boldsymbol{\omega} = \omega\hat{\mathbf{n}}_{OC}$ where $\hat{\mathbf{n}}_{OC} = \dfrac{a\hat{\mathbf{i}} + b\hat{\mathbf{j}} + c\hat{\mathbf{k}}}{d}$ is a unit vector along the diagonal OC of length $d = (a^2 + b^2 + c^2)^{1/2}$. Hence, the components of $\boldsymbol{\omega}$ along $Oxyz$ are

$$\omega_x = \frac{\omega a}{d}; \qquad \omega_y = \frac{\omega b}{d}; \qquad \omega_z = \frac{\omega c}{d}$$

Substituting these values and the moments and products of inertia into Eq. (16-2.1) will result in the following components of angular momentum:

$$H_x = \frac{m}{3}(b^2 + c^2)\left(\frac{\omega a}{d}\right) - \frac{mab}{4}\left(\frac{\omega b}{d}\right) - \frac{mca}{4}\left(\frac{\omega c}{d}\right) = \frac{m\omega a}{12d}(b^2 + c^2)$$

and similarly,

$$H_y = \frac{m\omega b}{12d}(c^2 + a^2); \qquad H_z = \frac{m\omega c}{12d}(a^2 + b^2)$$

16-2.2. Solve the previous problem by first computing centroidal components of angular momentum and then transferring these to the origin at O by means of Eq. (16-2.3).

Solution

For each of the three cases of the previous problem, the centroidal components of angular momentum are

$$\bar{H}_x = \bar{I}_x\omega_x; \qquad \bar{H}_y = \bar{I}_y\omega_y; \qquad \bar{H}_z = \bar{I}_z\omega_z$$

To apply them, we substitute the centroidal moments of inertia computed in the previous problem and use the components of $\boldsymbol{\omega}$ appropriate to each case.

Part (a)

For rotation about the edge OA, we note that $\omega_y = \omega$, while $\omega_x =$

$\omega_z = 0$. Hence, the components of $\bar{\mathbf{H}}$ are

$$\bar{H}_x = 0; \qquad \bar{H}_y = \frac{m\omega}{12}(c^2 + a^2); \qquad \bar{H}_z = 0 \tag{a}$$

The angular momentum about O can now be found by applying the transfer equation $\mathbf{H}_O = \bar{\mathbf{H}} + \boldsymbol{\rho}_{G/A} \times m\bar{\mathbf{v}}$ in which $\boldsymbol{\rho}_{G/A} = \frac{1}{2}(a\hat{\mathbf{i}} + b\hat{\mathbf{j}} + c\hat{\mathbf{k}})$ and the velocity of the mass center is $\bar{\mathbf{v}} = \boldsymbol{\omega} \times \bar{\boldsymbol{\rho}}$.

Noting that $\bar{\mathbf{v}} = \bar{\boldsymbol{\omega}} \times \bar{\boldsymbol{\rho}} = \omega\hat{\mathbf{j}} \times \frac{1}{2}(a\hat{\mathbf{i}} + b\hat{\mathbf{j}} + c\hat{\mathbf{k}}) = -\dfrac{\omega a}{2}\hat{\mathbf{k}} + \dfrac{\omega c}{2}\hat{\mathbf{i}}$

the value of the transfer term is

$$\bar{\boldsymbol{\rho}} \times m\bar{\mathbf{v}} = \frac{m\omega}{4}\begin{vmatrix} a & b & c \\ c & 0 & -a \\ \hat{\mathbf{i}} & \hat{\mathbf{j}} & \hat{\mathbf{k}} \end{vmatrix} = \frac{m\omega}{4}\begin{Bmatrix} \hat{\mathbf{i}}(-ab) \\ +\hat{\mathbf{j}}(c^2 + a^2) \\ +\hat{\mathbf{k}}(-bc) \end{Bmatrix} \tag{b}$$

Adding the components of Eq. (b) to those of Eq. (a) results in

$$H_x = -\frac{m\omega ab}{4}$$

$$H_y = \frac{m\omega}{12}(c^2 + a^2) + \frac{m\omega}{4}(c^2 + a^2) = \frac{m\omega}{12}(c^2 + a^2)$$

$$H_z = -\frac{m\omega bc}{4}$$

As expected, these results check those of the previous solution. Notice how the transfer term automatically accounted for the transfer of moments and products of inertia from G to O as well as eliminating the need for remembering or referring to the complex form of Eq. (16-2.1). These same comments will also be seen to apply to parts (b) and (c).

Part (b)

For fixed-axis rotation about the face diagonal OB, the angular velocity has the same components computed in the previous problem; namely,

$$\omega_x = \frac{\omega a}{d}; \qquad \omega_y = \frac{\omega b}{d}; \qquad \omega_z = 0$$

so that the centroidal components of $\bar{\mathbf{H}}$ become

$$\bar{H}_x = \frac{m\omega a}{12d}(b^2 + c^2); \qquad \bar{H}_y = \frac{m\omega b}{12d}(c^2 + a^2); \qquad \bar{H}_z = 0 \tag{c}$$

The components of $m\bar{\mathbf{v}} = m(\boldsymbol{\omega} \times \bar{\boldsymbol{\rho}})$, where $\bar{\boldsymbol{\rho}} = \frac{1}{2}(a\hat{\mathbf{i}} + b\hat{\mathbf{j}} + c\hat{\mathbf{k}})$ as before, are evaluated from the following determinant:

$$m\bar{v} = m(\boldsymbol{\omega} \times \bar{\boldsymbol{\rho}}) = \frac{m\omega}{2d}\begin{vmatrix} a & b & 0 \\ a & b & c \\ \hat{\mathbf{i}} & \hat{\mathbf{j}} & \hat{\mathbf{k}} \end{vmatrix} = \frac{m\omega}{2d}\begin{Bmatrix} \hat{\mathbf{i}}(bc) \\ +\hat{\mathbf{j}}(-ac) \\ +\hat{\mathbf{k}}(ab - ab) = 0 \end{Bmatrix}$$

We now insert these components in the following determinant form of $\bar{\boldsymbol{\rho}} \times m\bar{\mathbf{v}}$ to obtain

$$\bar{\boldsymbol{\rho}} \times m\bar{\mathbf{v}} = \frac{m\omega}{4d}\begin{vmatrix} a & b & c \\ bc & -ac & o \\ \hat{\mathbf{i}} & \hat{\mathbf{j}} & \hat{\mathbf{k}} \end{vmatrix} = \frac{m\omega}{4d}\begin{Bmatrix} \hat{\mathbf{i}}(ac^2) \\ +\hat{\mathbf{j}}(bc^2) \\ +\hat{\mathbf{k}}(-ca^2 - cb^2) \end{Bmatrix} \qquad (d)$$

Adding the components of Eq. (d) to those of Eq. (c) results in

$$H_x = \frac{m\omega a}{12d}(b^2 + c^2) + \frac{m\omega ac^2}{4d} = \frac{m\omega a}{12d}(b^2 + 4c^2) \qquad \textbf{\textit{Check}}$$

$$H_y = \frac{m\omega b}{12d}(c^2 + a^2) + \frac{m\omega bc^2}{4d^2} = \frac{m\omega b}{12d}(a^2 + 4c^2) \qquad \textbf{\textit{Check}}$$

$$H_z = -\frac{\omega mc}{4d}(a^2 + b^2) \qquad \textbf{\textit{Check}}$$

Part (c)

For fixed-axis rotation about the body diagonal OC, the angular velocity components, as computed in the previous problem, are

$$\omega_x = \frac{\omega a}{d}; \qquad \omega_y = \frac{\omega b}{d}; \qquad \omega_z = \frac{\omega c}{d}$$

so that the centroidal components of $\mathbf{H}$ now are

$$\bar{H}_x = \frac{m\omega a}{12d}(b^2 + c^2); \ \bar{H}_y = \frac{m\omega b}{12d}(c^2 + a^2); \ \bar{H}_z = \frac{m\omega c}{d}(a^2 + b^2) \quad (e)$$

If we now repeat the pattern of parts (a and b), we shall find that $\bar{\mathbf{v}} = 0$ so that the transfer term also is zero. Hence, the components given in Eq. (e) are also the desired components of angular momentum about O and agree with those found in the previous problem. Notice that angular momentum about a *fixed* axis of rotation has the same value about *any* point on the axis provided that the mass center G is also on this axis.

16-2.3. In the grinding mill shown in Fig. 16-2.5, centrifugal inertia causes the conical grinder to roll without slipping against the fixed vertical cylindrical surface as it also rotates about the axle AG of length L. The axle AG is hinged to and rotates with the vertical shaft AB at the constant rate of ω_1 rad/sec. Assuming that the conical grinder is equivalent to a thin disk of mass m and radius r, determine the components of angular momentum of the grinder about axes xyz at the fixed point A.

Solution

To simplify calculation, we shall first compute $\mathbf{H}_A$ about the principal axes $x_o y_o z_o$ at A and then convert back to the specified xyz axes. Recognize that angular momentum, like force or any other vector,

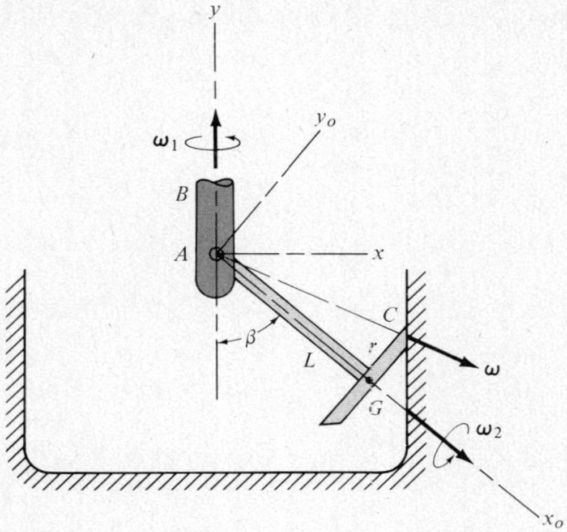

Figure 16-2.5 Grinding mill.

is a distinct entity whose various components depend on the choice of reference axes. With respect to the principal axes at the fixed point A, the components of $\mathbf{H}_A$ from Eq. (16-2.2) are

$$H_{x_o} = I_{x_o}\omega_x; \qquad H_y = I_{y_o}\omega_y; \qquad H_z = I_{z_o}\omega_z \tag{a}$$

where

$$I_{x_o} = \frac{1}{2}mr^2; \qquad I_{y_o} = I_{z_o} = \frac{1}{4}mr^2 + mL^2 = m\left(\frac{r^2}{4} + L^2\right)$$

To find the components of the absolute angular velocity $\boldsymbol{\omega}$ of the body axes, we first note that since point C has zero velocity from the pure rolling condition, AC is the instantaneous axis of rotation along which $\boldsymbol{\omega}$ is directed. But $\boldsymbol{\omega} = \boldsymbol{\omega}_1 + \boldsymbol{\omega}_2$ where $\boldsymbol{\omega}_2$ is the unspecified angular velocity of the disk about the axle AG. In terms of the unit vectors of the principal axes, we have

$$\boldsymbol{\omega} = \omega\hat{\mathbf{n}}_{AC} = \frac{\omega}{d}(L\hat{\mathbf{i}} + r\hat{\mathbf{j}})$$

$$\boldsymbol{\omega}_1 = -\omega_1 \cos\beta\,\hat{\mathbf{i}} + \omega_1 \sin\,\beta\,\hat{\mathbf{j}}$$

$$\boldsymbol{\omega}_2 = \omega_2\hat{\mathbf{i}}$$

so that on equating the coefficients of the unit vectors, we obtain

$$\frac{\omega L}{d} = -\omega_1 \cos\beta + \omega_2$$

$$\frac{\omega r}{d} = \omega_1 \sin\beta$$

from which, on multiplying the second of these by $-L/r$ and adding to the first, we find[1]

$$\omega_2 = \omega_1 \cos\beta + \omega_1 \frac{L}{r} \sin\beta$$

Hence,

$$\boldsymbol{\omega} = \boldsymbol{\omega}_1 + \boldsymbol{\omega}_2 = -\omega_1 \cos\beta\,\hat{\mathbf{i}} + \omega_1 \sin\beta\,\hat{\mathbf{j}} + \omega_1 \cos\beta\,\hat{\mathbf{i}} + \omega_1 \frac{L}{r} \sin\beta\,\hat{\mathbf{i}}$$

which reduces to

$$\boldsymbol{\omega} = \omega_1 \frac{L}{r} \sin\beta\,\hat{\mathbf{i}} + \omega_1 \sin\beta\,\hat{\mathbf{j}}$$

Returning to Eq. (*a*) and substituting the components of $\boldsymbol{\omega}$, we find the components of $\mathbf{H}_A$ along the principal axes are

$$H_{x_o} = \frac{1}{2} mr^2 \left(\omega_1 \frac{L}{r} \sin\beta \right) = \frac{1}{2} mrL\omega_1 \sin\beta$$

$$H_{y_o} = m\left(\frac{r^2}{4} + L^2\right)\omega_1 \sin\beta$$

$$H_{z_o} = 0$$

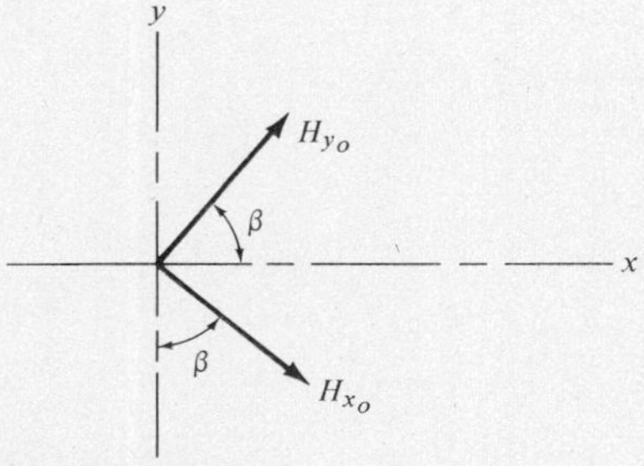

Figure 16-2.6

Finally, recalling that these are components of the entity $\mathbf{H}_A$, we find its components along the specified *xyz* axes by applying simple projection to Fig. 16-2.6 and obtain

$$H_x = H_{x_o} \sin\beta + H_{y_o} \cos\beta$$

$$= \frac{1}{2} mrL\omega_1 \sin^2\beta + m\left(\frac{r^2}{4} + L^2\right)\omega_1 \sin\beta\cos\beta \qquad \textbf{\textit{Ans.}}$$

$$H_y = H_{y_o} \sin\beta - H_{x_o} \cos\beta$$

$$= m\left(\frac{r^2}{4} + L^2\right)\omega_1 \sin^2\beta - \frac{1}{2} mrL\omega_1 \sin\beta\cos\beta \qquad \textbf{\textit{Ans.}}$$

PROBLEMS

16-2.4. As shown in Fig. P-16-2.4, a wheel, of weight 1610 lb, 6-in. thick, and 2-ft radius, rotates about the axis *OA* at the constant rate of

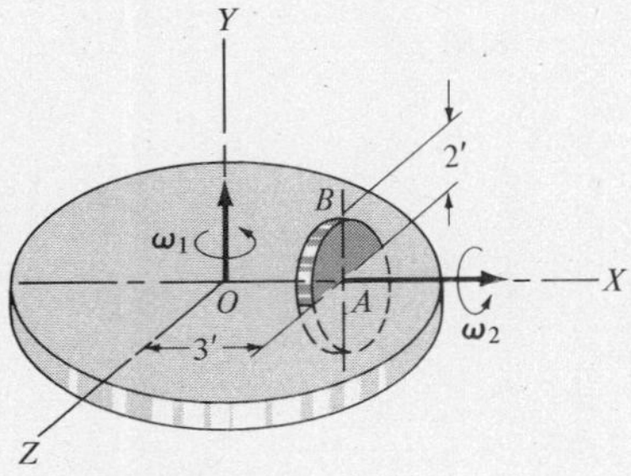

Figure P-16-2.4

[1] An alternate way of finding ω_2 in terms of ω_1 is to note that the velocity of point *C* is zero so that we write $\mathbf{v}_C = \boldsymbol{\omega} \times \boldsymbol{\rho}_{AC} = 0$, or

$$[\omega_2 - \omega_1 \cos\beta)\hat{\mathbf{i}} + (\omega_1 \sin\beta)\hat{\mathbf{j}}] \times (L\hat{\mathbf{i}} + r\hat{\mathbf{j}}) = 0$$

which reduces to

$$(\omega_2 - \omega_1 \cos\beta) r\hat{\mathbf{k}} - (\omega_1 L \sin\beta)\hat{\mathbf{k}} = 0$$

and hence, as before,

$$\omega_2 = \frac{\omega_2 L}{r} \sin\beta + \omega_1 \cos\beta$$

$\omega_2 = 6$ rad/sec as the platform carrying the wheel rotates about the fixed Y axis at a constant rate of $\omega_1 = 4$ rad/sec. Compute the angular momentum of the wheel about its mass center and about O.

$$\mathbf{H}_O = 600\hat{\mathbf{i}} + 2008\hat{\mathbf{j}} \text{ ft-lb-sec} \qquad \textbf{\textit{Ans.}}$$

16-2.5. A disk of mass m and radius r is fastened as shown in Fig. P-16-2.5 to a shaft which is rotating at a constant angular velocity ω. The plane of the disk is skewed at an angle θ with the normal to the shaft. Compute the angular momentum of the disk about xyz axes whose origin is at the mass center G.

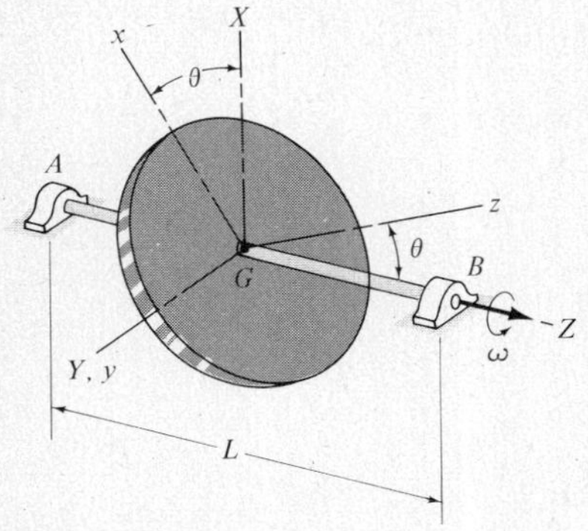

Figure P-16-2.5

16-2.6. The center A of the thin disk of mass m and radius r in Fig. P-16-2.6 has a constant speed of v fps. The disk rolls without slipping on a horizontal surface as it also rotates about the shaft axis AB. Determine the angular momentum of the disk about the fixed point B.

$$\mathbf{H}_B = -\frac{1}{2}mrv\hat{\mathbf{i}} + \frac{mv}{L}\left(\frac{r^2}{4} + L^2\right)\hat{\mathbf{j}} \qquad \textbf{\textit{Ans.}}$$

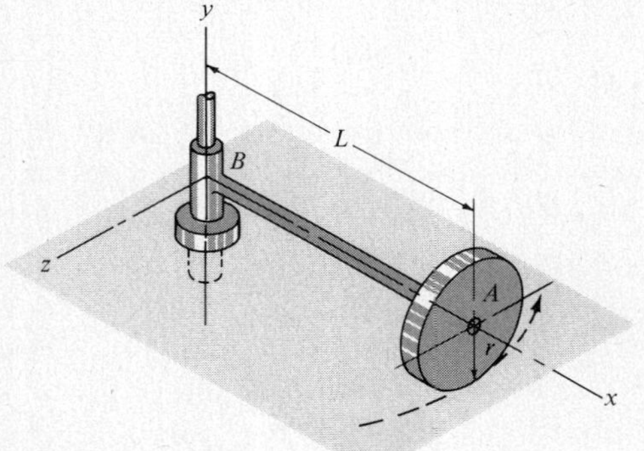

Figure P-16-2.6

16-2.7. The homogeneous solid cone of base radius $r = 7$ in., height $h = 24$ in., and weight 30 lb shown in Fig. P-16-2.7 rolls on a horizontal floor without slipping. The centerline OA rotates about the fixed Y axis at the constant rate of $\omega_1 = 2$ rad/sec. Compute the angular momentum of the cone about the XYZ axes.

$$H_X = -21.9 \text{ in.-lb-sec; } H_Y = 48.4 \text{ in.-lb-sec} \qquad \textbf{\textit{Ans.}}$$

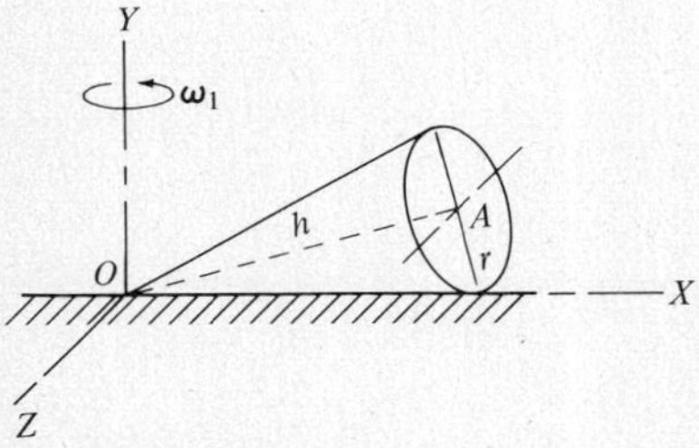

Figure P-16-2.7

16-2.8. A gyroscope consists of a thin disk weighing 30 lb and of 2-in. radius rotating at a constant rate of $\omega_2 = 40$ rad/sec about axis OA in Fig. P-16-2.8. Simultaneously, axis OA is precessing (i.e., rotating) about a fixed vertical Y axis at the constant rate of $\omega_1 = 2$ rad/sec. Assuming the angle ϕ is constant, find the angular momentum of the gyroscope about the fixed point O.

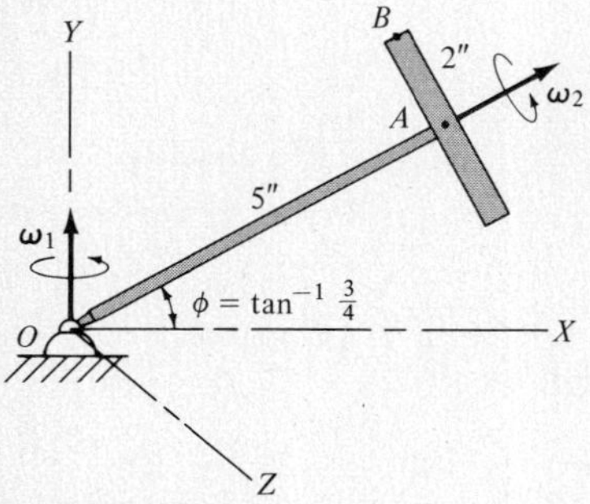

Figure P-16-2.8

16-2.9. Solve the preceding problem if, in addition to the given data, the angle ϕ of axis OA is increasing counterclockwise at the given instant at a constant rate of $\omega_3 = 1$ rad/sec.

$H_X = 0.288$ in.-lb-sec; $H_Y = 0.620$ in.-lb-sec; $H_Z = 0.202$ in.-lb-sec **_Ans._**

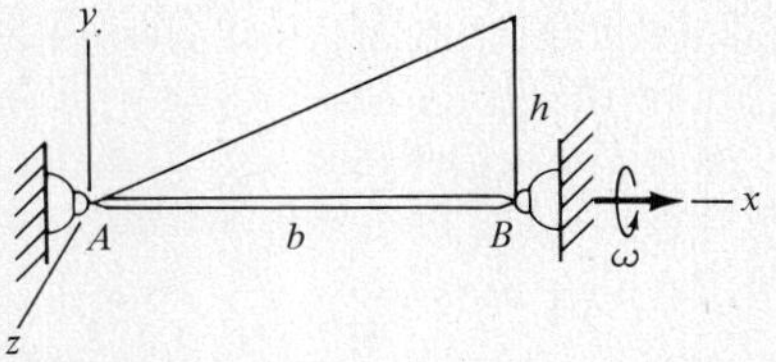

Figure P-16-2.10

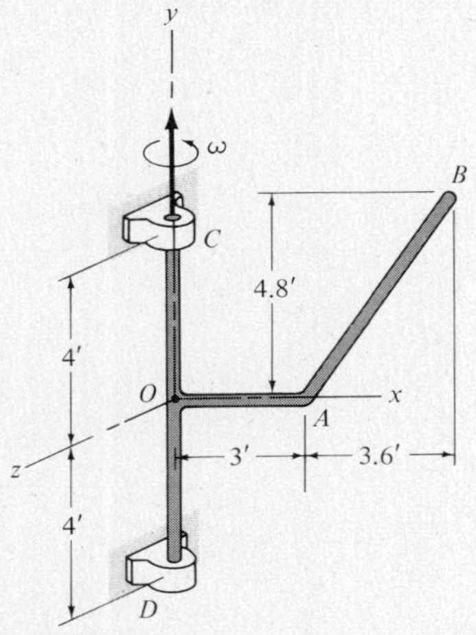

Figure P-16-2.11

16-2.10. A thin triangular plate of mass m is welded to a shaft rotating at the constant rate of ω rad/sec as shown in Fig. P-16-2.10. Determine the angular momentum of the plate about the ball-and-socket bearing at A and at B.

$$\text{At } A\text{: } \mathbf{H}_A = \tfrac{1}{12} mh\omega(2h\hat{\mathbf{i}} - 3b\hat{\mathbf{j}}) \qquad \textbf{\textit{Ans.}}$$

16-2.11. The bent bar OAB in Fig. P-16-2.11 is attached to a vertical shaft rotating about bearings at C and D at a constant rate of $\omega = 5$ rad/sec. Knowing that the bent bar always is in a vertical plane and that it weighs w lb/ft, determine its angular momentum about O.

$$\mathbf{H}_O = \frac{w}{g}(-389\hat{\mathbf{i}} + 769\hat{\mathbf{j}}) \text{ ft-lb-sec} \qquad \textbf{\textit{Ans.}}$$

16-2.12. As shown in Fig. P-16-2.12, a thin disk of mass m and radius r rolls without slipping on a flat horizontal surface as it also rotates about the axle AG. The axle AG is hinged to and rotates with the vertical shaft AB at the constant rate of ω_1 rad/sec. Determine the angular momentum of the disk about axes xyz at the fixed point A.

$$H_x = -\frac{1}{2} mrh\omega_1 \sin\beta;\ H_y = m\left(\frac{r^2}{4} + h^2\right)\omega_1 \sin\beta;\ H_z = 0 \qquad \textbf{\textit{Ans.}}$$

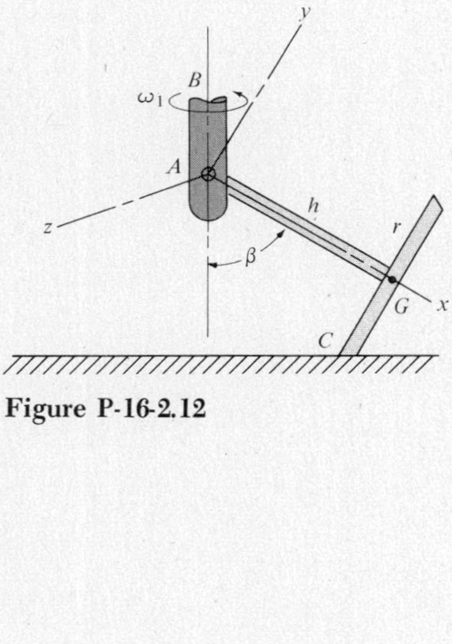

Figure P-16-2.12

* 16-3 INERTIA TENSOR

We include the following discussion as an introduction to the concepts of tensors and an index notation which is very useful in determining how the values of a tensor change with a rotation of the reference axes. It is appropriate at this place but is not essential.

If we rearrange Eq. (16-2.1) so that the coefficients of ω_x, ω_y, and ω_z are in the same vertical columns, we obtain

$$\left.\begin{aligned} H_x &= I_x\omega_x - P_{xy}\omega_y - P_{xz}\omega_z \\ H_y &= -P_{yx}\omega_x + I_y\omega_y - P_{yz}\omega_z \\ H_z &= -P_{zx}\omega_x - P_{zy}\omega_y + I_z\omega_z \end{aligned}\right\} \tag{16-3.1}$$

Lifting the inertia terms out of this arrangement and maintaining their order gives the following array which defines the *inertia tensor*.

$$\begin{pmatrix} I_x & -P_{xy} & -P_{xz} \\ -P_{yx} & I_y & -P_{yz} \\ -P_{zx} & -P_{zy} & I_z \end{pmatrix}$$

In order to determine how the values of the tensor change with a rotation of the reference axes (as part of the procedure to determine principal moments of inertia), it is advantageous to set $I_x = I_{11}$, $I_y = I_{22}$, $I_z = I_{33}$, and $-P_{xy} = I_{12}$, $-P_{xz} = I_{13}$, etc., thereby obtaining the following standard form of the inertia tensor:

$$\begin{pmatrix} I_{11} & I_{12} & I_{13} \\ I_{21} & I_{22} & I_{23} \\ I_{31} & I_{32} & I_{33} \end{pmatrix}$$

Notice that the subscripts x, y, and z have been replaced by the respective numerals 1, 2, and 3; also that the same symbol denotes both moments and products of inertia. Moments of inertia are identified by repeating the same subscript while products of inertia have differing subscripts.

Now by similarly denoting the components of **H** by H_1, H_2, and H_3, and the components of $\boldsymbol{\omega}$ by ω_1, ω_2, and ω_3, all the terms in Eq. (16-3.1) may be expressed compactly by the notation

$$H_i = \sum_{j=1}^{3} I_{ij}\omega_j, \qquad i = 1, 2, 3$$

in which each I term having differing indexes or subscripts represents the corresponding negative product of inertia term.

Index notation, which we shall not pursue further here,[2] is a compact method of keeping track of the many terms involved when deriving principal moments of inertia from a known set of inertia components. It is similarly useful in deformable body mechanics for determining principal stresses or strains from a known set of stress or strain components. In fact, the same procedure is used for transforming the components of any second-order tensor, such as the inertia, stress, or strain tensor, from one set of reference axes to another rotated set.

We shall not discuss such transformations, however, because in most practical applications in dynamics, the inertia components or their principal values are readily computed for insertion directly into Eqs. (16-2.1) or (16-2.2).

*16-4 EQUATIONS OF GENERAL SPATIAL MOTION

We now consider the spatial counterpart of plane dynamics. Recall from statics that any force system can be reduced to a resultant

[2] See, for example, Shames' *Engineering Mechanics* (Vol. II. *Dynamics*), Prentice-Hall (1966); or Yeh and Abrams' *Principles of Mechanics of Solids and Fluids* (Vol. 1. *Particle and Rigid-Body Mechanics*), McGraw-Hill (1960).

force-couple system. The value of the resultant force **R** at any instant is independent of a reference point whereas the resultant couple $\Sigma\mathbf{M}$, being the moment sum of the forces about the reference point, will depend on the location of the reference point. The correlation between the resultant force-couple and their dynamic effects were derived in Sections 10-5 and 10-6.

There we showed that the resultant of external forces on any body (rigid or nonrigid) is given by

$$\mathbf{R} = m\ddot{\bar{\mathbf{r}}} = m\bar{\mathbf{a}} \tag{10-5.2}$$

and that the moment of the external forces on the body is given by the general equation

$$\Sigma\mathbf{M}_A = \boldsymbol{\rho}_{G/A} \times m\mathbf{a}_A + \dot{\mathbf{H}}_A \tag{10-6.1}$$

where A is any reference point either in or outside the body and $\boldsymbol{\rho}_{G/A}$ is the position vector from A to the mass center G. If point A is fixed in space, $\mathbf{a}_A = 0$, or if A is at the mass center, $\boldsymbol{\rho}_{G/A} = 0$, so that with either of these restrictions, Eq. (10-6.1) reduces to the simpler forms

$$\Sigma\mathbf{M}_A = \dot{\mathbf{H}}_A \quad \text{or} \quad \Sigma\mathbf{M}_G = \dot{\mathbf{H}}_G \tag{10-6.2}$$

Usually we need not retain subscripts to distinguish between reference centers since each application of the moment equation clearly indicates which center is being used.

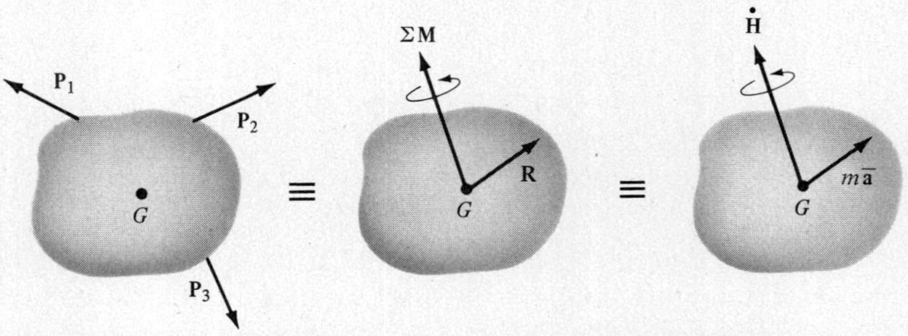

Figure 16-4.1 Equivalence of applied forces to their dynamic effects.

Our general approach to dynamics has been (and will continue to be) to reduce a given force system on a body to a resultant force-couple system acting at the mass center G. Under these conditions and in view of Eqs. (10-5.2) and (10-6.2) we had the equivalence shown in Fig. 10-6.2 which is repeated here as Fig. 16-4.1. When we use this concept, moment summations of the applied forces on a body may be taken about *any* point A in or fixed relative to the moving body by using the following general moment equation:

$$\Sigma \mathbf{M}_A = \boldsymbol{\rho}_{G/A} \times m\bar{\mathbf{a}} + \dot{\mathbf{H}}_G \tag{16-4.1}$$

where $\boldsymbol{\rho}_{G/A}$ is the position vector from A to the mass center G. Occasionally it may be convenient to use Eq. (10-6.1) as an alternate general moment equation which can serve to check results obtained by using Eq. (16-4.1).

We now consider how to evaluate the time derivative of angular momentum as a preliminary step to applying the general moment equations. Recall that expressions for **H** were determined with respect to body axes fixed in and rotating with the body at its absolute angular velocity $\boldsymbol{\omega}$. Note that the inertia integrals associated with the components of **H** are defined with respect to body axes fixed in and rotating with the body and hence have constant values invariant with time. However, **H** rotates with the body so that, using the body axes as a moving reference frame, the absolute time derivative of **H** with respect to inertial axes will involve both its change in magnitude as well as its change in direction. Both these changes in **H** are easily accounted for by applying the general omega theorem developed in Section 12-9. Using it, we obtain

$$\dot{\mathbf{H}} = \frac{\delta \mathbf{H}}{\delta t} + \boldsymbol{\omega} \times \mathbf{H} \tag{16-4.2}$$

where

$\dot{\mathbf{H}}$ = Absolute rate of change of **H** with respect to inertial axes XYZ

$\dfrac{\delta \mathbf{H}}{\delta t} = \dot{\mathbf{H}}_r$ = Relative rate of change of **H** with respect to the rotating body axes xyz; i.e., the change in magnitude of **H** relative to xyz

$\boldsymbol{\omega} \times \mathbf{H}$ = Rate of change of the direction of **H** due to rotation of the body axes xyz

We may evaluate the components of $\dot{\mathbf{H}}$ by formally expanding these terms, but it is conceptually more meaningful to indicate their components as shown in Fig. 16-4.2. These various components are directed as shown at the tips of the components (drawn dashed) of **H** along the rotating body axes xyz. Summations along the xyz directions then give the following components of the moment equation $\Sigma \mathbf{M} = \dot{\mathbf{H}}$:

$$\begin{aligned} \Sigma M_x &= \dot{H}_x + H_z\omega_y - H_y\omega_z \\ \Sigma M_y &= \dot{H}_y + H_x\omega_z - H_z\omega_x \\ \Sigma M_z &= \dot{H}_z + H_y\omega_x - H_x\omega_y \end{aligned} \tag{16-4.3}$$

To apply Eq. (16-4.3), we must substitute into it the general

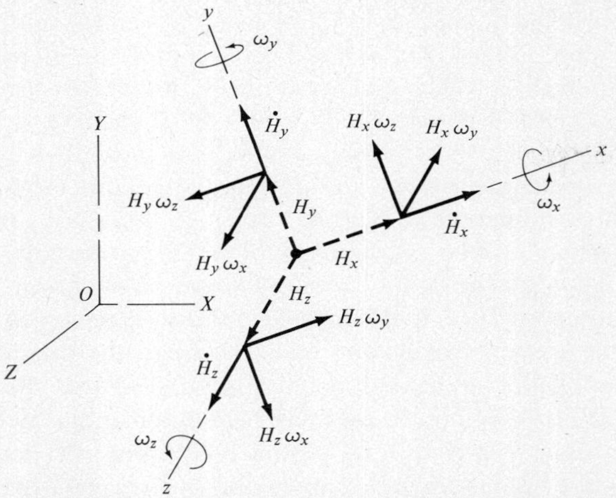

Figure 16-4.2 Components of $\dot{\mathbf{H}}_r$ and $\boldsymbol{\omega} \times \mathbf{H}$ along rotating body axes xyz.

expressions for H_x, H_y, and H_z given by Eq. (16-2.1) which are repeated here for reference:

$$\begin{aligned} H_x &= I_x\omega_x - P_{xy}\omega_y - P_{xz}\omega_z \\ H_y &= I_y\omega_y - P_{yz}\omega_z - P_{yx}\omega_x \\ H_z &= I_z\omega_z - P_{zx}\omega_x - P_{zy}\omega_y \end{aligned} \tag{16-2.1}$$

The result of substituting Eq. (16-2.1) into Eq. (16-4.3) is to get equations so lengthy and complex that they are almost unmanageable. The practical approach is to rewrite the general equations (16-2.1) and (16-4.3) using simplifying restrictions that are appropriate to each specific application. We now present several cases illustrating the manner by which this is accomplished.

EULER'S EQUATIONS

Considerable simplification of Eq. (16-2.1) is obtained by using body axes that coincide with the principal axes of inertia of the body. This makes the products of inertia zero so that the components of angular momentum in Eq. (16-2.1) reduce to

$$H_x = I_x\omega_x; \qquad H_y = I_y\omega_y; \qquad H_z = I_z\omega_z \tag{16-2.2}$$

Substituting these into Eq. (16-4.3) then gives

$$\begin{aligned} \Sigma M_x &= I_x\dot{\omega}_x + (I_z - I_y)\omega_y\omega_z \\ \Sigma M_y &= I_y\dot{\omega}_y + (I_x - I_z)\omega_z\omega_x \\ \Sigma M_z &= I_z\dot{\omega}_z + (I_y - I_x)\omega_x\omega_y \end{aligned} \tag{16-4.4}$$

These relations are known as Euler's equations of motion, named after Leonhard Euler (1707–1783) who ranks as one of the great mathematicians of all time. When you apply them, quite often the angular velocity $\boldsymbol{\omega}$ of the reference body axes will be the sum of several other angular velocities as in Illus. Probs. 16-2.3 and 16-4.1. In these cases, the time derivative of $\boldsymbol{\omega}$, i.e., $\dot{\boldsymbol{\omega}} = \boldsymbol{\alpha}$, will require using the kinematic analyses discussed in Section 12-8.

A modified version of Euler's equations is often used to describe the motion of bodies spinning at the rate $\boldsymbol{\omega}_s$ about an axis of symmetry as in a gyroscope. Then it is convenient to use a reference frame which is *not fixed* in the moving body, although the origin is still at the mass center or a fixed point. In this situation, the absolute velocity $\boldsymbol{\Omega}$ of the reference axes will be different from the absolute angular velocity $\boldsymbol{\omega} = \boldsymbol{\Omega} + \boldsymbol{\omega}_s$ of the body axes used to compute **H.** Since the orientation of the body now changes relative to the reference axes, the complications caused by time-dependent moments and products of inertia can be eliminated only if the reference axes always remain principal axes of inertia, even though they are not fixed in the body. This condition can exist only if one axis of the reference frame coincides with the axis of symmetry of a body of revolution (or more generally, a body in which two of the three principal moments of inertia are equal). Note that principal axes for a body of revolution consist of the axis of revolution together with *any* randomly selected orthogonal axes. Such axes, although not fixed in the body, will therefore always have constant principal moments of inertia.

When the general omega theorem is now applied to determine $\dot{\mathbf{H}}$, we shall have

$$\Sigma \mathbf{M} = \dot{\mathbf{H}} = \frac{\delta \mathbf{H}}{\delta t} + \boldsymbol{\Omega} \times \mathbf{H} \qquad (16\text{-}4.5)$$

where $\boldsymbol{\Omega}$ is now the absolute angular velocity of the reference axes. Remember, however, that $\boldsymbol{\omega}$, used to evaluate **H,** always is the absolute angular velocity of the body. Thus, assuming that the axis of symmetry of the body with its associated spin rate $\boldsymbol{\omega}_s$ coincides with the z axis of the reference frame, we have $\boldsymbol{\omega} = \boldsymbol{\Omega} + \boldsymbol{\omega}_s$ or

$$\omega_x = \Omega_x; \qquad \omega_y = \Omega_y; \qquad \omega_z = \Omega_z + \omega_s$$

and hence the components of **H,** expressed by Eq. (16-2.2), now become

$$H_x = I_x\Omega_x; \qquad H_y = I_y\Omega_y; \qquad H_z = I_z(\Omega_z + \omega_s)$$

Now carrying out the operations of Eq. (16-4.5), or using an equivalent form of Fig. 16-4.2, will result in the following *modified*

Euler's equations:

$$\begin{aligned} \Sigma M_x &= I_x\dot{\Omega}_x + (I_z - I_y)\Omega_y\Omega_z + I_z\Omega_y\omega_s \\ \Sigma M_y &= I_y\dot{\Omega}_y + (I_x - I_z)\Omega_z\Omega_x - I_z\Omega_x\omega_s \\ \Sigma M_z &= I_z(\dot{\Omega}_z + \dot{\omega}_s) + (I_y - I_x)\Omega_x\Omega_y \end{aligned} \qquad (16\text{-}4.6)$$

These equations will be especially useful in gyroscopic motion since they distinguish between the spin rate of the rotor and the rotation of the reference frame. They reduce to the original set of Euler's equations when $\boldsymbol{\Omega} + \boldsymbol{\omega}_s = \boldsymbol{\omega}$; that is, when the reference axes are fixed in the body. Also observe that letting the z axis of the reference frame coincide with the spin axis of the body results in $I_x = I_y$, and that all components of $\boldsymbol{\Omega}$ and $\boldsymbol{\omega}$ are identical except those along the z axis. If other than the z axis is chosen to coincide with the spin axis, Eq. (16-4.6) must be revised accordingly.

ROTATION ABOUT A FIXED AXIS

Here we extend fixed-axis rotation to include bodies that may not be symmetrical about the centroidal plane of motion as in Fig. 16-4.4 on p. 656 for Illus. Prob. 16-4.2. The origin of the body axes is fixed in space by locating it on the axis of rotation. Orient the body axes so that the z axis coincides with the axis of rotation. Doing this causes both ω_x and ω_y to become zero so that the components of angular momentum [from Eq. (16-2.1)] reduce to

$$H_x = -P_{zx}\omega_z; \qquad H_y = -P_{yz}\omega_z; \qquad H_z = I_z\omega_z$$

We may substitute these components of $\mathbf{H}$ directly into $\Sigma\mathbf{M} = \dot{\mathbf{H}}$ or, alternatively, substitute them into Eq. (16-4.3) which then yields the following equations:

$$\begin{aligned} \Sigma M_x &= -P_{zx}\dot{\omega}_z + P_{yz}\omega_z^{\,2} \\ \Sigma M_y &= -P_{yz}\dot{\omega}_z - P_{zx}\omega_z^{\,2} \\ \Sigma M_z &= I_z\dot{\omega}_z \end{aligned} \qquad (16\text{-}4.7)$$

Even with these simplifications, Eqs. (16-4.7) defining the equations of rotation for a general unsymmetrical body about its axis of rotation is complex. Fortunately, most practical applications of fixed-axis rotation involve rotation of a body that is *symmetrical* with respect to the plane of motion of its mass center. This is the situation previously discussed in Section 13-3 in which $P_{zx} = P_{yz} = 0$. Hence, Eq. (16-4.7) is reduced to the much simpler form:

$$\Sigma M_x = 0; \qquad \Sigma M_y = 0; \qquad \Sigma M_z = I_z\dot{\omega}_z = I_z\alpha \qquad (16\text{-}4.8)$$

Remember that I_z is always computed about the axis of rotation, whether the motion be noncentroidal or centroidal rotation. In the latter case, of course, I_z becomes $\bar{I}_z$.

GENERAL PLANE MOTION

Assume that the body is constrained to move parallel to a fixed XY plane. Here it is preferable to place the origin of the xyz body axes at the mass center of the body with the z axis directed perpendicular to the XY plane. This makes $\omega_x = \omega_y = 0$. These are the same restrictions that we just used for rotation about a fixed axis so that Eqs. (16-2.1) and (16-4.3) reduce to the same form as Eq. (16-4.7) except that here we use a bar superscript to emphasize that the origin is at the mass center. The equations become

$$\begin{aligned} \Sigma \bar{M}_x &= -\bar{P}_{zx}\dot{\omega}_z + \bar{P}_{yz}\omega_z^{\,2} \\ \Sigma \bar{M}_y &= -\bar{P}_{yz}\dot{\omega}_z - \bar{P}_{zx}\omega_z^{\,2} \\ \Sigma \bar{M}_z &= \bar{I}_z\dot{\omega}_z \end{aligned} \qquad (16\text{-}4.9)$$

For bodies which are also *symmetrical* with respect to the centroidal plane of motion so that $\bar{P}_{zx} = \bar{P}_{yz} = 0$, the above equations, as in the case of symmetrical body rotation about a fixed axis, further reduce to those discussed previously in Section 13-5; namely,

$$\Sigma \bar{M}_x = 0; \qquad \Sigma \bar{M}_y = 0; \qquad \Sigma \bar{M}_z = \bar{I}_z\dot{\omega}_z = \bar{I}_z\alpha \quad (16\text{-}4.10)$$

SUMMARY

As a guide to the solution of problems, start with a free-body diagram showing the isolated system under consideration. Observe that the force equation $\mathbf{R} = m\bar{\mathbf{a}}$ is always valid, regardless of the choice of a reference center for moment summations. Moment summations may be taken about any point A in or fixed relative to the moving body by using either of the following general equations:

$$\Sigma \mathbf{M}_A = \boldsymbol{\rho}_{G/A} \times m\bar{\mathbf{a}} + \dot{\mathbf{H}}_G \qquad (16\text{-}4.1)$$

or

$$\Sigma \mathbf{M}_A = \boldsymbol{\rho}_{G/A} \times m\mathbf{a}_A + \dot{\mathbf{H}}_A \qquad (10\text{-}6.1)$$

where $\dot{\mathbf{H}}_G$ and $\dot{\mathbf{H}}_A$ are the absolute time derivatives of the relative angular momentum about G or A respectively, and $\boldsymbol{\rho}_{G/A}$ is the position vector from A to G. Usually Eq. (16-4.1) is preferable since $\dot{\mathbf{H}}_G$ is easier to compute than $\dot{\mathbf{H}}_A$ and, furthermore, Eq. (16-4.1) may be applied to any of several moment centers once $\dot{\mathbf{H}}_G$ is known whereas $\dot{\mathbf{H}}_A$ is restricted to a single center.

You have probably noticed that moment equations for spatial motions involve both angular velocity $\boldsymbol{\omega}$ and angular acceleration $\dot{\boldsymbol{\omega}}$. When these quantities are specified, it is fairly simple to determine the corresponding moments. Indeed, this is usually the type of spatial problem considered in this book since the converse problem of finding $\boldsymbol{\omega}$ and $\dot{\boldsymbol{\omega}}$ when the moments are specified can be very difficult. A direct solution involves elliptic integrals which is beyond the level of this book. However, application of the work-energy method (Section 16-5) to find velocities may be a first step since then accelerations are not involved; after this, instantaneous accelerations may be determined. The general problem of computing spatial displacements requires an iterative numerical procedure which is best done by a digital computer. This step also is beyond the range of this book.

ILLUSTRATIVE PROBLEMS

16-4.1. A thin uniform disk of mass m and radius r spins at a uniform rate of ω_2 rad/sec about a horizontal axle in the yoke shown in Fig. 16-4.3. When the yoke is rotating about a fixed vertical axis at the uniform rate of ω_1 rad/sec, determine the moment exerted by the yoke upon the disk.

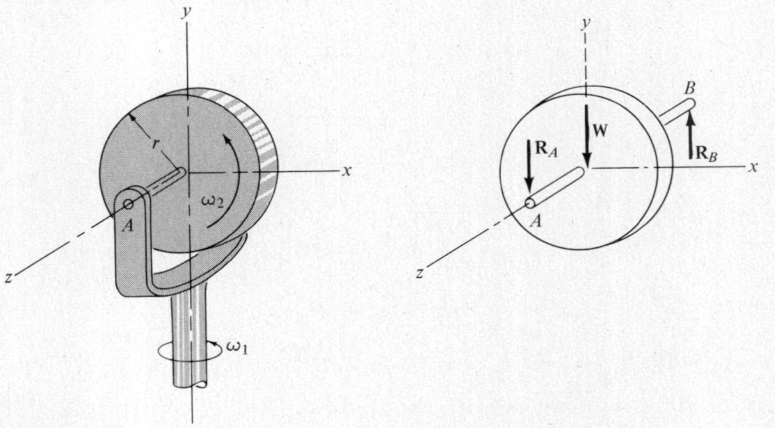

Figure 16-4.3

Solution

The FBD of the disk shows the forces $\mathbf{R}_A$ and $\mathbf{R}_B$ exerted by the yoke on the disk. Since the mass center of the disk is stationary in space, $\mathbf{R} = m\bar{\mathbf{a}} = \mathbf{0}$. Hence $\mathbf{R}_A = -\mathbf{R}_B$, thereby forming a couple $\mathbf{M}$ which the yoke exerts on the disk.

We shall solve this problem in three ways to compare the concepts discussed in this section. First, we shall assume the reference axes xyz in Fig. 16-4.3 are fixed in and rotating with the disk and apply the Euler equations (16-4.4). Next we shall assume the reference axes xyz are fixed in and rotate with the yoke and apply the modified Euler equations (16-4.6). Finally we shall apply Eq. (16-4.5) directly in its vector form.

Method 1: Reference axes fixed in the disk with origin at its mass center. The axes are principal axes about which the moments of inertia are

$$I_x = \tfrac{1}{4}mr^2; \qquad I_y = \tfrac{1}{4}mr^2; \qquad I_z = \tfrac{1}{2}mr^2$$

The absolute angular velocity of the body axes is

$$\boldsymbol{\omega} = \boldsymbol{\omega}_1 + \boldsymbol{\omega}_2 = \omega_1\hat{\mathbf{j}} + \omega_2\hat{\mathbf{k}}$$

and their absolute angular acceleration, by differentiation, is

$$\begin{aligned} \dot{\boldsymbol{\omega}} &= \dot{\omega}_1\hat{\mathbf{j}} + \omega_1\dot{\hat{\mathbf{j}}} + \dot{\omega}_2\hat{\mathbf{k}} + \omega_2\dot{\hat{\mathbf{k}}} \\ &= 0 + 0 + 0 + \omega_2(\boldsymbol{\omega}_1 \times \hat{\mathbf{k}}) = \omega_2(\omega_1\hat{\mathbf{j}}) \times \hat{\mathbf{k}}) = \omega_2\omega_1\hat{\mathbf{i}} \end{aligned}$$

As explained in Illus. Prob. 12-8.1, this result could also have been obtained directly by

$$\dot{\boldsymbol{\omega}} = \dot{\boldsymbol{\omega}}_1 + \dot{\boldsymbol{\omega}}_2 + \boldsymbol{\omega}_1 \times \boldsymbol{\omega}_2 = 0 + 0 + \omega_1\hat{\mathbf{j}} \times \omega_2\hat{\mathbf{k}} = \omega_1\omega_2\hat{\mathbf{i}}$$

Thus, the components of $\boldsymbol{\omega}$ and $\dot{\boldsymbol{\omega}}$ for use in Euler's equations (16-4.4) are

$$\begin{aligned} \omega_x &= 0 & \dot{\omega}_x &= \omega_1\omega_2 \\ \omega_y &= \omega_1 & \dot{\omega}_y &= 0 \\ \omega_z &= \omega_2 & \dot{\omega}_z &= 0 \end{aligned}$$

Substituting values in Euler's equations now yield

$$\begin{aligned} \Sigma M_x &= I_x\dot{\omega}_x + (I_z - I_y)\omega_y\omega_z \\ &= \tfrac{1}{4}mr^2(\omega_1\omega_2) + (\tfrac{1}{2}mr^2 - \tfrac{1}{4}mr^2)(\omega_1\omega_2) = \tfrac{1}{2}mr^2\omega_1\omega_2 \\ \Sigma M_y &= I_y\overset{=0}{\cancel{\dot{\omega}_y}} + (I_x - I_y)\omega_z\overset{=0}{\cancel{\omega_x}} = 0 \\ \Sigma M_z &= I_z\overset{=0}{\cancel{\dot{\omega}_z}} + (I_y - I_z)\overset{=0}{\cancel{\omega_x}}\omega_y = 0 \end{aligned}$$

Hence, the resultant moment exerted by the yoke on the disk is

$$\mathbf{M} = \tfrac{1}{2}mr^2\omega_1\omega_2\hat{\mathbf{i}} \qquad \textit{Ans.}$$

Method 2: Reference axes fixed in the yoke. The origin of these axes is a fixed point in space which happens to coincide with the mass center of the disk. We may use the modified Euler equations (16-4.6) since the spin axis of the disk coincides with the z axis of a reference frame rotating with the yoke. The angular velocity and acceleration components of $\boldsymbol{\Omega} + \boldsymbol{\omega}_s = \boldsymbol{\omega}$ are

$$\begin{aligned} \Omega_x &= 0 & \dot{\Omega}_x &= 0 & \omega_s &= \omega_2 \\ \Omega_y &= \omega_1 & \dot{\Omega}_y &= 0 & \dot{\omega}_s &= 0 \\ \Omega_z &= 0 & \dot{\Omega}_z &= 0 \end{aligned}$$

When these values are substituted in the modified Euler equations, we now obtain

$$
\begin{aligned}
\Sigma M_x &= I_x\dot{\Omega}_x + (I_z - I_y)\Omega_y\Omega_z + I_z\Omega_y\omega_s \\
&= 0 + 0 + \tfrac{1}{2}mr^2\omega_1\omega_2 \\
\Sigma M_y &= I_y\cancelto{=0}{\dot{\Omega}_y} + (I_x - I_y)\cancelto{=0}{\Omega_z}\Omega_x - I_z\cancelto{=0}{\Omega_x}\omega_s = 0 \\
\Sigma M_z &= I_z\cancelto{=0}{\dot{\Omega}_z} + \dot{\omega}_s) + (I_y - I_x)\cancelto{=0}{\Omega_z}\Omega_y = 0
\end{aligned}
$$

Hence, we verify that $\mathbf{M} = \frac{1}{2}mr^2\omega_1\omega_2\hat{\mathbf{i}}$.

Method 3: Direct application of Eq. (16-4.5). Recall that angular momentum is always computed using the absolute angular velocity of the body which here is $\boldsymbol{\omega} = \boldsymbol{\omega}_1 + \boldsymbol{\omega}_2$. Since $\boldsymbol{\omega}$ has the components

$$\omega_x = 0; \qquad \omega_y = \omega_1; \qquad \omega_z = \omega_2$$

which are directed along principal axes of inertia of the body, we use Eq. (16-2.2) to compute the following components of $\mathbf{H}$:

$$
\begin{aligned}
H_x &= I_x\omega_x = 0 \\
H_y &= I_y\omega_y = \tfrac{1}{4}mr^2\omega_1 \\
H_z &= I_z\omega_z = \tfrac{1}{2}mr^2\omega_2
\end{aligned}
$$

We now apply $\Sigma\mathbf{M} = \delta\mathbf{H}/\delta t + \boldsymbol{\Omega} \times \mathbf{H}$ where $\boldsymbol{\Omega} = \boldsymbol{\omega}_1 = \omega_1\hat{\mathbf{j}}$. Since ω_1 and ω_2 are constant in the rotating reference frame, $\mathbf{H}$ has a constant magnitude in this frame and hence $\delta\mathbf{H}/\delta t = 0$. Equation (16-4.5) therefore becomes

$$\mathbf{M} = 0 + \boldsymbol{\Omega} \times \mathbf{H} = \omega_1\hat{\mathbf{j}} \times (\tfrac{1}{4}mr^2\omega_1\hat{\mathbf{j}} + \tfrac{1}{2}mr^2\omega_2\hat{\mathbf{k}})$$

or

$$\mathbf{M} = \tfrac{1}{2}mr^2\omega_1\omega_2\hat{\mathbf{i}} \qquad \textit{Check}$$

Incidentally, the opposite of $\mathbf{M}$ is known as the gyroscopic moment exerted by the disk upon the yoke.

16-4.2. Two identical homogeneous triangular plates, each of mass m, are welded to the shaft in Fig. 16-4.4. Determine the dynamic bearing reactions acting at A and B when the shaft is rotating at a constant speed of ω rad/sec.

Solution

Since the weight of the system is balanced by the static reactions, the free-body diagram shows only the dynamic bearing reactions at the ball-and-socket bearings. To relate these six unknowns, there are available the three scalar components of $\mathbf{R} = m\bar{\mathbf{a}}$ and the three scalar components of $\Sigma\mathbf{M} = \dot{\mathbf{H}}$ taken about the mass center or a fixed point.

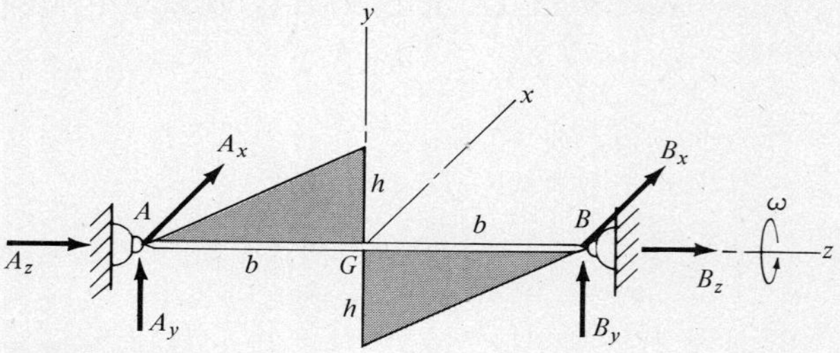

Figure 16-4.4

In this problem, the mass center G of the system happens to be on the fixed axis of rotation and hence $\mathbf{R} = m\bar{\mathbf{a}} = 0$. Its scalar components yield

$$[\Sigma X = 0] \qquad A_x + B_x = 0$$
$$[\Sigma Y = 0] \qquad A_y + B_y = 0$$
$$[\Sigma Z = 0] \qquad A_z + B_z = 0$$

so that the dynamic components of the bearing reactions are equal but oppositely directed.

Now consider the moment equation. Since the mass center of the system is on the fixed axis of rotation, recall, from part (c) of Illus. Prob. 16-2.2, that the system will have the same angular momentum about any origin on the axis of rotation. However, the directions of principal axes are unknown so Euler's equations cannot be used. Instead, we apply Eq. (16-4.7), selecting the origin of body-fixed reference axes at G because of the convenience this origin has for computing the product of inertia terms P_{zx} and P_{yz} appearing therein.

Since the yz plane is a plane of symmetry, $P_{zx} = \int zx\,dm = 0$. However, $P_{yz} = \int yz\,dm$ is not zero. By letting $dm = \gamma t\,dA$ where γ is the mass density and t the constant thickness of the plate, it is simple to show that

$$(P_{yz})_{\text{mass}} = (P_{yz})_{\text{area}}\left(\frac{m}{\text{area}}\right)$$

Hence, using the area product of inertia derived in Illus. Prob. 7-9.1, we have

$$P_{yz} = -\frac{b^2h^2}{24}\left(\frac{m}{bh/2}\right) = -\frac{mbh}{12}$$

where the minus sign appears because the face area of each plate is in a negative quadrant (refer to p. 231). Hence, for the system of two identical plates, the total mass product of inertia is their sum, or

$$P_{yz} = -\frac{mbh}{6}$$

Turn now to the components of the moment equation [Eq. (16-4.7)] and note that $\omega_z = \omega$ and $\dot{\omega}_z = 0$, whence we have

$$\left[\Sigma M_x = P_{yz}\omega^2 = -\frac{mbh}{6}\omega^2\right] \qquad b(A_y - B_y) = -\frac{mbh}{6}\omega^2$$

$$[\Sigma M_y = 0] \qquad b(B_x - A_x) = 0$$

$[\Sigma M_z = 0]$ Identically zero, since moment axis intersects all dynamic reaction components.

Finally, on combining all force and moment equations, we obtain

$$A_y = -\frac{mh\omega^2}{12} \qquad B_y = \frac{mh\omega^2}{12} \qquad \textit{Ans.}$$

$$A_x = B_x = 0$$

$$A_z = -B_z$$

Recognize that A_y and B_y form a couple, rotating with the shaft in the plane of the plates. A_z and B_z remain indeterminate but will each become zero if only one of the bearings (either A or B) can resist end thrust.

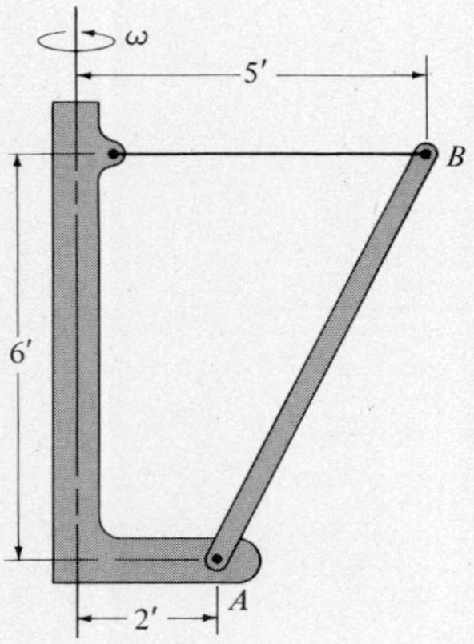

Figure 16-4.5

16-4.3. A uniform slender rod weighing 96.6 lb is fastened to the rotating frame in Fig. 16-4.5 by a smooth hinge at A and a horizontal cord at B. The frame rotates about its vertical axis at a constant speed of $\omega = 6$ rad/sec. Determine the tension in the cord and the horizontal and vertical components of the hinge reaction.

Solution

Considering the FBD of the bar and its equivalent dynamic effects shown in Fig. 16-4.6, we see that equating moment summations about A will lead directly to T. Then equivalence of horizontal and vertical force summations will determine A_h and A_v.

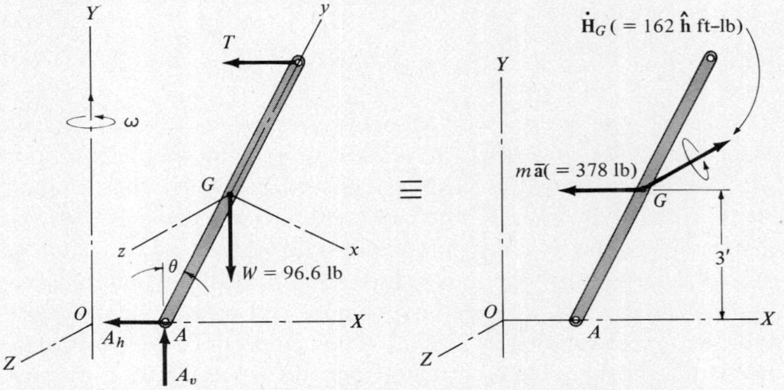

Figure 16-4.6 Equivalence of applied forces to dynamic effects.

Since G travels at constant speed in a circle of radius $(2 + \frac{3}{2})$ ft about the fixed Y axis of the frame, its acceleration is given by $\bar{a} = \bar{r}\omega^2$ and the effective force at G is $m\bar{a} = \dfrac{96.6}{32.2}(3.5)(6)^2 = 378$ lb.

To compute the effective couple $\dot{\mathbf{H}}_G$, we start with $\mathbf{H}_G$ using principal axes xyz with origin at G. Using Eq. (16-2.2) we have

$$\mathbf{H}_G = I_x\omega_x\hat{\mathbf{i}} + I_y\omega_y\hat{\mathbf{j}} + I_z\omega_z\hat{\mathbf{k}}$$

where

$$I_x = \tfrac{1}{12}mL^2; \qquad I_y = 0; \qquad I_z = \tfrac{1}{12}mL^2$$
$$\omega_x = -\omega\sin\theta; \qquad \omega_y = \omega\cos\theta; \qquad \omega_z = 0$$

and therefore

$$\mathbf{H}_G = \tfrac{1}{12}mL^2\omega\sin\theta\,\hat{\mathbf{i}}$$

Applying $\dot{\mathbf{H}}_G = \delta\mathbf{H}/\delta t + \boldsymbol{\omega}\times\mathbf{H}$, we note that since ω is constant, $\mathbf{H}_G$ does not change with respect to the rotating frame and hence $\delta\mathbf{H}/\delta t = 0$. But its direction does change, and hence

$$\dot{\mathbf{H}}_G = \boldsymbol{\omega}\times\mathbf{H}_G = (-\omega\sin\theta\,\hat{\mathbf{i}} + \omega\cos\theta\,\hat{\mathbf{j}})\times\left(-\frac{1}{12}mL^2\omega\sin\theta\,\hat{\mathbf{i}}\right)$$
$$= \frac{1}{12}mL^2\omega^2\sin\theta\cos\theta\,\hat{\mathbf{k}} = \frac{1}{12}\left(\frac{96.6}{32.2}\right)(L^2)(6)^2\left(\frac{3}{L}\right)\left(\frac{6}{L}\right)\hat{\mathbf{k}}$$
$$= 162\hat{\mathbf{k}}\text{ ft-lb}$$

which is equivalent to a counterclockwise couple of 162 ft-lb acting in the XY plane.

Equating the forces in the FBD to their equivalent dynamic effects, we have

$[\Sigma M_{Z_A} = (\Sigma M_{Z_A})_{\text{dyn}}]$ $\quad 6T - 96.6(1.5) = 378(3) + 162$

$T = 240$ lb $\quad$ ***Ans.***

$[\overset{+}{\leftarrow}\Sigma X = (\Sigma X)_{\text{dyn}}]$ $\quad A_h + 240 = 378 \qquad A_h = 138$ lb$\leftarrow$ $\quad$ ***Ans.***

$[+\uparrow\Sigma Y = (\Sigma Y)_{\text{dyn}}]$ $\quad A_v - 96.6 = 0 \qquad A_v = 96.6$ lb$\uparrow$ $\quad$ ***Ans.***

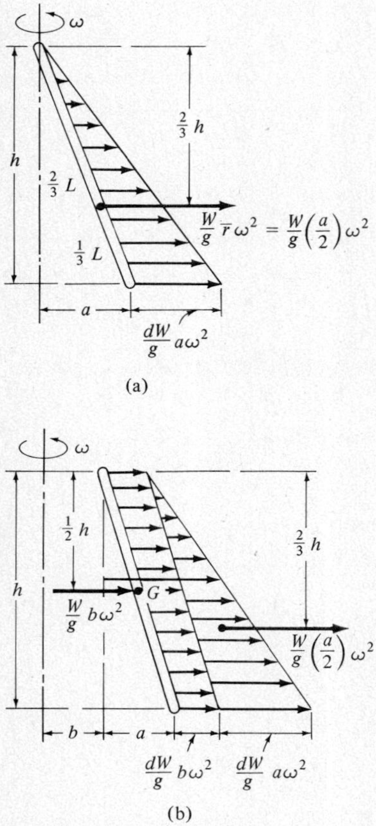

Figure 16-4.7 Centrifugal inertia forces acting on a rod inclined to the axis of rotation.

Let us now consider an alternate solution in terms of inertia effects which will give physical meaning to how the couple $\dot{\mathbf{H}}_G$ enters into the dynamic analysis. Although the *magnitude* of the resultant inertia force is always $m\bar{a}$, its line of action is *not* through the mass center when a body is unsymmetrical with respect to the plane of motion described by the mass center. One example of such unsymmetrical rotation is the case of a slender rod rotating at an angle with the axis of rotation as in Fig. 16-4.7. In part (a), the centrifugal inertia force acts radially outward from the axis of rotation through the centroid of the triangularly distributed inertia forces $dm\,r\omega^2$ and equals $m\bar{r}\omega^2 = \dfrac{W}{g}\left(\dfrac{a}{2}\right)\omega^2$. In part (b), the rod is displaced an additional distance b from the axis of rotation so that the centrifugal inertia force consists of two parts acting as shown and determined by

$$\frac{W}{g}\bar{r}\omega^2 = \frac{W}{g}\left(b + \frac{a}{2}\right)\omega^2 = \frac{W}{g}b\omega^2 + \frac{W}{g}\left(\frac{a}{2}\right)\omega^2$$

which is the sum of the uniformly distributed and triangularly distributed centrifugal inertia effects. Although not shown here, a similar analysis applies to tangential inertia effects caused by angular acceleration.

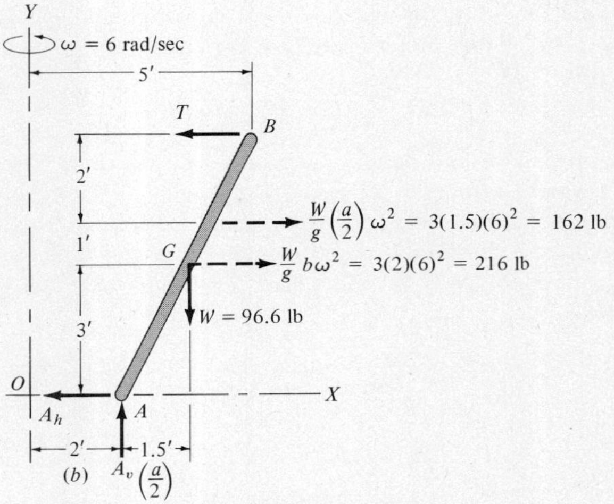

Figure 16-4.8 Dynamic equilibrium.

Applying the result of this discussion to the present problem, the FBD of the rod in dynamic equilibrium appears as shown in Fig. 16-4.8. Since there is no angular acceleration, all the forces are in the same plane so that the equations of dynamic equilibrium give

$[\Sigma M_A = 0]$	$6T - 96.6(1.5) - 216(3) - 162(4) = 0$	$T = 240$ lb
$[\Sigma X = 0]$	$162 + 216 - (T = 240) - A_h = 0$	$A_h = 138$ lb←
$[\Sigma Y = 0]$	$A_v - 96.6 = 0$	$A_v = 96.6$ lb↑

As expected, these results verify the original solution, but are much more direct. Furthermore, comparison with the original solution shows that replacing $\dfrac{W}{g}\left(\dfrac{a}{2}\right)\omega^2$ by a force at G and a couple results in $m\bar{a} = 216 + 162 = 378$ lb as before, while the couple $162(1) = 162$ ft-lb is equivalent to $\dot{\mathbf{H}}_G$.

PROBLEMS

16-4.4. Derive Eq. (16-4.7) for fixed-axis rotation of an unsymmetrical body by direct application of $\Sigma \mathbf{M} = \dot{\mathbf{H}}$. Assume the z axis is the axis of rotation and refer to Eq. (16-2.1) to determine the appropriate components of **H**.

16-4.5. Derive Euler's equations (16-4.4) analytically by substituting Eq. (16-2.2) into $\Sigma \mathbf{M} = \dot{\mathbf{H}}$.

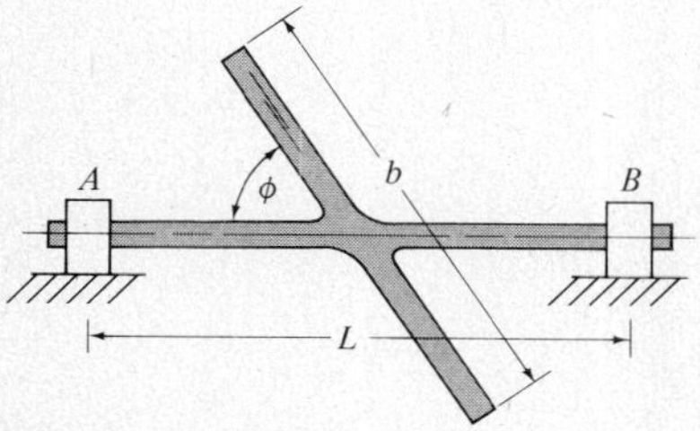

Figure P-16-4.6

16-4.6. A uniform slender rod b ft long and weighing w lb/ft is fastened at its midpoint to a horizontal shaft as shown in Fig. P-16-4.6. The rod is attached to the shaft midway between two bearings A and B a distance L ft apart. Compute the dynamic reactions at A and B when the shaft is rotating at a constant rate of ω rad/sec. Use Euler's equations and check by means of the dynamic equilibrium approach of Fig. 16-4.7.

$$R_B = -R_A = \frac{wb^2\omega^2}{24gL} \sin 2\phi \qquad \textbf{Ans.}$$

16-4.7. A disk of mass m and radius r is fastened as shown in Fig. P-16-4.7 to a shaft which is rotating at a constant angular velocity of ω rad/sec. The plane of the disk is skewed at an angle θ with the normal to the shaft. Determine the dynamic bearing reactions at A and B. Solve by direction application of $\Sigma \mathbf{M} = \dot{\mathbf{H}}$ and check by means of Euler's equations.

$$R_B = -R_A = \frac{mr^2\omega^2}{8L} \sin 2\theta \qquad \textbf{Ans.}$$

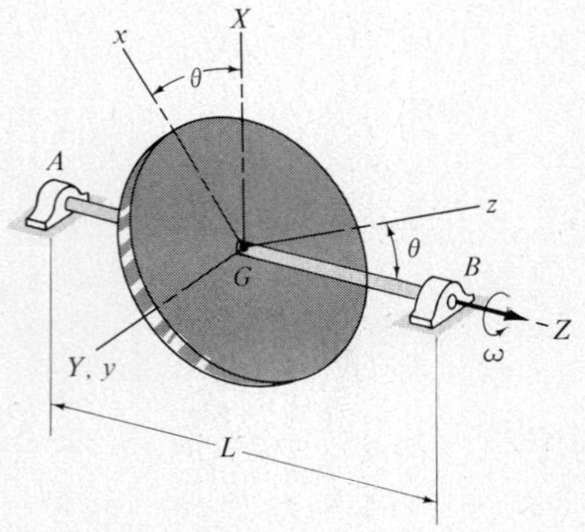

Figure P-16-4.7

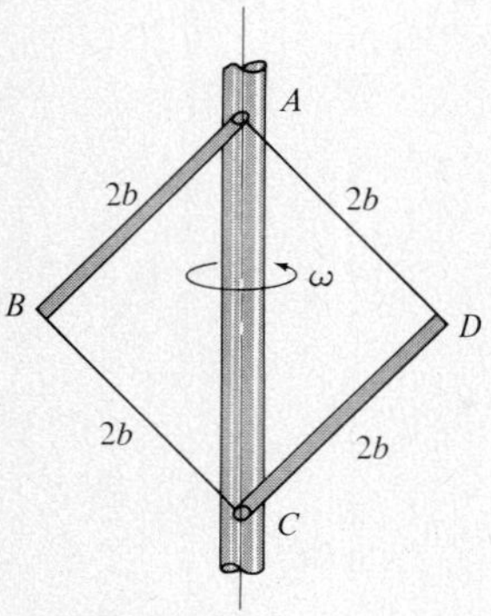

Figure P-16-4.8

16-4.8. Two slender uniform bars AB and CD, each of weight W and length $2b$, are pinned as shown in Fig. P-16-4.8 to the vertical shaft AC which rotates at a constant rate of ω rad/sec. Their outward swing is restrained by the cords AD and BC. Determine the magnitudes of the horizontal and vertical components of the pin reactions at A and C upon the bars.

16-4.9. In Fig. P-16-4.9, a disk weighing 1.61 lb and having a 2-in.

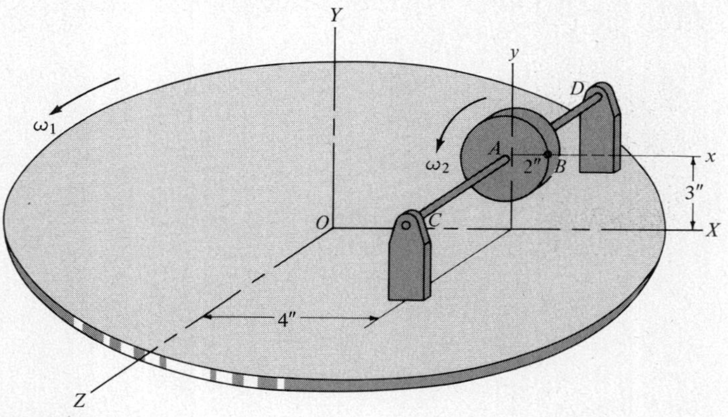

Figure P-16-4.9

radius rotates at the constant rate of $\omega_2 = 120$ rad/sec about the axle CD mounted on a platform which is itself rotating at the constant rate of $\omega_1 = 5$ rad/sec about a fixed vertical axis at O. Determine the moment exerted on shaft CD by the bearings at C and D. Solve by applying $\Sigma \mathbf{M}_A = \dot{\mathbf{H}}_A$ and check by using the gyroscopic equation (15-8.1).

16-4.10. In Fig. P-16-4.9, a disk weighing 9.66 lb rotates at the constant rate of $\omega_2 = 40$ rad/sec about the axle CD mounted on a wheel which is itself rotating at a constant rate of $\omega_1 = 5$ rad/sec about a vertical axis at O. If the distance between C and D is 8 in., find the dynamic reactions at C and D.

$$\mathbf{C} = (-1.25\hat{\mathbf{i}} - 1.25\hat{\mathbf{j}}) \text{ lb}; \ \mathbf{D} = (-1.25\hat{\mathbf{i}} + 1.25\hat{\mathbf{j}}) \text{ lb} \qquad \textit{Ans.}$$

16-4.11. The homogeneous rectangular plate of sides a and b and mass m in Fig. P-16-4.11 rotates about smooth ball-and-socket bearings at A and B at the constant rate of ω rad/sec. Determine the dynamic reactions at A and B.

$$R_B = -R_A = \frac{mab\omega^2}{12d^3}(b^2 - a^2) \qquad \textit{Ans.}$$

16-4.12. A thin triangular plate of mass m is welded to a shaft rotating at the constant rate of ω rad/sec as shown in Fig. P-16-4.12. Determine the dynamic bearing reactions at the ball-and-socket bearings at A and B.

$$\mathbf{R}_A = -\tfrac{1}{12}mh\omega^2\hat{\mathbf{j}} \text{ lb}; \ \mathbf{R}_B = -\tfrac{1}{4}mh\omega^2\hat{\mathbf{j}} \text{ lb} \qquad \textit{Ans.}$$

16-4.13. Two triangular plates, each weighing 15 lb, are welded as shown in Fig. P-16-4.13 to a shaft rotating at a constant rate of $\omega = 40$ rad/sec. Determine the dynamic reactions at the ball-and-socket bearings A and B.

$$\mathbf{R}_A = 108.7\hat{\mathbf{j}} \text{ lb}; \ \mathbf{R}_B = 139.8\hat{\mathbf{j}} \text{ lb} \qquad \textit{Ans.}$$

16-4.14. The center A of the thin disk of mass m and radius r in Fig. P-16-4.14 (p. 669) has a constant speed of v fps. The disk rolls without slip-

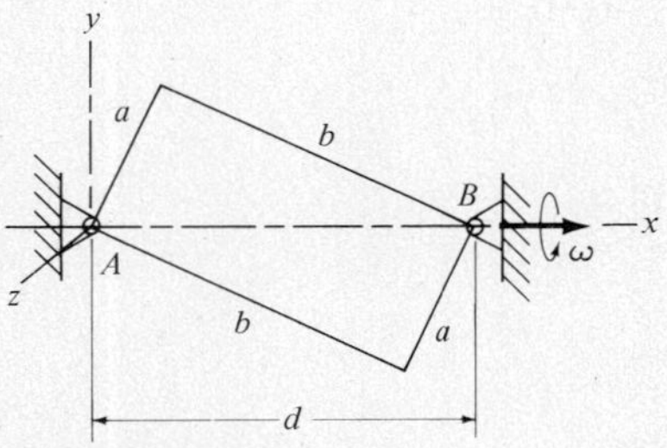

Figure P-16-4.11

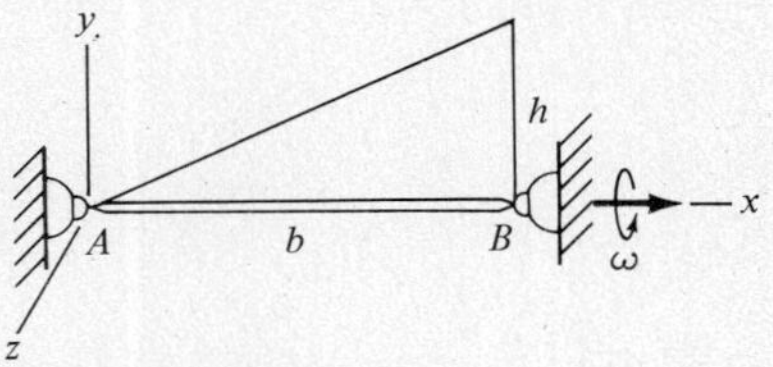

Figure P-16-4.12

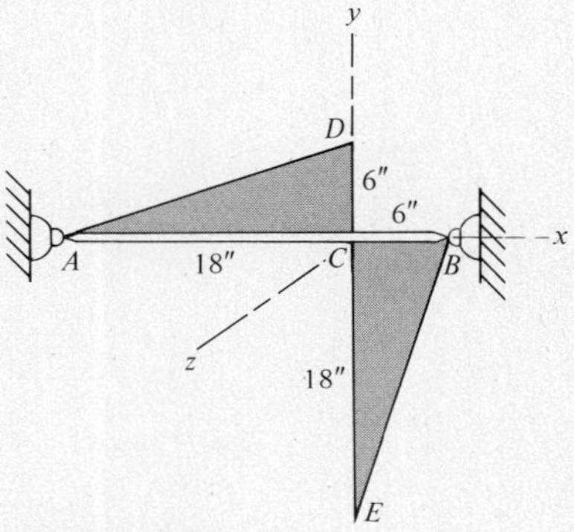

Figure P-16-4.13

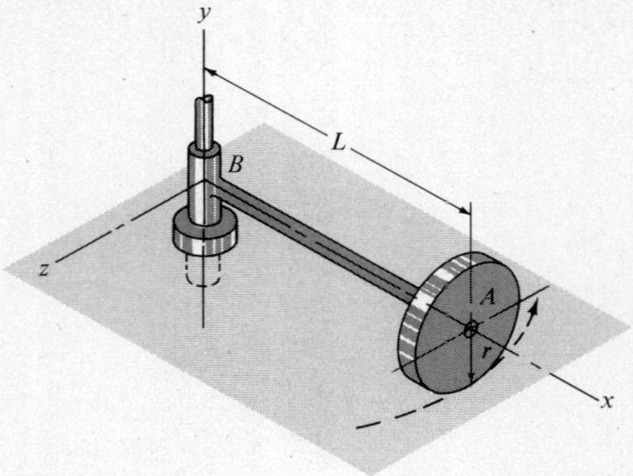

Figure P-16-4.14

ping on a horizontal surface as it also rotates about the shaft axis AB. Determine the normal and friction forces exerted by the surface on the disk.

$$N = W\left(1 + \frac{v^2 r}{2gL^2}\right); \; F = 0 \qquad \textbf{Ans.}$$

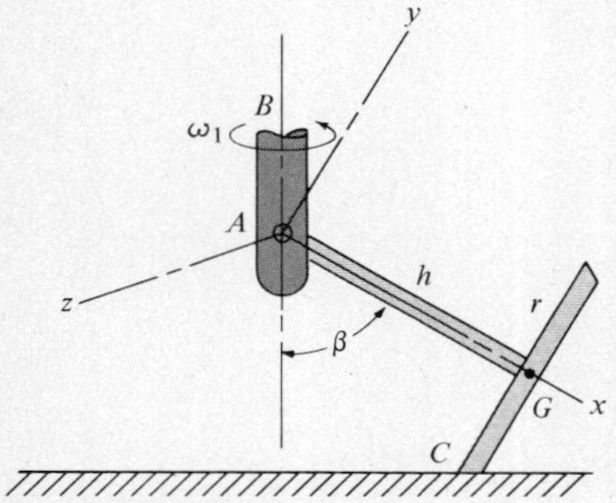

Figure P-16-4.15

16-4.15. As shown in Fig. P-16-4.15, a thin disk of weight W and radius r rolls without slipping on a flat horizontal surface as it also rotates about the axle AG. The axle AG is hinged to and rotates with the vertical shaft AB at the constant rate of ω_1 rad/sec. Assuming that $h = 5$ ft, $r = 2$ ft, and $\beta = 60°$, find the minimum value of ω_1 to reduce the reaction at C to zero.

$$\omega_1 = 4.31 \text{ rad/sec} \qquad \textbf{Ans.}$$

16-4.16. In the grinding mill shown in Fig. P-16-4.16, a disk of weight W and radius r rolls against the fixed vertical cylindrical surface as it also rotates about the axle AG of length L. The axle AG is hinged to and rotates with the vertical shaft AB at the constant rate of ω_1 rad/sec. Assuming

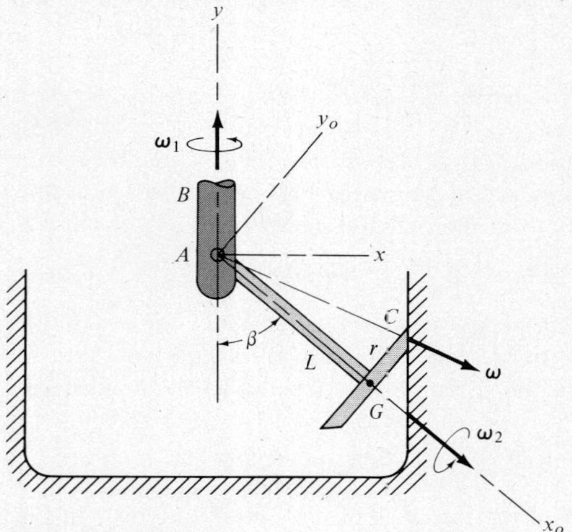

Figure P-16-4.16

that $r = 2$ ft, $L = 5$ ft, and $\beta = 60°$, determine the minimum value of ω_1 at which the disk loses contact with the cylinder.

$$\omega_1 = 3.05 \text{ rad/sec} \qquad \textbf{\textit{Ans.}}$$

* 16-5 MOMENTUM AND ENERGY METHODS IN SPATIAL MOTION

Now that we know how to express the angular momentum of a rigid body, we are ready to apply the impulse-momentum method to the spatial motion of rigid bodies. It is recommended that the impulse-momentum equation be expressed in the diagrammed form shown in Fig. 16-5.1. Such diagrams will indicate the most advantageous directions and centers about which to equate components of linear and angular impulse to the components of linear and angular momenta. For a body rotation about a fixed point A, the impulsive reaction at A can be eliminated by taking moments about A. A detailed explanation is given in Illus. Prob. 16-5.1.

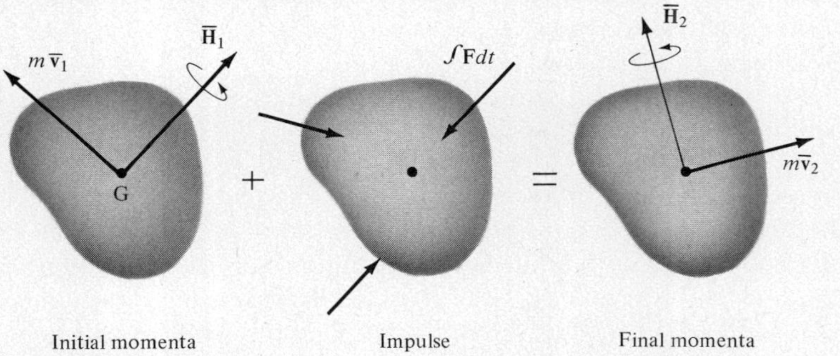

Figure 16-5.1

Let us now determine a general expression for the kinetic energy of a rigid body in spatial motion. Referring to the rigid body shown in Fig. 16-5.2, let a set of body axes xyz be fixed in it (and therefore rotating with the angular velocity $\boldsymbol{\omega}$ of the body) and let the origin of these axes be *any* point A in the body. Because the body is rigid, the position vector $\boldsymbol{\rho}$ from A to any other particle B of mass dm is of constant length.

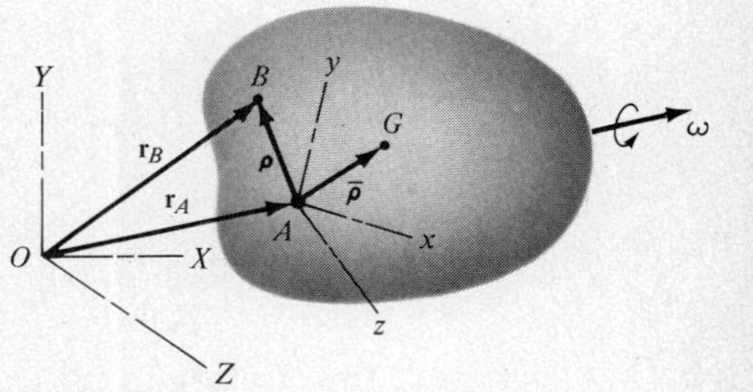

Figure 16-5.2

The total kinetic energy of the body is the sum of the kinetic energies of all particles in the body, or

$$\text{KE} = \int \tfrac{1}{2} v^2 \, dm \qquad (a)$$

But the velocity of any typical particle B is given by

$$\mathbf{v} = \dot{\mathbf{r}}_B = \dot{\mathbf{r}}_A + \dot{\boldsymbol{\rho}}$$

and therefore

$$v^2 = \mathbf{v} \cdot \mathbf{v} = (\dot{\mathbf{r}}_A + \dot{\boldsymbol{\rho}}) \cdot (\dot{\mathbf{r}}_A + \dot{\boldsymbol{\rho}}) = \dot{\mathbf{r}}_A \cdot \dot{\mathbf{r}}_A + \dot{\boldsymbol{\rho}} \cdot \dot{\boldsymbol{\rho}} + 2\dot{\mathbf{r}}_A \cdot \dot{\boldsymbol{\rho}} \qquad (b)$$

Note that $\dot{\mathbf{r}}_A \cdot \dot{\mathbf{r}}_A = v_A^2$ and that $\dot{\boldsymbol{\rho}} = \boldsymbol{\omega} \times \boldsymbol{\rho}$ so that $\dot{\boldsymbol{\rho}} \cdot \dot{\boldsymbol{\rho}}$ can be written as $(\boldsymbol{\omega} \times \boldsymbol{\rho}) \cdot \dot{\boldsymbol{\rho}}$ which becomes $\boldsymbol{\omega} \cdot (\boldsymbol{\rho} \times \dot{\boldsymbol{\rho}})$ by interchanging the cross and the dot in this scalar triple product (which does not alter its value, as shown on p. 46). Now on substituting Eq. (*b*) into Eq. (*a*), we obtain

$$\begin{aligned} \text{KE} &= \tfrac{1}{2}\int (v_A^2 + \boldsymbol{\omega} \cdot (\boldsymbol{\rho} \times \dot{\boldsymbol{\rho}}) + 2\dot{\mathbf{r}}_A \cdot \dot{\boldsymbol{\rho}})\, dm \\ &= \tfrac{1}{2}v_A^2 \int dm + \tfrac{1}{2}\boldsymbol{\omega} \cdot \underbrace{\int \boldsymbol{\rho} \times \dot{\boldsymbol{\rho}}\, dm}_{= \mathbf{H}_A} + \dot{\mathbf{r}}_A \cdot \underbrace{\int \dot{\boldsymbol{\rho}}\, dm}_{= \mathbf{v}_{G/A} m} \end{aligned} \qquad (c)$$

The equivalences indicated in Eq. (*c*) are now explained. Observe that $\dot{\boldsymbol{\rho}} = \mathbf{v}_{B/A}$ is the relative velocity of B about A so that the second integral is $\int \boldsymbol{\rho} \times \mathbf{v}_{B/A}\, dm$ which is recognized as the total relative moment of momentum about A, being denoted by $\mathbf{H}_A$. The third integral $\int \dot{\boldsymbol{\rho}}\, dm = \mathbf{v}_{G/A}\, m$ is obtained by first differentiating the expression $\int \boldsymbol{\rho}\, dm = \bar{\boldsymbol{\rho}} m$ (which relates the total moment of mass about A to the sum of the moments of component masses) to obtain $\int \dot{\boldsymbol{\rho}}\, dm = \dot{\bar{\boldsymbol{\rho}}} m$ and then denoting $\dot{\bar{\boldsymbol{\rho}}}$ (which is the velocity of the mass center relative to A) as $\mathbf{v}_{G/A}$. Thus, we obtain the following general expression for kinetic energy:

$$\text{KE} = \tfrac{1}{2}mv_A^2 + \tfrac{1}{2}\boldsymbol{\omega} \cdot \mathbf{H}_A + \mathbf{v}_A \cdot \mathbf{v}_{G/A} m \qquad (16\text{-}5.1)$$

Consider now several applications of this general expression for KE. For *rigid-body translation*, v_A is the same for all particles of the body and may be denoted by v, the velocity of translation of the body. Also $\boldsymbol{\omega} = 0$ and $\bar{\boldsymbol{\rho}}$ is constant in length and direction so that $\mathbf{v}_{G/A} = 0$. Thus, Eq. (16-5.1) reduces to the familiar expression for translation

$$\text{KE} = \tfrac{1}{2}mv^2 \qquad (16\text{-}5.2)$$

For *rotation* of a rigid body about a fixed point A, or any point which is instantaneously at rest as for an instant center of velocity, we have $v_A = 0$ and hence Eq. (16-3.1) reduces to

$$\text{KE} = \tfrac{1}{2}\boldsymbol{\omega} \cdot \mathbf{H}_A \qquad (16\text{-}5.3)$$

When the mass center G of a rigid body is selected as the reference point, v_A becomes $\bar{v}$, and also $\mathbf{v}_{G/A} = 0$ since the velocity of the mass center with respect to itself is zero. Thus, we obtain

$$\text{KE} = \tfrac{1}{2}m\bar{v}^2 + \tfrac{1}{2}\boldsymbol{\omega} \cdot \mathbf{H}_G \qquad (16\text{-}5.4)$$

This result may be interpreted as splitting KE into a translational component $\tfrac{1}{2}m\bar{v}^2$ (which results from considering the body to be translating with the velocity of the mass center) plus a rotational component $\frac{1}{2}\boldsymbol{\omega} \cdot \mathbf{H}_G$ (which represents KE due to rotation about the mass center). This interpretation of the total kinetic energy as due to translation of the mass center plus rotation about the mass center is consistent with our previous experience in Section 14-8.

EVALUATION OF KINETIC ENERGY

Equation (16-5.3) will reduce to a simple form if we are able to orient the body axes xyz so that they are principal axes of inertia. Then the components of $\mathbf{H}_A$ are expressed by

$$H_x = I_x\omega_x; \qquad H_y = I_y\omega_y; \qquad H_z = I_z\omega_z$$

and Eq. (16-5.3) reduces to

$$\text{KE} = \tfrac{1}{2}\omega \cdot \mathbf{H}_A = \tfrac{1}{2}\omega_x H_x + \tfrac{1}{2}\omega_y H_y + \tfrac{1}{2}\omega_z H_z$$

or

$$\text{KE} = \tfrac{1}{2}I_x\omega_x^{\,2} + \tfrac{1}{2}I_y\omega_y^{\,2} + \tfrac{1}{2}I_z\omega_z^{\,2} \tag{16-5.5}$$

On applying this result to bodies having plane motion, say parallel to the XY plane, we have $\omega_x = \omega_y = 0$ and the kinetic energy reduces to merely

$$\text{KE} = \tfrac{1}{2}I_z\omega_z^{\,2} = \tfrac{1}{2}I\omega^2 \tag{16-5.6}$$

where I is about the axis of rotation (either fixed or instantaneous). As is to be expected, this result agrees with those obtained previously in Sections 14-7 and 14-8.

A similar evaluation of the rotational component of kinetic energy in Eq. (16-5.4), i.e. $\frac{1}{2}\boldsymbol{\omega} \cdot \mathbf{H}_G$, is identical to Eq. (16-5.5) if we add bar signs over the I terms to denote body axes whose origin is at the mass center.

WORK DONE ON A RIGID BODY

The methods of computing work explained in Section 14-3 are also valid for spatial motion, but it may be instructive to expand that discussion. Consider the spatial motion of the rigid body in Fig. 16-5.3. Let the displacement of any reference point A in the body be $d\mathbf{r}_A$ while the body undergoes a differential rotation $d\boldsymbol{\theta}$. Then the differential displacement of any other point B is

$$d\mathbf{r}_B = d\mathbf{r}_A + d\boldsymbol{\theta} \times \boldsymbol{\rho} \tag{c}$$

and the differential work done by a force $\mathbf{P}_B$ applied at B is

$$\mathbf{P}_B \cdot d\mathbf{r}_B = \mathbf{P}_B \cdot d\mathbf{r}_A + \mathbf{P}_B \cdot (d\boldsymbol{\theta} \times \boldsymbol{\rho}) \tag{d}$$

Recalling that the dot and cross can be interchanged in a scalar triple product (as shown on p. 46), the last term in Eq. (d) can be written as $(\boldsymbol{\rho} \times \mathbf{P}_B) \cdot d\boldsymbol{\theta}$ whence the total work done by a system

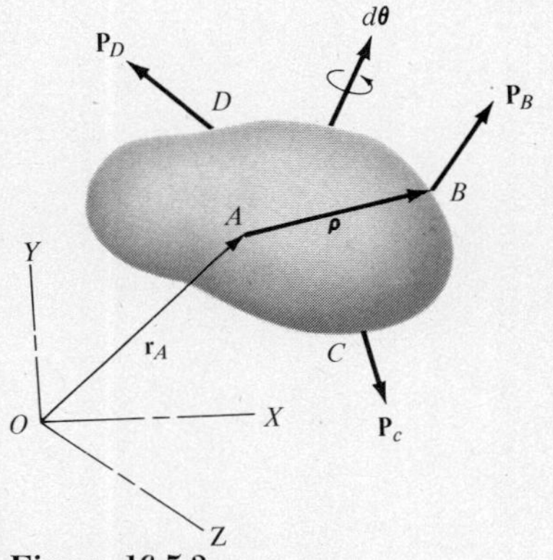

Figure 16-5.3

of forces is given by

$$\mathrm{RW} = \int \Sigma \mathbf{P} \cdot d\mathbf{r}_A + \int \Sigma \mathbf{M}_A \cdot d\boldsymbol{\theta} = \int \mathbf{R} \cdot d\mathbf{r}_A + \int \Sigma \mathbf{M}_A \cdot d\boldsymbol{\theta} \quad (16\text{-}5.7)$$

Notice that the resultant work is equivalent to that done by an equivalent resultant force-couple system at A.

ILLUSTRATIVE PROBLEMS

16-5.1. A heavy square plate of mass m and side a is suspended as shown in Fig. 16-5.4 from a ball-and-socket joint at A. It is struck at corner B by a particle of mass m_o moving with a velocity $\mathbf{v}_o$ directed perpendicular to the plane of the plate. Determine the angular velocity and instantaneous axis of rotation of the plate assuming that m_o becomes attached to the plate. What is the loss in kinetic energy due to the impact?

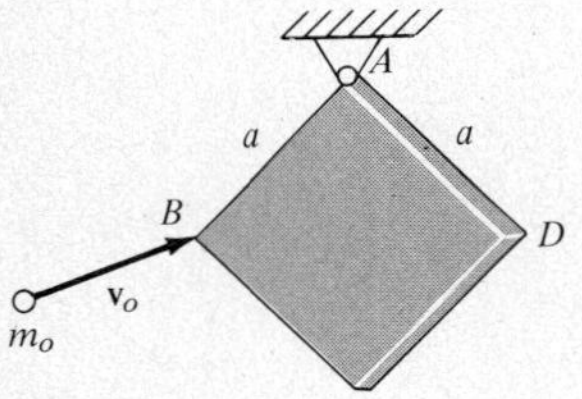

Figure 16-5.4

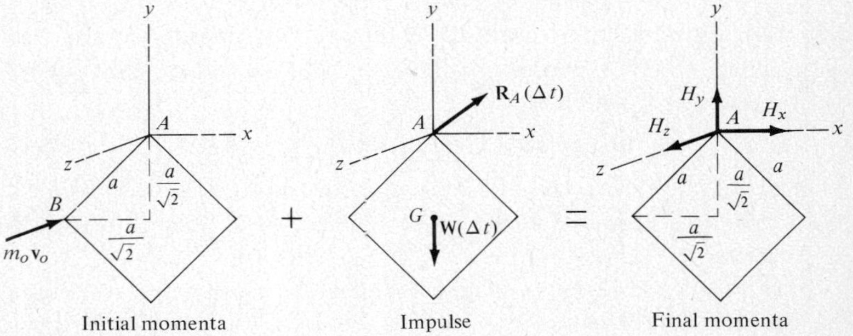

Figure 16-5.5

Solution

Using the diagrammed equation of the impulse-momentum relation shown in Fig. 16-5.5, we equate moment summations about A to eliminate the impulse at A. We obtain

$$\boldsymbol{\rho}_{AB} \times m_o\mathbf{v}_o = \mathbf{H}_A = H_x\hat{\mathbf{i}} + H_y\hat{\mathbf{j}} + H_z\hat{\mathbf{k}}$$

Evaluating $\boldsymbol{\rho}_{AB} \times m_o\mathbf{v}_o$, we have

$$\frac{a}{\sqrt{2}}(-\hat{\mathbf{i}} - \hat{\mathbf{j}}) \times m_o v_o(-\hat{\mathbf{k}}) = \frac{a m_o v_o}{\sqrt{2}}(-\hat{\mathbf{j}} + \hat{\mathbf{i}})$$

and therefore the components of $\mathbf{H}_A$ are

$$H_x = \frac{a m_o v_o}{\sqrt{2}}; \qquad H_y = -\frac{a m_o v_o}{\sqrt{2}}; \qquad H_z = 0$$

The axes at A are principal axes for which (neglecting m_o as negligible compared with m), we have

$$I_x = \frac{ma^2}{12} + m\left(\frac{a^2}{2}\right) = \frac{7}{12}ma^2$$

$$I_y = \frac{ma^2}{12}$$

Substituting these values in Eq. (16-2.2), we obtain

$$[H_x = I_x\omega_x] \qquad \frac{am_o v_o}{\sqrt{2}} = \frac{7}{12}ma^2\omega_x \qquad \omega_x = \frac{12m_o v_o}{7ma\sqrt{2}}$$

$$[H_y = I_y\omega_y] \qquad -\frac{am_o v_o}{\sqrt{2}} = \frac{ma^2}{12}\omega_y \qquad \omega_y = -\frac{12m_o v_o}{ma\sqrt{2}}$$

$$[H_z = I_z\omega_z] \qquad 0 = I_z\omega_z \qquad \omega_z = 0$$

and therefore

$$\boldsymbol{\omega} = \frac{12m_o v_o}{ma\sqrt{2}}\left(\frac{1}{7}\hat{\mathbf{i}} - \hat{\mathbf{j}}\right)$$

having a magnitude

$$\omega = \sqrt{\omega_x{}^2 + \omega_y{}^2} = \frac{12m_o v_o}{ma\sqrt{2}}\sqrt{\left(\frac{1}{7}\right)^2 + (1)^2} = \frac{60m_o v_o}{7ma}$$

and a direction with the x axis defined by

$$\phi = \tan^{-1}\frac{\omega_y}{\omega_x} = \tan^{-1} - \frac{1}{1/7} = \tan^{-1} - 7 = \searrow 81.87^\circ$$

which is also the direction of the instantaneous axis of rotation through A.

Using Eq. (16-5.5) and the above results, the final kinetic energy after impact is

$$\text{KE} = \frac{1}{2}I_x\omega_x{}^2 + \frac{1}{2}I_y\omega_y{}^2$$

$$= \frac{1}{2}\left(\frac{7}{12}ma^2\right)\left(\frac{12m_o v_o}{7ma\sqrt{2}}\right)^2 + \frac{1}{2}\left(\frac{ma^2}{12}\right)\left(\frac{12m_o v_o}{ma\sqrt{2}}\right)^2$$

which reduces to

$$\text{KE} = \frac{24m_o{}^2v_o{}^2}{7m}$$

Hence, the loss in kinetic energy is

$$\text{Loss in KE} = \frac{1}{2}m_o v_o{}^2 - \frac{24m_o{}^2v_o{}^2}{7m} = \frac{mv_o{}^2}{14}\left(7 - 48\frac{m_o}{m}\right) \qquad \textbf{\textit{Ans.}}$$

16-5.2. In the grinding mill shown in Fig. 16-5.6, a disk of weight W and radius r rolls against the fixed vertical cylindrical surface as it also rotates about the axle AG of length L. The axle AG is hinged to and rotates with the vertical shaft AB at the constant rate of ω_1 rad/sec. Assuming that $W = 400$ lb, $r = 2$ ft, $L = 5$ ft, $\omega_1 = 4$ rad/sec and $\beta = 60°$, determine the kinetic energy of the disk.

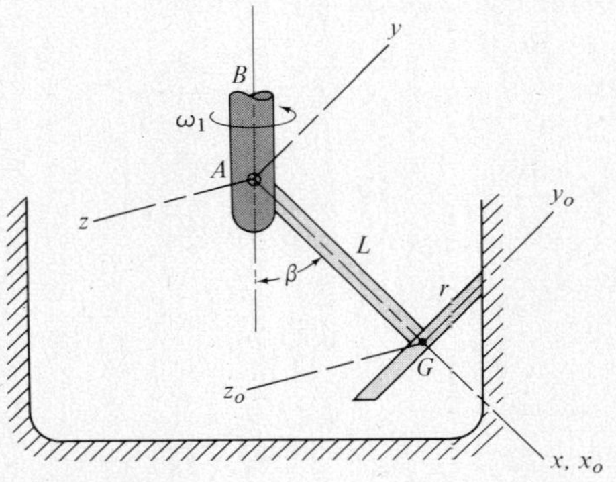

Figure. 16-5.6

Solution

We shall do this problem in two ways: first using Eq. (16-5.5) and considering the disk to be rotating about the fixed point A, and second using Eq. (16-5.4) and considering the disk to have the translational motion of the mass center G plus a rotational motion about G.

Part (a)

With respect to body-fixed principal axes xyz at A, we recall from Illus. Prob. 16-2.3 that the components of $\boldsymbol{\omega}$ are

$$\omega_x = \omega_1 \frac{L}{r} \sin\beta, \qquad \omega_y = \omega_1 \sin\beta, \qquad \omega_z = 0$$

Now using $\text{KE} = \frac{1}{2}I_x\omega_x^2 + \frac{1}{2}I_y\omega_y^2 + \frac{1}{2}I_z\omega_z^2$, we obtain

$$\frac{1}{2}I_x\omega_x^2 = \frac{1}{2}\left(\frac{1}{2}mr^2\right)\left(\omega_1\frac{L}{r}\sin\beta\right)^2 = \frac{1}{4}m\omega_1^2L^2\sin^2\beta$$

$$\frac{1}{2}I_y\omega_y^2 = \frac{1}{2}\left(\frac{1}{4}mr^2 + mL^2\right)(\omega_1\sin\beta)^2$$

$$= \frac{1}{8}m\omega_1^2r^2\sin^2\beta + \frac{1}{2}m\omega_1^2L^2\sin^2\beta$$

Summing up these components of kinetic energy results in

$$\text{KE} = \tfrac{1}{8}m\omega_1^2\sin^2\beta(6L^2 + r^2)$$

Part (b)

Using $KE = \frac{1}{2}m\bar{v}^2 + \frac{1}{2}\boldsymbol{\omega} \cdot \mathbf{H}_G$, the kinetic energy component caused by translation of the mass center G is

$$\tfrac{1}{2}m\bar{v}^2 = \tfrac{1}{2}m(\omega_1 L \sin\beta)^2 = \tfrac{1}{2}m\omega_1{}^2L^2\sin^2\beta$$

The rotational component of kinetic energy caused by rotation about body-fixed principal axes $x_o y_o z_o$ at the mass center is

$$KE = \tfrac{1}{2}\boldsymbol{\omega} \cdot \mathbf{H}_G = \tfrac{1}{2}\bar{I}_x\omega_x{}^2 + \tfrac{1}{2}\bar{I}_y\omega_y{}^2 + \tfrac{1}{2}\bar{I}_z\omega_z{}^2$$

where as before

$$\omega_x = \omega_1\frac{L}{r}\sin\beta, \qquad \omega_y = \omega_1\sin\beta, \qquad \omega_z = 0$$

Then

$$\frac{1}{2}\bar{I}_x\omega_x{}^2 = \frac{1}{2}\left(\frac{1}{2}mr^2\right)\left(\omega_1\frac{L}{r}\sin\beta\right)^2 = \frac{1}{4}m\omega_1{}^2L^2\sin^2\beta$$

$$\frac{1}{2}\bar{I}_y\omega_y{}^2 = \frac{1}{2}\left(\frac{1}{4}mr^2\right)(\omega_1\sin\beta)^2 = \frac{1}{8}m\omega_1{}^2r^2\sin^2\beta$$

$$\frac{1}{2}\bar{I}_z\omega_z{}^2 = 0$$

Summing up all the components of kinetic energy results, as before, in

$$KE = \tfrac{1}{8}m\omega_1{}^2\sin^2\beta(6L^2 + r^2)$$

Comparing parts (a) and (b), notice that the transfer term mL^2 for I_y in part (a) automatically included the translational component of KE in part (b). Finally, on substituting numerical data, we have

$$KE = \frac{1}{8}\left(\frac{400}{32.2}\right)(4)^2\sin^2 60°[6(5)^2 + (2)^2] = 2870 \text{ ft-lb} \qquad \textbf{\textit{Ans.}}$$

PROBLEMS

16-5.3. The thin rectangular plate of sides a and b and mass m in Fig. P-16-5.3 rotates about smooth ball-and-socket bearings at A and B at the constant rate of ω rad/sec. Determine the kinetic energy of the plate.

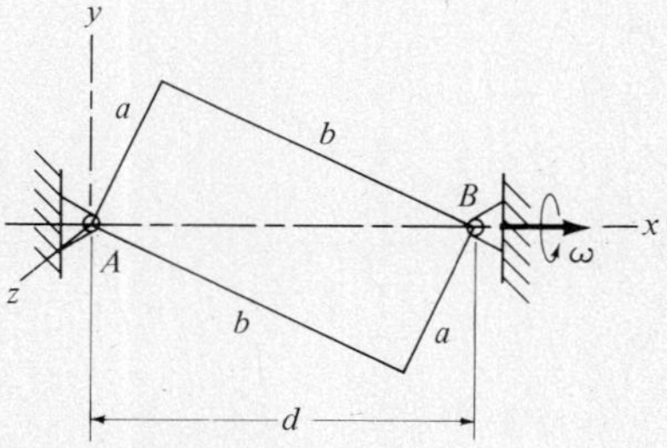

Figure P-16-5.3

16-5.4. Determine the kinetic energy of the homogeneous rectangular block of mass m in Fig. P-16-5.4 (p. 670) which is rotating at a constant rate of ω rad/sec about a shaft whose axis passes through the opposite corners A and B.

$$KE = \frac{m\omega^2(a^2b^2 + b^2c^2 + c^2a^2)}{12(a^2 + b^2 + c^2)} \qquad \textbf{\textit{Ans.}}$$

16-5.5. In Fig. P-16-5.5 (p. 670), a wheel, of weight 1610 lb, 6 in. thick, with a 2-ft radius, rotates about axis OA at a constant rate of $\omega_2 = 6$ rad/sec as the platform carrying the wheel rotates about the fixed

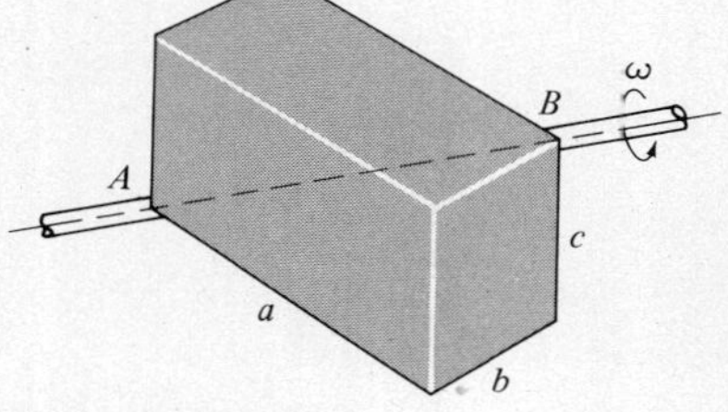

Figure P-16-5.4

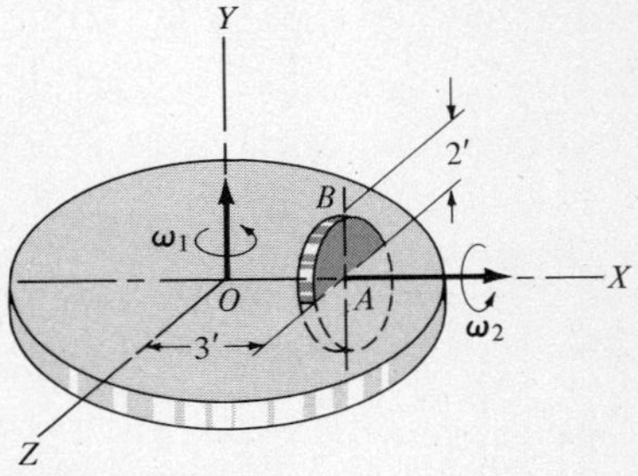

Figure P-16-5.5

Y axis at a constant rate of $\omega_1 = 4$ rad/sec. Compute the KE of the wheel.

$$\text{KE} = 5808 \text{ ft-lb}$$ ***Ans.***

16-5.6. The center A of the thin disk of mass m and radius r in Fig. P-16-5.6 has a constant speed of v fps. The disk rolls without slipping on a horizontal surface as it also rotates about the shaft axis AB. Determine the kinetic energy of the disk.

$$\text{KE} = \frac{mv^2}{8}\left(6 + \frac{r^2}{L^2}\right)$$ ***Ans.***

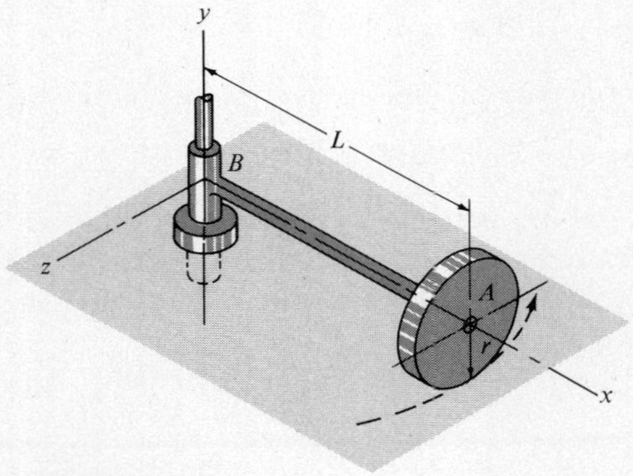

Figure P-16-5.6

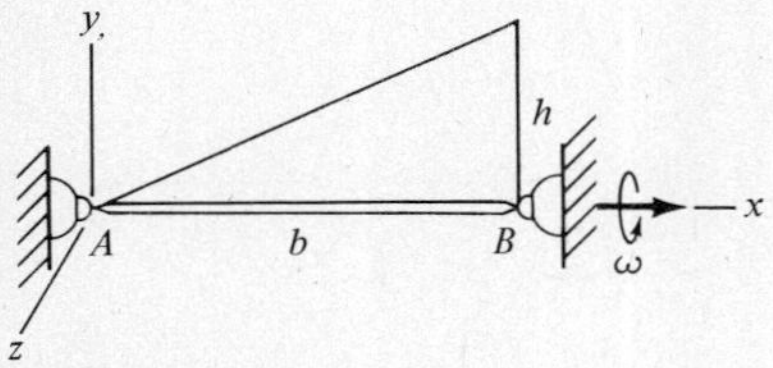

Figure P-16-5.7

16-5.7. A thin triangular plate weighing 30 lb is welded to a shaft as shown in Fig. P-16-5.7. The plate and shaft are free to rotate about smooth ball-and-socket bearings at A and B. If $b = 3$ ft, $h = 1$ ft, and a slight push starts the plate from rest at the given position, determine its angular velocity after it has rotated through 90°.

$$\omega = 11.35 \text{ rad/sec}$$ ***Ans.***

16-5.8. If the plate described in the previous problem has an angular velocity of 4 rad/sec at the given highest position, determine its angular velocity as it passes through its lowest position.

$$\omega = 16.54 \text{ rad/sec}$$ ***Ans.***

16-5.9. In Fig. P-16-5.9, a disk weighing 1.61 lb and of 2-in. radius rotates at the constant rate of $\omega_2 = 4$ rad/sec about the axle CD mounted

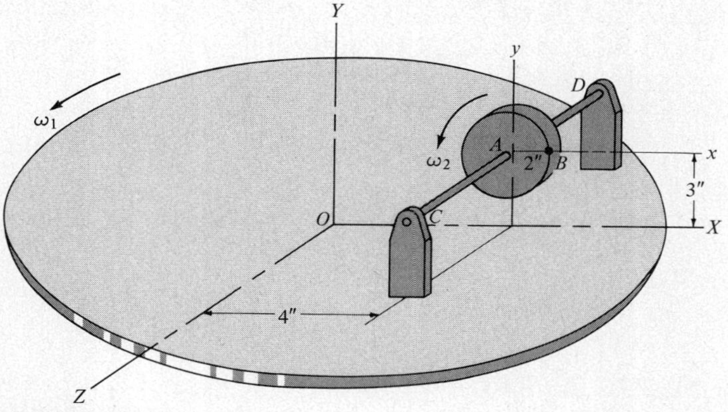

Figure P-16-5.9

on a platform which is itself rotating at the constant rate of $\omega_1 = 5$ rad/sec about a fixed vertical axis at O. Compute the kinetic energy of the disk.

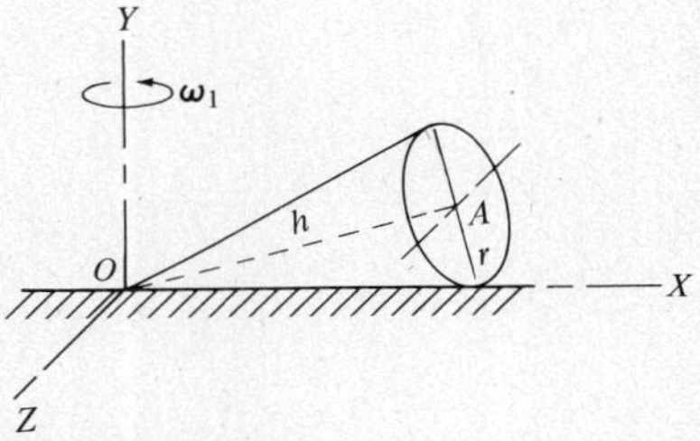

Figure P-16-5.10

16-5.10. The homogeneous solid cone of base radius $r = 7$ in., height $h = 24$ in., and weight $W = 30$ lb shown in Fig. P-16-5.10 rolls without slipping on a horizontal floor. The centerline OA rotates about the fixed Y axis at the constant rate of $\omega_1 = 2$ rad/sec. Compute the kinetic energy of the cone.

KE = 75.2 in.-lb ***Ans.***

16-5.11. Refer to the data of Illus. Prob. 16-5.2. When $\omega_1 = 4$ rad/sec, an additional constant torque $M = 200$ ft-lb is applied to shaft AB for 2 revolutions and then removed. What will then be the angular velocity of the disk about the axle AG?

$\omega_2 = 14.6$ rad/sec ***Ans.***

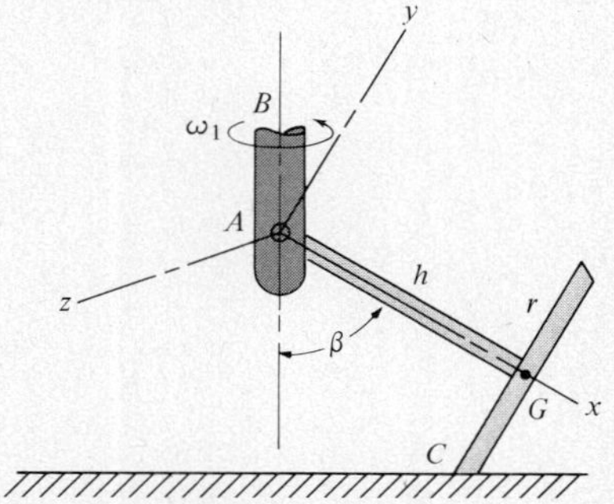

Figure P-16-5.12

16-5.12. As shown in Fig. P-16-5.12, a thin disk of weight W and radius r rolls without slipping on a flat horizontal surface as it also rotates about the axle AG. The axle AG is hinged to and rotates with the vertical shaft AB at the constant rate of $\omega_1$rad/sec. Assuming that $W = 300$ lb, $r = 2$ ft, $h = 3$ ft, $\omega_1 = 5$ rad/sec, and $\beta = 60°$, compute the kinetic energy of the disk.

KE = 1266 ft-lb ***Ans.***

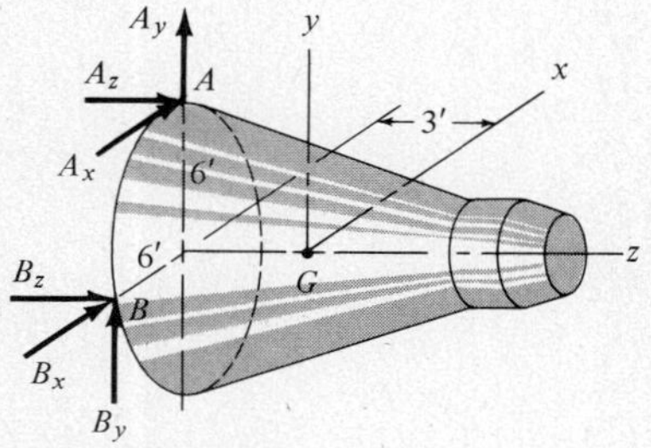

Figure P-16-5.13

16-5.13. The motion of a space capsule is controlled by four jets situated at quarter points around its bell bottom. As indicated by the components at A and B in Fig. P-16-5.13, each jet can exert a force in any direction. As the results of jerks on the tether of an astronaut during a space walk, the capsule acquires an angular velocity $\boldsymbol{\omega} = 0.12\hat{\mathbf{i}} + 0.15\hat{\mathbf{j}}$ rad/sec. The capsule has a mass of 60 slugs and centroidal radii of gyration $\bar{k}_z = 4$ ft, $\bar{k}_x = \bar{k}_y = 5$ ft. Assuming that the operating components of the jets at A and B are $\mathbf{A} = 15\hat{\mathbf{k}}$ lb and $\mathbf{B} = 15\hat{\mathbf{k}}$ lb, determine (a) the operating time for each jet that will reduce the angular velocity of the capsule to zero and (b) the resulting change in the velocity of the mass center G.

$t_A = 2.0$ sec; $t_B = 2.5$ sec; $\Delta v = 1.125$ fps ***Ans.***

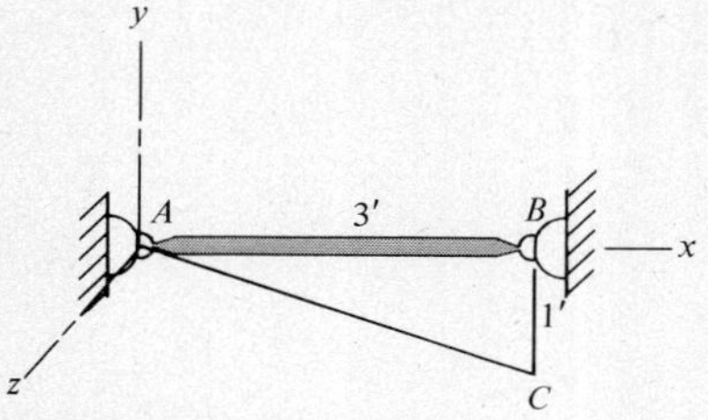

Figure P-16-5.14

16-5.14. A thin triangular plate weighing 30 lb is welded to a shaft from which it hangs at rest as shown in Fig. P-16-5.14. A sudden blow exerts an impulse of $-15\hat{k}$ lb-sec at C. Determine the impulsive reactions at the ball-and-socket bearings A and B.

$A = 32.5$ lb-sec; $B = 12.5$ lb-sec **Ans.**

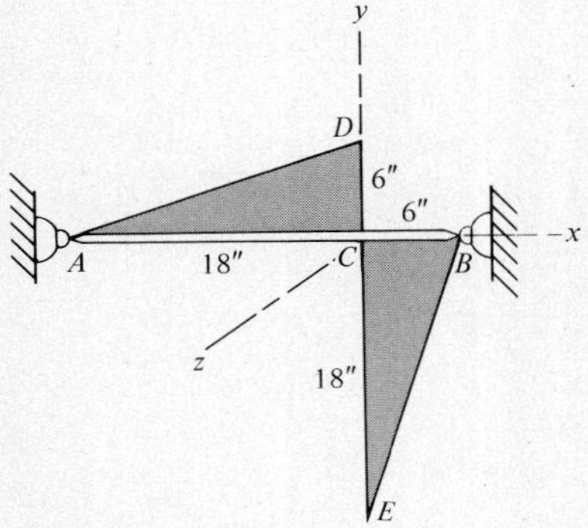

Figure P-16-5.15

16-5.15. Two triangular plates, each weighing 15 lb, are welded as shown in Fig. P-16-5.15 to a shaft supported by ball-and-socket bearings at A and B. The plates are at rest in the given vertical position when a sudden blow exerts an impulse of $-24\hat{k}$ lb-sec at E. Determine the impulsive reactions at A and B.

* 16-6 EULER'S ANGLES

The modified Euler equations [Eq. (16-4.6)] were developed for the motion of an axisymmetrical body spinning about an axis which is itself rotating with respect to fixed inertial axes. Euler's angles are ideal for defining the spatial orientation of such a body. To understand how Euler's angles do this, let us start with a gyroscope mounted in a Cardan suspension as shown in Fig. 16-6.1. Here the gyro rotor is free to spin about the axis A-A' in the inner gimbal ring. The inner gimbal ring is free to rotate about the axis B-B' relative to the outer gimbal ring. Finally, the outer gimbal ring is free to rotate about the fixed axis C-C'. As a result, the mass center G of the gyro rotor remains fixed in space although the rotor can have any spatial orientation.

Starting from a reference position of the gyroscope in Fig. 16-6.1

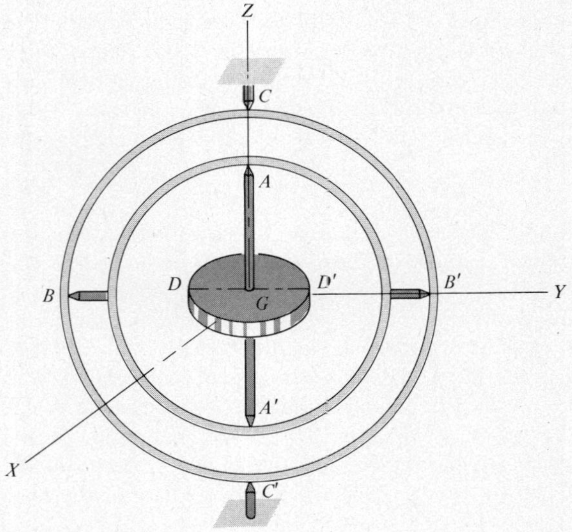

Figure 16-6.1

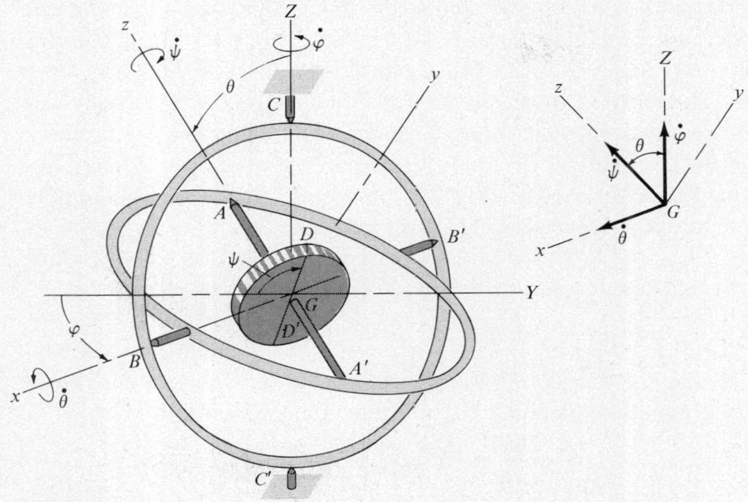

Figure 16-6.2

in which the two gimbal rings and a diameter D-D' of the gyro rotor are in the fixed YZ plane, Fig. 16-6.2 shows how the rotor can assume any spatial orientation. First let the outer gimbal ring rotate through an angle φ about the Z axis. Next let the plane of the inner gimbal ring rotate through the angle θ about the axis B-B'. Finally, let the gyro rotor rotate through the angle ψ about the axis A-A'. The result of these three rotations, each of which can occur independently of the others, is to determine a unique position of the gyro rotor while maintaining a fixed position of its mass center G with respect to XYZ axes. Note that since angular displacements are not vectors (see footnote on p. 426), the rotations must occur in the order φ, θ, and ψ to obtain a unique orientation of the gyro rotor. However, the time derivatives of these angles, i.e., $\dot{\varphi}$, $\dot{\theta}$, and $\dot{\psi}$, are vectors shown directed respectively along the positive senses of the C-C' axis, the B-B' axis, and the A-A' axis. These time derivatives define respectively what are known as the rate of *precession*, the rate of *nutation*, and the rate of *spin* at any given instant.

We have used the gimbal suspension of a gyroscope to help us visualize the Eulerian angles. Regardless of how a body is supported, however, these angles may be used to define the spatial orientation of any rigid body with respect to axes centered at any point of the body.

* 16-7 GYROSCOPIC PHENOMENA

For the analysis of gyroscopic phenomena, we use Euler's angles and the modified Euler equations [Eq. (16-4.6)] since both apply to bodies rotating in a coordinate system which is itself rotating. The coordinate frame $Gxyz$ used here is attached to the inner gimbal ring as shown in Fig. 16-6.2 with the x axis along B-B' and the z axis along A-A'. The y axis is always in the plane defined by the Z-z axes. These axes are always principal axes of inertia for the gyro rotor in which $I_x = I_y$. They follow the rotor in its precession and nutation. Thus, we have a set of axes exactly like those used to obtain the modified Euler equations except that here $\dot{\boldsymbol{\varphi}} + \dot{\boldsymbol{\theta}} = \boldsymbol{\Omega}$ and $\dot{\boldsymbol{\psi}} = \boldsymbol{\omega}_s$. Then, using Fig. 16-6.2, the components of $\boldsymbol{\Omega}$ and $\boldsymbol{\omega}_s$ along $Gxyz$ which are to be substituted into the modified Euler equations are

$$\Omega_x = \dot{\theta}; \qquad \Omega_y = \dot{\varphi} \sin \theta; \qquad \Omega_z = \dot{\varphi} \cos \theta; \qquad \omega_s = \dot{\psi}$$

Doing this and noting that $I_x = I_y$, we obtain the following three general equations defining the motion of a gyroscope:[3]

$$\begin{aligned}
\Sigma M_x &= I_x(\ddot{\theta} - \dot{\varphi}^2 \sin \theta \cos \theta) + I_z \dot{\varphi} \sin \theta (\dot{\varphi} \cos \theta + \dot{\psi}) \\
\Sigma M_y &= I_x(\ddot{\varphi} \sin \theta + 2\dot{\varphi}\dot{\theta} \cos \theta) - I_z \dot{\theta} (\dot{\varphi} \cos \theta + \dot{\psi}) \qquad (16\text{-}7.1) \\
\Sigma M_z &= I_z(\ddot{\varphi} \cos \theta - \dot{\varphi}\dot{\theta} \sin \theta + \ddot{\psi})
\end{aligned}$$

These three nonlinear differential equations have been studied extensively. For almost any situation, numerical solutions can be found using digital computers although no general analytical solution has yet been obtained. Fortunately, however, most technical applications involve motion with constant angular velocities so that, with zero angular accelerations, considerable simplification is achieved. We shall consider two important cases: first, steady precession in which θ is constant and second, torque-free motion in which the moment sum of external forces about the mass center is zero. These two cases have many technical applications. They depend essentially on the resistance to a change in the direction of the spin axis of a rapidly spinning axisymmetrical body. Some examples are the gyrocompass, the rate-of-turn gyro and an artificial horizon for aircraft, gyroscopic stabilization of such vehicles as ships, torpedos, and rockets. The literature on gyroscopes abounds in these and still additional applications, but a discussion of their details is beyond the scope of this book.

STEADY PRECESSION

The most common gyroscopic applications, which were treated in Section 15-8, are those in which the spin, moment, and precession axes are mutually perpendicular. Let us now extend that discussion to the case where the spin axis is not perpendicular to the precession axis as for the spinning top shown in Fig. 16-7.1. Our objective is

[3] These equations assume the mass of the gimbals is negligibly small compared to that of the gyro rotor.

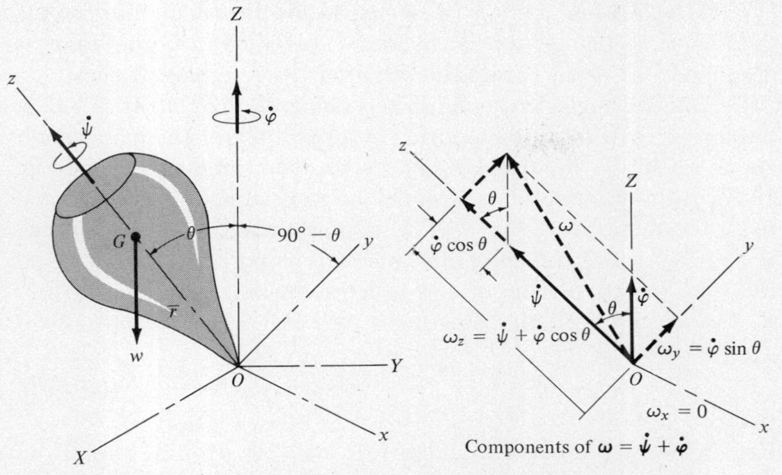

Figure 16-7.1 Steady precession of a spinning top with constant θ.

to determine the conditions for a steady motion in which the inclination θ of the spin axis, the rate of precession $\dot{\varphi}$, and the spin rate $\dot{\psi}$ all remain constant.

The top is spinning at the constant rate $\dot{\psi}$ about the z axis of the reference frame $Oxyz$ which is itself rotating about the fixed Z axis at the constant rate $\dot{\varphi}$. Under these conditions, the equations of motion are obrained by substituting $\ddot{\theta} = \dot{\theta} = 0$ and $\ddot{\varphi} = \ddot{\psi} = 0$ into Eqs. (16-7.1). Doing this yields

$$\begin{aligned} \Sigma M_x &= [I_z(\dot{\psi} + \dot{\varphi}\cos\theta)\dot{\varphi} - I_x\dot{\varphi}^2\cos\theta]\sin\theta \\ \Sigma M_y &= 0 \\ \Sigma M_z &= 0 \end{aligned} \tag{16-7.2}$$

The applied moment ΣM_x is due to the weight W located a distance $\bar{r}$ from O along the z axis. Since W is always in the rotating yz plane, its moment about the x axis is $W\bar{r}\sin\theta$. Hence, the equation of motion becomes

$$W\bar{r} = Iz(\dot{\psi} + \dot{\varphi}\cos\theta)\dot{\varphi} - I_x\dot{\varphi}^2\cos\theta \tag{16-7.3}$$

from which we now determine the rate of precession $\dot{\varphi}$. To simplify matters, refer to Fig. 16-7.1 and replace $\dot{\psi} + \dot{\varphi}\cos\theta$ by ω_z where ω_z is the z component of the top's absolute angular velocity $\boldsymbol{\omega} = \dot{\boldsymbol{\psi}} + \dot{\boldsymbol{\varphi}}$. Then we obtain the following quadratic equation in $\dot{\varphi}$

$$W\bar{r} = I_z\omega_z\dot{\varphi} - I_x\dot{\varphi}^2\cos\theta \tag{16-7.4}$$

from which the rate of precession is found to be

$$\dot{\varphi} = \frac{I_z\omega_z}{2I_x\cos\theta}\left[1 \pm \sqrt{1 - \frac{4W\bar{r}I_x\cos\theta}{I_z^2\omega_z^2}}\right] \tag{16-7.5}$$

This equation will also determine the precession rate of the gyroscope of Fig. 16-6.2 if the mass center of the gyro rotor is displaced a distance $\bar{r}$ from the origin of the reference frame.

Consider angle θ as lying in the range 0° to 180°. If $\theta > 90°$, the top or gyro rotor is supported at a point above the mass center and is known as a *pendulous gyroscope*. The grinding mill of Fig. 16-2.5 is an example. For a pendulous gyroscope (with $\cos\theta$ negative), the radicand in Eq. (16-7.5) is positive and $\dot{\varphi}$ is always real. For the two possible values of $\dot{\varphi}$, one will be positive and one negative. The negative value of $\dot{\varphi}$ is called *retrograde precession*.

When $\theta < 90°$, the precession rate $\dot{\varphi}$ has real values only if

$$1 \geqq \frac{4W\bar{r}I_x \cos\theta}{I_z{}^2\omega_z{}^2} \quad \text{or} \quad \omega_z{}^2 \geqq \frac{4W\bar{r}I_x \cos\theta}{I_z{}^2} \tag{16-7.6}$$

from which we observe that the top must be spinning rapidly to achieve steady precession. As ω_z increases, we may replace the radicand in Eq. (16-7.5) by using only the first two terms of a binomial expansion to obtain

$$\dot{\varphi} = \frac{I_z\omega_z}{2I_x \cos\theta}\left[1 \pm \left(1 - \frac{2W\bar{r}I_x \cos\theta}{I_z{}^2\omega_z{}^2}\right)\right]$$

from which we see that the two roots of Eq. (16-7.5) approach the limiting values

$$\dot{\varphi} = \frac{I_z\omega_z}{I_x \cos\theta} \quad \text{and} \quad \dot{\varphi} = \frac{W\bar{r}}{I_z\omega_z} \tag{16-7.7}$$

These are known as the *rapid* and *slow* rates of precession, respectively. The slow rate is the one usually occurring. It becomes the rate specified in Eq. (15-8.1) when $\theta = 90°$. Notice, however, that while simple steady precession *can* occur, it will do so only under appropriate initial conditions. The gyroscope must be started with $\dot{\theta} = 0$ *and* one of the precessional rates specified by Eq. (16-7.5). If the spinning top or gyro rotor with its mass center above O merely is released from rest, then the spin axis will dip downward initially, and the precessional motion will be complicated by alternate increases and decreases in θ known as nutational motion.

TORQUE-FREE MOTION

We now consider the motion of a spinning body moving with no external moment about its mass center. Typical examples are orbiting space vehicles, the gimbal-supported gyroscope of Fig. 16-6.2 with frictionless bearings, or any axisymmetrical spinning body whose point of support coincides with its mass center.

Since $\Sigma\mathbf{M}_G = 0$, it follows from $\Sigma\mathbf{M}_G = \dot{\mathbf{H}}_G$ that the angular momentum $\mathbf{H}_G$ is constant. The constant direction of $\mathbf{H}_G$ establishes what is called the *invariable line* and is analogous to the fixed Z axis of the spinning top. Also, with zero external moment, no rotational work is done so that the rotational kinetic energy remains constant. Hence, using Eq. (16-5.4), we have

$$\boldsymbol{\omega} \cdot \mathbf{H}_G = \text{Constant}$$

which means that the projection of $\boldsymbol{\omega}$ upon the invariable line remains constant. Consequently, the tip of $\boldsymbol{\omega}$ moves in a plane, known as the *invariable plane*, which is perpendicular to the invariable line as shown in Fig. 16-7.2.

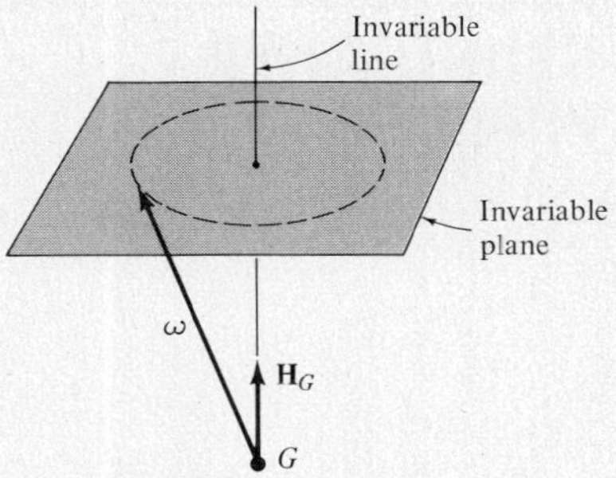

Figure 16-7.2

If the absolute angular velocity $\boldsymbol{\omega}$ of the body is directed along either the axis of symmetry or a transverse axis through G, then $\boldsymbol{\omega}$ and $\mathbf{H}_G$ are collinear and the body is spin stabilized. Frequently, however, $\boldsymbol{\omega}$ and $\mathbf{H}_G$ are not collinear as in Fig. 16-7.3 where the rotating $Gxyz$ axes show the z axis along the axis of symmetry and

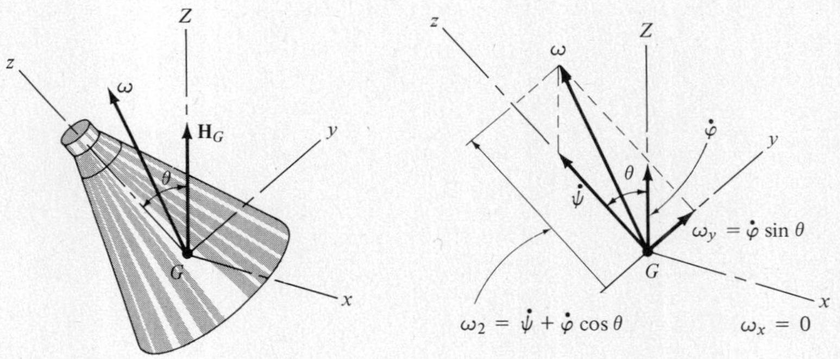

Figure 16-7.3

the y axis in the plane defined by the Z-z axes. The decomposition of $\boldsymbol{\omega}$ into components $\dot{\varphi}$ and $\dot{\psi}$ along the Z and z axes is exactly like that in Fig. 16-7.1. From this analogy, we conclude that steady precession will occur about the Z axis at a rate found by setting $\Sigma M_x = 0$ in Eq. (16-7.2). This gives

$$\dot{\varphi} = \frac{I_z \dot{\psi}}{(I_x - I_z)\cos\theta} \tag{16-7.8}$$

which, as you can easily verify, is exactly the same as the fast precession rate of Eq. (16-7.7).

Notice that the direction of the precession depends on the relative magnitudes of I_x and I_z. For an elongated body resembling a long cylinder in which $I_x > I_z$, the positive value of $\dot{\varphi}$ produces what is called direct precession. For a relatively flattened body like that of a disk, $I_x < I_z$ and the resulting negative value of $\dot{\varphi}$ produces what is called retrograde precession.

ILLUSTRATIVE PROBLEM

16-7.1. In the grinding mill shown in Fig. 16-7.4, a disk of weight $W = 30$ lb and radius $r = 2$ ft rolls against the fixed vertical cylindrical surface as it also rotates about the axle AG of length $L = 5$ ft. Knowing that $\beta = 60°$, determine the normal force exerted by the vertical wall on the disk when the shaft AB is rotating at $\dot{\varphi} = 6$ rad/sec. Also determine the components of the hinge force at A. Neglect the weight of the axle AG.

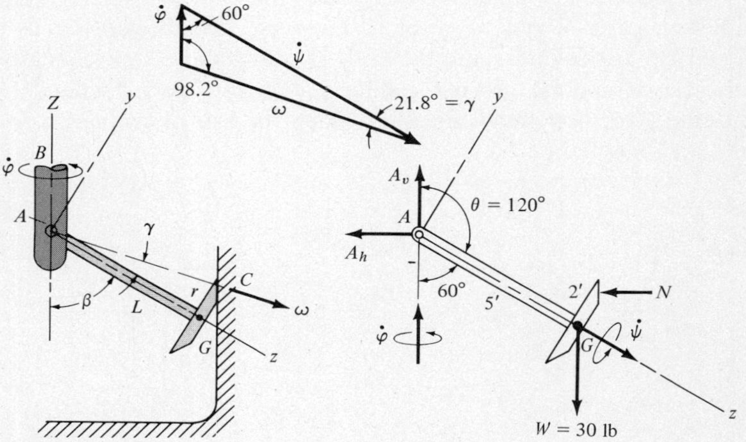

Figure 16-7.4

Solution

By using a rotating coordinate system $Axyz$ in which the y and z axes are always in the same vertical plane as the fixed Z axis, this problem becomes equivalent to a pendulous gyroscope in which $\theta = 120°$. The free-body diagram discloses the normal reaction N and the components at A which all rotate with the axle in the Z-y-z plane. Using Eq. (16-7.2) for a moment summation about the x axis at A will determine N, after which force summations from $\mathbf{R} = (W/g)\bar{\mathbf{a}}$ will determine A_h and A_v.

Equation (16-7.2) is

$$\Sigma M_x = [I_z\dot{\varphi}(\dot{\psi} + \dot{\varphi}\cos\theta) - I_x\dot{\varphi}^2\cos\theta]\sin\theta$$

Before using this equation, some preliminaries are required. The unknown spin rate $\dot{\psi}$ is easily found in terms of the known precession rate $\dot{\varphi}$ by applying the sine law to the vector polygon of $\boldsymbol{\omega} = \dot{\boldsymbol{\varphi}} + \dot{\boldsymbol{\psi}}$. The direction of $\boldsymbol{\omega}$ relative to $\dot{\boldsymbol{\psi}}$ is found by noting that it must pass through points A and C which have zero velocity. Then $\gamma = \tan^{-1}\frac{2}{5} = 21.8°$, and

$$\frac{\dot{\psi}}{\sin 98.2°} = \frac{\dot{\varphi}}{\sin 21.8°} \quad \text{or} \quad \dot{\psi} = 2.67\dot{\varphi}$$

from which

$$\dot{\psi} + \dot{\varphi}\cos\theta = 2.67\dot{\varphi} + \dot{\varphi}(\cos 120°) = 2.17\dot{\varphi}$$

Alternatively, this sum can be found directly by recalling that $\dot{\psi} + \dot{\varphi}\cos\theta = \omega_z$ where ω_z is the projection of $\boldsymbol{\omega}$ upon the spin axis. Thus, applying the sine law to the velocity polygon to find $\boldsymbol{\omega}$ in terms of $\dot{\boldsymbol{\varphi}}$, we have

$$\frac{\omega}{\sin 60°} = \frac{\dot{\varphi}}{\sin 21.8°}$$

whence the component of $\boldsymbol{\omega}$ along the spin axis is

$$\omega_z = \omega\cos 21.8° = \frac{\dot{\varphi}\sin 60°}{\sin 21.8°}\cos 21.8° = 2.17\dot{\varphi}$$

as before.

The moments of inertia are

$$I_z = \frac{1}{2}\frac{W}{g}r^2 = \frac{1}{2}\frac{30}{g}(2)^2 = \frac{60}{g} \text{ ft-lb-sec}^2$$

$$I_x = \frac{1}{4}\frac{W}{g}r^2 + \frac{W}{g}L^2 = \frac{30}{g}\left(\frac{2^2}{4} + 5^2\right) = \frac{780}{g} \text{ ft-lb-sec}^2$$

We now substitute these preliminary results into the moment equation. Note that the positive x axis is directed into the plane of the paper so that positive moments are clockwise. Resolving N into components perpendicular to and along AG, we obtain

$$5(N\cos 60°) - 2(N\sin 60°) + 30(5)\sin 60°$$
$$= \left[\frac{60}{g}\dot{\varphi}(2.17\dot{\varphi}) - \frac{780}{g}\dot{\varphi}^2\cos 120°\right]\sin 120° = \frac{450}{g}\dot{\varphi}^2$$

whence on substituting the given value of $\dot{\varphi} = 6$ rad/sec, we find

$$N = 486 \text{ lb}$$ ***Ans.***

The components at A are now obtained by using the components of $\mathbf{R} = \frac{W}{g}\bar{\mathbf{a}}$.

$$\left[\overset{+}{\leftarrow} \Sigma F_h = \frac{W}{g}\bar{a}_h\right] \qquad A_h + 486 = \frac{30}{g}(5\sin 60°)(6)^2$$

$$A_h = -341 \text{ lb}$$ ***Ans.***

$$\left[+\uparrow \Sigma F_v = \frac{W}{g}\bar{a}_v\right] \qquad A_v - 30 = 0 \qquad A_v = 30 \text{ lb}$$ ***Ans.***

We deliberately left the right side of the moment equation as $\frac{450}{g}\dot{\varphi}^2$ so that we could use it to determine the minimum speed at which N is zero. With $N = 0$, the moment equation is

$$30(5) \sin 60° = \frac{450}{g}\dot{\varphi}^2 \quad \text{or} \quad \dot{\varphi} = 3.05 \text{ rad/sec}$$

which verifies the answer to Prob. 16-4.16.

PROBLEMS

16-7.2. A spinning gyro rotor is supported by two gimbal rings. When the inner gimbal ring is horizontal, the outer gimbal ring is vertical as shown in Fig. P-16-7.2. Describe what will happen if the outer gimbal ring is rotated.

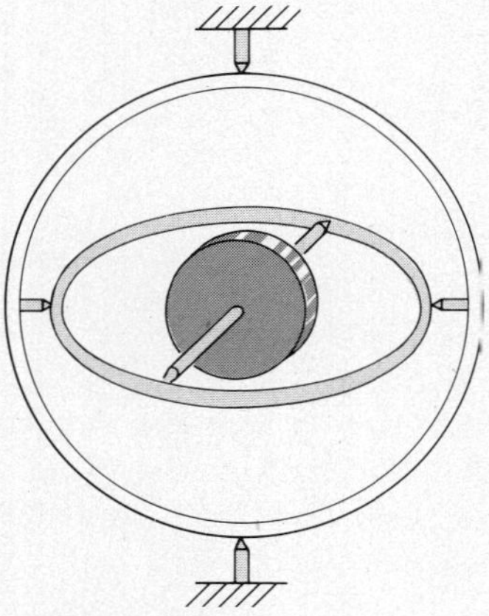

Figure P-16-7.2

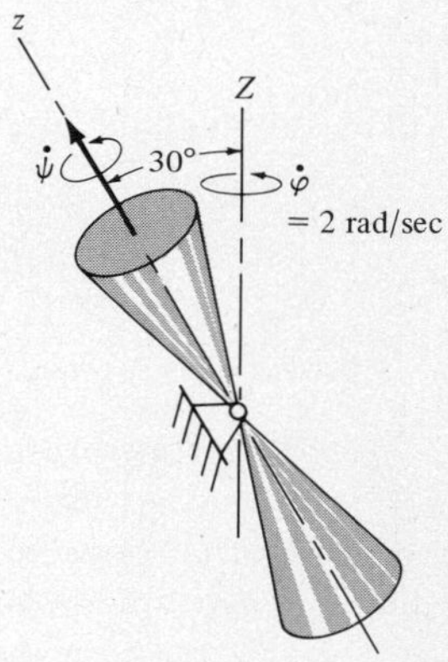

Figure P-16-7.3

16-7.3. The two cones in Fig. P-16-7.3 form a single body which is spinning about its axis of symmetry. If this axis is precessing at the constant rate of 2 rad/sec about the fixed vertical Z axis with which it remains inclined at 30°, determine the spin rate $\dot{\psi}$. The altitude of each cone is three times its base radius.

$\dot{\psi} = 41$ rad/sec **Ans.**

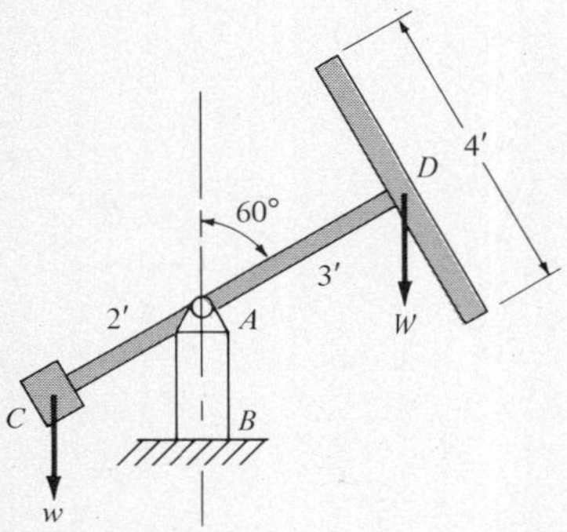

Figure P-16-7.4

16-7.4. The assembly in Fig. P-16-7.4 consists of a 4-ft diameter disk of weight $W = 25$ lb which is spinning about the shaft CD to which a weight $w = 20$ lb is attached at C. The shaft is supported by a ball-and-socket joint at A. Find the spin rate of the disk about CD if the assembly is precessing at a constant rate of 10 rpm about the vertical axis AB with which it makes a constant angle of 60°. Neglect the weight of the shaft.

16-7.5. If the disk in the previous problem is spinning at a constant rate of 40 rad/sec about the shaft *CD*, determine the constant rate at which the assembly precesses about *AB* at a constant angle of 60°

$\dot{\varphi} = 0.84$ rad/sec **Ans.**

16-7.6. A top spinning in an upright position is called a "sleeping" top. Determine the minimum spin rate of such a top having the shape of a right circular cone of altitude 3 in. and base radius 1 in.

$\dot{\psi} = 463$ rad/sec **Ans.**

16-7.7. Determine the spin rate $\dot{\psi}$ of the top shown in Fig. P-16-7.7. Its axis of symmetry remains at 30° to the vertical through *O* about which it precesses at a constant rate of 2 rad/sec. The radii of gyration of the top about its axis of symmetry and about a transverse axis at *O* are respectively 0.866 in. and 1.732 in. The distance of its gravity center *G* from *O* is 2 in.

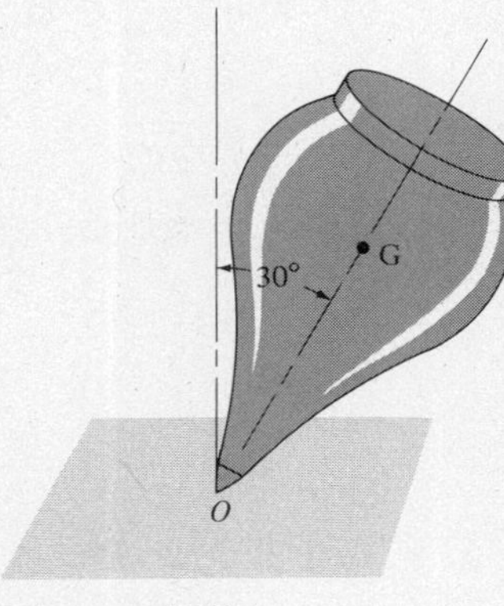

Figure P-16-7.7

16-7.8. Find the rate of precession of the top in the previous problem if it is spinning about its axis of symmetry at 1500 rpm. Assume the axis of the top precesses at a constant angle of 30° to the vertical through *O*.

$\dot{\varphi} = 7.5$ rad/sec (slow rate) **Ans.**

16-7.9. In the grinding mill of Illus. Prob. 16-7.1, assume $r = 1$ ft, $L = 4$ ft, and $\beta = 30°$. Find the normal force exerted by the vertical surface of the cylinder upon the 30-lb disk when the disk is spinning about the axle *AG* at $\dot{\psi} = 15$ rad/sec.

$N = 44.3$ lb **Ans.**

16-7.10. The disk of weight $W = 30$ lb and radius $r = 2$ ft in Fig. P-16-7.10 rolls without slipping on the flat horizontal surface as it also rotates about the axle *AG* whose length is $L = 3$ ft. Knowing that $\beta = 60°$, determine the normal reaction at *C* when the shaft *AB* to which *AG* is attached by a hinge is rotating at $\omega_1 = 5$ rad/sec. Neglect the weight of the axle *AG*.

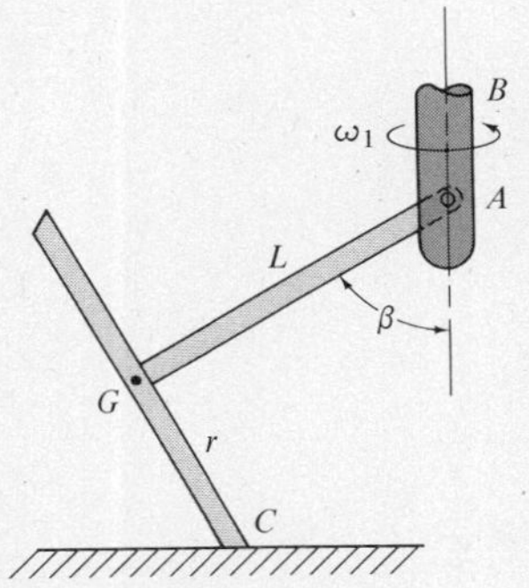

Figure P-16-7.10

$N = 18.43$ lb **Ans.**

16-7.11. Determine the minimum value of ω_1 in the previous problem that will reduce the normal reaction at *C* to zero.

16-7.12. The space capsule of Fig. P-16-7.12 has no angular velocity when the jet at *A* is actuated for 1 sec with a value of $\mathbf{A} = 15\hat{\mathbf{i}}$ lb. Determine the resulting axis of precession for the capsule and the precession and spin rates. The mass of the capsule is 60 slugs and the centroidal radii of gyration are $k_x = k_y = 5$ ft and $k_z = 4$ ft.

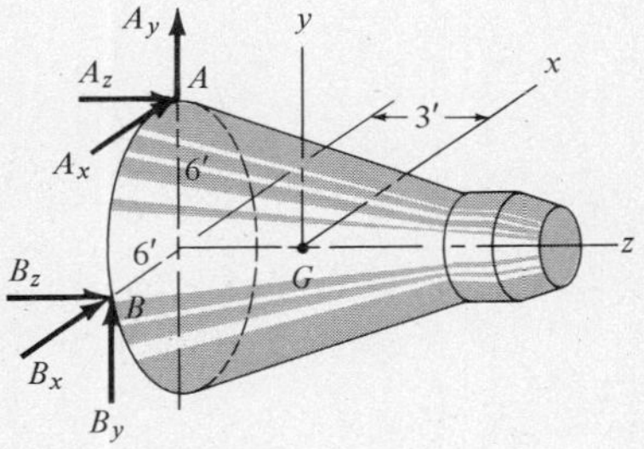

Figure P-16-7.12

$\dot{\varphi} = -0.067$ rad/sec (retrograde precession); $\dot{\psi} = -0.0339$ rad/sec **Ans.**

16-7.13. The impact of a meteorite on the space capsule of Prob. 16-7.12 causes it to acquire a spin velocity of 0.5 rad/sec about its axis of symmetry. If the spin axis is known to generate a cone with a total vertex angle of 10° as it then precesses about the axis of total angular momentum, find the period τ of each full precession.

$\tau = 7.04$ sec **Ans.**

SUMMARY

For application of $\mathbf{R} = m\bar{\mathbf{a}}$ and $\Sigma\mathbf{M} = \dot{\mathbf{H}}$ to spatial dynamics, we start with the development of general expressions for $\mathbf{H}$ and then find its time derivation by applying the general omega theorem $d\mathbf{H}/dt = \delta\mathbf{H}/\delta t + \boldsymbol{\omega} \times \mathbf{H}$.

With respect to either a fixed point A or the mass center G of the moving body, the components of H are

$$\begin{aligned} H_x &= I_x\omega_x - P_{xy}\omega_y - P_{xz}\omega_z \\ H_y &= I_y\omega_y - P_{yz}\omega_z - P_{yx}\omega_x \\ H_z &= I_z\omega_z - P_{zx}\omega_x - P_{zy}\omega_y \end{aligned} \tag{16-2.1}$$

or with respect to principal axes at A or G, these components of $\mathbf{H}$ reduce to

$$H_x = I_x\omega_x; \qquad H_y = I_y\omega_y; \qquad H_z = I_z\omega_z \tag{16-2.2}$$

A transfer formula relating angular momenta about A and G is given by

$$\mathbf{H}_A = \mathbf{H}_G + \boldsymbol{\rho}_{G/A} \times m\bar{\mathbf{v}} \tag{16-2.3}$$

in which $\boldsymbol{\rho}_{G/A}$ is the position vector from A to G.

Euler's equations express the components of $\Sigma\mathbf{M} = \dot{\mathbf{H}}$ in terms of body axes that coincide with the principal axes of the body. We obtain

$$\begin{aligned} \Sigma M_x &= I_x\dot{\omega}_x + (I_z - I_y)\omega_y\omega_z \\ \Sigma M_y &= I_y\dot{\omega}_y + (I_x - I_z)\omega_z\omega_x \\ \Sigma M_z &= I_z\dot{\omega}_z + (I_y - I_x)\omega_x\omega_y \end{aligned} \tag{16-4.4}$$

A modified set of Euler's equations is useful to describe the motion of bodies spinning at the rate $\boldsymbol{\omega}_s$ about an axis of symmetry. Here we use reference axes which are *not* fixed in the body but have an absolute angular velocity $\boldsymbol{\Omega}$. Assuming the axis of symmetry of the body to coincide with the z axis of the reference frame, we use

$$\dot{\mathbf{H}} = \frac{\delta\mathbf{H}}{\delta t} + \boldsymbol{\Omega} \times \mathbf{H} \tag{16-4.5}$$

to obtain the following set of modified Euler's equations:

$$\begin{aligned} \Sigma M_x &= I_x\dot{\Omega}_x + (I_z - I_y)\Omega_y\Omega_z + I_z\Omega_y\omega_s \\ \Sigma M_y &= I_y\dot{\Omega}_y + (I_x - I_z)\Omega_z\Omega_x - I_z\Omega_x\omega_s \\ \Sigma M_z &= I_z(\dot{\Omega}_z + \dot{\omega}_s) + (I_y - I_x)\Omega_x\Omega_y \end{aligned} \tag{16-4.6}$$

Observe that by letting the z axis of the reference frame coincide with the axis of symmetry of the body, we shall have $I_x = I_y$ and that all components of $\mathbf{\Omega}$ for the reference frame and $\boldsymbol{\omega}$ of body-fixed axes are identical except for those along the z axis.

For fixed-axis rotation of bodies which are *unsymmetrical* about the plane of motion of the mass center, the components of $\mathbf{H}$ about body axes whose z axis coincides with the axis of rotation become [from Eq. (16-2.1)]

$$H_x = -P_{zx}\omega_z; \qquad H_y = -P_{yz}\omega_z; \qquad H_z = I_z\omega_z$$

so that the general omega theorem gives the following components of $\Sigma\mathbf{M} = \dot{\mathbf{H}}$:

$$\begin{aligned} \Sigma M_x &= -P_{zx}\dot{\omega}_z + P_{yz}\omega_z^2 \\ \Sigma M_y &= -P_{yz}\dot{\omega}_z - P_{zx}\omega_z^2 \\ \Sigma M_z &= I_z\dot{\omega}_z \end{aligned} \tag{16-4.7}$$

The kinetic energy of a body in spatial motion is expressed by

$$\text{KE} = \tfrac{1}{2}m\bar{v}^2 + \tfrac{1}{2}\boldsymbol{\omega}\cdot\mathbf{H}_G \tag{16-5.4}$$

It represents the sum of the translational component $\frac{1}{2}m\bar{v}^2$ resulting from considering the body to be translating with the velocity of the mass center, plus the rotational component $\frac{1}{2}\boldsymbol{\omega}\cdot\mathbf{H}_G$ resulting from rotation of the body about the mass center. If the body axes are principal axes of inertia, the rotational component of kinetic energy reduces to

$$(\text{KE})_{\text{rot}} = \tfrac{1}{2}I_x\omega_x^2 + \tfrac{1}{2}I_y\omega_y^2 + \tfrac{1}{2}I_z\omega_z^2 \tag{16-5.5}$$

The resultant work done is found by reducing the external forces on a body to a resultant force-couple system acting at any point A of the body and applying

$$\text{RW} = \int \mathbf{R}\cdot d\mathbf{r}_A + \int \Sigma\mathbf{M}_A \cdot d\boldsymbol{\theta} \tag{16-5.7}$$

Euler's angles are ideal for defining the spatial orientation of a body rotating in a coordinate system which is itself rotating. Using the time derivatives of Euler's angles and the modified Euler equations, we obtain the general equations [Eq. (16-7.1)] for the motion of a gyroscope. Because of their complexity, only two special but technically important cases are discussed: that of steady precession under constant moment and that of torque-free motion. The rates of precession for these motions are

$$\dot{\varphi}_1 = \frac{I_z\omega_z}{I_x\cos\theta} \quad \text{and} \quad \dot{\varphi}_2 = \frac{W\bar{r}}{I_z\omega_z} \tag{16-7.7}$$

where the fast rate $\dot{\varphi}_1$ applies to torque-free motion and the slow rate $\dot{\varphi}_2$ applies to devices like a spinning top.

17

MECHANICAL VIBRATIONS

17-1 INTRODUCTION. DEFINITIONS AND CONCEPTS

Vibrations in an elastic structure are caused by disturbing forces which create a displacement of the structure from its position of static equilibrium. Such displacements create elastic forces which tend to restore the body to its original condition of equilibrium. When the disturbing force is removed, the elastic forces cause the body to accelerate back to its equilibrium position. However, the body will now possess some velocity as it passes through its equilibrium position, and this will cause it to overshoot. Thus, vibrations of it are set up which may or may not diminish, depending on whether any resisting forces are present.

When these vibrations are caused by an initially applied force which is then removed from the body, there result what are known as free vibrations. If resisting forces also act on the vibrating body, the motion is known as a damped free vibration. When the disturbing force continues to act at periodic intervals upon the body, the result is a forced vibration which may or may not be damped.

Generally speaking, a *vibration* is a periodic motion which repeats itself after a definite interval of time. This time interval is called the *period* of the vibration and will be designated by the symbol τ, usually measured in seconds. Each repetition of the motion is called a *cycle*. The *frequency* (f) of the vibration is the reciprocal of the period, i.e., $f = 1/\tau$, and is measured in cycles per second (also called Hertz and abbreviated as Hz). The maximum displacement of the body from its equilibrium position is known as the *amplitude* of the vibration.

17-2 SIMPLE HARMONIC MOTION. FREE VIBRATIONS

Simple harmonic motion (shm) is the name used to describe the straight-line motion of a body whose acceleration is proportional to the displacement from a fixed origin and is always directed toward the origin. This motion is a special case of rectilinear motion with variable acceleration. The motion is essentially a vibratory displacement such as that described by a weight which is attached to one end of a spring and is allowed to vibrate freely.

The mathematical description of the motion is given by the equation

$$a = -Ks$$

where K is the constant of proportionality and s is the displacement from the origin. It is preferable, however, to denote this constant by ω^2, thus making

$$a = -\omega^2 s \tag{17-2.1}$$

the mathematical statement of simple harmonic motion. The negative sign indicates that the direction of the acceleration is always opposite to that of the displacement.

The equations for simple harmonic motion may be derived from the kinematic differential equations by means of the method developed in Section 9-3. Another simpler method is to describe the motion graphically by considering a point that moves with constant speed around a circular path. The projection of this moving point upon a diameter of the circle possesses simple harmonic motion.

Consider, for example, a circle whose radius r is rotating at a constant rate of ω radians per second. Choose a diameter of the circle as the X axis (Fig. 17-2.1), and let the initial position of a point on the circle be at A. The speed of this point will be constant at the value $v = r\omega$. Its acceleration is a normal one due only to the change in the direction of the velocity because the magnitude of the latter is constant. In Section 12-2 this acceleration was given by $a_n = r\omega^2$. After a time interval of t sec, point A reaches position C. The velocity and acceleration of the point in this position are $r\omega$ and $r\omega^2$, directed as shown.

During this time the projection of the point on the X axis moves from O to B through a displacement x. From the figure arc AC

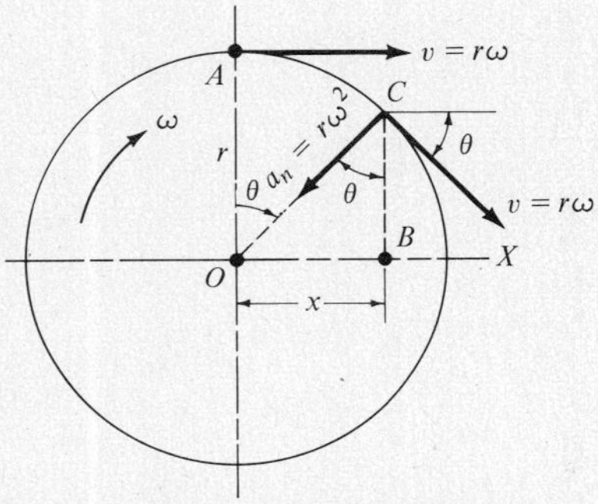

Figure 17-2.1 Auxiliary circle.

projected upon the diameter is

$$x = r \sin \theta \qquad (a)$$

If Eq. (*a*) is differentiated with respect to time, we obtain

$$\frac{dx}{dt} = r \cos \theta \frac{d\theta}{dt}$$

or

$$v_x = r\omega \cos \theta \qquad (b)$$

since $\omega = d\theta/dt$. Observe that $r\omega \cos \theta$ is the diametral projection of the velocity of point C.

If Eq. (*b*) is differentiated with respect to time, the following expression for acceleration results:

$$\frac{dv_x}{dt} = -r\omega \sin \theta \frac{d\theta}{dt}$$

or

$$a_x = -r\omega^2 \sin \theta \qquad (c)$$

The minus sign indicates that the acceleration is directed toward the origin O. Notice that $r\omega^2 \sin \theta$ is the diametral projection of the acceleration of point C. This projection is also directed toward the origin.

However, $r \sin \theta = x$ from Eq. (*a*); hence, Eq. (*c*) may be rewritten as

$$a_x = -\omega^2 r \sin \theta$$

or

$$a_x = -\omega^2 x$$

This is the mathematical description of simple harmonic motion given in Eq. (17-2.1). Hence, we may conclude that the displacement, velocity, and acceleration of a simple harmonic motion are described by the diametral projection of the properties of a point moving with constant speed around a circular path. It is evident from Fig. 17-2.1 that the diametral projection of such a point oscillates back and forth as the point rotates around the circle;[1] also that ω is the constant angular velocity of the radius. A convenient name for ω is *natural circular frequency*. This is sometimes written as ω_n, the subscript being added to avoid confusion when ω is used in its usual meaning of angular velocity.

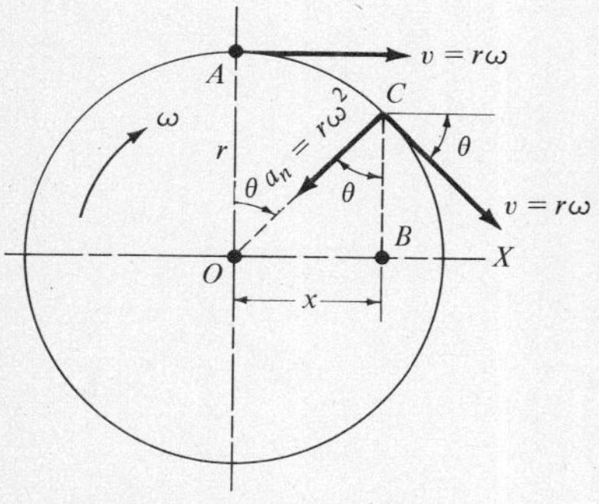

Figure 17-2.1 Auxiliary circle.

The circle in this discussion is known as the auxiliary circle. Its radius is called the *amplitude* of the motion. The time required to complete one oscillation backward and forward is called the *period*

[1] A physical analogy is to whirl a stone in a horizontal circle. Look at the edge view of the plane described by the rotating stone. The horizontal movement of the stone is describing simple harmonic motion.

of the motion. The period is determined by the time required for the radius of the auxiliary circle to complete one revolution, or

$$\tau = \frac{2\pi}{\omega} \tag{17-2.2}$$

The *frequency* is the number of oscillations per second, or

$$f = \frac{1}{\tau} = \frac{\omega}{2\pi} \tag{17-2.3}$$

ILLUSTRATIVE PROBLEMS

17-2.1. If a particle in simple harmonic motion has an amplitude of 1 ft and a period of 1 sec, determine the displacement, velocity, and acceleration after 0.4 sec from when the particle was at the right end of its path.

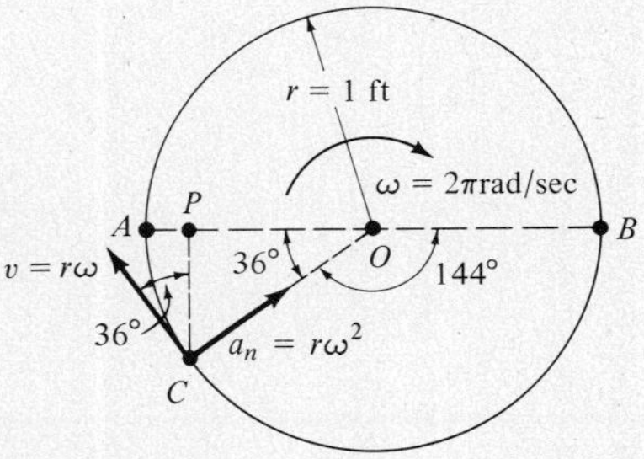

Figure 17-2.2

Solution

Construct an auxiliary circle having a radius equal to the amplitude, as shown in Fig. 17-2.2. Choose diameter AOB as the path of the simple harmonic motion. The natural circular frequency is determined from Eq. (17-2.2).

$$\left[\tau = \frac{2\pi}{\omega}\right] \qquad 1 = \frac{2\pi}{\omega} \qquad \omega = 2\pi \text{ rad per sec}$$

After 0.4 sec the radius will have swept through an angular distance

$$[\theta = \omega t] \qquad \theta = 2\pi \times (0.4) = 0.8\pi \text{ radians}$$
$$= 0.8\pi \times 57.3 = 144°$$

The moving point will then be at position C on the auxiliary circle. The properties of its diametral projection P are given by

$$[s = r\cos\theta] \qquad s = -1 \times \cos 36° = -0.809 \text{ ft}$$
$$[v = r\omega \sin\theta] \qquad v = -1 \times 2\pi \sin 36° = -3.70 \text{ ft per sec}$$
$$[a = r\omega^2 \cos\theta] \qquad a = 1 \times (2\pi)^2 \cos 36° = 31.9 \text{ ft per sec}^2$$

The signs are determined by inspection of Fig. 17-2.2, rightward values being plus and leftward minus.

17-2.2. An elastic string of length $2L$ tightly stretched between two rigid supports, as in Fig. 17-2.3, carries a small ball of weight W at its midpoint. Show that for small displacements, the ball will have a simple harmonic motion; compute the period.

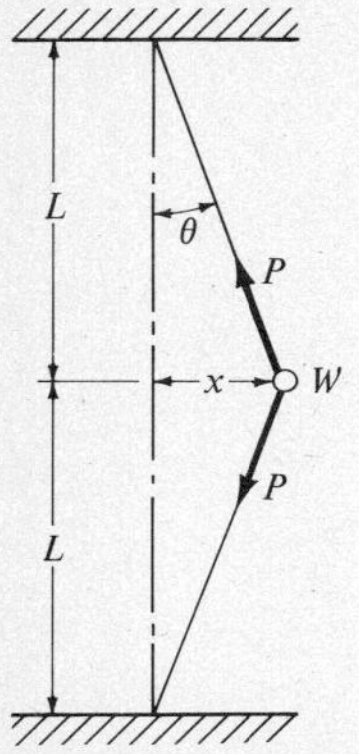

Figure 17-2.3

Solution

We shall assume that the initial tension P in the string is so large that we can neglect any increase in it caused by the small displacement of the ball. If the rightward direction from the equilibrium position is considered positive, the unbalanced force acting on the ball is

$$\Sigma X = -2P \sin\theta \approx -2P\frac{x}{L}$$

since $\sin\theta \approx \tan\theta$ for small values of θ.

Substituting this value in the equation for translation, Eq. (11-2.1), we obtain

$$\left[\Sigma X = \frac{W}{g}a\right] \qquad -2P\frac{x}{L} = \frac{W}{g}a$$

or

$$a = -\frac{2Pg}{WL}x$$

which is in the form of a simple harmonic motion ($a = -\omega^2 s$) where $\omega^2 = \dfrac{2Pg}{WL}$. Hence, from Eq. (17-2.2) the period is

$$\left[\tau = \frac{2\pi}{\omega}\right] \qquad \tau = 2\pi\sqrt{\frac{WL}{2Pg}} \qquad \textit{Ans.}$$

PROBLEMS

17-2.3. A simple harmonic motion is defined by the relation $a = -36s$. Determine its period and frequency.

$\tau = 1.047$ sec, $f = 0.957$ osc per sec *Ans.*

17-2.4. The amplitude of a simple harmonic motion is 2 ft and the period is 1 sec. Determine the maximum velocity and the maximum acceleration.

17-2.5. A particle in simple harmonic motion has an amplitude of 15 in. and a period of $\pi/2$ sec. Find the velocity and acceleration of the particle when it has traveled 9 in. to the right of the center of its path. What time is required for this displacement?

$v = 4$ fps; $a = -12$ fps^2; $t = 0.161$ sec *Ans.*

17-2.6. A particle has a simple harmonic motion defined by $a = -9s$ and an amplitude of 10 in. Find the velocity and acceleration of the particle when it is 6 in. away from the center of its path. What is its period?

17-2.7. A particle moving with simple harmonic motion has a maximum velocity of 20 fps and a maximum acceleration of 40 fps^2. Determine the velocity and acceleration of the particle when it is midway between the center and the right end of its path. How long does it take to move from the center to the specified position?

$$v = 17.32 \text{ fps};\ a = -20 \text{ fps}^2;\ t = 0.261 \text{ sec} \qquad \textit{Ans.}$$

17-2.8. The amplitude of a particle moving with simple harmonic motion is 20 in. When the particle is 10 in. from the extreme left position, its acceleration is 160 ips^2. What is its velocity at that position? How many seconds are required to move 10 in. from the extreme left position?

17-2.9. A particle moving with simple harmonic motion is observed to have a velocity of 16 ips when it is 6 in. from the center of its path and a velocity of 12 ips when it is 8 in. from the center. Determine the maximum velocity, maximum acceleration, and the frequency of vibration.

$$v_{max} = 20 \text{ ips};\ a_{max} = 40 \text{ ips}^2;\ f = 0.318 \text{ vib per sec} \qquad \textit{Ans.}$$

17-2.10. A weight of W lb is attached to one end of a vertical spring (Fig. P-17-2.10) whose scale is k lb per ft. The weight is displaced vertically from its equilibrium position by an external force. It is then released and vibrates freely. By means of the equation $\Sigma X = (W/g)a$, show that the vibration is in the form of a simple harmonic motion and that the period of vibration is given by $\tau = 2\pi\sqrt{W/kg}$.

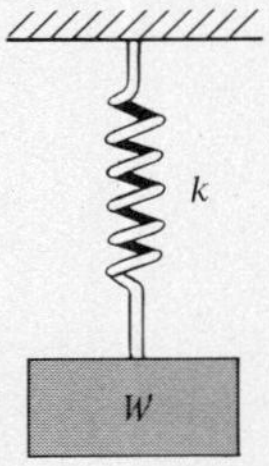

Figure P-17-2.10

17-2.11. A 161-lb weight is suspended vertically from the end of a spring whose modulus is 20 lb per in. The weight is pulled down 6 in. below its equilibrium position and then released. Compute the velocity of the weight as it passes through the equilibrium position and also its frequency of vibration. Check the velocity using the work-energy method.

$$v = 3.47 \text{ ft per sec};\ f = 1.104 \text{ vib per sec} \qquad \textit{Ans.}$$

17-2.12. The average depth of immersion of a ship is h and its area at the water line is A. Using the equation $\Sigma X = (W/g)a$, show that if the ship is displaced slightly downward and then released, the buoyant force of the water will cause vertical oscillations that have a simple harmonic motion. Determine the period of oscillation.

$$\tau = 2\pi\sqrt{\frac{h}{g}} \qquad \textit{Ans.}$$

17-3 SIMPLE PENDULUM

A simple pendulum is defined as a particle at the end of a weightless cord that is allowed to oscillate in a vertical arc of a circle under the influence of gravity and the tension in the cord. This motion is approximated by suspending a small weight at the end of a thread, as shown in Fig. 17-3.1.

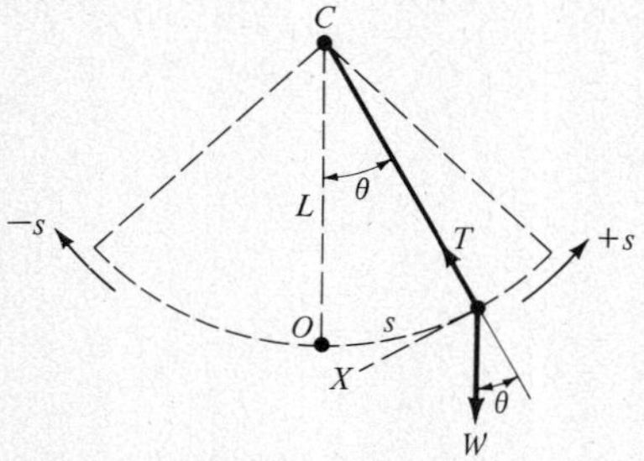

Figure 17-3.1 Simple pendulum.

Taking the displacement as positive in the direction shown, we have

$$\left[\Sigma X = \frac{W}{g}a\right] \qquad -W\sin\theta = \frac{W}{g}a \qquad (a)$$

If the oscillation is restricted to small angles of vibration, $\sin\theta$ may be considered practically equivalent to θ measured in radians. From the figure, $\theta = s/L$, wherefore

$$-W\frac{s}{L} = \frac{W}{g}a \qquad (b)$$

so that

$$a = -\frac{g}{L}s \qquad (c)$$

This is the equation of simple harmonic motion ($a = -\omega^2 s$) in which $\omega^2 = \dfrac{g}{L}$. From Section 17-2, the period is given by

$$\tau = \frac{2\pi}{\omega} = 2\pi\sqrt{\frac{L}{g}} \qquad (17\text{-}3.1)$$

and the frequency by

$$f = \frac{\omega}{2\pi} = \frac{1}{2\pi}\sqrt{\frac{g}{L}} \qquad (17\text{-}3.2)$$

These results show that for small angles of vibration the period and frequency depend only on the length of the cord of the simple pendulum.

PROBLEMS

17-3.1. Determine the length of a simple pendulum whose period is 1 sec.

17-3.2. Find the period of a simple pendulum whose length is 50 in.

$\tau = 2.26$ sec ***Ans.***

17-3.3. A simple pendulum is suspended from the roof of an elevator which is accelerating at a fps^2. Assuming that the vibrations are small, determine the period of the pendulum when the elevator is (a) accelerating upward, (b) accelerating downward, (c) falling freely.

(a) $\tau = 2\pi\sqrt{\dfrac{L}{g+a}}$; (b) $\tau = 2\pi\sqrt{\dfrac{L}{g-a}}$;

(c) $\tau = \infty$, pendulum stationary ***Ans.***

17-4 COMPOUND PENDULUM

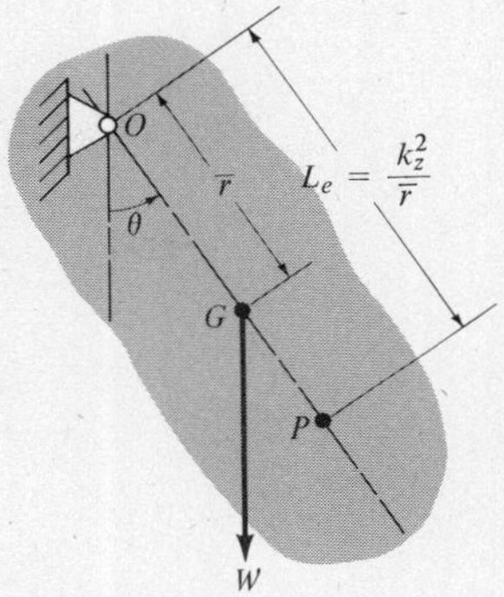

Figure 17-4.1 Compound pendulum.

A compound pendulum is a rigid body that is free to rotate about a horizontal axis and that oscillates under the influence of gravity. Such a body is represented in Fig. 17-4.1. Point O is called the *center of suspension*. The position of the pendulum is defined by its angular displacement θ from the equilibrium position.

Applying the equation of rotation about a fixed axis and noting that moments are positive in the sense of angular displacement, we obtain (neglecting friction at the axis)

$$[\Sigma M_z = I_z\alpha] \qquad -W\bar{r}\sin\theta = I_z\alpha \tag{a}$$

Expressing I_z in terms of its mass $\frac{W}{g}$ and radius of gyration k_z gives

$$-W\bar{r}\sin\theta = \frac{W}{g}k_z{}^2\alpha \tag{b}$$

If the oscillations are restricted to small angles for which $\sin\theta$ is practically equal to θ, Eq. (b) may be written in the form

$$\alpha = -\frac{g\bar{r}}{k_z{}^2}\theta \tag{c}$$

which is similar to a simple harmonic motion in which $\omega^2 = \frac{g\bar{r}}{k_z{}^2}$. Therefore, the period is given by $\tau = 2\pi/\omega$ (Section 17-2), or

$$\tau = 2\pi\sqrt{\frac{k_z{}^2}{g\bar{r}}} = 2\pi\sqrt{\frac{L_e}{g}} \tag{17-4.1}$$

Observe that the period of a compound pendulum is the same as that of a simple pendulum (Section 17-3) that has an equivalent length L_e equal to $\frac{k_z{}^2}{\bar{r}}$.

From the transfer formula relating radii of gyration, Eq. (7-4.2),

$$k_z{}^2 = \bar{k}^2 + \bar{r}^2$$

Hence the equivalent length of a compound pendulum may be written as

$$L_e = \frac{\bar{k}^2}{\bar{r}} + \bar{r} \tag{d}$$

This may be interpreted as meaning that the period is the same for all parallel axes of suspension located at the same distance from the center of gravity.

Point P on line OG at the distance $L_e = \frac{k_z{}^2}{\bar{r}}$ from O is called the *center of oscillation*. This point is also the center of percussion (Illus. Prob. 13-3.2.). Huygens showed that if the center of oscillation is made the center of rotation, the period will be unchanged, i.e., *the centers of suspension and oscillation are interchangeable.* Thus since $GO = \bar{r}$ and $GP = L_e - \bar{r}$, Eq. (d) gives

$$(GO)(GP) = \bar{k}^2 \tag{e}$$

If the pendulum is suspended from a horizontal axis through P and the new center of oscillation is denoted by O', then from Eq. (e) we have

$$(GO')(GP) = \bar{k}^2 \tag{f}$$

or O' and O must coincide. We conclude that if P becomes the center of suspension, O becomes the center of oscillation.

Equation (17-4.1) is sometimes used in experiments to determine the radius of gyration or moment of inertia of a body that can be swung as a pendulum. For example, if a connecting rod is swung from a knife edge that passes through the hole for the wrist pin or crank pin and the duration of the period is observed, the radius of gyration and hence the moment of inertia about the axis of suspension can be computed. The center of gravity can be determined by balancing the connecting rod on a knife edge, and the moment of inertia about any parallel axis can then be found by applying the transfer formula.

PROBLEMS

17-4.1. A circular ring is suspended from a knife edge as shown in Fig. P-17-4.1. The ring weighs 64.4 lb, its outer radius is 3 ft, and its inner radius is 2 ft. Determine the period for small oscillations.

$\tau = 2.54$ sec **_Ans._**

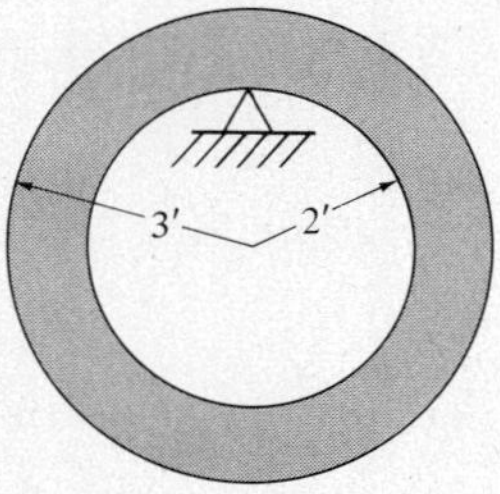

Figure P-17-4.1

17-4.2. The compound pendulum in Fig. P-17-4.2 consists of a slender rod 2 ft long weighing 6 lb to which is attached a solid circular disk of 12-in. diameter that weighs 8 lb. Compute the period of small oscillations.

$\tau = 1.67$ sec **_Ans._**

17-4.3. A connecting rod weighing 9.66 lb is swung as a pendulum

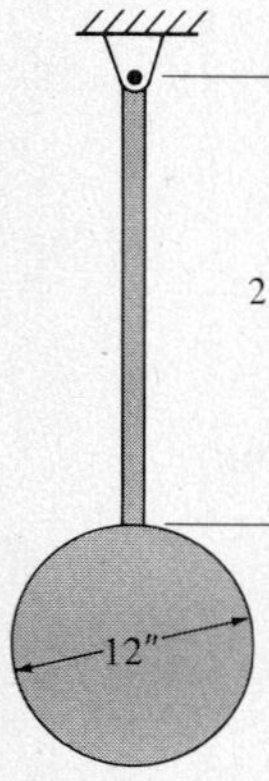

Figure P-17-4.2

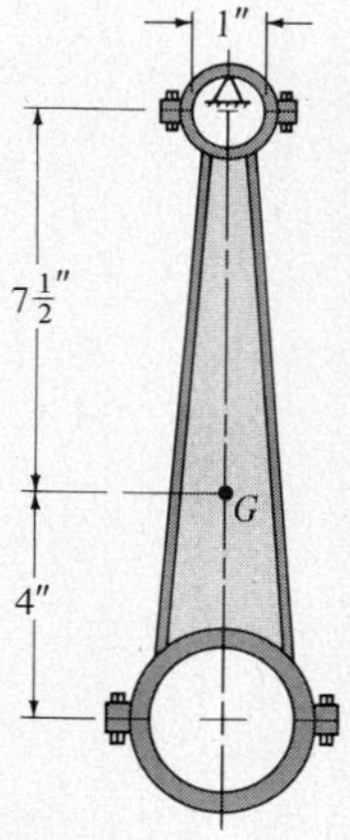

Figure P-17-4.3

from a horizontal knife edge resting on the inner face of the wrist pin bearing. The dimensions are given in Fig. P-17-4.3. If the observed period of oscillation is 1 sec, compute the moment of inertia with respect to the axis of the crank bearing.

$$I = 0.0633 \text{ ft-lb-sec}^2 \qquad \textbf{\textit{Ans.}}$$

17-5 TORSION PENDULUM

A torsion pendulum consists of a body that is fastened to the lower end of an elastic wire or rod whose upper end is clamped. The body is assumed to be fastened so that the axis of the rod passes through the center of gravity of the body, as in Fig. 17-5.1.

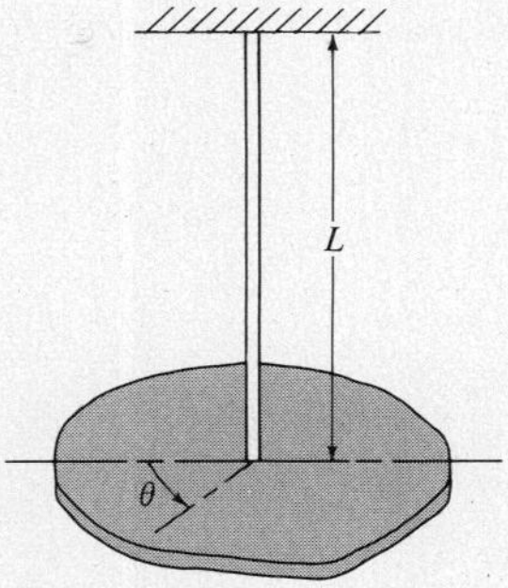

Figure 17-5.1 Torsion pendulum.

Let an external moment be applied to the body that rotates it (and the lower end of the rod) through an angular displacement θ from the equilibrium position. When this moment is removed, the body undergoes an oscillatory rotation known as torsional vibration. During it, the only external moment acting on the body is supplied by the resisting moment M exerted by the rod. From strength of materials, it is known that the relation between the twisting moment and the angle of twist for a circular cylindrical rod is

$$M = \frac{JG}{L}\theta \qquad (a)$$

in which J is the polar moment of inertia of the cross-sectional area of the rod,[2] G the shear modulus of elasticity of the material, and L the length. The product $\frac{JG}{L}$ is sometimes called the torsional spring constant of the rod, i.e., the torque required to produce an angle of twist of 1 rad.

Applying the equation of rotation, we have

$$[\Sigma M = I\alpha] \qquad -M = I\alpha$$

in which I represents the mass moment of inertia of the body about the axis of rotation. From Eq. (a) this becomes

$$-\frac{JG}{L}\theta = I\alpha$$

or

$$\alpha = -\frac{JG}{IL}\theta \qquad (b)$$

This is in the form of a simple harmonic motion in which $\omega^2 = \frac{JG}{IL}$.[3]

[2] For a rod of diameter d, $J = \pi d^4/32$.

[3] Observe that I should be expressed in in.-lb-sec²; also that L should be expressed in inches.

In Section 17-2 the period was given by $\tau = 2\pi/\omega$, or

$$\tau = 2\pi \sqrt{\frac{IL}{JG}} \tag{17-5.1}$$

If the axis of the rod does not pass through the center of gravity of the attached body, the rod tends to be thrown out of plumb. However, if it is supported in bearings at the lower end to prevent this, the attached body will undergo torsional vibrations whose period is defined by Eq. (17-5.1). Under these conditions, observing the period of vibration is an easy method of determining the mass moment of inertia of any body with respect to an axis coinciding with that of the rod.

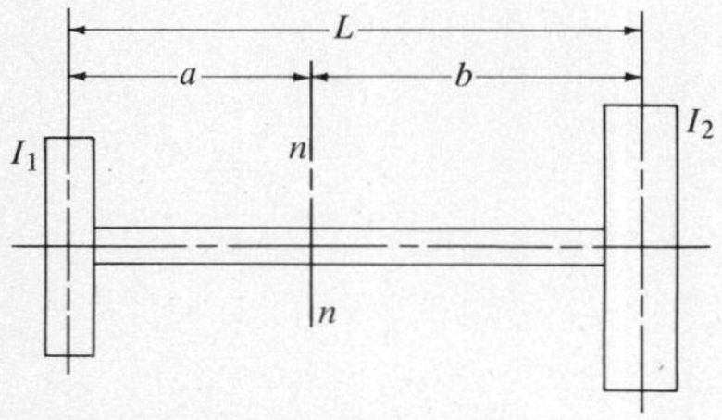

Figure 17-5.2 Nodal section.

Many practical applications of torsional vibration occur when the load or power transmitted along a shaft between two heavy masses is suddenly released. An example is the shaft connecting the engine and propeller of a vessel that is pitching in a heavy sea. If Fig. 17-5.2 represents two heavy masses having moments of inertia I_1 and I_2 between which a transmitted torque is suddenly released, the elastic energy stored in the shaft will cause the masses to rotate in opposite directions. (See Prob. 15-6.10.) One section *n-n* of the shaft, called the nodal section, will remain stationary. The motion of each mass may be considered equivalent to a torsion pendulum on a shaft fixed at the nodal section.

The position of the nodal section is located from the condition that each segment of the shaft has the same period of oscillation. Hence by applying Eq. (17-5.1) we obtain

$$I_1 a = I_2 b \tag{c}$$

from which, since $a + b = L$

$$a = \frac{I_2 L}{I_1 + I_2} \qquad b = \frac{I_1 L}{I_1 + I_2} \tag{d}$$

Since each segment has the same period, we may substitute either of these lengths in Eq. (17-5.1) to obtain the following expression for the period of free torsional vibration for the system:

$$\left[\tau = 2\pi \sqrt{\frac{IL}{JG}}\right] \qquad \tau = 2\pi \sqrt{\frac{I_1 I_2 L}{JG(I_1 + I_2)}} \tag{e}$$

ILLUSTRATIVE PROBLEM

17-5.1. A solid circular disk weighing 32.2 lb and having a diameter of 3 ft is suspended at its midpoint horizontally from the end of a vertical steel wire 3 ft long and $\frac{1}{8}$ in. in diameter. The shear modulus of elasticity for the wire is $G = 12 \times 10^6$ lb per sq in. Determine the period of torsional vibration.

Solution

The moment of inertia of the disk with respect to the axis of the wire is

$$\left[I = \frac{1}{2}\frac{W}{g}r^2\right] \qquad I = \frac{1}{2} \times \frac{32.2}{32.2} \times (1.5)^2 = 1.125 \text{ ft-lb-sec}^2$$

$$= 1.125 \times 12 = 13.5 \text{ in.-lb-sec}^2$$

The polar moment of inertia of the wire is

$$\left[J = \frac{\pi d^4}{32}\right] \qquad J = \frac{\pi \times (\frac{1}{8})^4}{32} = 24 \times 10^{-6} \text{ in.}^4$$

Substituting these values in Eq. (17-5.1) gives the period

$$\left[\tau = 2\pi\sqrt{\frac{IL}{JG}}\right] \qquad \tau = 2\pi\sqrt{\frac{13.5 \times (3 \times 12)}{24 \times 10^{-6} \times 12 \times 10^6}} = 8.17 \text{ sec} \qquad \textit{Ans.}$$

PROBLEMS

17-5.2. A horizontal bar 1 ft long that weighs 16.1 lb is suspended at its midpoint by a vertical steel rod 4 ft long and $\frac{1}{4}$ in. in diameter. If the shear modulus of elasticity is 12×10^6 lb per sq in., compute the frequency of torsional vibration.

$f = 2.21$ osc per sec *Ans.*

17-5.3. A certain body has a moment of inertia of 0.5 ft-lb-sec^2 with respect to an axis coincident with the center line of a wire from which the body is suspended. The period of torsional vibration is 0.6 sec. When another body is suspended from the same wire in place of the first body, the period is 0.9 sec. Determine the moment of inertia of the second body.

$I = 1.125$ ft-lb-sec^2 *Ans.*

17-5.4. A shaft 3 in. in diameter and 10 ft long connects the rotor of a steam turbine with the armature of a generator. The rotor weighs 1932 lb and has a radius of gyration of 2 ft. The armature weighs 966 lb and has a radius of gyration of 3 ft. If $G = 12 \times 10^6$ lb per sq in., compute the frequency of free torsional vibration of the system.

$f = 3.64$ osc per sec *Ans.*

17-6 GRAPHICAL REPRESENTATION OF SIMPLE HARMONIC MOTION

As discussed in Section 17-2, the displacement-time relation defining a simple harmonic motion may be written in the form

$$x = r \sin \omega t \tag{a}$$

Differentiating this equation with respect to the time determines the velocity-time relation

$$v = r\omega \cos \omega t \tag{b}$$

The maximum value of velocity is $v_m = r\omega$. The acceleration-time relation is obtained by another differentiation with respect to the time; this gives

$$a = -r\omega^2 \sin \omega t \tag{c}$$

from which the maximum value of acceleration is found to be $a_m = r\omega^2$.

One method of visualizing the variations in these equations is by means of the auxiliary circle described in Section 17-2. Another method, to be described, permits the displacement-time, velocity-time, and acceleration-time curves to be plotted easily. This technique is useful when analyzing mechanical vibrations, especially when combining vibrations of the same frequency.

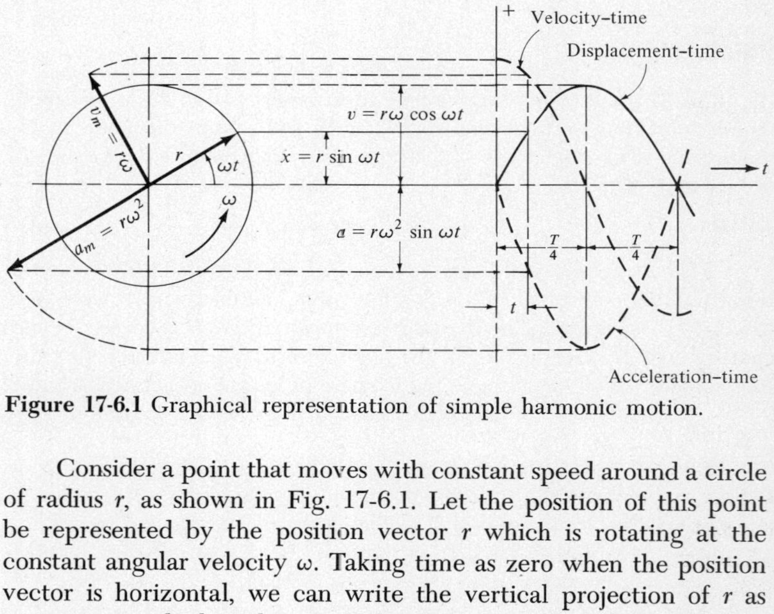

Figure 17-6.1 Graphical representation of simple harmonic motion.

Consider a point that moves with constant speed around a circle of radius r, as shown in Fig. 17-6.1. Let the position of this point be represented by the position vector r which is rotating at the constant angular velocity ω. Taking time as zero when the position vector is horizontal, we can write the vertical projection of r as $x = r \sin \omega t$, which is the simple harmonic motion expressed by Eq. (a). Similarly the velocity can be represented by the vertical projection of a vector of length $r\omega$ that rotates with the same angular velocity ω as the position vector but is always 90° *ahead* of it. The acceleration is represented by the vertical projection of a vector of length $r\omega^2$ that rotates with the same angular velocity ω as the position and velocity vectors but is 180° *ahead* of the position vector and 90° *ahead* of the velocity vector. Figure 17-6.1 indicates how the projections of these rotating vectors are used to plot the values of displacement, velocity, and acceleration against time.

17-7 FREE VIBRATIONS WITHOUT DAMPING. GENERAL CASE

Let us consider a spring-suspended weight which is displaced from its equilibrium position and allowed to vibrate freely. This is described in Fig. 17-7.1. It is assumed that all resistances such as air resistance and internal friction in the spring are neglected. This means that the spring will vibrate indefinitely under the action of the variable spring force exerted upon the weight. This condition is called free vibration.

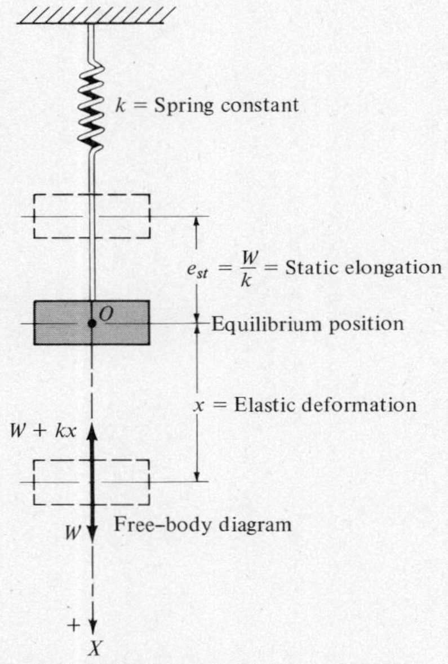

Figure 17-7.1 Free vibration of a spring-suspended weight.

The differential equation of motion for this case is found by applying the relation between the unbalanced force and the resulting acceleration. Considering downward forces and displacements as positive, we obtain

$$\left[\Sigma X = \frac{W}{g}a\right] \qquad -kx = \frac{W}{g}a$$

or, since $a = \dfrac{d^2x}{dt^2}$

$$\frac{d^2x}{dt^2} = -\frac{kg}{W}x = -\omega^2 x \tag{a}$$

where

$$\omega = \sqrt{\frac{kg}{W}} \tag{b}$$

Equation (*a*) is the differential equation of a free vibration. It indicates that the displacement x is a function of time such that, upon being differentiated twice with respect to time, it equals itself multiplied by a negative constant of value ω^2. The sine and cosine functions repeat themselves in this manner. Substitution of the relations $x = \sin \omega t$ and $x = \cos \omega t$ in Eq. (*a*) shows that they are solutions of that equation. A more general solution is obtained by multiplying these solutions by arbitrary constants C_1 and C_2 which are evaluated to fit the actual condition of the vibration. With this procedure, the complete solution of Eq. (*a*) is given by the sum of these solutions, or

$$x = C_1 \sin \omega t + C_2 \cos \omega t \tag{c}$$

To determine the constants C_1 and C_2 it is necessary to understand that Eq. (*c*) describes all the possible motions that the system of weight and spring shown in Fig. 17-7.1 can have. If we now specify that the weight has a velocity v_o when given an initial displacement x_o, there result two conditions which can be used to determine the constants of integration. Measuring the time as zero under the given conditions, at $t = 0$ we obtain

$$x = x_o \quad \text{and} \quad v = v_o$$

Substituting the first condition in Eq. (*c*) yields

$$x_o = C_1(0) + C_2(1) \quad \text{or} \quad C_2 = x_o \tag{d}$$

The second condition is substituted in the velocity-time relation obtained by differentiating Eq. (*c*) with respect to the time. Thus,

$$\left[v = \frac{dx}{dt}\right] \qquad v = \omega C_1 \cos \omega t - \omega C_2 \sin \omega t$$

Setting $v = v_o$ when $t = 0$, we have

$$v_o = \omega C_1(1) - \omega C_2(0) \quad \text{or} \quad C_1 = \frac{v_o}{\omega} \tag{e}$$

Under the specified conditions, the vibration is described by the equation

$$x = \frac{v_o}{\omega} \sin \omega t + x_o \cos \omega t \tag{17-7.1}$$

The period is the time required for the position vector describing the vibration (Section 17-6) and moving with a constant angular velocity ω, to sweep through one revolution, or

$$\tau = \frac{2\pi}{\omega} = 2\pi \sqrt{\frac{W}{kg}} = 2\pi \sqrt{\frac{e_{st}}{g}} \tag{17-7.2}$$

where the static elongation is

$$e_{st} = \frac{W}{k} \tag{f}$$

and represents the elongation produced by the weight when hanging freely from the spring. The frequency is

$$f = \frac{1}{\tau} = \frac{\omega}{2\pi} = \frac{1}{2\pi} \sqrt{\frac{kg}{W}} = \frac{1}{2\pi} \sqrt{\frac{g}{e_{st}}} \tag{17-7.3}$$

It will be observed that Eq. (17-7.1) really shows the combined effect of two vibrations of the same frequency. If the position vectors

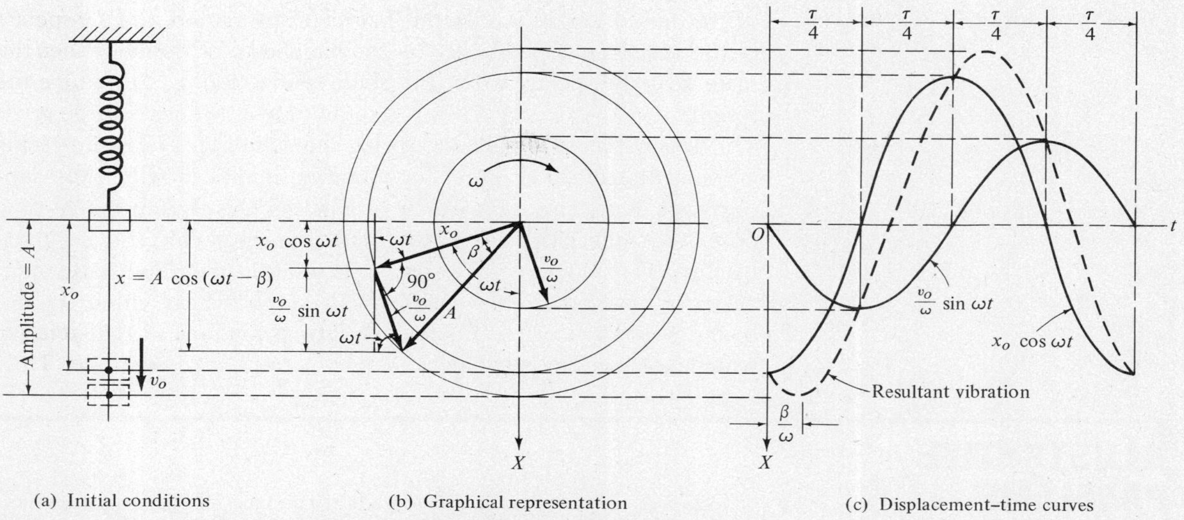

Figure 17-7.2 Combination of vibrations showing phase angle and resultant vibration.

of these vibrations are plotted 90° apart, as in Fig. 17-7.2, the sum of their projections on the vertical X axis will be

$$x = x_o \cos \omega t + \frac{v_o}{\omega} \sin \omega t = A \cos(\omega t - \beta) \tag{17-7.4}$$

The figure shows that the sum of two simple harmonic motions of the same frequency is another simple harmonic motion of the same frequency. The amplitude of the resultant vibration is

$$A = \sqrt{x_o^2 + \left(\frac{v_o}{\omega}\right)^2} \tag{g}$$

which lags behind vibration $x_o \cos \omega t$ by the phase angle β, the value of which is determined from the condition

$$\tan \beta = \frac{v_o}{\omega x_o} \tag{h}$$

The phase angle β indicates that the resultant vibration reaches its maximum values of displacement, velocity, and acceleration at β/ω sec behind vibration $x_o \cos \omega t$. The construction shown in Fig. 17-7.2 may be used to represent the displacement curves of each vibration. The resultant vibration is determined by adding these displacements either geometrically or directly from the position vector A.

The following conclusion can be drawn from this discussion: The period and frequency of free vibration of a body are independent of the initial conditions of the motion. The period and frequency of the vibration depend only on the weight W of the body and the scale k of the spring, or on the static elongation e_{st} caused by the weight.

The system of weight and spring shown in Fig. 17-7.1 represents a conventionalized diagram for vibrating bodies in which the supports act like springs, regardless of whether this is their actual purpose. For such cases the period and frequency can also be given by Eqs. (17-7.2) and (17-7.3) by substituting in them the static elongation e_{st} caused by the weight of the body whose vibrations are being considered or the equivalent spring constant of that weight, whichever is more convenient.

ILLUSTRATIVE PROBLEMS

17-7.1. The deflection produced at the free end of the cantilever beam in Fig. 17-7.3 by a static load of 200 lb is 0.20 in. If a weight $W = 300$ lb is dropped on the free end of the beam from a height $h = 2$ in., compute the frequency of vibration. What will be the maximum deflection? Assume that the weight stays in contact with the beam after striking it.

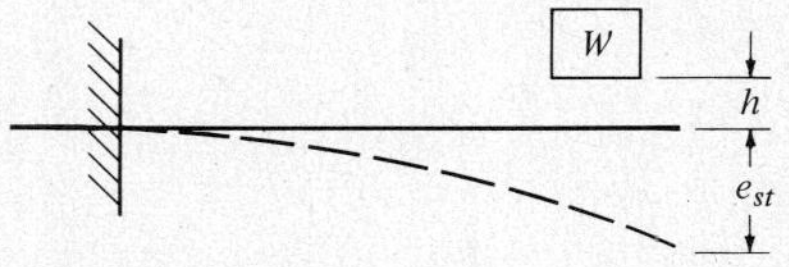

Figure 17-7.3

Solution

Using Eq. (f), we find that the spring constant, determined from the deflection produced by the static load, is

$$\left[k = \frac{W}{e_{st}}\right] \qquad k = \frac{200}{0.20} = 1000 \text{ lb per in.}$$

The frequency with which the 300-lb weight will vibrate is determined from Eq. (17-7.3).

$$\left[f = \frac{1}{2\pi}\sqrt{\frac{kg}{W}}\right] \qquad f = \frac{1}{2\pi}\sqrt{\frac{(1000 \times 12) \times 32.2}{300}}$$

$$= 5.72 \text{ vib per sec} \qquad \textit{Ans.}$$

The amplitude of vibration is given by Eq. (g). Since the initial conditions of the motion are $x_o = -W/k$ and $v_o = \sqrt{2gh}$, and also since $\omega = \sqrt{kg/W}$, Eq. (g) may be rewritten

$$\left[A = \sqrt{x_o^2 + \left(\frac{v_o}{\omega}\right)^2}\right] \qquad A = \sqrt{\left(\frac{W}{k}\right)^2 + \frac{2Wh}{k}}$$

Since the amplitude is measured from the position of static equilibrium, the maximum deflection will be the sum of the static deflection plus the amplitude of vibration. Hence,

$$\left[\begin{aligned} e_{max} &= e_{st} + A \\ &= \frac{W}{k} + \sqrt{\left(\frac{W}{k}\right)^2 + \frac{2Wh}{k}} \end{aligned}\right]$$

$$\begin{aligned} e_{max} &= \frac{300}{1000} + \sqrt{\left(\frac{300}{1000}\right)^2 + \frac{2 \times 300 \times 2}{1000}} \\ &= 0.30 + 1.136 = 1.436 \text{ in.} \end{aligned}$$ ***Ans.***

A check on this result, as well as a more direct determination of the maximum deflection, is afforded by the work-energy method. The weight has zero velocity when initially dropped, and zero velocity when the cantilever beam is deflected through e_{max} in. Equating the resultant work on W to the zero change in kinetic energy, we have

$$[W(h + e_{max}) - \tfrac{1}{2}ke_{max}^2 = 0]$$

$$300(2 + e_{max}) - \tfrac{1}{2} \times 1000 \times e_{max}^2 = 0 \qquad e_{max} = 1.436 \text{ in.}$$ ***Ans.***

17-7.2. Consider a body of weight W supported by a system of two springs, as shown in Fig. 17-7.4. Determine the equivalent spring constant for each arrangement.

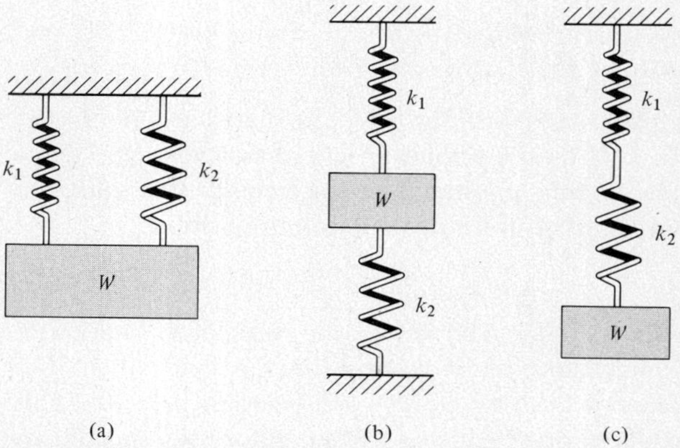

Figure 17-7.4 Springs in parallel and in series.

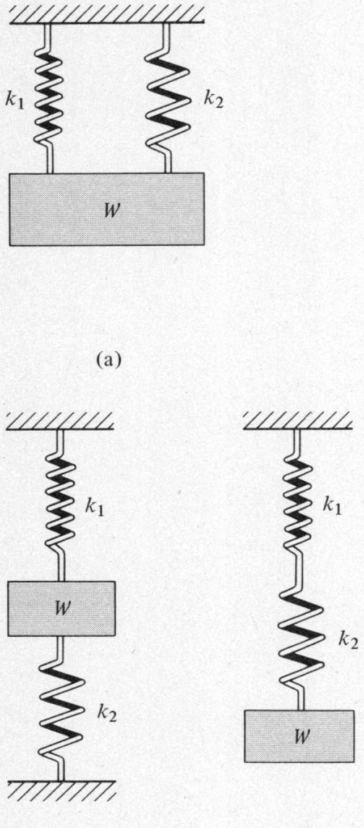

Figure 17-7.4 Springs in parallel and in series.

Solution

In Fig. 17-7.4a, it is evident that a displacement of the body through a unit distance stretches each spring a unit distance. Hence the equivalent spring constant for the system is

$$k = k_1 + k_2 \tag{a}$$

This arrangement is known as springs in parallel.

A variation of it is shown in Fig. 17-7.4b. At first it is not so evident that the spring constant here is given by Eq. (*a*), but it will be observed that a unit displacement of the weight will stretch one spring a unit amount while compressing the other a unit amount. Hence, the spring force exerted on the weight when displaced a unit distance is also

$$k = k_1 + k_2 \tag{a}$$

The arrangement in Fig. 17-7.4c is known as springs in series. A unit force applied to W will be transmitted through both springs. Since the displacement of the weight is the sum of the deflection of both springs, a unit force will cause a displacement of the amount

$$s = \frac{1}{k_1} + \frac{1}{k_2}$$

From Hooke's law applied to a spring, $P = ks$, we find $k = P/s$, or for a unit force

$$k = \frac{1}{\dfrac{1}{k_1} + \dfrac{1}{k_2}} \tag{b}$$

Equation (*b*) can be rewritten in the form

$$\frac{1}{k} = \frac{1}{k_1} + \frac{1}{k_2} \tag{c}$$

This form shows clearly that the reciprocal of the equivalent spring constant is the sum of the reciprocals of the springs in series.

17-7.3. Consider a U tube (Fig. 17-7.5) of cross section A that contains a length L of fluid weighing w lb per unit volume. Determine the equivalent spring constant and the frequency of vibration when the fluid is disturbed from its equilibrium position.

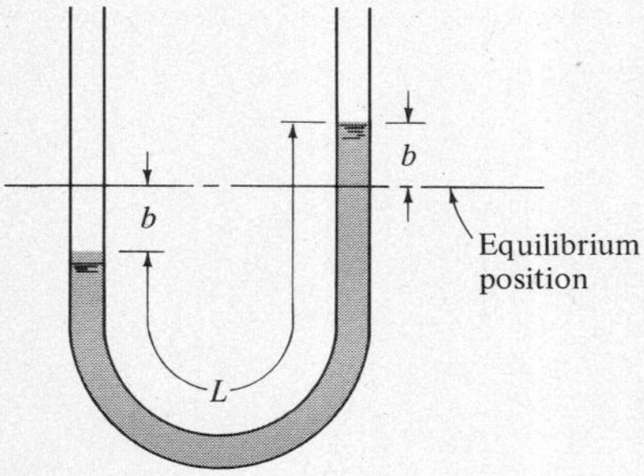

Figure 17-7.5

Solution

If the fluid oscillates back and forth, the weight in motion is $W = wAL$. The unbalanced force acting on the fluid when it is displaced a distance b from its equilibrium position is $2wAb$. This results in a "gravity spring" which tends to restore the fluid to its equilibrium position. The spring constant or force required to displace the fluid through a unit distance is therefore

$$k = 2wA$$

Applying Eq. (17-7.3) gives the frequency

$$\left[f = \frac{1}{2\pi}\sqrt{\frac{kg}{W}}\right] \qquad f = \frac{1}{2\pi}\sqrt{\frac{2wAg}{wAL}} = \frac{1}{2\pi}\sqrt{\frac{2g}{L}} \qquad \textbf{\textit{Ans.}}$$

Observe that the frequency of vibration is independent of the fluid. A column of mercury, for example, vibrates with the same frequency as a column of water.

17-7.4. The system shown in Fig. 17-7.6 consists of a stiff rod of length L that has a body of weight W at its free end and is supported by a spring having a constant k_s. Determine the frequency of vibration of the weight W if it is displaced vertically from its position of equilibrium.

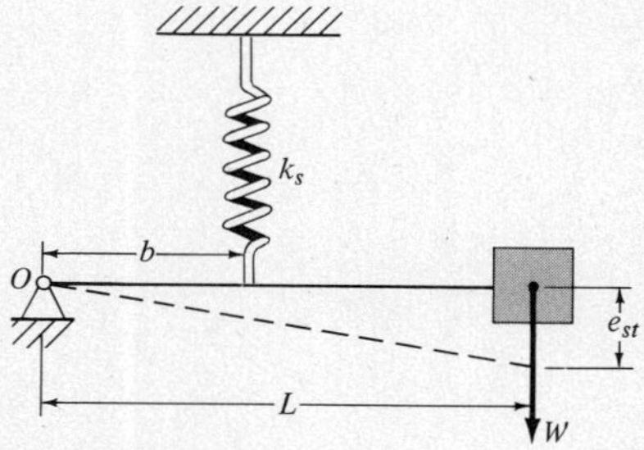

Figure 17-7.6

Solution

It was shown in Eq. (17-7.3) that the frequency of vibration can be expressed in terms of the static deflection produced by the body whose vibrations are being considered.

From the static equilibrium of the rod, moments about O determine the force exerted on the spring to be $P = WL/b$. This stretches the spring an amount

$$e = \frac{P}{k_s} = \frac{WL}{k_s b}$$

From the geometry in Fig. 17-7.6, the corresponding deflection of the weight is

$$e_{st} = e\frac{L}{b} = \frac{W}{k_s}\left(\frac{L}{b}\right)^2$$

Applying Eq. (17-7.3), we find the frequency of vibration to be

$$\left[f = \frac{1}{2\pi}\sqrt{\frac{g}{e_{st}}}\right] \qquad f = \frac{1}{2\pi}\frac{b}{L}\sqrt{\frac{gk_s}{W}} \qquad \textbf{\textit{Ans.}}$$

PROBLEMS

17-7.5. A weight $W = 100$ lb is suspended from two springs in series, as shown in Fig. P-17-7.5 on p. 704. If $k_1 = 30$ lb per in. and $k_2 = 20$ lb per in., compute the period of free vibration when W is placed vertically from its equilibrium position.

$\tau = 0.922$ sec **Ans.**

17-7.6. Compute the period of vibration if the springs in Prob. 17-7.5 are arranged in parallel.

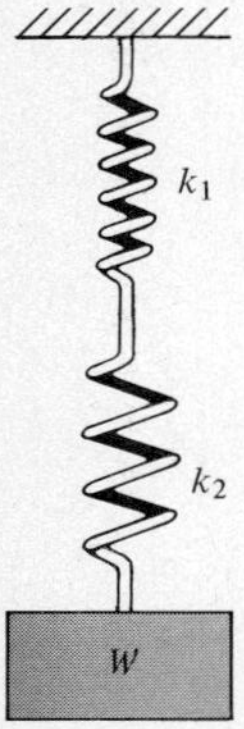

Figure P-17-7.5

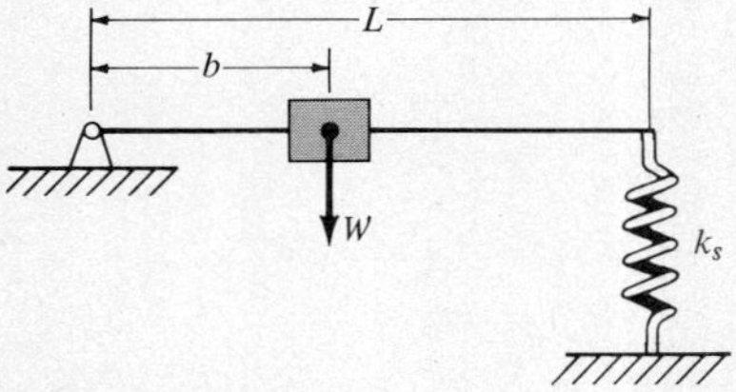

Figure P-17-7.8

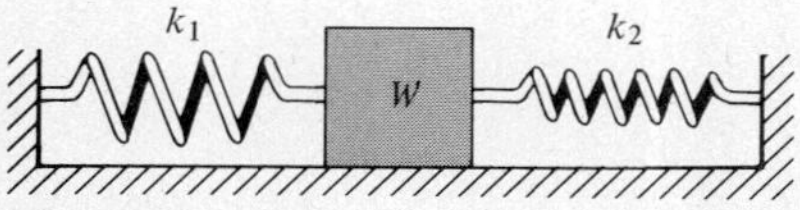

Figure P-17-7.9

17-7.7. Assume that the weight in Prob. 17-7.5 has a velocity of 10 ft per sec as it passes through its equilibrium position. Compute the maximum tension produced in the springs. By how much will this tension be changed if the lower spring is omitted?

Max tension = 311 lb; increase in tension = 122 lb ***Ans.***

17-7.8. A stiff bar of length $L = 4$ ft and negligible weight is supported by a frictionless hinge at its left end, as in Fig. P-17-7.8. It carries a body of weight $W = 10$ lb at a distance $b = 1.5$ ft from the hinge. The bar is supported at the right end by a spring having a constant $k_s = 40$ lb per in. Determine the frequency of vibration of W if it is displaced vertically from its equilibrium position.

$f = 16.7$ vib per sec ***Ans.***

17-7.9. As shown in Fig. P-17-7.9, a weight $W = 4$ lb rests upon a frictionless surface between two springs having constants $k_1 = 1.5$ lb per in. and $k_2 = 2.5$ lb per in. The weight is released from rest when it is $\frac{3}{4}$ in. from its equilibrium position. Determine its period of vibration and velocity as it passes through the equilibrium position.

$\tau = 0.32$ sec; $v = 14.7$ ips ***Ans.***

17-7.10. An elevator weighing 4000 lb is being lowered at the constant rate of 10 ft per sec when the hoisting drum is stopped suddenly. If the elastic properties of the cable are such that 1000 lb stretches it 0.025 in., compute the frequency of free vibration of the elevator. Determine the maximum tension produced in the cable. Neglect the weight of the cable.

$f = 9.9$ vib per sec; max tension = 81,200 lb ***Ans.***

17-7.11. In Prob. 17-7.10, assume that a spring which elongates 1 in. per 2000 lb is inserted between the elevator and the end of the cable. Determine the frequency of vibration and the maximum tension produced in the cable.

Max tension = 20,900 lb ***Ans.***

17-8 FREE VIBRATIONS ANALYZED BY WORK-ENERGY METHOD

The work-energy method provides an alternative technique for determining the frequency and period of vibrating systems. Stated briefly, it consists of determining the elastic energy stored in the system at the instant of maximum displacement and equating this energy to the kinetic energy at the instant the system passes through the position of static equilibrium.

As an example of this method, consider the spring-suspended weight in Fig. 17-7.1. The resultant work is done by the unbalanced spring force, which is zero at the position of static equilibrium and maximum at the position of maximum displacement x_m. The velocity is maximum as the body passes through the position of static equilibrium, and zero at the position of maximum displacement when the body is instantaneously at rest. Expressing the work-energy relation between these two positions as an equation, we obtain

$$\left[\mathrm{RW} = \frac{W}{2g}(v^2 - v_o^2)\right] \qquad -\frac{1}{2}kx_m^2 = \frac{W}{2g}(0 - v_m^2) \qquad (a)$$

Since the restoring force kx is always directed toward the origin and is directly proportional to the displacement, we can assume that the motion is harmonic and is defined by the relation

$$x = x_m \sin \omega t$$

The velocity, obtained by differentiating the displacement with respect to the time, is given by

$$v = \omega x_m \cos \omega t$$

Obviously the maximum value of the velocity is

$$v_m = \omega x_m$$

Substituting this value in Eq. (a) gives

$$\frac{1}{2}kx_m^2 = \frac{1}{2}\frac{W}{g}\omega^2 x_m^2$$

from which

$$\omega^2 = \frac{kg}{W} \qquad (b)$$

This agrees with the value of natural circular frequency as determined in Section 17-7. Since x_m^2 cancels out, we conclude as before that the frequency is independent of the amplitude.

ILLUSTRATIVE PROBLEMS

17-8.1. The stiff bar of length L and weight W in Fig. 17-8.1 is hinged freely at one end and supported at the other by a spring for which the constant is k. Determine the period of vibration, using the work-energy method.

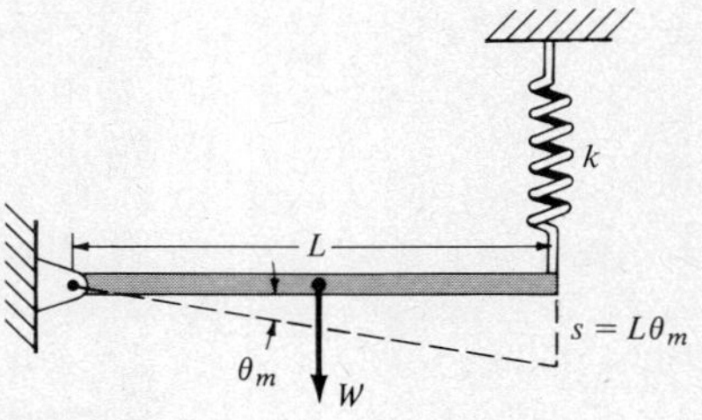

Figure 17-8.1

Solution

Let θ_m be the angle through which the bar rotates from its equilibrium position to reach the position of maximum displacement. Assume that the angular displacement at any instant is defined by the harmonic motion

$$\theta = \theta_m \sin \omega t$$

The angular velocity[4] at any instant, found by differentiation, is

$$\Omega = \frac{d\theta}{dt} = \omega\theta_m \cos \omega t$$

from which the maximum angular velocity is seen to be $\Omega_m = \omega\theta_m$.

When the bar is in an extreme position, the spring is stretched an amount $s = L\theta_m$. Equating the elastic energy at an extreme position to the kinetic energy as the bar passes through its equilibrium position, we obtain

$$\frac{1}{2}k(L\theta_m)^2 = \frac{1}{2}I\Omega_m{}^2 = \frac{1}{2}\left(\frac{1}{3}\frac{W}{g}L^2\right)(\omega\theta_m)^2$$

Therefore,

$$k = \frac{1}{3}\frac{W}{g}\omega^2$$

or

$$\omega = \sqrt{\frac{3kg}{W}}$$

The period is therefore expressed by

$$\tau = \frac{2\pi}{\omega} = 2\pi\sqrt{\frac{W}{3kg}}$$

Comparing this result with that in Prob. 17-2.10, we see that the period is the same as though a body equal to one-third the weight of the bar were applied directly to the spring.

17-8.2. A uniform rod of weight w and length L is welded to the inside of a hoop of negligible weight and radius L as shown in Fig. 17-8.2. Find the frequency of small oscillations if the assembly is displaced slightly from its equilibrium position and rolls without slipping on the horizontal surface.

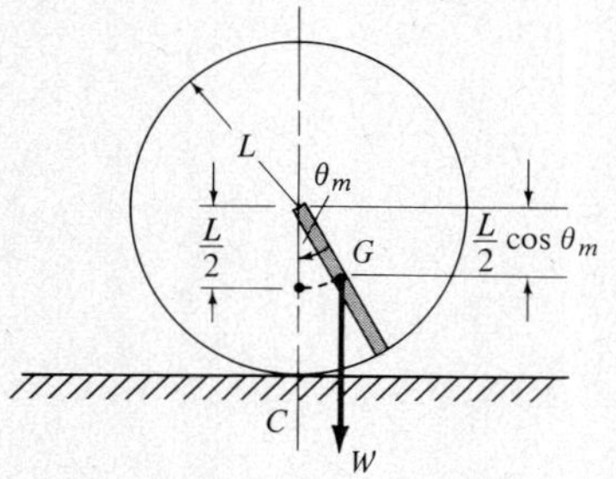

Figure 17-8.2

Solution

As in the previous problem, let θ_m represent the maximum angular displacement from the equilibrium position. Here the system is in plane motion and we use Eq. (14-8.2) to express the kinetic energy in terms of rotation about the instant center C. Equating the work

[4] To avoid confusion with natural circular frequency ω, the angular velocity is here denoted by Ω.

done in moving from the extreme position to the kinetic energy as the system moves through its equilibrium position, we obtain

$$\left[RW = \frac{1}{2}I_c\omega^2\right] \qquad W\left(\frac{L}{2} - \frac{L}{2}\cos\theta_m\right) = \frac{1}{2}\left(\frac{1}{3}\frac{W}{g}L^2\right)\Omega_m{}^2$$

or

$$(1 - \cos\theta_m) = \frac{L}{2g}\Omega_m{}^2 \qquad (a)$$

Assuming harmonic motion, we have $\Omega_m{}^2 = \omega^2\theta_m{}^2$ as in the previous problem. Also, by series expansion for small angles, $\cos\theta_m = 1 - \theta_m{}^2/2$. Substituting these values in Eq. (a) gives

$$\frac{\theta_m}{2} = \frac{L}{3g}\omega^2\theta_m{}^2 \quad \text{or} \quad \omega^2 = \frac{3g}{2L}$$

whence the frequency is

$$\left[f = \frac{\omega}{2\pi}\right] \qquad f = \frac{1}{2\pi}\sqrt{\frac{3g}{2L}} \qquad \textbf{Ans.}$$

PROBLEMS

17-8.3. In Illus. Prob. 17-8.1, assume $W = 60$ lb and $k = 30$ lb per in. What additional weight Q applied at the right end of the bar will produce a frequency of 2 vib per sec?

17-8.4. A stiff weightless bar of length L (Fig. P-17-8.4) is hinged at its upper end and carries a weight W at its lower end. Two springs, each having the constant k, are attached to the bar at a distance b from the hinge. Compute the frequency of vibration for small oscillations.

$$f = \frac{1}{2\pi}\sqrt{\frac{g}{L} + \frac{2kgb^2}{WL^2}} \qquad \textbf{Ans.}$$

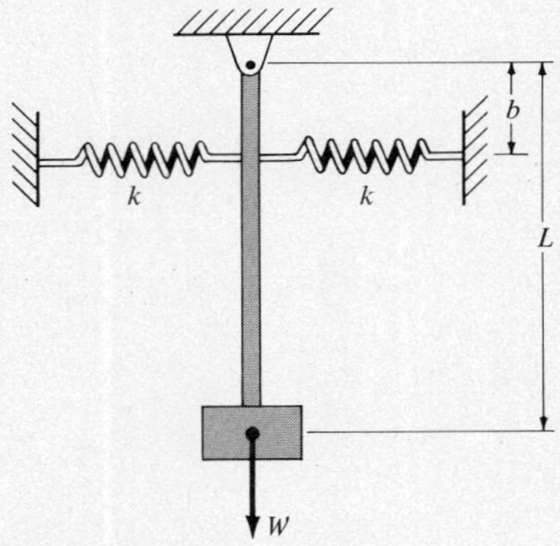

Figure P-17-8.4

17-8.5. If the bar in Fig. P-17-8.4 weighs Q lb and the weight W is removed, determine the frequency of vibration, assuming small oscillations.

$$f = \frac{1}{2\pi}\sqrt{\frac{3g}{2L} + \frac{6kgb^2}{QL^2}} \qquad \textbf{Ans.}$$

17-8.6. Let the bar in Fig. P-17-8.4 weigh $Q = 30$ lb and the weight $W = 16$ lb. If $L = 2$ ft, $b = \frac{1}{2}$ ft, and $k = 20$ lb per in., compute the period of vibration.

$$\tau = 0.837 \text{ sec} \qquad \textbf{Ans.}$$

17-8.7. In Fig. P-17-8.7, a homogeneous solid cylinder of radius r and mass m is displaced slightly from its equilibrium position at the bottom of a curved path of radius R. Find the period of small oscillations if the cylinder rolls without slipping.

$$\tau = \pi\sqrt{6(R - r)/g} \qquad \textbf{Ans.}$$

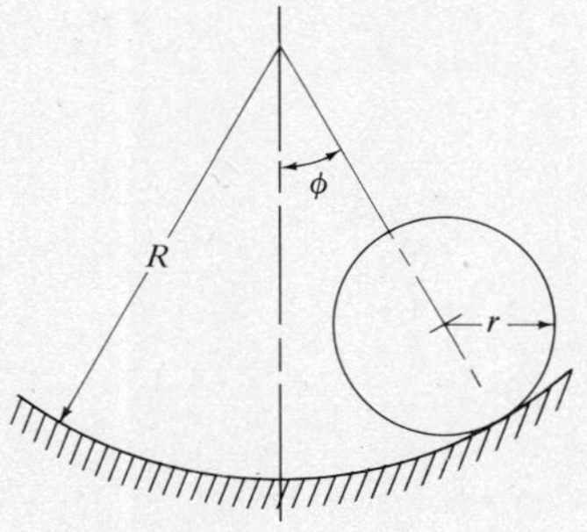

Figure P-17-8.7

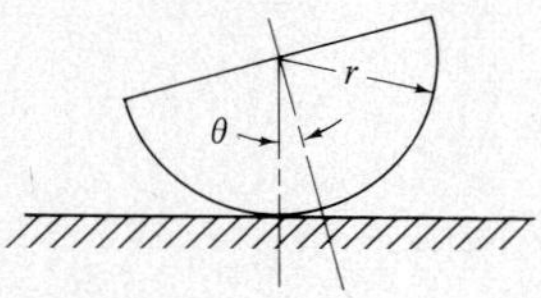

Figure P-17-8.8

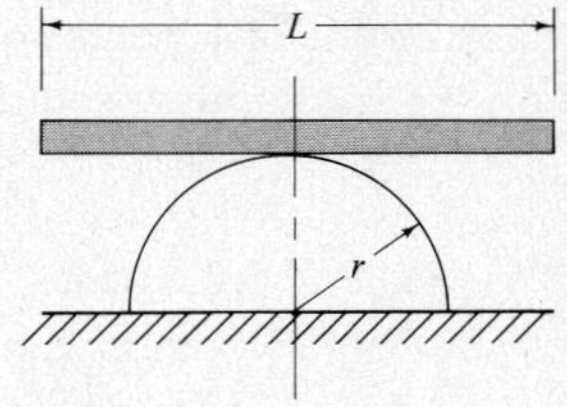

Figure P-17-8.9

17-8.8. A uniform semicircular cylinder of radius r and weight W is displaced through a small angle θ from its equilibrium position as shown in Fig. P-17-8.8. Find the period of its oscillations of it rolls without slipping.

$$\tau = \sqrt{\frac{(9\pi - 16)r}{2g}} \qquad \textit{Ans.}$$

17-8.9. A uniform board of length L and weight W is balanced on a fixed semicircular cylinder of radius r as shown in Fig. P-17-8.9. If the plank is tilted slightly from its equilibrium position, determine the period of its oscillation.

$$\tau = \pi L/\sqrt{3gr} \qquad \textit{Ans.}$$

17-9 FORCED VIBRATIONS

Thus far we have considered free vibrations in which the amplitude depends only on the initial conditions of the motion. The free vibration is maintained by the action of an unbalanced variable spring force created by the motion itself. The vibration exists only after it is begun by an external disturbing force which, after displacing the body from its position of equilibrium, ceases to act on the vibrating body.

If the disturbing force is not removed but continues to act upon the body at periodic intervals, *forced vibrations* are created in which the frequency and amplitude of vibration are affected by the frequency and magnitude of the disturbing force. A typical example of a disturbing force is the force caused by lack of balance in a rotating machine. If the frequency of the disturbing force is the same as the machine's natural frequency of free vibration, a condition known as *resonance* arises which results in vibrations of dangerously large amplitudes.

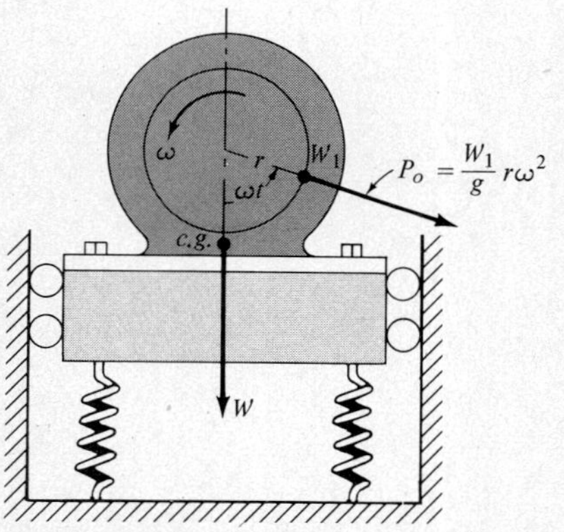

Figure 17-9.1 Forced vibration caused by an unbalanced rotating weight.

In most practical cases the disturbing force varies with the time according to a sine or cosine law. For instance, the motor in Fig. 17-9.1 has a weight W and is mounted on a spring-supported base which permits only vertical movement. Denote the equivalent spring constant of the support by k. A small unbalanced weight W_1 rotates with the faceplate of the motor, and creates a centrifugal inertia force $P_o = (W_1/g)r\omega^2$. The vertical component of this force varies harmonically with the time, its magnitude being $P_o \cos \omega t$.

If displacements from the position of static equilibrium are taken as positive downward, the equation of motion is

$$\left[\Sigma X = \frac{W}{g}a\right] \qquad W + P_o \cos \omega t - (W + kx) = \frac{W}{g}\frac{d^2x}{dt^2}$$

Multiplying this by $\frac{g}{W}$ gives

$$\frac{d^2x}{dt^2} = -\frac{kg}{W}x + \frac{gP_o}{W}\cos \omega t \tag{a}$$

Before solving this equation, consider another example of forced vibration illustrated by the spring-suspended weight shown in Fig. 17-9.2. The upper end of the spring is connected by an inextensible cord to a disk of radius r that rotates with an angular velocity ω. Assuming the cord to be long compared with r, this creates in the upper end of the spring the simple harmonic motion

$$b = r \cos \omega t$$

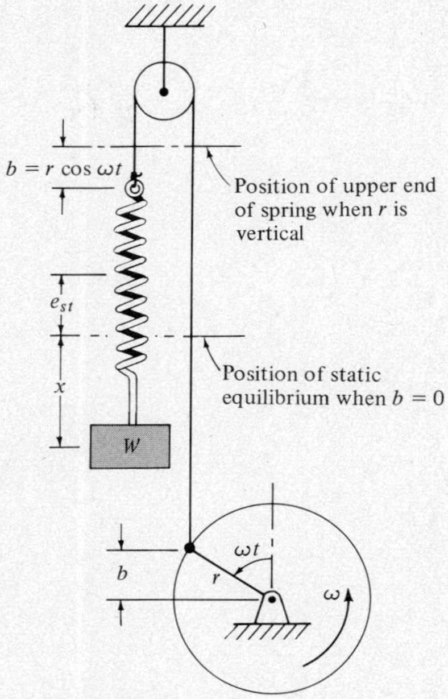

Figure 17-9.2 Forced vibration of a spring-suspended weight due to motion of the point of support.

The time t is measured from the instant radius r is vertically up, and b is measured from the center of the disk. The spring extension is $x - b$ because the downward movement of the top of the spring tends to overcome the downward stretching movement x caused by the vibration of the weight. Hence, at any instant the unbalanced upward spring force is $k(x - b)$.

Forces, displacements, and accelerations being taken as positive downward, the equation of motion is

$$\left[\Sigma X = \frac{W}{g}a\right] \qquad -k(x - b) = \frac{W}{g}\frac{d^2x}{dt^2}$$

If we multiply this by $\frac{g}{W}$ and replace b by $r \cos \omega t$, this becomes

$$\frac{d^2x}{dt^2} = -\frac{kg}{W}x + \frac{krg}{W}\cos \omega t \tag{b}$$

It will be noticed that Eqs. (a) and (b) are identical except for the form of the constant term preceding the expression $\cos \omega t$ in each equation. The only difference between the two cases is that in the second case the disturbing force is transmitted to the weight through the spring instead of being applied directly to it. Denoting the coefficient of $\cos \omega t$ by h, we have in the first case

$$h = \frac{gP_o}{W}$$

and in the second case

$$h = \frac{krg}{W}$$

Also, as in Section 17-7, the expression $\frac{kg}{W}$ can be replaced by ω_n^2. Note that ω_n is the natural circular frequency of a free vibration, whereas in each of Eqs. (a) and (b) above, ω is the circular frequency of the disturbing force. Thus, Eqs. (a) and (b) may be put into the form

$$\frac{d^2x}{dt^2} = -\omega_n^2 x + h \cos \omega t \qquad (c)$$

The forced vibration represented by Eq. (c) may be visualized as a combination of a free vibration of the type discussed in Section 17-7 plus a new motion caused by the disturbing force. Thus, if x_1 represents the displacement caused by the free vibration (which is independent of the disturbing force) and x_2 is the additional displacement caused by the disturbing force, the actual displacement is

$$x = x_1 + x_2$$

With this assumption, Eq. (c) may be considered to be composed of the following parts:

$$\frac{d^2x_1}{dt^2} = -\omega_n^2 x_1 \qquad (d)$$

$$\frac{d^2x_2}{dt^2} = -\omega_n^2 x_2 + h \cos \omega t \qquad (e)$$

Adding these equations gives the original equation (c); Eq. (d) is similar to Eq. (a) in Section 17-7. Hence the value of x_1 is given by

$$x_1 = C_1 \sin \omega_n t + C_2 \cos \omega_n t \qquad (f)$$

It seems reasonable to suppose that a tentative solution of Eq. (e) is $x_2 = A \cos \omega t$, in which the constant A, representing the amplitude of forced vibration, is selected to satisfy the equation. Substituting this solution in Eq. (e) yields

$$-A\omega^2 \cos \omega t = -A\omega_n^2 \cos \omega t + h \cos \omega t$$

from which the value of A is

$$-A\omega^2 + A\omega_n^2 = h$$

or

$$A = \frac{h}{\omega_n^2 - \omega^2} \qquad (17\text{-}9.1)$$

Hence, a solution of Eq. (*e*) is

$$x_2 = A \cos \omega t = \frac{h}{\omega_n^2 - \omega^2} \cos \omega t \qquad (17\text{-}9.2)$$

The complete solution of Eq. (*c*) is the sum of the solutions of its parts, or

$$x = x_1 + x_2 = C_1 \sin \omega_n t + C_2 \cos \omega_n t + \frac{h}{\omega_n^2 - \omega^2} \cos \omega t \qquad (17\text{-}9.3)$$

This result is shown graphically in Fig. 17-9.3. Part (a) represents the displacement-time diagram for the free vibration (assuming $C_1 = 0$) and (b) is the diagram of the forced vibration. The summation of these two harmonic motions (which have different periods) is shown in (c).

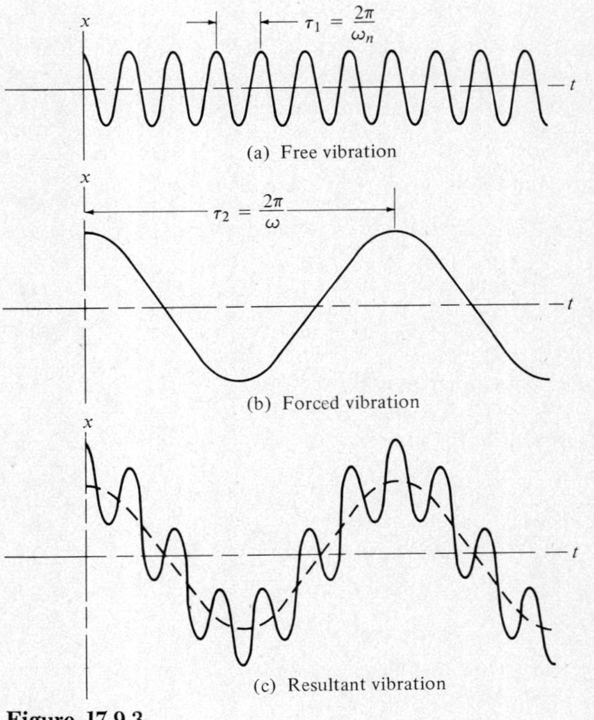

Figure 17-9.3

Since some damping of the free vibration will always exist, the free vibration will die out, as shown in Fig. 17-9.4a. Combining the damped free vibration with the forced vibration results in the steady

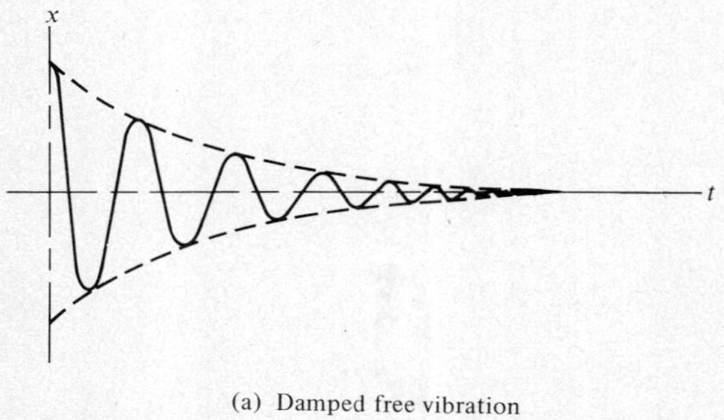

(a) Damped free vibration

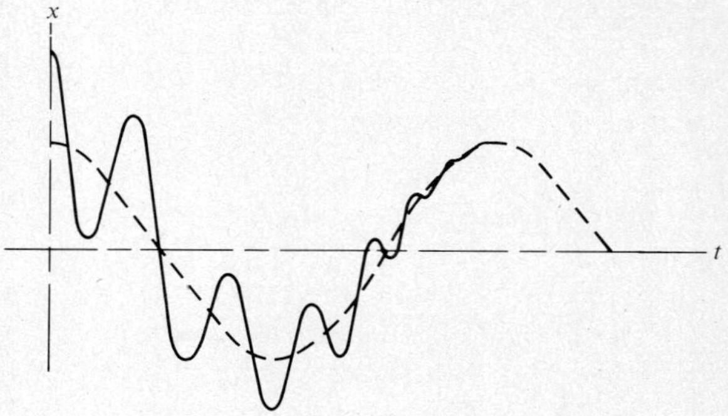

(b) Resultant steady state of forced vibration

Figure 17-9.4

state of forced vibration shown in Fig. 17-9.4b which is maintained indefinitely by the periodic disturbing force.

MAGNIFICATION FACTOR

Because of the damping of the free vibration, it is the motion of forced vibration represented in Fig. 17-9.4b that is of practical importance. If the free vibrations which soon die out are neglected, the maximum displacement (or amplitude) of forced vibration will be given by Eq. (17-9.1).

For the spring-mounted motor shown in Fig. 17-9.1, $\omega_n{}^2 = \dfrac{kg}{W}$ and $h = \dfrac{gP_o}{W}$. Hence the amplitude of forced vibration reduces to

$$A = \frac{P_o}{k}\left(\frac{1}{1 - \dfrac{\omega^2}{\omega_n{}^2}}\right) \tag{g}$$

Similarly in the spring-suspended weight in Fig. 17-9.2, if h is replaced by $\dfrac{krg}{W}$, the amplitude of forced vibration becomes

$$A = r\left(\frac{1}{1 - \dfrac{\omega^2}{\omega_n{}^2}}\right) \tag{h}$$

The factor $\dfrac{P_o}{k}$ in Eq. (g) represents the maximum displacement caused by the static action of the disturbing force P_o. Also r in Eq. (h) represents the maximum displacement of W in Fig. 17-9.2 when the disk revolves very slowly. In general,

$$A = x_{st}\left(\frac{1}{1 - \dfrac{\omega^2}{\omega_n{}^2}}\right) \tag{17-9.4}$$

where x_{st} represents the displacement of the body caused by the static application of the disturbing force. We conclude that the dynamic effect of the disturbing force on the maximum displacement is given by the expression

$$\frac{1}{1 - \frac{\omega^2}{\omega_n{}^2}} \tag{i}$$

This is known as the *magnification factor.*

The effect of this factor is shown in Fig. 17-9.5, in which the factor has been plotted in terms of the ratio ω/ω_n. Thus, when the disturbing force alternates slowly compared with the natural frequency (ω/ω_n less than $\frac{1}{2}$), the magnification factor is only slightly larger than unity. Hence, the dynamic deflection is only slightly greater than that produced if the disturbing force were applied statically.

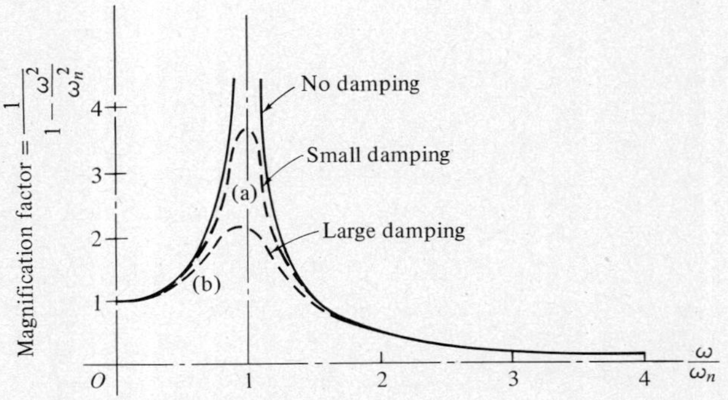

Figure 17-9.5 Magnification factor.

As the impressed frequency approaches the natural frequency, i.e., as ω becomes equal to ω_n, the magnification factor increases rapidly, reaching infinity when ω equals ω_n. Physically this means that when $\omega/\omega_n = 1$, the frequency of the disturbing force coincides exactly with the natural frequency; i.e., the disturbing force tends to push the body always at the right time in the right direction so that the amplitude increases without limit. The condition existing when the impressed frequency is equal to the natural frequency is known as *resonance.* In a machine, the speed of rotation at which resonance occurs is called the *critical speed.* Obviously a machine should be designed to operate well above or below this speed.

As ω is increased above resonance, the magnification factor becomes negative. Physically this means that the disturbing force is 180° out of phase with the motion; i.e., it is of opposite sense to the displacement. Usually this phase relation is of only little interest and the sign of the magnification factor is neglected. It will be observed that when the impressed frequency is well above the natural frequency the magnification factor approaches zero and the body tends to stand still in space. This is because the disturbing force moves up and down so rapidly that the body cannot follow it because of its inertia.

It can be shown that the effect of a relatively small and a relatively large viscous damping can be represented by curves (*a*) and (*b*) in Fig. 17-9.5. Note that there is practically no change in the magnification factor except near resonance and that the resonant frequency is practically unchanged.

For torsional forced vibrations, equations similar to those discussed above can be obtained. To do this, we replace force by torque, mass by mass moment of inertia, and linear values of displacement, velocity, and acceleration by corresponding angular values.

PROBLEMS

17-9.1. A horizontal shaft rotates in bearings at its ends. At its midpoint is keyed a disk weighing 40 lb, whose center of gravity is 0.1 in. from the axis of rotation. If a static force of 200 lb deflects the shaft and disk through 0.1 in., determine the critical speed of rotation of the shaft.

$\omega = 547$ rpm **Ans.**

17-9.2. The motor in Fig. 17-9.1 has a weight $W = 1000$ lb and is supported by four springs each having a constant of 250 lb per in. When the motor is rotating at 1800 rpm, the centrifugal force caused by the eccentric weight is $P_o = 200$ lb. Compute the maximum amplitude of forced vibration.

$A = 0.0022$ in. **Ans.**

17-9.3. The spring-suspended weight in Fig. 17-9.2 weighs 60 lb and the spring constant is $k = 100$ lb per in. The disk of radius $r = 1$ in. rotates at 24 rad per sec. Compute the maximum tension produced in the spring.

1008 lb **Ans.**

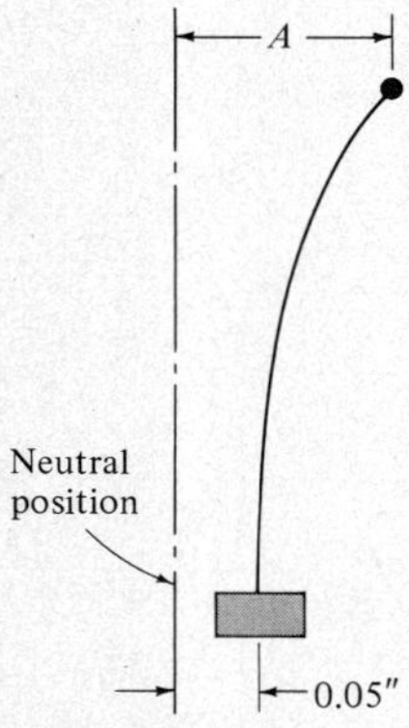

Figure P-17-9.4

17-9.4. The bob in Fig P-17-9.4 weighs 8.05 lb and is fastened to the upper end of a flexible rod. The lower end of the rod is embedded in a block that oscillates horizontally with an amplitude of 0.05 in. and a frequency of 1 vib per sec. The elastic properties of the rod are such that a horizontal force of 10 lb applied at the upper end deflects the bob through 1 in. Determine the amplitude of forced vibration of the bob.

SUMMARY

Vibrations are essentially periodic motions which repeat themselves after a definite time interval called the *period*. The *frequency* is the reciprocal of the period and represents the number of vibrations per second.

Most vibrations have simple harmonic motion. The mathematical description of simple harmonic motion is

$$a = -\omega^2 s \tag{17-2.1}$$

in which ω is the natural circular frequency of the motion. If a motion can be expressed in this form, the period and frequency are respectively

$$\tau = \frac{2\pi}{\omega} \tag{17-2.2}$$

$$f = \frac{\omega}{2\pi} \tag{17-2.3}$$

For purposes of comparison, the period and frequency for various types of pendulums are summarized in the following table:

Pendulum	Period	Frequency	Reference
Simple	$2\pi\sqrt{\frac{L}{g}}$	$\frac{1}{2\pi}\sqrt{\frac{g}{L}}$	Section 17-3
Compound	$2\pi\sqrt{\frac{k_z^2}{g\bar{r}}}$	$\frac{1}{2\pi}\sqrt{\frac{g\bar{r}}{k_z^2}}$	Section 17-4
Torsion	$2\pi\sqrt{\frac{IL}{JG}}$	$\frac{1}{2\pi}\sqrt{\frac{JG}{IL}}$	Section 17-5

The manner in which displacement, velocity, and acceleration vary with time can be visualized by projecting upon a diameter the length of the position, velocity, and acceleration vectors. These vectors rotate about a common axis with the same angular velocity ω, which is the natural circular frequency of the vibration. The magnitude of the position vector is r, the amplitude of the vibration. The magnitude of the velocity vector is $r\omega$ located 90° ahead of the position vector; the magnitude of the acceleration vector is $r\omega^2$ located 90° ahead of the velocity vector.

The general solution of the differential equation of the free vibration of a spring-suspended weight is

$$x = \frac{v_o}{\omega}\sin\omega t + x_o\cos\omega t \tag{17-7.1}$$

in which x_o is the initial displacement from the position of static equilibrium and v_o is the initial velocity at x_o.

The combined effect of the two vibrations of the same frequency expressed in Eq. (17-7.1) is equivalent to

$$x = A\cos(\omega t - \beta) \tag{17-7.4}$$

in which A is the amplitude of a resultant vibration that has the same frequency as each of the combined vibrations. The magnitude of the amplitude A and of the phase angle β is determined from

$$A = \sqrt{x_o^2 + \left(\frac{v_o}{\omega}\right)^2} \qquad \text{and} \qquad \tan\beta = \frac{v_o}{\omega x_o}$$

Section 17-7 shows that the period and frequency of a free vibration are independent of the initial conditions of the motion. They depend only on the weight W and the constant k of the spring support or on the static elongation e_{st} caused by the weight as expressed by

$$\tau = 2\pi\sqrt{\frac{W}{kg}} = 2\pi\sqrt{\frac{e_{st}}{g}} \tag{17-7.2}$$

$$f = \frac{1}{2\pi}\sqrt{\frac{kg}{W}} = \frac{1}{2\pi}\sqrt{\frac{g}{e_{st}}} \tag{17-7.3}$$

Hence, the period or frequency of a body of weight W which is part of a vibrating system may be found by determining the static elongation caused by that weight or computing the equivalent spring constant of that weight, whichever is more convenient.

An alternative technique for determining the frequency and period of vibrating systems is the work-energy method. In this, the elastic energy in the system at the instant of maximum displacement is equated to its kinetic energy at the instant the system passes through its mid-position. This procedure determines the natural circular frequency of the system.

In a forced vibration, the maximum displacement (i.e., amplitude) is given by

$$A = x_{st}\left(\frac{1}{1 - \dfrac{\omega^2}{\omega_n^{\,2}}}\right) \tag{17-9.5}$$

where x_{st} is the displacement caused by the static application of the disturbing force, ω is the circular frequency of the disturbing force, and ω_n is the natural circular frequency of free vibration of the system.

Moments of Inertia for Geometric Shapes

Shape	Moment of inertia	Radius of gyration
Rectangle 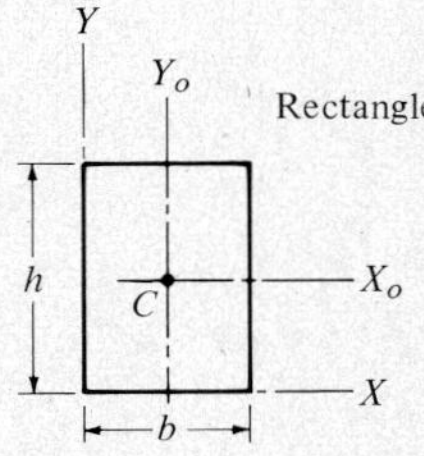	$\bar{I}_x = \dfrac{bh^3}{12}$ $I_x = \dfrac{bh^3}{3}$	$\bar{k}_x = \dfrac{h}{\sqrt{12}}$ $k_x = \dfrac{h}{\sqrt{3}}$
Any triangle 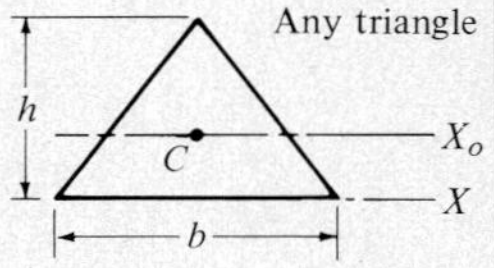	$\bar{I}_x = \dfrac{bh^3}{36}$ $I_x = \dfrac{bh^3}{12}$	$\bar{k}_x = \dfrac{h}{\sqrt{18}}$ $k_x = \dfrac{h}{\sqrt{6}}$
Circle 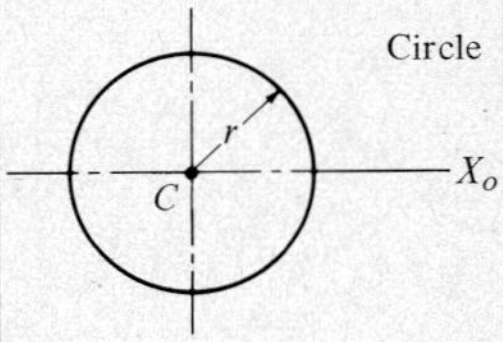	$\bar{I}_x = \dfrac{\pi r^4}{4}$ $\bar{J} = \dfrac{\pi r^4}{2}$	$\bar{k}_x = \dfrac{r}{2}$ $\bar{k}_z = \dfrac{r}{\sqrt{2}}$
Semicircle 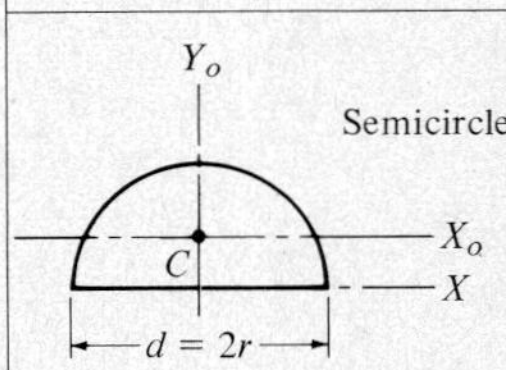	$I_x = \bar{I}_y = \dfrac{\pi r^4}{8}$ $\bar{I}_x = 0.11r^4$	$k_x = \bar{k}_y = \dfrac{r}{2}$ $\bar{k}_x = 0.264r$
Quarter circle 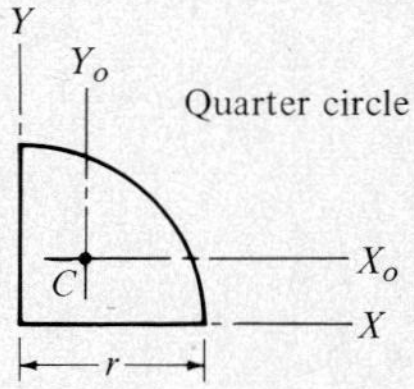	$I_x = I_y = \dfrac{\pi r^4}{16}$ $\bar{I}_x = \bar{I}_y = 0.055r^4$	$k_x = k_y = \dfrac{r}{2}$ $\bar{k}_x = \bar{k}_y = 0.264r$
Ellipse 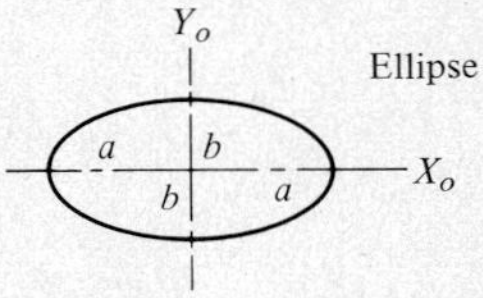	$\bar{I}_x = \dfrac{\pi ab^3}{4}$ $\bar{I}_y = \dfrac{\pi ba^3}{4}$	$\bar{k}_x = \dfrac{b}{2}$ $\bar{k}_y = \dfrac{a}{2}$
Spandrel 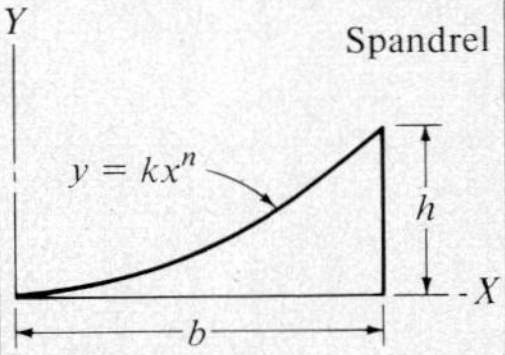	$I_x = \dfrac{bh^3}{3(3n + 1)}$ $I_y = \dfrac{hb^3}{n + 3}$	

Mass Moments of Inertia for Homogeneous Bodies

Description of Body		Axis	Moment of Inertia
Solid right circular cylinder or radius r and length L	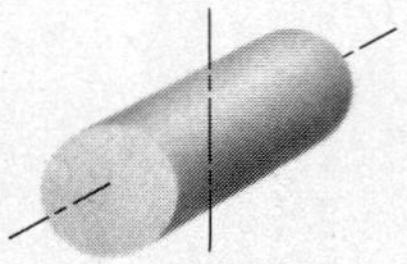	Longitudinal axis	$\frac{1}{2}mr^2$
		Through center, perpendicular to longitudinal axis	$\frac{1}{12}m[L^2 + 3r^2]$
Hollow right circular cylinder of outer radius R and inner radius r, and length L	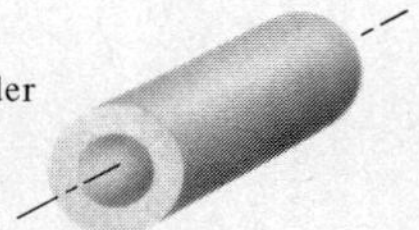	Longitudinal axis	$\frac{1}{2}m(R^2 + r^2)$
		Through center, perpendicular to longitudinal axis	$\frac{1}{12}m[L^2 + 3(R^2 + r^2)]$
Sphere of radius r		Any diameter	$\frac{2}{5}mr^2$
Spherical shell of mean radius r	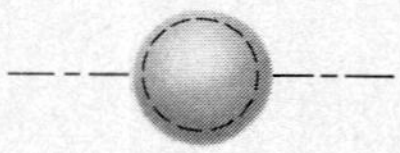	Any diameter	$\frac{2}{3}mr^2$
Uniform slender rod of length L		At end, perpendicular to rod	$\frac{1}{3}mL^2$
		Through center, perpendicular to rod	$\frac{1}{12}mL^2$
Rectangular prism	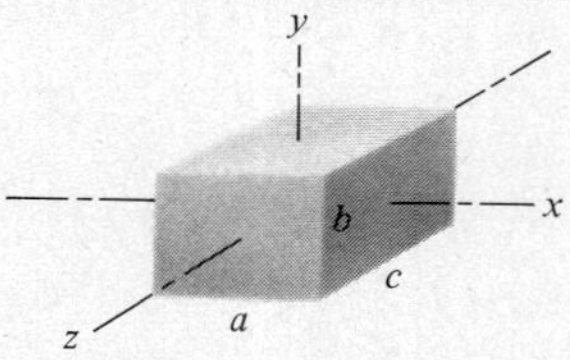	Centroidal axes	$\bar{I}_x = \frac{1}{12}m(b^2 + c^2)$ $\bar{I}_y = \frac{1}{12}m(c^2 + a^2)$ $\bar{I}_z = \frac{1}{12}m(a^2 + b^2)$
Right circular cone of base radius r and altitude h	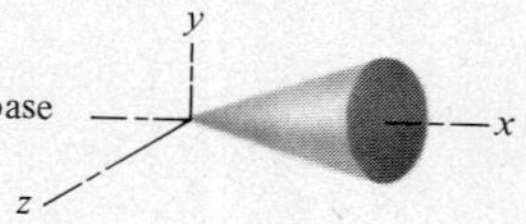	Axis of revolution	$\frac{3}{10}mr^2$
		Through vertex, perpendicular to axis of symmetry	$I_y = I_z = \frac{3}{5}m\left(\frac{r^2}{4} + h^2\right)$

INDEX

75 76 77 9 8 7 6 5 4 3 2 1